Climate Science: Concepts and Applications

Climate Science: Concepts and Applications

Editor: Malena Parker

Callisto Reference
www.callistoreference.com

Callisto Reference,
118-35 Queens Blvd., Suite 400,
Forest Hills, NY 11375, USA

Visit us on the World Wide Web at:
www.callistoreference.com

ISBN: 978-1-63239-928-1 (Hardback)

The publisher's policy is to use permanent paper from mills that operate a sustainable forestry policy. Furthermore, the publisher ensures that the text paper and cover boards used have met acceptable environmental accreditation standards.

Printed in China.

Cataloging-in-Publication Data

Climate science : concepts and applications / edited by Malena Parker.
p. cm.
Includes bibliographical references and index.
ISBN 978-1-63239-928-1
1. Climatology. 2. Climatic changes. I. Parker, Malena.
QC861.3 .C55 2018
551.6--dc23

Table of Contents

Preface

This book has been a concerted effort by a group of academicians, researchers and scientists, who have contributed their research works for the realization of the book. This book has materialized in the wake of emerging advancements and innovations in this field. Therefore, the need of the hour was to compile all the required researches and disseminate the knowledge to a broad spectrum of people comprising of students, researchers and specialists of the field.

Climate science helps in understanding the climate of our planet. Understanding the climate helps in forecasting the weather by using techniques such as the Madden–Julian oscillation, the Northern Annular Mode, El Niño–Southern Oscillation, etc. Climatology also helps in understanding the impact of human activities on the atmosphere and the climate. The various sub-fields of climate science along with technological progress that have future implications are glanced at in this book. It is an essential guide for both academicians and those who wish to pursue this discipline further.

At the end of the preface, I would like to thank the authors for their brilliant chapters and the publisher for guiding us all-through the making of the book till its final stage. Also, I would like to thank my family for providing the support and encouragement throughout my academic career and research projects.

Editor

Clouds enhance Greenland ice sheet meltwater runoff

K. Van Tricht[1], S. Lhermitte[1], J.T.M. Lenaerts[2], I.V. Gorodetskaya[1], T.S. L'Ecuyer[3], B. Noël[2], M.R. van den Broeke[2], D.D. Turner[4] & N.P.M. van Lipzig[1]

The Greenland ice sheet has become one of the main contributors to global sea level rise, predominantly through increased meltwater runoff. The main drivers of Greenland ice sheet runoff, however, remain poorly understood. Here we show that clouds enhance meltwater runoff by about one-third relative to clear skies, using a unique combination of active satellite observations, climate model data and snow model simulations. This impact results from a cloud radiative effect of 29.5 ($\pm$5.2) W m^{-2}. Contrary to conventional wisdom, however, the Greenland ice sheet responds to this energy through a new pathway by which clouds reduce meltwater refreezing as opposed to increasing surface melt directly, thereby accelerating bare-ice exposure and enhancing meltwater runoff. The high sensitivity of the Greenland ice sheet to both ice-only and liquid-bearing clouds highlights the need for accurate cloud representations in climate models, to better predict future contributions of the Greenland ice sheet to global sea level rise.

[1] KU Leuven, Department of Earth and Environmental Sciences, Celestijnenlaan 200E, Leuven 3001, Belgium. [2] Institute for Marine and Atmospheric research Utrecht—Utrecht University, Utrecht 3584CC, The Netherlands. [3] Atmospheric and Oceanic Sciences, University of Wisconsin-Madison, Madison, Wisconsin 53706, USA. [4] National Severe Storms Laboratory, NOAA, Norman, Oklahoma 73072, USA. Correspondence and requests for materials should be addressed to K.V.T. (email: Kristof.VanTricht@ees.kuleuven.be).

Regionally amplified[1] anthropogenic climate change[2] in combination with anomalous large-scale atmospheric circulations[3] have made the Greenland ice sheet (GrIS) a dominant contributor[4,5] to recent global sea level rise (0.63 ± 0.07 mm per year during 2000–2012; ref. 6). This increasing contribution is primarily driven by a reduction in GrIS surface mass balance (SMB)[6–8], related to a significant increase in GrIS surface melt extent[9] and volume[10]. Surface melt is determined by the surface energy balance (SEB)[11], the sum of incoming and outgoing energy fluxes at the surface.

Clouds are known to play a pivotal role in regulating the local SEB, with competing warming and cooling effects on the surface[12–14]. The dominating effect depends strongly on cloud properties such as vertically integrated ice and liquid water contents that determine cloud optical depth and emissivity, in addition to cloud temperature, sun position and surface albedo[13–15].

Despite their importance, the distribution of clouds over the GrIS, their phase partitioning (for example, ice-only clouds versus liquid-bearing or mixed-phase clouds) and their precise impact on the SEB and SMB are poorly understood, owing to the limited availability of measurements. Sensitivity studies based on observations have been limited to particular events or cloud types such as the role of low-level liquid-bearing clouds[16] together with the enhanced poleward transport of humid and warm air[17] in the July 2012 extreme GrIS melt event, whereas ice sheet-wide assessments of cloud SEB impact relied on climate model output to overcome the limited availability of observations[18–20].

However, analysis of the recent Coupled Model Intercomparison Project phase 5 model ensemble and a regional climate model shows a significant inter-model spread in the estimates of both ice-only and liquid-bearing cloud frequency and water paths over the GrIS (Supplementary Fig. 1 and Supplementary Methods). This highlights the need for an observation-based assessment of the cloud distribution over the GrIS.

In this study, we use a unique combination of state-of-the-art satellite remote sensing, ground-based observations and a regional climate model, to quantify the impact of clouds on the SEB over the entire GrIS. The GrIS shows a strong sensitivity to cloud presence, where both ice-only and liquid-bearing clouds are warming the ice sheet surface. Using an advanced snow model, we find that this warming enhances GrIS meltwater runoff in response to reduced refreezing rates at night, when cloud warming is highest compared with clear skies.

Results

Cloud properties from active satellite remote sensing. In this study, a total of ~6.3 million satellite cloud observations were compiled over the GrIS for the period 2007–2010 from the CloudSat and CALIPSO (Cloud-Aerosol Lidar and Infrared Pathfinder Satellite Observation) satellites. These missions carry the first complementary collocated spaceborne radar and lidar sensors[21,22], providing information on vertical distributions of cloud ice and liquid water contents, cloud phase and effective radii, supplemented by cloud optical depths derived from MODerate resolution Imaging Spectroradiometer (MODIS)[23].

These observation-based vertical profiles of cloud properties provide the main input for radiative flux calculations in the level-2 'Fluxes and Heating Rates' product (2B-FLXHR-LIDAR)[24]. An updated refined version of the 2B-FLXHR-LIDAR algorithm that includes low-level supercooled liquid-bearing clouds detected using a combination of CloudSat and CALIPSO observations[25] is used in this study. This new data set, which will form the basis of the Release 05 version of the CloudSat 2B-FLXHR-LIDAR product, provides for the first time a robust partitioning of ice-only and liquid-bearing clouds that does not suffer from errors typically present in conventional satellite products[25,26], and provides an unprecedented three-dimensional view of clouds over the GrIS and their microphysical properties[22].

Figure 1 shows that clouds occur frequently over the GrIS, where the annual mean cloud cover is 67% (Table 1). The CloudSat and CALIPSO data sets employed here suggest that liquid-bearing clouds that contain both ice and (supercooled) liquid water are present 28% of the time, consistent with other work showing that such clouds are prevalent throughout the Arctic[27]. There were almost no (<1%) detections of liquid-only clouds over the GrIS. Whereas the average total cloud fraction is fairly stable throughout the year ranging from 63 to 70%, the satellites observe strong spatial (Fig. 1) and seasonal (Table 1) variations in ice-water path (IWP), liquid-water path (LWP) and liquid/ice partitioning. Liquid-bearing clouds occur predominantly during summer (46% of the time), associated with the presence of warmer and moister air[28], whereas winter is characterized by a much higher frequency of ice-only clouds (55%). Ice clouds occur frequently over the entire GrIS, whereas liquid-bearing clouds are much more abundant in the coastal regions. Annual mean cloud LWP in all conditions varies between 3 and 33 $g\,m^{-2}$ over the entire GrIS with highest values in the coastal zones, whereas IWP is lowest (<7 $g\,m^{-2}$) in the northern part of the interior and highest (>100 $g\,m^{-2}$) in the high-snowfall coastal zones of the southeastern GrIS.

Cloud radiative effect. The satellite-based cloud observations allow to estimate the cloud impact on the SEB using a two-stream broadband radiative transfer model (see Methods). Cloud impact is defined here as the net cloud radiative effect (CRE)[5] at the surface, which is the difference between the all-sky and clear-sky surface radiation budgets[12].

The spatial distribution of the annual mean CRE is positive over the entire GrIS and is strongly correlated with LWP and IWP (Fig. 2a). This indicates that longwave (LW) warming dominates shortwave (SW) cooling, which can be explained by the high solar zenith angles (SZAs) and surface albedo[29].

Averaged over the GrIS, clouds reduce annual mean surface radiative heat loss by 29.5 (±5.2) $W\,m^{-2}$, a factor of 2.7 decrease relative to clear skies. Cloud liquid water contributes only 11.2 (±2.4) $W\,m^{-2}$ to the total radiative effect. This is explained by the lower occurrence frequency and strong seasonal cycle of liquid-bearing clouds with highest frequencies in summer, when both LW warming and SW cooling constitute the CRE. Ice-only clouds occur most frequently in winter, when only LW warming contributes to the radiative effect.

Translating SEB to SMB using a snow model. The annual mean CRE of 29.5 $W\,m^{-2}$ provides enough energy to melt ~90 Gt of ice in the GrIS ablation area during July and August. However, part of this excess energy is absorbed by the snowpack, warming the surface and enhancing outgoing LW radiation. Warmer snow promotes snow metamorphism, which lowers the albedo of the snowpack[30], whereas multiple reflections of SW radiation between clouds and the surface in combination with spectrally dependent cloud absorption generally increase it[31,32].

To account for these processes, which are not all included in the satellite product, the one-dimensional physical snow model SNOWPACK[33] was used. The cloud impact on the GrIS SMB is assessed by forcing SNOWPACK with different cloud inputs (all-sky, clear-sky and no-liquid). As our SNOWPACK simulations require hourly meteorological input data, which is much more frequent than the satellite overpasses (on average CloudSat

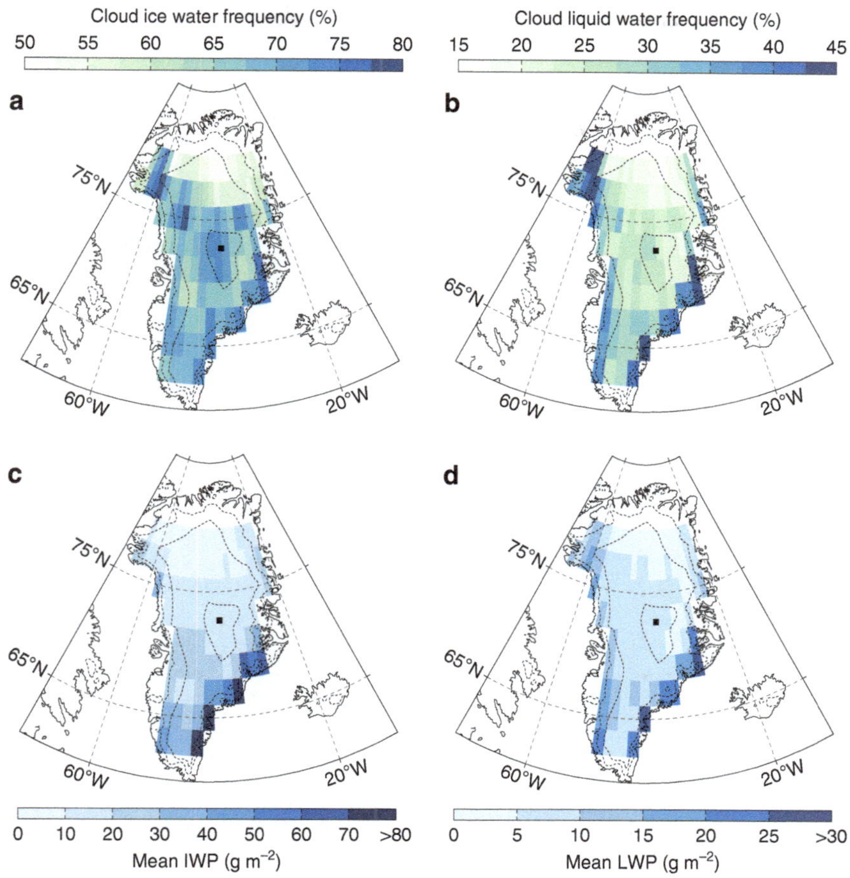

Figure 1 | Retrieved cloud properties over the GrIS by satellite remote sensing. These retrievals result from joint CloudSat, CALIPSO and MODIS observations. Both (**a**) cloud ice and (**b**) cloud liquid water occur frequently over the GrIS, although strong spatial variations exist. (**c**) Mean IWP in all conditions and (**d**) mean LWP in all conditions also show spatial variability in cloud water contents. Dashed curves indicate 1,000 m height contours and the black dot represents the location of the Summit station.

Table 1 | Seasonal and yearly cloud frequencies and mean water paths.

	All clouds (%)	Ice-only (%)	Liquid-bearing (%)	Mean IWP (g m^{-2})	Mean LWP (g m^{-2})
Spring (MAM)	63	41	22	20	6
Summer (JJA)	66	20	46	19	16
Autumn (SON)	68	41	27	25	8
Winter (DJF)	70	55	15	23	5
Year	67	39	28	21	9

GrIS, Greenland ice sheet; IWP, ice-water path; LWP, liquid-water path. MAM, March-April-May; JJA, June-July-August; SON, September-October-November; DJF, December-January-February.
The observed total cloud frequency (in %) over the GrIS (2007-2010) is fairly constant year round, but the partitioning and water paths of ice and liquid water exhibit strong seasonal variations. It is noteworthy that in the event of multiple detected cloud layers in a vertical profile, these are combined and reported as a single cloudy profile in these statistics.

and CALIPSO observe each 2° by 2° grid box once every 1–7 days, depending on the location), we constructed a hybrid satellite-climate model data set, combining the benefits of the observationally based estimates of CREs with the hourly temporal resolution of a regional climate model (see Methods).

The snow model simulations, which capture the evolution of the GrIS SMB from 2007 to 2010 (Fig. 3a, black dashed curve), indicate that clouds warm the GrIS surface by 1.2 (±0.1) °C on average over the entire period. The changes in the snowpack as a result of this cloud-induced warming, however, in combination with increased albedo under cloudy conditions, lower the annual mean CRE (Fig. 2a, orange line) by 8.4 W m^{-2} (~30%) relative to the satellite estimate (Fig. 2a, purple line). The spatial distribution of this updated CRE is shown in Fig. 2b.

Enhanced meltwater runoff. Although GrIS surface melt is nearly identical under all-sky and clear-sky conditions (Fig. 3b), clouds enhance GrIS meltwater runoff by 56 ± 20 Gt per year (32 ± 12 %), with similar contributions from cloud ice (25 Gt) and cloud liquid (31 Gt) water. This suggests that the primary influence of clouds is by reducing meltwater refreezing; in clear-sky conditions, about 58% of the meltwater refreezes, but this fraction decreases to 45% in the presence of clouds.

The governing mechanisms that explain this effect are illustrated in Fig. 4, for a case study during 16–21 June 2008 at 67°N-49°E (ablation zone), during an albedo transition period. In the presence of a thick cloud (Fig. 4a), strong SW cooling and LW warming partly compensate during daytime (Fig. 4b). Surface melt occurs during and shortly after solar noon, when incoming SW radiation peaks (Fig. 4c, red curves). During the subsequent

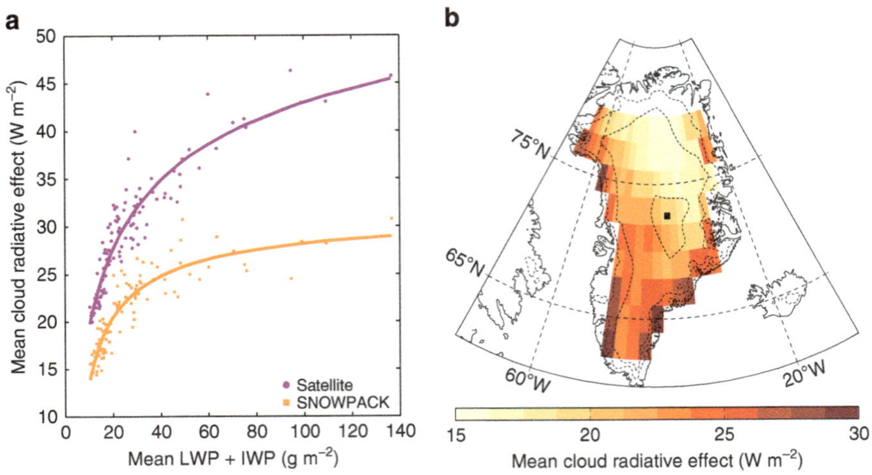

Figure 2 | CRE over the GrIS. (**a**) The average relationship between mean LWP + IWP and mean annual CRE is positive at each location, with higher radiative effects from optically thicker (more liquid and/or ice) or more frequent clouds. The purple circles were derived from satellite observations, whereas the orange squares result from the SNOWPACK runs. The satellite-retrieved CRE estimates assume fixed surface conditions (albedo and surface temperature). In reality, these conditions may change in the absence of the detected clouds. The entire snowpack changes as a result of cloud warming, reducing the initial CRE (lower values in the orange curve). (**b**) Yearly mean CRE over the GrIS (2007-2010) from the SNOWPACK output. This estimate takes into account changing surface conditions (albedo and surface temperature) and snowpack with changing energy inputs due to the presence/absense of clouds. Dashed lines indicate 1,000 m height contours and the black dot represents the location of the Summit station.

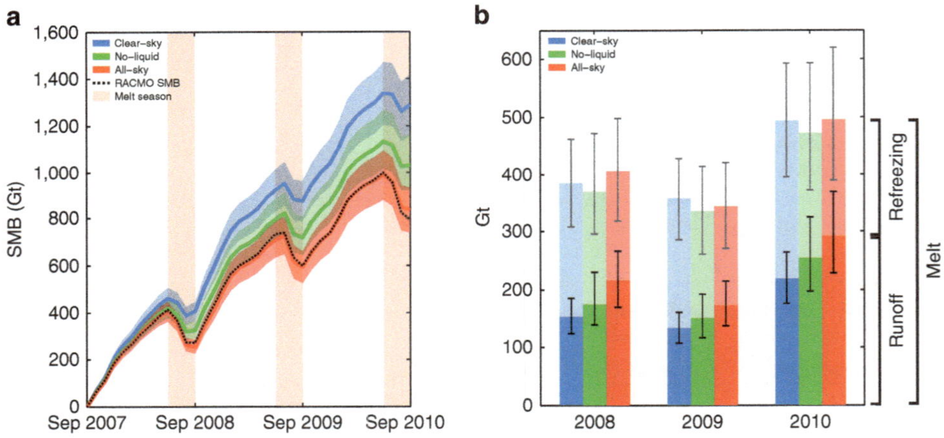

Figure 3 | The effects of clouds on the SMB during the period September 2007-September 2010. (**a**) Evolution of GrIS SMB indicates that the cloudy simulations have a lower SMB. Uncertainties are shown by the shaded areas. The black curve represents the regional climate model RACMO2.3 SMB values and show the performance of SNOWPACK in simulating the GrIS SMB. (**b**) Yearly GrIS melt, refreezing and runoff. Despite negligible differences in melt, 58% of the meltwater refreezes in clear-sky conditions, whereas only 45% refreezes in all-sky conditions. Annual meltwater runoff is therefore about one-third higher in the presence of clouds, with a slightly higher contribution of liquid-bearing clouds. The whiskers indicate an inherent SNOWPACK uncertainty and the sensitivity to the amount of LWP/IWP in the SNOWPACK simulations.

nighttime hours, when incoming SW is small (Fig. 4b, grey zones), clear-sky conditions favour strong surface radiative cooling, extracting heat from the warmer subsurface snow by conduction, (partly) refreezing the meltwater (Fig. 4c, blue curves). This radiative cooling is reduced under cloudy conditions, impeding the refreezing mechanism. The additional meltwater that does not refreeze fills up the available pore space until it runs off[34] (Fig. 4d), whereas a persistent positive CRE leads to firn warming, further enhancing melt and subsequent runoff[35]. The combined effect causes an earlier exposure of bare ice and slush under cloudy conditions, which amplifies these mechanisms through enhanced SW absorption with lower albedo values (albedo feedback, Fig. 4e). During the last 2 days of the case study, when the cloud has disappeared in the all-sky simulation, this albedo feedback is strengthened due to increased SW insolation in the absence of cloud SW blocking, enhancing melt relative to the clear-sky simulation (Fig. 4c, red curves).

Along with meltwater runoff, our results suggest that clouds enhance GrIS sublimation. In summary, liquid-bearing and ice-only clouds contribute equally to a GrIS SMB decrease during 2007–2010 of $35 \pm 5\%$ relative to clear-sky conditions by increasing meltwater runoff and sublimation, which is the average equivalent of 152 ± 20 Gt per year (Fig. 3a, red curve).

Discussion

Our results highlight the strong sensitivity of the GrIS SEB and SMB to clouds. This analysis overcomes many of the current issues that exist in observing and modelling of mixed-phase clouds by creating a hybrid satellite-climate model data set that

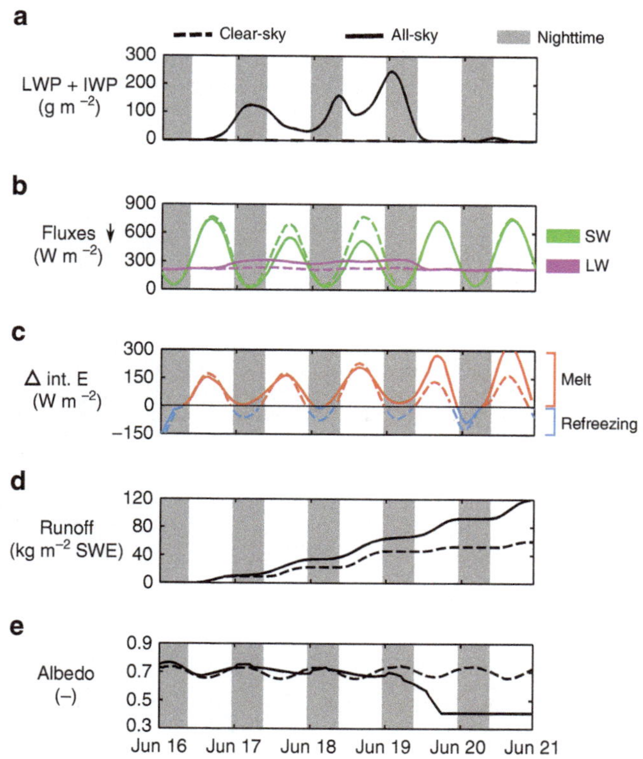

Figure 4 | Case study showing cloud impacts on melt/refreezing during 16–21 June 2008 at 67°N–49°E. Solid curves represent all-sky conditions and dashed curves represent clear-sky conditions. Grey zones are characterized by a SZA > 70°, considered to be nighttime. All variables as shown in the figure have been smoothed using a 6-hourly moving-average window. (**a**) Total (liquid + ice) water path, showing a cloud in the all-sky simulation. (**b**) Downwelling SW (green) and LW (purple) radiative fluxes. In the presence of a cloud, SW cooling (solid curve below dashed curve) and LW warming (solid curve above dashed curve) occur simultaneously during daytime, whereas LW warming dominates nighttime. (**c**) Internal energy change of the snowpack due to melting (positive) and refreezing (negative) processes. Melt rates are highest at solar noon when SW insolation peaks, whereas refreezing rates are highest at night when strong surface radiative cooling dominates. In the presence of clouds, this radiative cooling is reduced, impeding the refreezing mechanism. (**d**) Cumulative meltwater runoff in snow water equivalent (SWE) is higher under cloudy conditions, due to limited meltwater refreezing and earlier bare-ice exposure. (**e**) Surface albedo as simulated by SNOWPACK. Persistent warming by clouds enhances meltwater runoff, leading to an earlier exposure of bare ice and slush that have a much lower albedo than (fresh) snow (from 19 June onwards). At this point, the warming is amplified due to a much higher absorption of SW radiation in the all-sky simulation, as opposed to the clear-sky simulation.

closely matches satellite retrievals and ground-based observations with hourly temporal resolution. The strong positive CRE over the GrIS is found to warm the surface, enhance meltwater runoff in the GrIS ablation area and increase sublimation in the GrIS accumulation area.

However, a substantial uncertainty is associated to the sublimation term, as SNOWPACK was used in an offline configuration and does not account for any feedbacks to the atmosphere in response to changes in the snowpack. In reality, changes in surface temperature induced by clouds will modify turbulent fluxes and thus the boundary-layer air temperature and humidity. Here we did not include these second-order effects and focused exclusively on the sensitivity of the GrIS to cloud-related changes in the SEB. This constitutes a first important step in assessing the impact of clouds on the GrIS SMB.

These results further indicate that not only liquid-bearing clouds[16] but also clouds composed exclusively of ice significantly increase radiative fluxes into the surface and decrease GrIS SMB. This underscores the need for continued research into the factors that govern the formation and maintenance of these distinct cloud regimes, and their evolution in a future warmer and wetter Arctic[36]. Evidence of the large spread in cloud cover and liquid/ice partitioning over the GrIS in current state-of-the-art climate models, in combination with our limited understanding of the interaction between clouds, circulation and climate[37], suggests that improved cloud representations in climate models could significantly increase the fidelity of future projections of GrIS SMB and subsequent global sea level rise.

Methods

General approach. State-of-the-art satellite cloud observations have been used in a refined multi-stage hybrid model-observation algorithm, to estimate surface radiative fluxes. To overcome the limited temporal resolution of these satellite observations, a regional climate model has been incorporated to interpolate the observations to hourly resolution. Ground-based observations from Summit[28,38] and several automatic weather stations (AWSs) have been used to evaluate the resulting hybrid satellite-climate model data set. This new hybrid data set was finally used to force a snow model that simulates the impact of the CRE on the GrIS SMB.

CloudSat/CALIPSO satellite observations. Although radiative fluxes over the GrIS are available from the European Centre for Medium-Range Weather Forecasts (ECMWF) atmospheric re-analysis (ERA), the uncertainty in these fluxes is large, as they are calculated based on modelled clouds, excluding any cloud observational data[39]. However, with the launch of the CloudSat and CALIPSO satellites, radiative flux calculations can now be constrained by a comprehensive set of cloud observations from space[21–23]. The Cloud-Aerosol Lidar with Orthogonal Polarization (CALIOP) sensor aboard CALIPSO was designed to focus on optically thin clouds and the 94-GHz Cloud Profiling Radar (CPR) aboard CloudSat probes optically thicker clouds and precipitation. CALIOP and CPR measure the backscattered energy by cloud particles, which can be converted into vertical distributions of cloud ice and liquid water contents, and effective radii, filled in by MODIS radiance information when the retrieval algorithms of the active sensors fail to converge. Despite their limited spatiotemporal sampling along specific tracks and limited collocated availability from 2007 to 2010, these active satellite cloud observations have been proven to be of great relevance for polar cloud studies[39,40].

The 2B-FLXHR-LIDAR algorithm. The primary satellite product used in this study is the refined CloudSat 2B-FLXHR-LIDAR[24]. This product combines (i) cloud and aerosol observations from CloudSat, CALIPSO and MODIS; (ii) ancillary temperature and humidity profiles from ECMWF atmospheric re-analyses; and (iii) International Geosphere-Biosphere Programme surface albedo/emissivity data to constrain the two-stream radiative transfer model (RTM) BugsRad[24], which calculates broadband radiative fluxes. The CloudSat footprint of 1.4 km × 1.8 km provides the basis for the radiative transfer calculations, determining the horizontal resolution of the product, whereas the vertical resolution is 240 m.

We found that in the presence of low-level supercooled liquid-bearing clouds that occur frequently over the Arctic[27], the original 2B-FLXHR-LIDAR product was severely underestimating downwelling LW fluxes because of the inability of the existing algorithm to properly handle these clouds, which were falsely identified as ice-only clouds. This was attributed to the fact that the CloudSat CPR is insensitive to small supercooled liquid droplets, while an exact retrieval of water path in this case was often not available from the CALIOP sensor. In such an event, a simple linear temperature-based relationship was used in the original version, to decide on the portion of ice and liquid water to which the CALIOP backscattered energy was translated, not allowing any supercooled liquid water to occur below −20 °C, while it can physically occur down to −40 °C. In this study, we therefore refined the 2B-FLXHR-LIDAR product by introducing the use of the CloudSat level-2 'Combined Radar and Lidar Cloud Scenario Classification' product[25,41] as input for the refined 2B-FLXHR-LIDAR algorithm, which combines information from the CloudSat CPR and CALIPSO CALIOP to identify cloud phase. When low-level supercooled liquid-bearing clouds are detected and no exact retrieval is available for determining the LWP that is assigned to these clouds, we rely on a fill value based on the median LWP value retrieved by ground-based observations at Summit (see below).

A total of ~6.3 million refined 2B-FLXHR-LIDAR flux profiles were aggregated in a 2° by 2° grid, to ensure a sufficient amount of satellite profiles within each grid box. Ice and liquid water occurrence frequencies and water paths were extracted from the refined 2B-FLXHR-LIDAR product and represent the combined CloudSat/CALIPSO/MODIS observations that were used in the radiative flux retrievals.

The radiative fluxes of the original 2B-FLXHR-LIDAR product have been evaluated[24] through comparisons with Clouds and the Earth's Radiant Energy System (CERES) top-of-atmosphere observations and surface flux retrievals on a global scale. To assess the quality of the refined 2B-FLXHR-LIDAR product over the GrIS, the satellite LWP retrievals were first evaluated with ground-based LWP measurements[42], retrieved by a microwave radiometer (MWR) installed at Summit, Greenland, as part of the ICECAPS project[28], during July 2010–December 2010. Despite the limited available overlap in time of 6 months, relative frequency histograms of LWP show a strong agreement between MWR-retrieved and satellite-retrieved LWP values (Supplementary Fig. 2). Furthermore, cloud macrophysical properties such as cloud fraction agree well (not shown) with the available ground-based observations[28,38].

An additional evaluation was performed by comparing the retrieved downwelling LW/SW radiative fluxes from the satellite product with ground-based flux measurements from 11 AWSs on the GrIS (IMAU[29] and PROMICE, http://www.promice.org/). Monthly averages were computed for satellite overpasses within 100 km from the AWS locations, whereas for consistency with the original overpasses, monthly averages for the AWS-measured fluxes were based on interpolated flux measurements to the times of these overpasses. AWS stations that are located in the same 2° by 2° grid box were averaged to yield one representative flux observation for that grid box. The resulting eight comparisons (Supplementary Fig. 3) show the good performance of the 2B-FLXHR-LIDAR surface radiative fluxes, as there is in general a close agreement between satellite-retrieved and AWS-measured LW and SW fluxes. Statistics on bias and root-mean-square errors (RMSEs) are given in Supplementary Table 1. Although there is a considerable range in these statistics, the overall agreement is good, with a mean bias of -4 and 7.6 W m^{-2} for downwelling LW and SW, respectively, well below the estimated monthly mean measurement uncertainty of $\sim$5% in the radiation instruments[29,43]. RMSEs equal 13.2 W m^{-2} for LW fluxes and 24.2 W m^{-2} for SW fluxes; these RMSE values are very similar to radiative flux closure results shown at Barrow, Alaska, where the cloud properties were specified by ground-based radars, lidars and radiometers[44]. Local disagreement between satellite and AWS can be explained by heterogeneous terrain within the 100-km radius of the AWS, AWS sensor uncertainties, AWS sensor tilt affecting SW measurements and low sampling rate for some AWS locations to compare 2B-FLXHR-LIDAR and AWS observations. Despite this, our evaluation indicates that the refined 2B-FLXHR-LIDAR product provides reliable estimates of radiative fluxes over the GrIS.

Cloud radiative effect. CRE at the surface (W m^{-2}) is defined by equation (1), where $F^{\downarrow\uparrow}_{\text{SW, all-sky}}$ and $F^{\downarrow\uparrow}_{\text{LW, all-sky}}$ are the net surface radiative SW and LW fluxes in all-sky conditions, and $F^{\downarrow\uparrow}_{\text{SW, clear-sky}}$ and $F^{\downarrow\uparrow}_{\text{LW, clear-sky}}$ are the net surface radiative fluxes that would occur in the absence of clouds[12].

$$\text{CRE} = F^{\downarrow\uparrow}_{\text{SW, all-sky}} + F^{\downarrow\uparrow}_{\text{LW, all-sky}} - (F^{\downarrow\uparrow}_{\text{SW, clear-sky}} + F^{\downarrow\uparrow}_{\text{LW, clear-sky}}) \quad (1)$$

A positive effect indicates net cloud warming at the surface, whereas a negative effect is indicative of net cloud cooling. The magnitude of the radiative effect is intimately connected to the amount of ice and liquid water in the cloud, its temperature, SZA and surface albedo[14].

All analyses of CRE over the GrIS were focused on three different cloud scenarios. In the 'all-sky' simulation (that is, control run), radiative fluxes were calculated based on the satellite-observed clouds and their LWP and IWP. In the 'clear-sky' simulation, all cloud ice and liquid water was removed (that is, IWP and LWP were set to zero), to simulate the radiative fluxes that would occur in the absence of clouds. In the 'no-liquid' simulation, only the retrieved cloud liquid water was removed (LWP set to zero), effectively eliminating the impact of (supercooled) liquid water in liquid-bearing clouds on the calculated radiative fluxes, whereas IWP was kept unchanged. All-sky CRE is defined as the difference in radiative fluxes between the 'all-sky' and the 'clear-sky' simulations, whereas liquid-bearing CRE is defined as the difference in radiative fluxes between the 'all-sky' and the 'no-liquid' simulations.

To determine an observationally based CRE estimate, the 2B-FLXHR-LIDAR RTM was used for the three different cloud scenarios described above. Because of the large computational requirements of running the RTM on all satellite observations between 2007 and 2010, the 'clear-sky' and 'no-liquid' simulations were limited to 4 months, distributed over the year 2010 to be able to capture seasonal variability. For this purpose, all available observations from March, June, September and December 2010 were used.

Downwelling LW radiation is described by equation (2)[45], in which $\text{LW}_{\text{as}}\downarrow$ and $\text{LW}_{\text{cs}}\downarrow$ are the downwelling LW radiation in all-sky and clear-sky conditions, respectively, ϵ_{cs} is the clear-sky atmospheric emissivity, T is the temperature (K) at a reference height, σ is the Stefan–Boltzmann constant and F is the cloud enhancement factor.

$$\text{LW}_{\text{as}}\downarrow = \epsilon_{cs} F \sigma T^4 = F \times \text{LW}_{\text{cs}}\downarrow \quad (2)$$

The cloud enhancement factor F (≥ 1) describes the increase in downwelling LW radiation relative to clear sky due to clouds[46]. In this study we estimate F for all clouds and for liquid water in liquid-bearing clouds using the previously described RTM runs, in which respectively all clouds ('clear-sky' simulation) and liquid water in liquid-bearing clouds ('no-liquid' simulation) were removed from the satellite profiles, while keeping all other parameters unchanged.

Subsequently, $F_{\text{all, LW}}$ and $F_{\text{liq, LW}}$, that is, the LW cloud enhancement factors F for all clouds and liquid-bearing clouds, respectively, can be calculated by equations (3) and (4).

$$F_{\text{all, LW}} = \frac{\text{LW}_{\text{as}}\downarrow}{\text{LW}_{\text{cs}}\downarrow} \quad (3)$$

$$F_{\text{liq, LW}} = \frac{\text{LW}_{\text{as}}\downarrow}{\text{LW}_{\text{noliq}}\downarrow} \quad (4)$$

CRE in the LW, and therefore the LW cloud enhancement factor, is primarily a function of cloud temperature (often approximated by cloud-base temperature (CBT)), cloud height and emissivity[14]. The latter is mainly determined by the cloud LWP and IWP.

In an analogous manner, we can define an SW cloud enhancement factor that describes the decrease in downwelling SW radiation due to clouds relative to clear-sky conditions. Similar to equations (3) and (4), cloud enhancement factors $F_{\text{all, SW}}$ and $F_{\text{liq, SW}}$ (≤ 1) that account for the decrease in SW radiation due to all clouds and liquid-bearing clouds, respectively, can be defined by equations (5) and (6).

$$F_{\text{all, SW}} = \frac{\text{SW}_{\text{as}}\downarrow}{\text{SW}_{\text{cs}}\downarrow} \quad (5)$$

$$F_{\text{liq, SW}} = \frac{\text{SW}_{\text{as}}\downarrow}{\text{SW}_{\text{noliq}}\downarrow} \quad (6)$$

The SW cloud enhancement factor mainly depends on cloud microphysical properties (LWP and IWP), surface albedo and SZA[13].

To extend the satellite CRE data set from the four available months to the full 2007–2010 period, we trained neural networks to predict $F_{\text{all, LW}}$, $F_{\text{all, SW}}$, $F_{\text{liq, LW}}$ and $F_{\text{liq, SW}}$ (ref. 47).

Inputs for the neural networks to predict the cloud enhancement factors were LWP, IWP and near-surface temperature for LW and LWP, IWP, SZA and surface albedo for SW. LWP, IWP, SZA and surface albedo are readily available in our data sets. However, probably the best temperature-related predictor of LW CRE is cloud temperature or CBT. Yet, this variable is not readily available from the ERA-Interim product used by 2B-FLXHR-LIDAR ('ECMWF-AUX') and it is not available in the RACMO model outputs that were used in a later stage (see below). Therefore, near-surface temperature (2 m), originating from ECMWF ERA-Interim reanalysis data, was used instead of CBT, to train the neural networks. As such, the method only requires a good correlation between near-surface temperature and CBT. We therefore investigated the relationship between both temperatures, with CBT approximated here by the temperature of the satellite bin containing the cloud base (as detected by the combined CloudSat/CALIPSO observations), for the 4 months explicitly calculated by the RTM, and found a high correlation ($r = 0.83$) for cases over the GrIS with LWP + IWP ≥ 10 g m^{-2}. From this we expect the near-surface temperature to be a good proxy to predict LW CRE.

Hence, LWP, IWP, near-surface temperature, SZA and surface albedo were used as inputs for the neural networks and the cloud enhancement factors F as targets. Of the available satellite-retrieved CRE estimates, 70% was used for training the neural networks, whereas 15% was employed to avoid overfitting and 15% was used as independent validation set to assess the neural network performance. As the range of albedo values that are used in the 2B-FLXHR-LIDAR algorithm is relatively small, we performed additional offline RTM simulations with a variable surface albedo, to extend the training data set for a wider range of albedo values. The neural networks for both SW and LW consist of one hidden layer with ten neurons. Supplementary Fig. 4 shows the error histograms of the neural networks, expressed as the difference between the F factors predicted by the neural networks, and the F factors calculated from the RTM runs. For all F factors, biases are small (0 ± 0.03 for $F_{\text{liq, SW}}$, 0 ± 0.07 for $F_{\text{liq, LW}}$, 0 ± 0.03 for $F_{\text{all, SW}}$ and 0 ± 0.05 for $F_{\text{all, LW}}$), confirming the good skill of the neural networks to predict CRE. The errors on the F factors were used to calculate the uncertainty on the estimated CRE over the GrIS (see 'Uncertainty derivations').

The final all-year CRE as estimated by the neural networks, agrees very well with ground-based observations atop the GrIS at Summit[48] (30 versus 33 W m^{-2}), whereas in the ablation zone of the GrIS this estimate is substantially higher (33 versus 23 W m^{-2}), based on AWS stations S6 and S9 (ref. 29). This disparity is almost exclusively attributed to a difference in LW CRE, most probably due to the absence of cloud observations in the ground-based analysis and related assumptions for cloud optical depth retrievals[29]. In addition, we have calculated the monthly mean CRE over 2007–2010 for Summit, to ensure that the seasonal cycle of CRE is captured well. The seasonal pattern shown in Supplementary Fig. 5 agrees well with the results of a previous study[48], where their retrieval uncertainty spans the observed differences with our results. One clear difference that stands out is a lower LW CRE in our numbers compared with theirs in wintertime. Two possible explanations are (i) the difference in years that are used for the sampling and (ii) the ground-based instruments that are more sensitive to very small amounts of LWP/IWP that are missed in the CloudSat/CALIPSO estimate. Overall, the satellite-based estimates of CRE agree well with ground-based observations.

Regional climate model RACMO2.3. As the temporal resolution of satellite observations is limited by the amount of overpasses, a hybrid approach was designed (see below) to combine the observational constraints of the satellite data with the temporal resolution of the regional climate model RACMO2.3.

RACMO version 2.3 (ref. 49) combines the atmospheric dynamics from HIRLAM with the physical processes described by ECMWF Integrated Forecast System Cycle 33R1 (ref. 50). RACMO2.3 is specifically adapted for use over polar ice sheets, as it is interactively coupled to a multi-layer snow model[51], which includes an advanced snow albedo scheme[52,53], accounts for drifting snow[54] and contains several improvements in the representation of atmospheric physics compared with RACMO2.1. In particular, changes were introduced in the cloud scheme, which now includes a new ice super-saturation parameterization, improving the representation of supercooled liquid-bearing clouds, and also in the turbulence and radiation parameterization schemes[55,56]. The original RACMO data were aggregated in the same 2° by 2° grid that was used for the satellite observations.

Part of the results in this study strongly depend on the performance of the RACMO2.3 model. As this study focuses in particular on GrIS melt, it is of utmost importance that the climate is well simulated for reproducing melt events. The RACMO2.3 regional climate model has been extensively evaluated for simulating the near-surface climate, showing a very good performance in representing air temperature and specific and relative humidity[51,57], wind speed[58,59] and surface albedo[52]. With regard to SEB, however, significant biases remain in modelled surface radiative fluxes, with, in particular, an underestimation of the downwelling LW radiation and an overestimation of the downwelling SW radiation, due to an underestimated cloud optical depth[49]. These findings are confirmed in our comparisons between the original RACMO2.3-modelled radiative fluxes and retrieved radiative fluxes by the 2B-FLXHR-LIDAR algorithm (red dots in Supplementary Fig. 6). Downwelling LW radiation is systematically underestimated with a mean bias of $-6.2\,\mathrm{W\,m^{-2}}$ and an RMSE value of $8.8\,\mathrm{W\,m^{-2}}$, whereas downwelling SW radiation is slightly overestimated with a mean bias of $2.1\,\mathrm{W\,m^{-2}}$ and an RMSE value of $5.5\,\mathrm{W\,m^{-2}}$. The earlier hypothesis[49] is supported by our IWP/LWP comparisons in the Supplementary Methods. To improve the modelled IWP/LWP values and reduce the corresponding downwelling SW/LW biases, we therefore constructed a hybrid satellite-climate model data set.

Hybrid satellite-climate model data set. The accuracy of the satellite observations and high temporal resolution of RACMO2.3 were combined by fitting 3-day moving averages through both the satellite and model LWP/IWP values and calculating an additive correction factor CF (that is, the difference between the satellite and model moving average value) at each model output timestep for each 2° by 2° grid box. The CF factor is exponentially weighted with respect to the original model values, to avoid that too much LWP/IWP is added when originally no cloud was present in the model. The resulting exponentially weighted factor was used to rescale the LWP/IWP model values, whereas the original model values were retained when no satellite retrieval was available within the moving average window. The correction procedure is mathematically shown in equation (7) for the example of LWP. The factor P determines the shape of the exponential function. Both the width of the moving average window and the exponential factor P were chosen in such a way that the resulting corrected radiative fluxes resemble the satellite-retrieved radiative fluxes as closely as possible (see below). The factor P ranges from 0.03 in summer to 1,000 in winter. The resulting differences between corrected and original IWP values range from -69 to $203\,\mathrm{g\,m^{-2}}$, whereas the difference in LWP values range from -13 to $82\,\mathrm{g\,m^{-2}}$ (99 percentiles).

$$\mathrm{LWP}_{\mathrm{i,corrected}} = \mathrm{LWP}_{\mathrm{i,original}} + CF\times(1-\exp(-P\times\mathrm{LWP}_{\mathrm{i,original}})) \quad (7)$$

There are no significant lags in time and place for RACMO2.3 simulations compared with satellite observations. RACMO2.3 is forced by ERA-Interim, giving very good correlations with observations for daily and subdaily variations in near-surface climate (temperature, wind speed and humidity)[51,57]. We also investigated the possible occurrence of lags in time by performing cross-correlation tests between RACMO2.3 LWP/IWP simulations and the satellite LWP/IWP retrievals, and found no significant lag (not shown).

Comparisons of the original RACMO and hybrid satellite-climate model LWP histograms with the ground-based retrievals by the MWR at Summit are shown in Supplementary Fig. 7. These histograms confirm that the hybrid LWP values agree significantly better with ground-based observations, although LWP values remain somewhat underestimated. Moreover, for the five available months (July to December 2010) with concurrent ground-based observations, the difference in median LWP value has decreased from an underestimation of 5.3 to $1.7\,\mathrm{g\,m^{-2}}$.

After constraining the original model LWP/IWP by satellite observations, the trained neural networks were used to scale the modelled radiative fluxes accordingly. This was done by first converting the original RACMO fluxes to clear-sky fluxes and subsequently converting these clear-sky fluxes to radiative fluxes that match the hybrid LWP/IWP data set. In this framework, clear-sky radiative fluxes were calculated by obtaining cloud enhancement factors $F_{\mathrm{all,\ LW,\ original}}$ and $F_{\mathrm{all,\ SW,\ original}}$ from the neural networks with the original LWP/IWP/T2m/SZA/albedo model values as inputs (equation (8) for the LW example).

$$\mathrm{LW_{cs}}\downarrow = \frac{\mathrm{LW_{original}}\downarrow}{F_{\mathrm{all,\ LW,\ original}}} \quad (8)$$

Next, a new set of cloud enhancement factors $F_{\mathrm{all,\ LW,\ hybrid}}$ and $F_{\mathrm{all,\ SW,\ hybrid}}$ was calculated by using the hybrid LWP/IWP data set as inputs for the neural networks, in addition to updated albedo values based on a physical albedo parameterization[60], which match the corrected optical depths, as albedo generally increases with optical depth[32]. Finally, these adjusted cloud enhancement factors were used to calculate the hybrid all-sky $\mathrm{LW_{hybrid}}\downarrow/\mathrm{SW_{hybrid}}\downarrow$ fluxes, as shown in equation (9) for the LW.

$$\mathrm{LW_{hybrid}}\downarrow = F_{\mathrm{all,LW,hybrid}}\times\mathrm{LW_{cs}}\downarrow \quad (9)$$

The LWP/IWP correction parameters were chosen in such a way that the corrected downwelling LW/SW fluxes over the entire GrIS match the satellite-retrieved radiative fluxes as closely as possible (Supplementary Fig. 6, blue squares). This reduced the original downwelling LW bias of $-6.2\,\mathrm{W\,m^{-2}}$ and RMSE of $8.8\,\mathrm{W\,m^{-2}}$ to $-0.9\,\mathrm{W\,m^{-2}}$ and $6.1\,\mathrm{W\,m^{-2}}$, respectively. Downwelling SW bias and RMSE were reduced from 2.1 and $5.5\,\mathrm{W\,m^{-2}}$ to -0.3 and $3.7\,\mathrm{W\,m^{-2}}$, respectively.

Finally, we independently evaluated both the original and hybrid downwelling LW/SW fluxes in RACMO by comparisons with 11 AWSs on the GrIS. The results in Supplementary Fig. 8 show that the systematic underestimation of downwelling LW fluxes at the surface in the original RACMO2.3 model has been greatly reduced in the hybrid satellite-climate model data set. Mean bias in the LW fluxes is reduced from -9.9 to $-3.2\,\mathrm{W\,m^{-2}}$, with a reduced RMSE from 10.9 to $6.8\,\mathrm{W\,m^{-2}}$. Mean bias in the SW fluxes is reduced from 4.4 to $1.8\,\mathrm{W\,m^{-2}}$, with no change in RMSE.

CRE estimates for all clouds in the hybrid satellite-climate model data set follow directly from the previous results, whereas an additional set of $F_{\mathrm{liq,\ LW,\ hybrid}}$ and $F_{\mathrm{liq,\ SW,\ hybrid}}$ factors was calculated for estimating CRE by cloud liquid water in liquid-bearing clouds. Surface albedo values from RACMO2.3 were updated for this purpose to include the effect of changing cloud optical depth on the albedo values, following a physical albedo parameterization[60].

Simulating the GrIS SMB using SNOWPACK. To couple the effect of clouds on the SEB to the SMB, SNOWPACK simulations were performed for the 'all-sky', 'no-liquid' and 'clear-sky' scenarios on the 2° by 2° grid for the three hydrological years from September 2007 to September 2010, after a 10-year spinup. SNOWPACK[33] is a one-dimensional physical snow model, which, driven by time series of standard meteorological observations, models the stratigraphy, snow microstructure, snow metamorphism, snow temperature profile and settlement, as well as surface energy exchange and mass balance of a snowpack. SNOWPACK can be run with a variety of schemes for albedo, metamorphism and water balance. The output of the model is a set of time series that describe the snow profile (albedo, temperature, grain size, density and water content) and its processes (refreezing, water retention and so on). SNOWPACK was developed for seasonal snow, but it has recently been successfully applied to Antarctica as well[61].

The all-sky simulation (that is, control run) was forced by the hybrid satellite-climate model radiative fluxes and RACMO2.3 meteorological input data (2 m air temperature, 10 m wind speed, 2 m relative humidity and total precipitation), linearly interpolated to hourly values, and represents the most realistic estimate of the actual GrIS SMB. In the 'clear-sky' simulation the LW and SW CRE of all clouds was removed to simulate the cloud impact (both ice-only and liquid-bearing) on the SMB, whereas in the 'no-liquid' simulation LW and SW CRE by cloud (supercooled) liquid water was excluded to distinguish between the ice-only and liquid-only radiative impacts. As we focus solely on the radiative effect of clouds on the SMB, their role as precipitation source is not considered.

These simulations were performed using a physically based broadband albedo parameterization[53,60], which was implemented in SNOWPACK and accounts for albedo variations in function of the specific surface area of the snow/ice profile, cloud optical depth, SZA and concentration of light-absorbing carbon in the snow/ice, but with a lower limit of albedo values that corresponds to the spatially variable ice-albedo values derived from MODIS[53]. The specific surface area of the snow/ice profile was derived from SNOWPACK geometric grain size per layer, whereas the cloud optical depth was calculated from the hybrid satellite-climate model LWP/IWP data set. Finally, the concentration of light-absorbing carbon in the snow/ice was set to a fixed value per layer together with the other SNOWPACK model parameters (Supplementary Table 2). SNOWPACK was run in an offline configuration, that is, identical atmospheric temperature, wind speed, relative humidity and precipitation were used for all three simulations. This implies that we do not allow the atmosphere to react to changes in the snowpack.

The performance of SNOWPACK to simulate GrIS surface conditions was assessed based on the comparison of SNOWPACK SMB with RACMO SMB and stake measurements, SNOWPACK melt with satellite-observed melt and SNOWPACK refreezing rates with refreezing rates from literature. First, the yearly average SMB for each grid box was calculated, showing a close correspondence between SNOWPACK and RACMO SMB (Supplementary Fig. 9). We would like to stress here that this comparison is not aimed at an absolute evaluation of GrIS SMB estimates, but merely serves as a check that SNOWPACK simulates a similar SMB when forced by the same meteorological inputs as RACMO2.3. Integrated over the

GrIS, this is also confirmed by the strong agreement between the red and black curves in Fig. 3a. The limited amount of SMB observations through yearly stake measurements hinder a Greenland-wide observation-based evaluation of SMB. Yet, we compared the SNOWPACK-simulated SMB with observations from the K-transect[62] in the same period. The spatial resolution of the SNOWPACK runs prevents individual comparisons for all stations of the K-transect. However, the S10 station is representative for an accumulation pixel, whereas the other stations cover an ablation pixel. The yearly mean recorded SMB at S10 is 203 $kg\,m^{-2}$, whereas SNOWPACK indicates 243 $kg\,m^{-2}$. The other stations have a mean yearly SMB of $-1{,}714\,kg\,m^{-2}$, whereas the corresponding grid box of SNOWPACK has an SMB of $-1{,}708\,kg\,m^{-2}$. This comparison gives a good indication that the SNOWPACK estimates are sufficiently in line with the available SMB observations in the field. In addition, melt occurrence was calculated, to assess the ability of SNOWPACK to reproduce observed melt events from spaceborne passive microwave data[63]. In this framework, melt days in the SNOWPACK simulations were defined as days during which hourly surface temperatures reached the melting point within the window of overpass of the satellite[63] to detect surface melt. The original daily melt detections[63] were aggregated to the 2° by 2° grid for comparison with SNOWPACK. The yearly amount of melt days for each grid box agree reasonably well between observations and SNOWPACK simulations (Supplementary Fig. 10). Some scatter exists due to the definition of melt days, slight timing offsets and relatively coarse resolution of the SNOWPACK simulations. Nonetheless, the general agreement with observations is good and also the spatial patterns are captured well (Supplementary Fig. 11). Based on our definition of melt days, we find a mean yearly melt index (number of days melt is observed, multiplied by the area where the melt occurs) of $3.45 \times 10^7\,km^2$ days, whereas the melt index calculated from the observed melt data[63] is $3.60 \times 10^7\,km^2$ days, a difference of $<5\%$. In addition, also the timing of simulated melt events was compared with observed melt events, using the same data set[63]. The result shows a very good agreement between both simulated SNOWPACK melt events and observed melt events (Supplementary Fig. 12). Finally, we quantified the mean refreezing rate in SNOWPACK in the all-sky simulation as 45%, whereas another study[64] found 42 ± 4 %. The SMB, melt and refreezing rate comparisons illustrate the good performance of SNOWPACK for simulating the GrIS SMB.

Uncertainty derivations. Radiative fluxes retrieved from satellite observations carry an inherent uncertainty due to many different factors. In the retrieval of CRE from satellite observations, we have accounted for the following sources of uncertainty: (i) LW/SW radiative flux retrievals from 2B-FLXHR-LIDAR, (ii) cloud enhancement factors F predicted by the neural networks and (iii) assumed surface albedo values in 2B-FLXHR-LIDAR. To incorporate uncertainty estimates based on the LW/SW radiative flux retrievals by 2B-FLXHR-LIDAR, we used the comparisons with ground-based radiative flux measurements from AWS observations. In this framework, a general RMSE value for both LW fluxes (9.3 $W\,m^{-2}$) and SW fluxes (22.5 $W\,m^{-2}$) was derived from the individual comparisons with the stations. These RMSE values represent the mean uncertainty of the satellite-retrieved radiative fluxes over the GrIS and were subsequently used to perturb the obtained satellite LW/SW fluxes (100 iterations), by introducing random errors from a normal distribution $N(\mu = 0; \sigma = \text{RMSE})$. These were then used as inputs to the neural networks, where an additional uncertainty was added by perturbing the predicted cloud enhancement factors F using random errors from a normal distribution with means and s.d. based on the network performance results (Supplementary Fig. 4). Finally, a last source of uncertainty was added to the 100 sets of radiative fluxes, based on the surface albedo assumption. Comparison of the albedo values used in the 2B-FLXHR-LIDAR algorithm with the physically based albedo retrievals in the RACMO model resulted in a mean bias of 0.04 ± 0.05. A random error was introduced based on these characteristics, from which updated radiative fluxes were calculated. The final 100 sets of radiative fluxes were used to calculate corresponding sets of CRE and include the three most important sources of uncertainty. The s.d. of this data set provides the combined uncertainty estimate of CRE values that are reported in this study.

SMB terms in SNOWPACK contain a combination of uncertainties that accumulated in earlier steps of our methodology. These uncertainties were estimated using a combination of (i) perturbed CRE to force SNOWPACK and (ii) inherent SNOWPACK model uncertainty. As the SMB terms for the different simulations depend strongly on the microphysical characteristics of clouds, we quantified the LWP/IWP uncertainty of the hybrid satellite-climate model data set by comparing these values with the satellite observations. We found a mean LWP uncertainty of 35% and IWP uncertainty of 12%. Therefore, we performed four additional SNOWPACK runs using all possible combinations of adding/removing 35% of the LWP and 12% of the IWP at each timestep, and calculating the corresponding radiative fluxes. The range of resulting outputs for the all-sky and no-liquid simulations were used as uncertainty due to the amount of cloudiness in the SNOWPACK runs. Inherent SNOWPACK uncertainty was estimated by comparing SNOWPACK and RACMO estimates of the SMB terms (Fig. 3a). Results of this comparison showed that the uncertainty on the total GrIS SMB is $\sim 10\%$, whereas the uncertainty on the individual components is $\sim 50\%$ for sublimation and $\sim 20\%$ for runoff. Finally, both sources of uncertainty (inherent model uncertainty and cloudiness) on SNOWPACK SMB and individual mass terms were combined (whiskers and shaded areas in Fig. 3, and all mass-related uncertainties in the text).

Code availability. All codes that have contributed to the results reported in this study are available on request. The 2B-FLXHR-LIDAR data are available at http://www.cloudsat.cira.colostate.edu/.

References

1. Box, J. E. *et al.* Greenland ice sheet albedo feedback: thermodynamics and atmospheric drivers. *Cryosphere* **6,** 821–839 (2012).
2. Fyke, J. G., Vizcaíno, M. & Lipscomb, W. H. The pattern of anthropogenic signal emergence in Greenland Ice Sheet surface mass balance. *Geophys. Res. Lett.* **41,** 6002–6008 (2014).
3. Fettweis, X. *et al.* Brief communication 'Important role of the mid-tropospheric atmospheric circulation in the recent surface melt increase over the Greenland ice sheet'. *Cryosphere* **7,** 241–248 (2013).
4. Shepherd, A. *et al.* A reconciled estimate of ice-sheet mass balance. *Science* **338,** 1183–1189 (2012).
5. IPCC. *Climate Change 2013—The Physical Science Basis* (eds Stocker, TF. *et al.*) pp. 1535 (Cambridge Univ. Press, Cambridge, UK and New York, doi:10.1017/CBO9781107415324, 2014).
6. Enderlin, E. M. *et al.* An improved mass budget for the Greenland ice sheet. *Geophys. Res. Lett.* **41,** 866–872 (2014).
7. Hanna, E. *et al.* Increased runoff from melt from the Greenland Ice Sheet: a response to global warming. *J. Clim.* **21,** 331–341 (2008).
8. Andersen, M. *et al.* Basin-scale partitioning of Greenland ice sheet mass balance components (2007-2011). *Earth Planet Sci. Lett.* **409,** 89–95 (2015).
9. Fettweis, X., van Ypersele, J.-P., Gallée, H., Lefebre, F. & Lefebvre, W. The 1979–2005 Greenland ice sheet melt extent from passive microwave data using an improved version of the melt retrieval XPGR algorithm. *Geophys. Res. Lett.* **34,** L05502 (2007).
10. Rignot, E. & Kanagaratnam, P. Changes in the velocity structure of the Greenland Ice Sheet. *Science* **311,** 986–990 (2006).
11. van den Broeke, M. *et al.* Partitioning recent Greenland mass loss. *Science* **326,** 984–986 (2009).
12. Ramanathan, V. *et al.* Cloud-radiative forcing and climate: results from the earth radiation budget experiment. *Science* **243,** 57–63 (1989).
13. Curry, J. A., Schramm, J. L., Rossow, W. B. & Randall, D. Overview of Arctic cloud and radiation characteristics. *J. Clim.* **9,** 1731–1764 (1996).
14. Shupe, M. D. & Intrieri, J. M. Cloud radiative forcing of the Arctic surface: the influence of cloud properties, surface albedo, and solar zenith angle. *J. Clim.* **17,** 616–628 (2004).
15. Turner, D. D. *et al.* Thin Liquid water clouds: their importance and our challenge. *Bull. Am. Meteorol. Soc.* **88,** 177–190 (2007).
16. Bennartz, R. *et al.* July 2012 Greenland melt extent enhanced by low-level liquid clouds. *Nature* **496,** 83–86 (2013).
17. Neff, W., Compo, G. P., Martin Ralph, F. & Shupe, M. D. Continental heat anomalies and the extreme melting of the Greenland ice surface in 2012 and 1889. *J. Geophys. Res.-Atmos.* **119,** 6520–6536 (2014).
18. Ettema, J., van den Broeke, M. R., van Meijgaard, E. & van de Berg, W. J. Climate of the Greenland ice sheet using a high-resolution climate model—Part 2: Near-surface climate and energy balance. *Cryosphere* **4,** 529–544 (2010).
19. Franco, B., Fettweis, X. & Erpicum, M. Future projections of the Greenland ice sheet energy balance driving the surface melt. *Cryosphere* **7,** 1–18 (2013).
20. Vizcaíno, M., Lipscomb, W. H., Sacks, W. J. & van den Broeke, M. Greenland surface mass balance as simulated by the community earth system model. Part II: twenty-first-century changes. *J. Clim.* **27,** 215–226 (2014).
21. Stephens, G. L. *et al.* CloudSat mission: performance and early science after the first year of operation. *J. Geophys. Res.* **113,** D00A18 (2008).
22. L'Ecuyer, T. S. & Jiang, J. H. Touring the atmosphere aboard the A-train. *Phys. Today* **63,** 36–41 (2010).
23. Platnick, S. *et al.* The MODIS cloud products: algorithms and examples from terra. *IEEE Trans. Geosci. Remote Sens.* **41,** 459–473 (2003).
24. Henderson, D. S., L'Ecuyer, T., Stephens, G., Partain, P. & Sekiguchi, M. A multisensor perspective on the radiative impacts of clouds and aerosols. *J. Appl. Meteorol. Climatol.* **52,** 853–871 (2013).
25. Zhang, D., Wang, Z. & Liu, D. A global view of midlevel liquid-layer topped stratiform cloud distribution and phase partition from CALIPSO and CloudSat measurements. *J. Geophys. Res.* **115,** D00H13 (2010).
26. Kay, J. E., L'Ecuyer, T., Gettelman, A., Stephens, G. & O'Dell, C. The contribution of cloud and radiation anomalies to the 2007 Arctic sea ice extent minimum. *Geophys. Res. Lett.* **35,** L08503 (2008).
27. Cesana, G., Kay, J. E., Chepfer, H., English, J. M. & de Boer, G. Ubiquitous low-level liquid-containing Arctic clouds: new observations and climate model constraints from CALIPSO-GOCCP. *Geophys. Res. Lett.* **39,** L20804 (2012).
28. Shupe, M. D. *et al.* High and dry: new observations of tropospheric and cloud properties above the Greenland ice sheet. *Bull. Am. Meteorol. Soc.* **94,** 169–186 (2013).
29. van den Broeke, M., Smeets, P., Ettema, J. & Munneke, P. K. Surface radiation balance in the ablation zone of the west Greenland ice sheet. *J. Geophys. Res.* **113,** D13105 (2008).

30. Flanner, M. G. & Zender, C. S. Linking snowpack microphysics and albedo evolution. *J. Geophys. Res.* **111,** D12208 (2006).
31. Yamanouchi, T. Variations of incident solar flux and snow albedo on the solar zenith angle and cloud cover, at Mizuho Station, Antarctica. *J. Meteorol. Soc. Jpn* **61,** 879–893 (1983).
32. Niwano, M. *et al.* Numerical simulation of extreme snowmelt observed at the SIGMA-A site, northwest Greenland, during summer 2012. *Cryosphere* **9,** 971–988 (2015).
33. Lehning, M., Bartelt, P., Brown, B., Fierz, C. & Satyawali, P. A physical SNOWPACK model for the Swiss avalanche warning: part II. Snow microstructure. *Cold Regions Sci. Technol.* **35,** 147–167 (2002).
34. Harper, J., Humphrey, N., Pfeffer, W. T., Brown, J. & Fettweis, X. Greenland ice-sheet contribution to sea-level rise buffered by meltwater storage in firn. *Nature* **491,** 240–243 (2012).
35. de la Peña, S. *et al.* Changes in the firn structure of the western Greenland Ice Sheet caused by recent warming. *Cryosphere* **9,** 1203–1211 (2015).
36. Boisvert, L. N. & Stroeve, J. C. The Arctic is becoming warmer and wetter as revealed by the atmospheric infrared sounder. *Geophys. Res. Lett.* **42,** 4439–4446 (2015).
37. Bony, S. *et al.* Clouds, circulation and climate sensitivity. *Nat. Geosci.* **8,** 261–268 (2015).
38. Van Tricht, K. *et al.* An improved algorithm for polar cloud-base detection by ceilometer over the ice sheets. *Atmos. Meas. Tech.* **7,** 1153–1167 (2014).
39. Kay, J. E. & L'Ecuyer, T. Observational constraints on Arctic Ocean clouds and radiative fluxes during the early 21st century. *J. Geophys. Res.-Atmos.* **118,** 7219–7236 (2013).
40. Liu, Y., Key, J. R., Ackerman, S. A., Mace, G. G. & Zhang, Q. Arctic cloud macrophysical characteristics from CloudSat and CALIPSO. *Remote Sens. Environ.* **124,** 159–173 (2012).
41. Sassen, K., Wang, Z. & Liu, D. Global distribution of cirrus clouds from CloudSat/Cloud-Aerosol Lidar and Infrared Pathfinder Satellite Observations (CALIPSO) measurements. *J. Geophys. Res.* **113,** D00A12 (2008).
42. Turner, D. D. *et al.* Retrieving liquid water path and precipitable water vapor from the Atmospheric Radiation Measurement (ARM) microwave radiometers. *IEEE Trans. Geosci. Remote Sens.* **45,** 3680–3690 (2007).
43. van den Broeke, M., van As, D., Reijmer, C. & van de Wal, R. Assessing and improving the quality of unattended radiation observations in Antarctica. *J. Atmos. Ocean. Technol.* **21,** 1417–1431 (2004).
44. Shupe, M. D. *et al.* Deriving Arctic cloud microphysics at Barrow, Alaska: algorithms, results, and radiative closure. *J. Appl. Meteorol. Climatol.* **54,** 1675–1689 (2015).
45. Unsworth, M. H. & Monteith, J. L. Long-wave radiation at the ground I. Angular distribution of incoming radiation. *Q. J. R. Meteorol. Soc.* **101,** 13–24 (1975).
46. Sicart, J. E., Pomeroy, J. W., Essery, R. L. H. & Bewley, D. Incoming longwave radiation to melting snow: observations, sensitivity and estimation in Northern environments. *Hydrol. Process.* **20,** 3697–3708 (2006).
47. Chen, Y., Aires, F., Francis, J. A. & Miller, J. R. Observed relationships between Arctic longwave cloud forcing and cloud parameters using a neural network. *J. Clim.* **19,** 4087–4104 (2006).
48. Miller, N. B. *et al.* Cloud radiative forcing at Summit, Greenland. *J. Clim.* **28,** 6267–6280 (2015).
49. Noël, B. *et al.* Evaluation of the updated regional climate model RACMO2.3: summer snowfall impact on the Greenland Ice Sheet. *Cryosphere* **9,** 1831–1844 (2015).
50. ECMWF-IFS: Part IV : PHYSICAL PROCESSES (CY33R1). Technical Report (2008).
51. Ettema, J. *et al.* Climate of the Greenland ice sheet using a high-resolution climate model—Part 1: evaluation. *Cryosphere* **4,** 511–527 (2010).
52. Kuipers Munneke, P. *et al.* A new albedo parameterization for use in climate models over the Antarctic ice sheet. *J. Geophys. Res.* **116,** D05114 (2011).
53. van Angelen, J. H. *et al.* Sensitivity of Greenland Ice Sheet surface mass balance to surface albedo parameterization: a study with a regional climate model. *Cryosphere* **6,** 1175–1186 (2012).
54. Lenaerts, J. T. M. *et al.* Modeling drifting snow in Antarctica with a regional climate model: 1. Methods and model evaluation. *J. Geophys. Res.* **117,** D05108 (2012).
55. Van Wessem, J. M. *et al.* Updated cloud physics in a regional atmospheric climate model improves the modelled surface energy balance of Antarctica. *Cryosphere* **8,** 125–135 (2014).
56. Lenaerts, J. T. M. *et al.* Extreme precipitation and climate gradients in Patagonia revealed by high-resolution regional atmospheric climate modeling. *J. Clim.* **27,** 4607–4621 (2014).
57. Lenaerts, J. T. M. *et al.* Drifting snow measurements on the Greenland Ice Sheet and their application for model evaluation. *Cryosphere* **8,** 801–814 (2014).
58. Sanz Rodrigo, J., Buchlin, J.-M., van Beeck, J., Lenaerts, J. T. M. & van den Broeke, M. R. Evaluation of the antarctic surface wind climate from ERA reanalyses and RACMO2/ANT simulations based on automatic weather stations. *Clim. Dyn.* **40,** 353–376 (2012).
59. Gorter, W., van Angelen, J. H., Lenaerts, J. T. M. & van den Broeke, M. R. Present and future near-surface wind climate of Greenland from high resolution regional climate modelling. *Clim. Dyn.* **42,** 1595–1611 (2013).
60. Gardner, A. S. & Sharp, M. J. A review of snow and ice albedo and the development of a new physically based broadband albedo parameterization. *J. Geophys. Res.* **115,** 1–15 (2010).
61. Groot Zwaaftink, C. D. *et al.* Event-driven deposition of snow on the Antarctic Plateau: analyzing field measurements with SNOWPACK. *Cryosphere* **7,** 333–347 (2013).
62. van de Wal, R. S. W. *et al.* Twenty-one years of mass balance observations along the K-transect, West Greenland. *Earth Syst. Sci. Data* **4,** 31–35 (2012).
63. Tedesco, M. *Greenland Daily Surface Melt 25 km EASE-Grid, 2007-2010.* City College of New York, The City University of New York Digital Media (2014).
64. van Angelen, J. H., van den Broeke, M. R., Wouters, B. & Lenaerts, J. T. M. Contemporary (1960-2012) evolution of the climate and surface mass balance of the Greenland Ice Sheet. *Surv. Geophys.* **35,** 1155–1174 (2013).

Acknowledgements

K.V.T. and S.L. are funded by the Research Foundation Flanders (FWO). I.V.G. was supported via the project HYDRANT funded by the Belgian Science Policy Office under grant number EN/01/4B. The contributions of T.S.L. to this project were supported by CloudSat/CALIPSO science team grant NNX14AB35G. This study is partly funded by Utrecht University through its strategic theme Sustainability, sub-theme Water, Climate and Ecosystems, and the Netherlands Polar Programme of NWO. Part of this work was carried out under the programme of the Netherlands Earth System Science Centre (NESSC), financially supported by the Ministry of Education, Culture and Science (OCW). This work is further supported by the Belgian Federal Science Policy Office project AEROCLOUD (BR/143/A2). The Summit data were recorded in the frame of the ICECAPS project, which is supported by the US National Science Foundation under Grants ARC-0856773, 0904152 and 0856559 as part of the Arctic Observing Network (AON) programme, with additional instrumentation provided by the NOAA Earth System Research Laboratory, US Department of Energy ARM Program and Environment Canada. We thank the CloudSat Data Processing Center, in particular Phil Partain, for their help with the 2B-FLXHR-LIDAR algorithm runs.

Author contributions

K.V.T. conceived this study, constructed the hybrid satellite-climate model data set, performed most of the analyses and wrote the manuscript. S.L. implemented the SNOWPACK snow model and conducted all simulations, and participated in writing the manuscript. J.T.M.L. performed the Coupled Model Intercomparison Project phase 5 analyses and theoretical calculations of SMB impacts, provided many insights in the coupling between SEB and SMB, and participated in writing the manuscript. I.V.G. worked on the theoretical basis of the study and helped with the interpretation of results. T.S.L. was responsible for the CloudSat/CALIPSO/MODIS satellite products and has overseen the conducted 2B-FLXHR-LIDAR runs. B.N. performed RACMO2.3 runs over Greenland and supplied the necessary climate model data to construct the hybrid satellite-climate model data set. M.R.v.d.B. is responsible for the use of RACMO2.3 over Greenland and has overseen its use in this study and helped interpreting the results. D.D.T. performed the all-sky retrievals to derive LWP from the microwave radiometer at Summit and provided theoretical insights in clouds over the GrIS. N.P.M.v.L. developed the methodology for constraining the original RACMO model output by the available satellite observations and has worked on the theoretical part of the study and result interpretations. All authors commented on the manuscript.

Additional information

Competing financial interests: The authors declare no competing financial interests.

2

A global reanalysis of storm surges and extreme sea levels

Sanne Muis[1], Martin Verlaan[2,3], Hessel C. Winsemius[2], Jeroen C.J.H. Aerts[1] & Philip J. Ward[1]

Extreme sea levels, caused by storm surges and high tides, can have devastating societal impacts. To effectively protect our coasts, global information on coastal flooding is needed. Here we present the first global reanalysis of storm surges and extreme sea levels (GTSR data set) based on hydrodynamic modelling. GTSR covers the entire world's coastline and consists of time series of tides and surges, and estimates of extreme sea levels. Validation shows that there is good agreement between modelled and observed sea levels, and that the performance of GTSR is similar to that of many regional hydrodynamic models. Due to the limited resolution of the meteorological forcing, extremes are slightly underestimated. This particularly affects tropical cyclones, which requires further research. We foresee applications in assessing flood risk and impacts of climate change. As a first application of GTSR, we estimate that 1.3% of the global population is exposed to a 1 in 100-year flood.

[1]Institute for Environmental Studies (IVM), Vrije Universiteit Amsterdam, 1081 HV Amsterdam, The Netherlands. [2]Deltares, 2600 MH Delft, The Netherlands. [3]Mathematical Physics, Electrical Engineering, Mathematics and Computer Science (EEMCS), TU Delft, 2628 CD Delft, The Netherlands. Correspondence and requests for materials should be addressed to S.M. (email: sanne.muis@vu.nl).

Storm surge, a rise in sea level due to low atmospheric pressure and strong winds, is the main driver of coastal flood events[1]. The most extreme of these events are caused by tropical cyclones[2], but extra-tropical storms can also produce high sea levels, especially when they coincide with high tide[3]. With over 600 million people living in low-lying coastal areas[4], coastal floods can have devastating societal impacts. It is estimated that on average 0.8–1.1 million people per year are flooded globally[5]. This is reflected by disasters such as the flooding of The Netherlands and the United Kingdom in 1953, which resulted in over 2,000 fatalities[6], and led to the construction of a series of flood protection works along the Dutch coast and the Thames Barrier in London. Another more recent catastrophe was the flooding of New Orleans in 2005 due to tropical cyclone Katrina, which resulted in around 1,100 fatalities[7].

In recent years, coastal flood risk has increased due to population, economic growth[8] and land subsidence[9,10]. To date, socioeconomic development has been the main driver of increasing risk[11], but in the future sea level rise will be an important driver of increasing risk of coastal floods[5,12–14]. To analyse spatial patterns and temporal trends in coastal flood risk, several continental to global scale studies have been carried out[5,8,12,15–17] based on the extreme sea levels in the Dynamic Interactive Vulnerability Assessment (DIVA) input database[18,19]. The extreme sea levels in the DIVA database have been instrumental in many risk assessments, and have provided important insights into which areas face the highest risk, as well as the potential effects of sea level rise and adaptation. However, some applications require time series of sea levels (instead of extreme values), such as: assessing interannual variability; assessing the impact of changes in storm regimes; and the modelling of past events. Such time series can be obtained from tide gauge observations, but many regions at risk have insufficient numbers of tide gauges and/or record lengths available to reliably estimate extreme sea levels. Extreme sea levels vary significantly along the coast due to variability in storminess, coastline shape, and bathymetry[1]. Hence, interpolation between different stations will not accurately capture the spatial variability. Recently, more advanced techniques based on altimetry data have been developed[20], but the limited length of altimetry records prevent their application to low-probability extreme events[21,22]. Due to these data limitations, there is still limited understanding of the global coastal flood hazard, even under current (stationary) climate conditions.

At the regional scale, hydrodynamic modelling is the state-of-art approach to develop consistent and complete sea level reanalyses that can be applied to flood risk assessments[18,19]. As the modelling of surges in shallow coastal areas requires a high resolution, generally such a modelling approach has been computationally too costly to apply on the global-scale. The application of unstructured grids (or 'flexible mesh') in hydrodynamic models make it possible to have a sufficient resolution in shallow coastal areas[23], while maintaining computational efficiency[24]. These developments in hydrodynamic modelling combined with the increasing availability of global data sets on climate and elevation, make it possible to upscale the hydrodynamic modelling to the global scale.

Here we apply such a hydrodynamic approach for the first time at the global scale and present the Global Tide and Surge Reanalysis (GTSR) data set. To obtain the first near-coast global reanalysis of storm surges (1979–2014), we force the newly developed Global Tide and Surge Model (GTSM)[24] with wind speed and atmospheric pressure from the ERA-Interim global atmospheric reanalysis[25]. Tides are modelled separately using a recent update of the Finite Element Solution (FES2012) hydrodynamic model (Supplementary Fig. 1 and Supplementary Note 1)[26]. To validate the GTSR data set, we compare the results with observed sea levels from a global data set from University of Hawaii Sea Level Center (UHSLC). Results show that the GTSR time series agree very well with observed sea levels. Extreme sea levels are slightly underestimated, especially those induced by tropical cyclones; this is mainly due to the resolution of the meteorological forcing. To illustrate one of the potential applications of the GTSR data set, we assess global exposure to coastal flooding using a simple inundation model, and find that 1.3% of the global population, equal to 76 million people, is living in the 1 in 100-year floodplain. We foresee various other applications of the GTSR data set, such as: assessing flood risk at the global scale; assessing interannual variability of storminess; providing warning thresholds for operational forecasting models; and assessing the impacts of climate change.

Results

Validation of time series. For the total sea level, there is generally a good agreement between modelled and observed sea levels. The root mean square errors (RMSE), based on the time series with a 10 minute temporal resolution, are lower than 0.20 m for 80% of the stations (Fig. 1a). The average RMSE across all validation sites is 0.17 m (s.d. is 0.15 m). The performance of the surge levels is even better, with 95% of the stations having a RMSE lower than 0.2 m. The average RMSE is 0.11 m (s.d. is 0.05 m). For illustration, Fig. 2a– shows the modelled and observed surge levels for selected sites around the world. For tidal levels, about 85% of the stations have a RMSE lower than 0.2 m, but the maximum RMSE value is greater than 1 m. The average RMSE is 0.15 m (s.d. is 0.42 m). The large errors (RMSE >1 m) for the tide gauge stations Windham, Victoria and Puerto Montt (red dots in Fig. 1a) are caused by an over- or underestimation of the tidal amplitude. Supplementary Fig. 2a–b shows the RMSEs for the surge levels and tide levels for all observation stations.

To assess whether sea level extremes are also adequately represented in the GTSR time series, we calculate the performance based on daily maxima. Figure 3a–k shows scatter density plots for modelled and observed daily maxima for 12 selected locations. The majority of the daily maxima (orange to red areas in Fig. 3a–k) are close to the perfect-fit line, indicating a good performance. The performance for the majority of these 12 stations decreases for more extreme sea levels (the least-squares line diverges from the best-fit line). The underestimation of extreme sea levels is primarily an issue of resolution (see Discussion section). The average Pearson correlation coefficient is 0.83 (s.d. is 0.14), indicating a good representation of the maxima by the model. Over 75% of the stations have a correlation coefficient higher than 0.75. In regions prone to tropical cyclones (Supplementary Fig. 3a–b and Supplementary Note 2), such as the Caribbean Sea, we obtain correlation coefficients lower than 0.5. The average correlation coefficient in these tropical regions is 0.77, which is significantly lower than the average correlation coefficient of 0.87 in extra-tropical regions (two-tailed Student's *t*-test, $P<0.05$).

Validation of extremes. To obtain extreme sea levels for various return periods, we fit a Gumbel distribution to the annual maxima using the maximum-likelihood method. The Gumbel plot is shown in Fig. 4a–e for five selected stations. For all stations shown, the annual maxima follow a relatively straight line, indicating a good fit of the Gumbel distribution. Average relative errors for return periods from 5 to 100 years are in the range of 11–14%, increasing with higher return periods. As there is a limited number of stations with observation records longer than

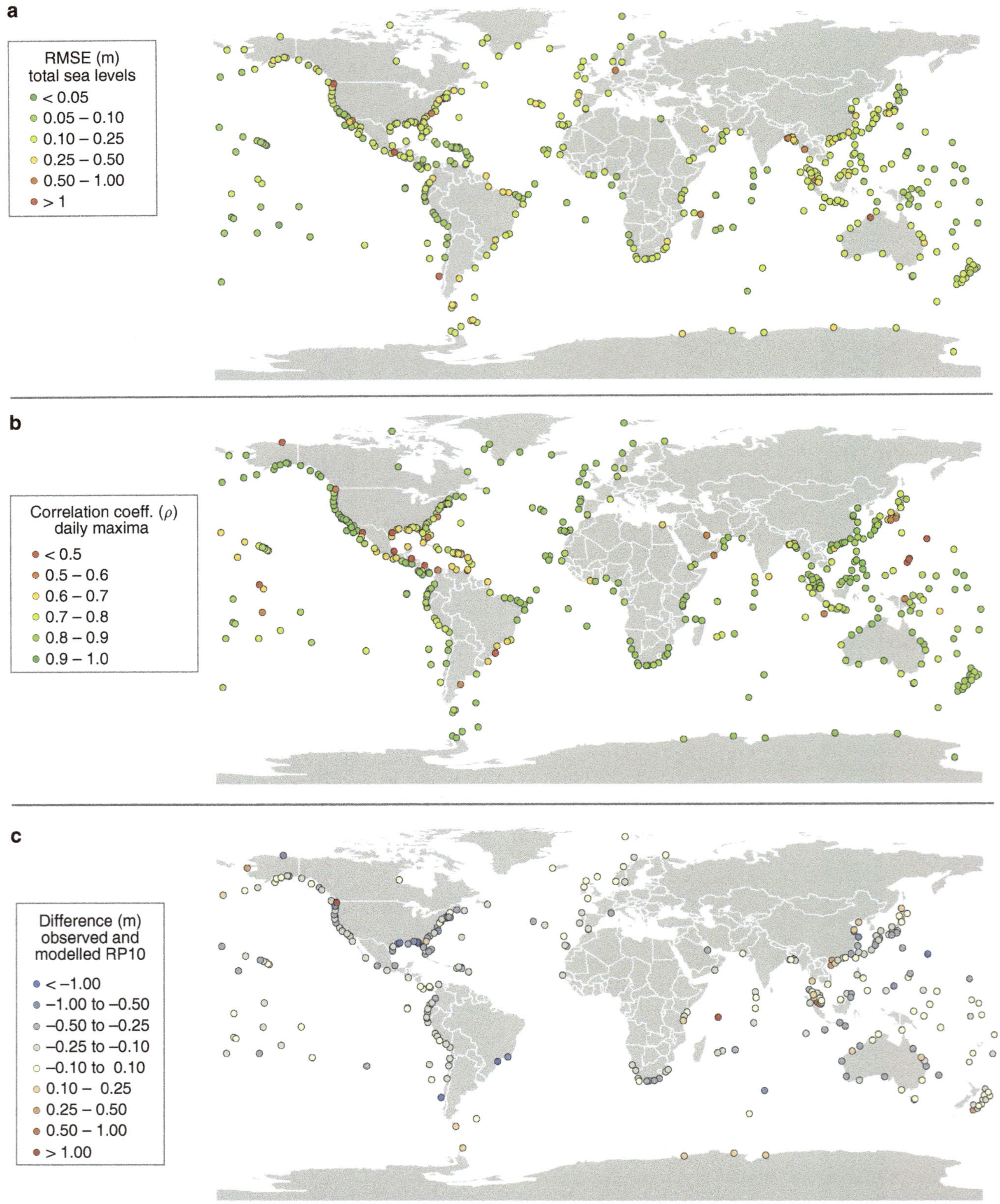

Figure 1 | Maps showing the performance of GTSR against observed sea levels. The performance of GTSR shown as (**a**) the RMSE (m) between modelled and observed sea level time series; (**b**) the Pearson correlation coefficient (ρ) between modelled and observed daily maximum sea levels; and (**c**) the bias (m) calculated by subtracting the modelled extreme sea levels from the observed extreme sea levels with a return period of 10 years.

the 30 years, we focus on 1 in 10-year extreme sea levels and use all stations that have an observation record longer than 10 years. For 75% of these 144 stations, the absolute error for the 1 in 10-year sea level is lower than 0.3 m (Fig. 1c). On average, the extremes are underestimated by −0.14 m (s.d. is 0.39). Supplementary Fig. 4 shows the results for the 1 in 100-year sea levels. The performance of GTSR is best in areas where extremes are dominated by large, extra-tropical storms, such as

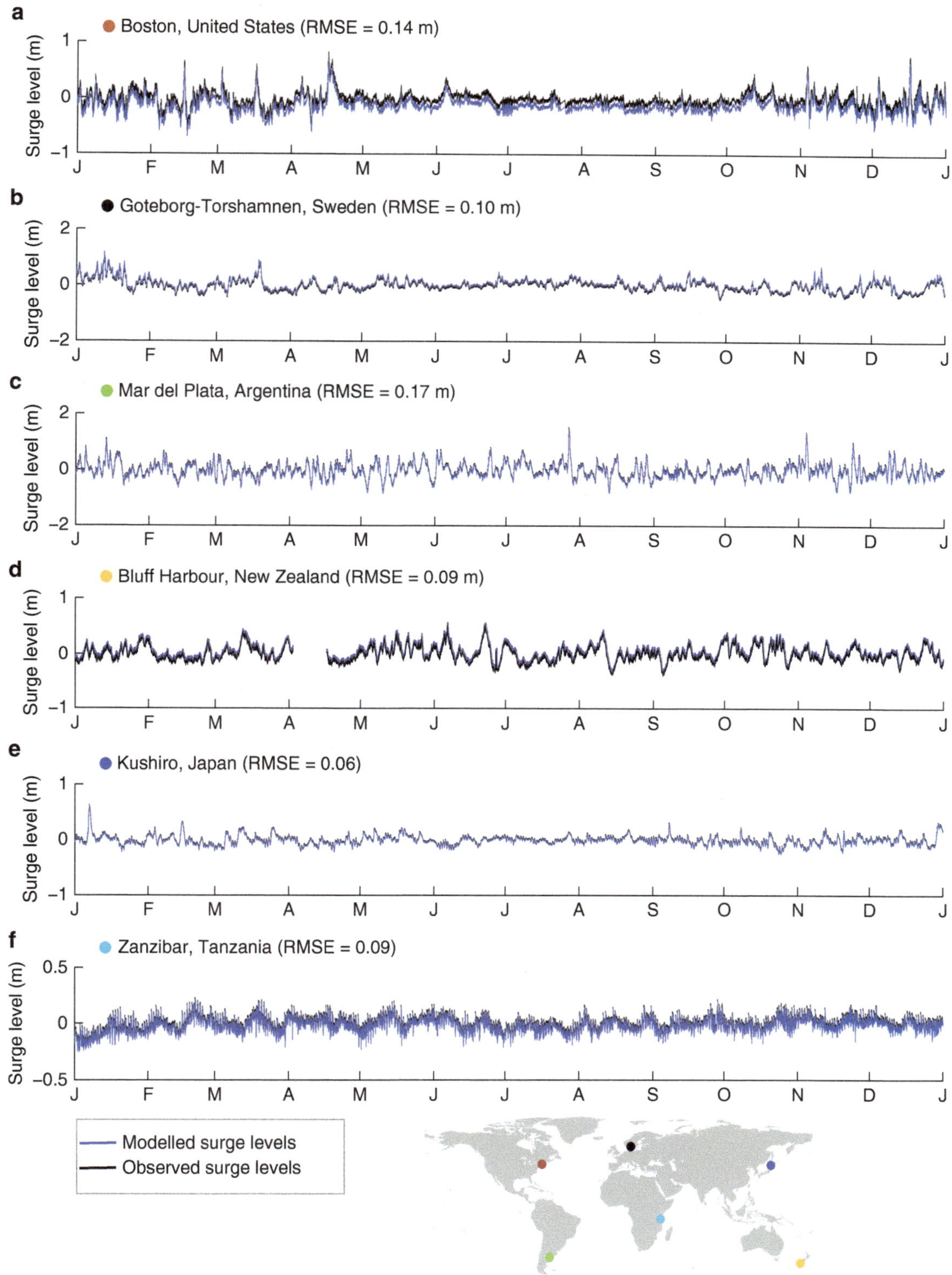

Figure 2 | Comparison of the modelled and observed surge levels at selected stations. Comparison of the modelled and observed surge levels for 2007 at six selected stations around the world, **(a)** Boston, United States; **(b)** Goteborg-Torshamn, Sweden; **(c)** Mar de la Plata, Argentina; **(d)** Bluff Harbour, New Zealand; **(e)** Kushiro, Japan; and **(f)** Zanzibar, Tanzania. The coloured dots in the world map indicate the location of the observation stations.

Europe, southeast Australia, eastern South-America and northwest North-America. In regions where storm surges are largely induced by tropical cyclones, the mean absolute difference between the extreme sea levels with a return period of 10 years is 0.23 m, which is significantly higher than in extra-tropical regions, where the mean absolute difference is 0.08 m (two-tailed Student's t-test, $P < 0.05$).

Application to assess flood exposure. To illustrate a first application of the GTSR extremes, we calculate the flood hazard (that is, inundation extent) and flood exposure (that is, exposed people) based on 1 in 100-year extreme sea levels. This assessment is based on SRTM elevation[27] and GRUMP population density for the year 2000 (refs 28,29). For this first demonstration, we assume no protection from coastal flooding and planar flood levels.

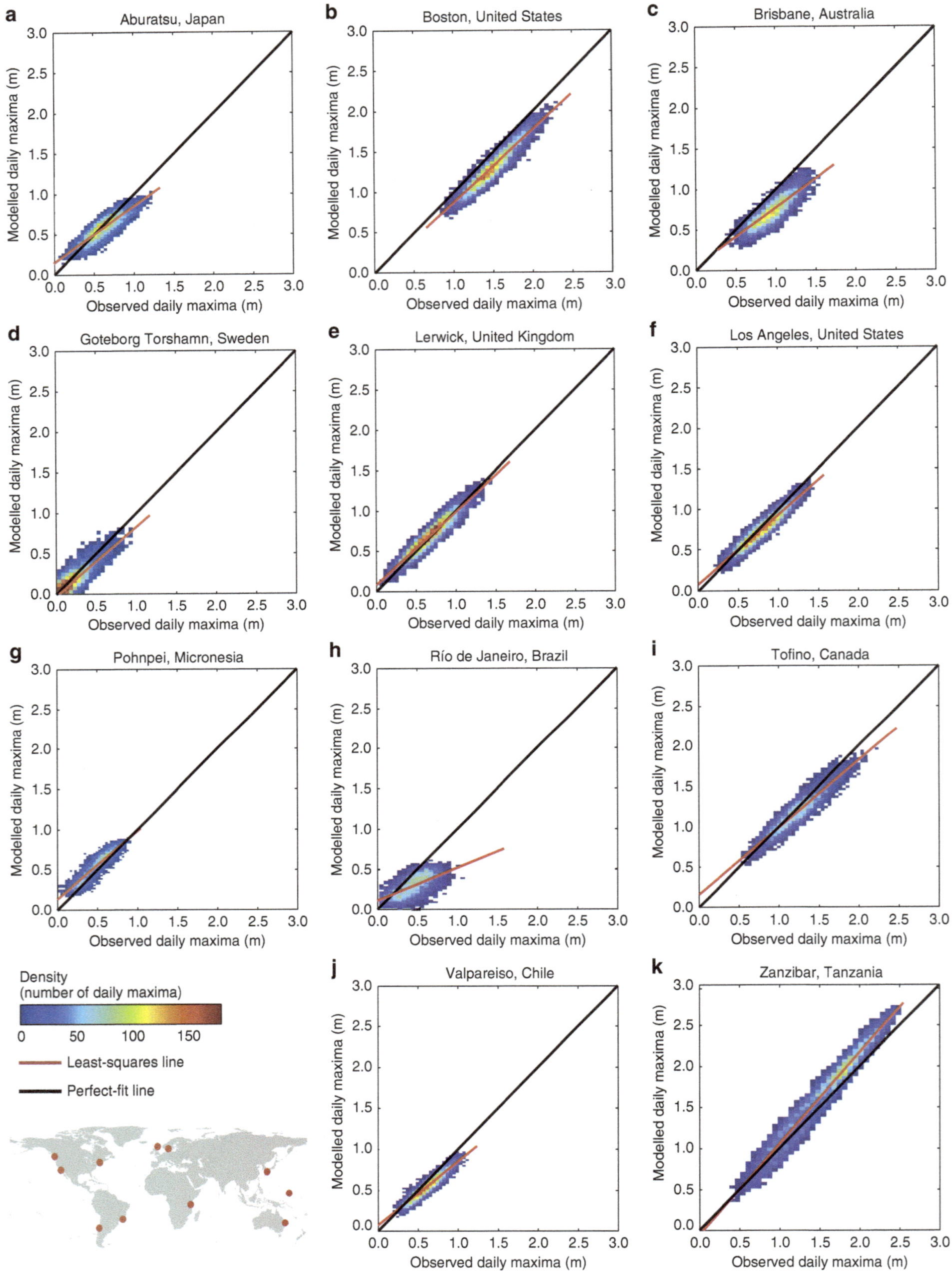

Figure 3 | Scatter plots of modelled and observed daily maxima at selected stations. Scatter plots for (**a**) Abaratsu, Japan; (**b**) Boston, United States; (**c**) Brisbane, Australia; (**d**) Goteborg-Torshamn, Sweden; (**e**) Lerwick, United Kingdom; (**f**) Los Angeles, United States; (**g**) Pohnpei, Micronesia; (**h**) Río de Janeiro, Brazil; (**i**) Tofino, Canda; (**j**) Valpareiso, Chile; and (**k**) Zanzibar, Tanzania. Colours indicate the data density for bins with a 5 cm × 5 cm size. The red dots in the world map indicate the location of the observations stations.

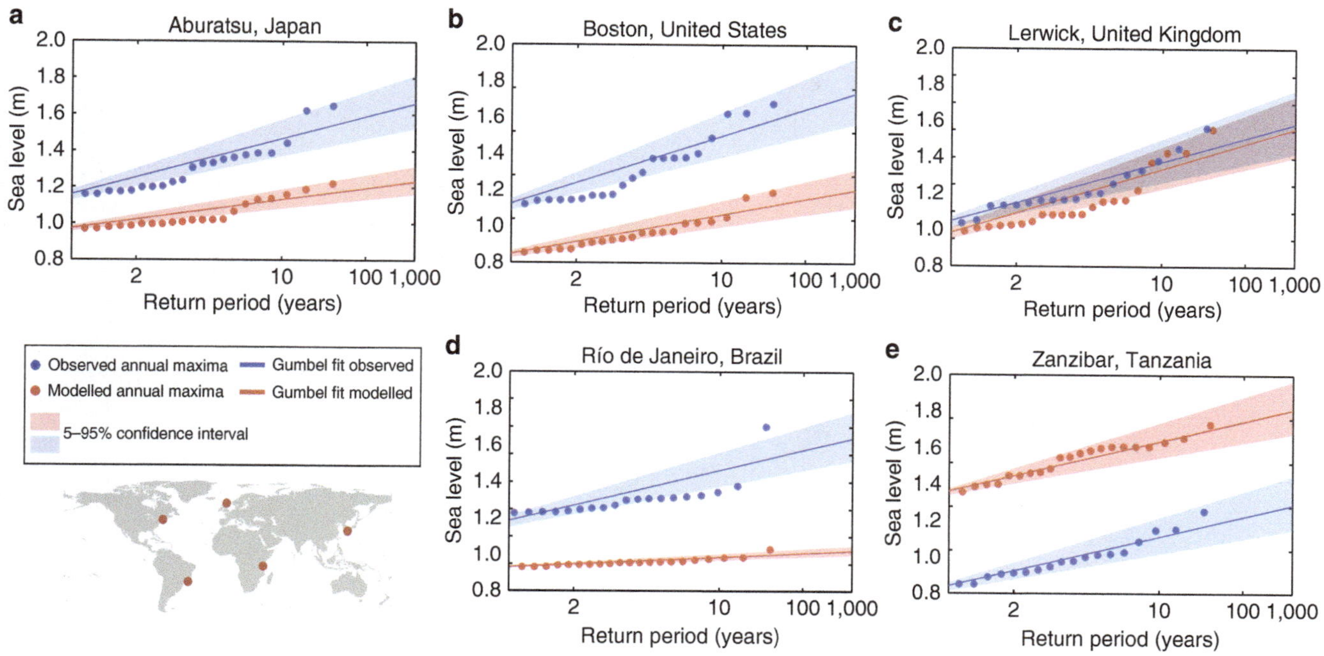

Figure 4 | Gumbel plot for the modelled and observed annual maxima at selected stations. Gumbel plots for five selected stations around the world, (**a**) Aburatsu, Japan; (**b**) Boston, United States; (**c**) Lerwick, United Kingdom; (**d**) Río de Janeiro, Brazil; and (**e**) Zanzibar, Tanzania. The annual maxima are plotted against the Gumbel variate, which is equal to $x = -\ln(-\ln(G(y)))$. The red dots in the world map indicate the location of the observation stations.

Results show that there are large regional variations in the 1 in 100-year extreme sea levels (Fig. 5a)[30]. Relatively high extreme sea levels are found in areas with high tidal amplitude, such as northwest Europe, south Argentina, China and Bangladesh. This spatial pattern agrees with other global data sets[19,20,31], and does not change for higher return periods (Supplementary Fig. 5a–b). Combining extreme sea levels with elevation, and assuming no flood protection, shows that major inundation occurs particularly in delta areas in Europe and Asia (Supplementary Fig. 6). The global population exposed to a 1 in 100-year flood is 76 million, which is equal to 1.3% of the total world population (Table 1). The flood exposure map in Fig. 5b shows that China has a particularly large exposure, with 37 million people, which is 3% of the country's population and 53% of the global exposure (Supplementary Fig. 7a shows the absolute exposure). Other countries that have a relatively high exposure include The Netherlands (8.4 million people, 53% of total population), Vietnam (4.7 million people, 6% of global exposure) and Egypt (4.3 million people, 6% of global exposure). Relative to the total national population, The Netherlands and Greenland stand out with 53% and 31% of the population exposed to a 1 in 100-year flood (Supplementary Fig. 7b), respectively; in both these countries the highest concentration of people is found near the coast[28].

We assess the uncertainty in the extreme value statistics by calculating the 5 and 95% confidence bounds of the Gumbel fit, and express the uncertainty as the percentage difference compared to extreme sea level based on the best-fit. For extreme sea levels with a return period of 100 years, the uncertainty of the Gumbel fit is below 10% for half of the world's coastline (Fig. 6a). The uncertainty is greater than 25% for only 4% of the world's coastline. The largest uncertainty is seen in regions where the extreme values are relatively low, such as the Mediterranean Sea and Caribbean Sea, where a small change in extreme sea level leads to a large increase in relative terms. To assess how sensitive the exposure estimates are to the uncertainty in extreme values statistics, we use the 5 and 95% confidence bounds as input to the inundation and impact model. Globally, the sensitivity to the uncertainty in extreme values is relatively small, with a range of −8 to +21% around the best-fit for exposed population (that is, exposure ranges between 70 and 92 million people). The results per country are shown in Fig. 6, which shows that the results for the USA, Thailand, Mauritania and several countries along the Baltic Sea are particularly sensitive, with uncertainty values greater than 50% of the best-fit. Of the top 10 countries listed in Table 1, the results for China and Vietnam have the largest uncertainty, with exposure estimates ranging from −34% to +39% and −28% to +4% respectively. While the largest uncertainty in extreme sea levels is found along the north coast of Russia, this does not lead to large uncertainty in exposure because the area has a very low population density. On the other hand, an uncertainty in extreme sea levels of 10–50% along the east coast of the USA and <10% along the west coast of the USA leads to uncertainty in the country's exposure estimates between −40% and +49%.

Discussion

The performance of the GTSR time series is similar to the performance of other hydrodynamic models that cover a large geographic domain. For example, Cid *et al.*[32] reported a mean RMSE of 0.08–0.10 m for surge levels in the Mediterranean Sea. For a hydrodynamic model covering the entire Australian coastline, Haigh *et al.*[33] reported mean RMSEs of 0.14 m and 0.05 m for total sea level and surge, respectively. The validation of GTSR shows that extreme sea levels are generally underestimated but that the differences with observed extreme sea levels are rather small (that is, <0.45 m for 90% of the observation stations for the 1 in 10-year sea level). For applications to risk assessment, which make use of data from global digital elevation models (DEMs), this underestimation is reasonable, since the vertical resolution of such DEMs is much greater. For example, the global SRTM DEM has a vertical resolution of 1 m, but its uncertainty is up to several metres[27]. Hence, while acknowledging that the

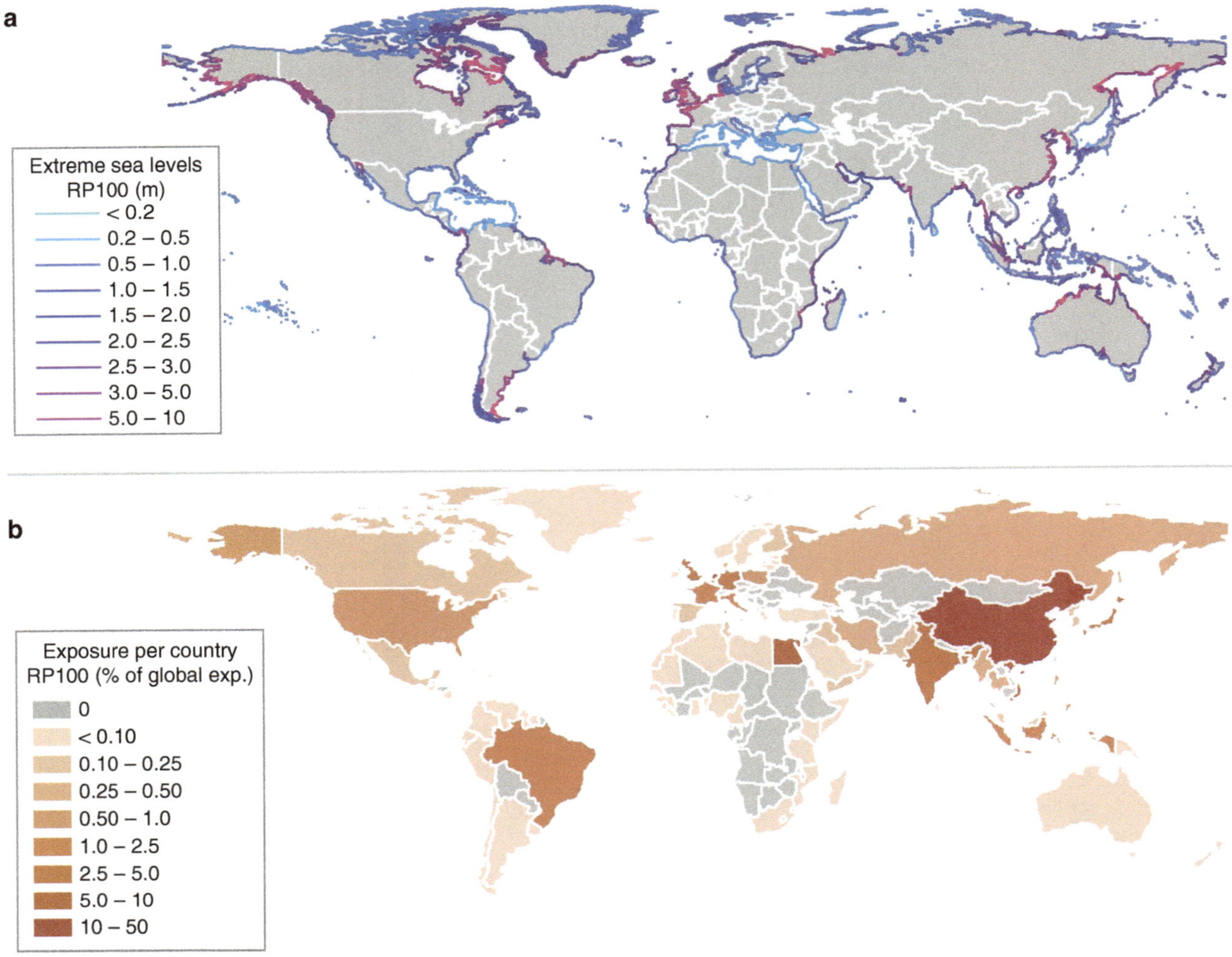

Figure 5 | Extreme sea levels with a return period of 100 years and the exposed population. Maps showing (**a**) the height of extreme sea levels with a return period of 100 years (based on the best Gumbel fit) around the entire world's coastline; and (**b**) the estimated exposed population estimates per country (relative to the global exposure) with return period of 100 years.

Table 1 | Absolute and relative exposed population to a 1 in 100-year flood for the 10 most exposed countries.

Country	Absolute exposure (in millions)	Relative exposure (% of population)
China	36	2.9
The Netherlands	8.4	53
Vietnam	4.7	6.0
Egypt	4.3	6.4
Germany	2.9	3.5
India	2.5	0.3
United Kingdom	2.4	4.1
Japan	2.0	1.6
Bangladesh	1.4	1.1
Belgium	1.1	11
World	**76**	**1.3**

errors can potentially be large in specific locations, we are confident that the GTSR time series (that is, surge, tide and total sea levels) and extremes are a very valuable addition to the global data sets that are currently available. Further benchmarking may be performed by comparing the GTSR extremes with other modelled data sets, such as extreme sea levels in the DIVA database[19], and regional hydrodynamic models[23,34].

There are several limitations to the GTSR time series and extremes. We aim to update the data set in the future, addressing some of the issues described here. First, the validation shows that extreme values are slightly underestimated. This is an inevitable result of the relatively coarse resolution of the model grid, bathymetry and meteorological forcing (compared with point observations). Extremes in wind speed and atmospheric pressure are smoothed in the ERA-Interim data set due to the temporal (6 h) and spatial (0.75°) resolution. This is particularly problematic for tropical cyclones, which are characterized by strong gradients in atmospheric pressures and wind fields both in time and in space. Our validation results show that the underestimation of extreme sea levels is more severe in tropical areas. However, because of the sparseness and shortness of the available records, the largest tropical cyclone-induced surges are not all included in the available observations[2]. Hence, if we compare the GTSR extremes against reported sea levels induced by tropical cyclones we see larger deviations. For example, the maximum sea levels during tropical cyclone Katrina in New Orleans (2005), and during Typhoon Haiyan (2013), in the Philippines exceeded 8 m (ref. 35) and 4–5 m (refs 36,37), respectively, while our extreme sea level estimates do not exceed 2–3 m for a return period of 1,000 year. This illustrates that accurately modelling these intense storms requires a much higher resolution than atmospheric

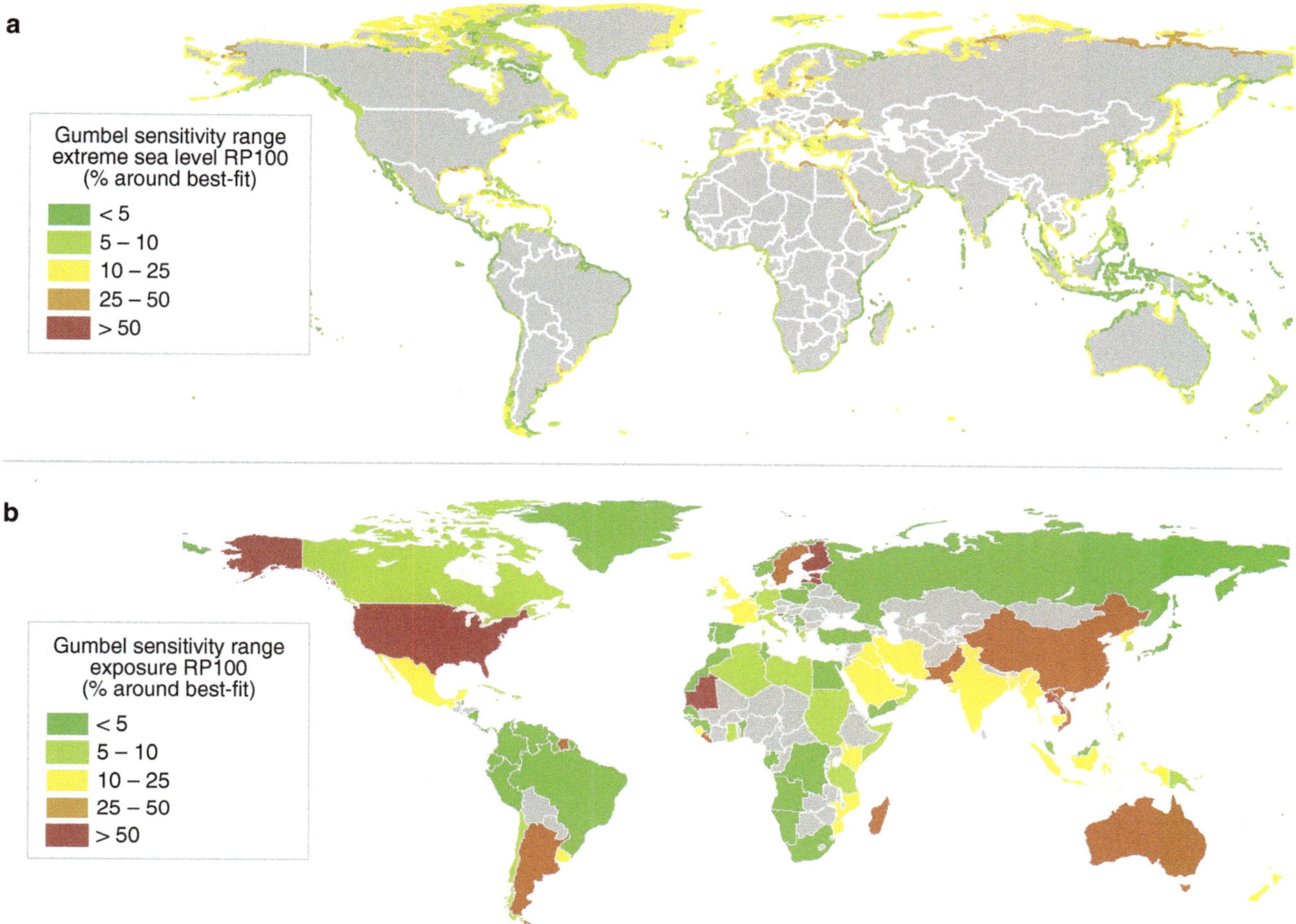

Figure 6 | Gumbel sensitivity range for extreme sea levels and flood exposure. Maps showing the uncertainty of the extreme values statistics for the 1 in 100-year return period. The values shown is the range for the 5-95% confidence bounds expressed as a percentage of the value for the best-fit for (**a**) the height of extreme sea around the entire world's coastline; and (**b**) the estimated exposed population estimates per country.

reanalysis data can deliver at present day. For updated versions of the GTSR data set, this issue could be resolved by generating localized wind fields based on storm track data and the parametric model of Holland[38]. This is expected to more accurately simulate surge levels resulting from tropical cyclones. However, even when the extreme sea levels are adequately modelled, time series of 36 years contain insufficient number of tropical cyclones to obtain reliable statistics of extreme values. Hence, synthetic resampling techniques are needed to extend the tropical cyclone record to a longer record[33,39,40]. Including tropical cyclones will lead to higher coastal flood risk, as the most damaging storm surges are often induced by tropical cyclones[41].

Second, we apply the annual maxima method to obtain the GTSR extremes. Although the method is widely applied, it neglects the fact that sea levels are composed of two independent processes, namely a tide-driven (deterministic) process and a surge-driven (stochastic) process. In addition, data are used inefficiently, as extreme values are estimated based on annual maxima only. To address these limitations, more sophisticated statistical methods, such as the *r* largest or joint probability, could be applied[42]. However, these methods are more sensitive to timing errors and/or temporal or spatial variation. Assuming that tides and storm surges behave independently, the estimates of extreme values could also be made more robust by resampling the surge and tide levels in time to obtain longer time series.

Third, the sea level variations in the GTSR time series are due to gravitational tides and barotropic changes (changes in wind and pressure). Baroclinic effects (density differences) are not considered. While in most parts of the world the generation of extreme sea levels is dominated by tide and surge, in some regions the variations in mean sea level are relatively large[43] and thus affect the total sea level. For example, this is the case in parts of the Australian coastline[23]. Also, non-linear interactions between storm surges and tides, the effects of waves, and precipitation and river flow are not considered. In this version of GTSR, tides and surges are modelled separately and surge-tide interactions are thus not included, although they are known to be important in shallow water areas with a large tidal range[44,45]. Including the surge-tide interaction has improved the model performance for regional hydrodynamic models[46]. For example, in the case of the North Sea, the RMSE was lowered by ca. 40% (ref. 47). However, at this stage this does not apply to GTSM (Supplementary Note 1), as the current version of GTSM is not capable of adequately reproducing tidal characteristics in all coastal regions. Wave setup may increase total sea levels considerably near the coast with the largest contribution in regions with steep slopes[48]. In deltas and estuaries, precipitation and river flow may also contribute to coastal flooding[49]. Including all these processes on a global scale is not feasible at present, but it is important to note that these processes may lead to significantly different extreme sea levels locally.

Also, the impact modelling has some limitations. Aside from limitations of the simple inundation model (see Methods section), the most important limitation is that flood protection is not included in this analysis, and many exposed countries are protected by dikes and storm surge barriers up to a certain design standard. For example, despite its high exposure rates, The Netherlands has very-low flood probabilities as all low-lying areas are protected by coastal defences, some with very high protection standards (1 in 10,000-year return period)[50]. However, Hallegatte *et al.*[12] estimated protection standards for major cities and found that a city such as Jakarta has a protection standard that equals a 10-year return period. Hence, for many countries listed in Table 1, the protection standard will be lower than a return period of 100 years. Using a similar methodology (planar inundation and no flood protection), but an alternative population database, Jongman *et al.*[8] estimated that 271 million people are exposed to 1 in 100-year coastal flooding in 2010. On the basis of the similar elevation and population data, but using a somewhat different methodology that includes flood protection, Hinkel *et al.*[5] estimated that between 160 million people are exposed to a 1 in 100-year flood. The differences in methodology and data make a direct comparison difficult, but our exposure estimate of 70–92 million people seems low compared with previous studies[5,8]. In countries prone to tropical cyclones, such as Bangladesh, exposure is underestimated as the tropical storms are responsible for the largest events[2]. Nevertheless, previous studies of global flood risk, including the studies of Jongman *et al.*[4] and Hinkel *et al.*[7], are based on the extreme sea levels in the DIVA database[19]. This data set has not been validated at the global scale, but compared to observed sea levels, Muis *et al.*[51] found that the DIVA 1 in 10-year extreme sea levels in Indonesia are overestimated with on average 0.8 m. Further research could assess differences in flood hazard between DIVA and GTSR and its influence on the estimated flood exposure.

In future research, the GTSR data set could be applied to assess global flood risk, both under current and future climate conditions. Coastal floods may become more severe due to sea level rise and changes in storminess[14,52]. While sea level rise may be directly added to the return levels[5], this is not valid for all regions[53], and the advantage of the physical based modelling approach is that the sensitivity of this assumption can be tested. Another advantage is that changes in storminess can be assessed using a dynamical approach by forcing the model with future global climate model simulations[54,55]. Furthermore, changes in storm duration can be propagated into the full dynamics of storm surge-induced flooding. Such climate change assessments can be used to identify areas that face rapidly increasing risks, which is important for planning disaster risk reduction efforts and to prioritize adaptation efforts[56]. Aside from flood risk applications, the GTSR data set can be used for a variety of other applications. For example, the GTSR time series may be used to: assess changes in storminess[57]; correct for meteorological effects in mean sea level[58]; disentangle the drivers of extreme events in different areas (tidal versus surge dominated)[20]; or assess the influence of interannual variability on risk[59]. As our new time series also includes the duration of flood events, another potential research direction may be the improvement of inundation modelling on a global scale by including flood duration in the inundation modelling. Recent research demonstrated the importance of assessing compound flooding in major US coastal cities[60]. In other delta regions, including compound flood events may also be critical for correctly assessing flood risk[61–64]. In combination with time series of precipitation or discharge, the GTSR time series could be used to assess compound floods on a global scale. Another application of the GTSM model that is currently being developed is the global operational forecasting system called GLOSSIS[65], which produces 10-day forecasts of coastal storm surges worldwide. The GTSR extremes could be used to raise warning flags and to identify potential flood hazards. In the future this should not only be based on exceeding a physical threshold, but also include potential impacts of a flood.

Methods

General approach. The method to develop the GTSR time series is based on two global hydrodynamic models: GTSM for storm surges, and FES2012 for tides (see Supplementary Fig. 8 for a flowchart). To simulate water levels in all coastal areas, while not generating huge amounts of data, model output is produced for 16,395 locations along the coastline based on the centroids of the DIVA segments database[19] and the locations of observation stations used for validation. Reduced data sets, useful for extreme value analyses, were generated by post-processing the results into daily maxima as well as annual maxima, for each of the output locations.

Modelling tides with FES2012. FES2012 is a global tidal model, which assimilates satellite altimetry data. The model is available for download at: http://www.avi-so.altimetry.fr. Tidal elevations are distributed on a regular grid of 1/16°. Tides are simulated here with a 10-min time interval. A review on the performance of global tide models by Stammer[66] shows that FES2012 performs relatively well compared to other global tidal models in coastal areas.

Modelling storm surge with GTSM. GTSM is based on the Delft3D Flexible Mesh software developed by Deltares[24]. To accurately resolve hydrodynamic equations in topographically complex areas, such as coastal regions, while not decreasing the computational efficiency, it is desirable to locally refine the computational grid[24]. Delft3D FM enables this by allowing the use of unstructured grids. The cell size of the computational grid is dependent on the bathymetry and increases from 1/2° ($\sim$50 km) in deeper parts of the ocean towards 1/20° ($\sim$5 km) in shallow coastal areas (Supplementary Fig. 9). The bathymetric data with a resolution of 1/60° are collected from the General Bathymetric Chart of Oceans[67] and are interpolated onto the computational grid.

To simulate the sea levels resulting from storm surges, the model was forced with 10 m wind speed and atmospheric pressure obtained from the ERA-Interim data set developed by The European Centre For Medium-Range Weather Forecasts (ECMWF)[25]. The ERA-Interim data set is a global atmospheric reanalysis from 1979 to present day. The meteorological fields are available every 6 h and have a spatial resolution of $0.75 \times 0.75°$. The meteorological fields are temporally downscaled to 10 min using linear interpolation. The 10-m wind speed is translated into wind stress using a drag coefficient based on Charnock[68]. For consistency with the ECMWF climate model, we applied a Charnock parameter of 0.041. The sensitivity of applying different values for the Charnock parameter is discussed in the Supplementary Note 1.

The model simulations were carried out for each year separately, using a spin-up time of 11 days. Using a parallel setup with 16 cores, each simulation of one year takes approximately 30 h. The different runs were started in parallel, so in theory without other users on the computer cluster, the whole computation can be completed in 30 h.

Data used for validation. A global data set with observed sea levels is used to validate GTSR. Hourly levels from 472 stations (Supplementary Fig. 10) over the period 1980–2011 are obtained from the archives of the UHSLC (data set is available at http://uhslc.soest.hawaii.edu). All stations are quality-checked by tidal analysis using TideMAT 1.05 (ref. 69). The records of each station are only used when $<20\%$ of the data are missing. Each year is analysed separately and mean annual sea level is used as reference date; as such the sea levels are detrended. We subtract the tidal component from the total sea level to obtain the residual level. This component primarily contains the meteorological contribution to sea level (that is, surge level), but may also contain harmonic prediction errors or timing errors. However, we consider these errors as negligible as each station has been inspected visually. Decomposing the total sea level into a tide and surge component enables the separate validation of the two components. For each station with a record longer than 25 years, we also estimate the extreme sea levels for different return periods and compare those with modelled sea levels (see next paragraph).

Extreme value statistics. To estimate the probabilities of extreme sea levels, we apply extreme value statistics using the annual maxima method[70,71]. For each output location, we extract the annual maximum for the calendar years 1979–2014 and fit a Gumbel distribution using the maximum-likelihood method. From the parameterised distribution, we can obtain estimates of sea levels corresponding to selected return periods. We also obtain uncertainty estimates of the parameterised extreme value distribution (that is, 5 and 95% confidence bounds). The Gumbel distribution is often a good approximation of observed extreme sea levels and is frequently applied to estimate return periods[42]. While more advanced statistical methods are available, such as peak-over-threshold or joint probability, the annual maxima method is more robust to temporal and spatial variations, and is thus relatively easy to apply on a global scale.

Modelling inundation and flood exposure. On the basis of the extreme sea levels, coastal inundation is calculated using a GIS-based planar approach, which uses the extreme sea level and a DEM as input. Inundated areas are defined as areas that have an elevation lower than the water level, and have a direct connection to the sea[5,8]. The DEM is obtained during NASA's Shuttle Radar Topography Mission and its original resolution is $1'' \times 1''$ ($\sim 30 \times 30$ m at the equator)[27]. We use a $30'' \times 30''$ ($\sim 1 \times 1$ km at the equator) average to estimate inundation. Impacts of coastal floods are measured in terms of exposed population using the GRUMPv1 population counts maps for 2000 (refs 28,29). Flood protection is not included in the analysis and will lead to an overestimation of the flood hazard. Furthermore, the planar approach assumes that a maximum sea level during any event will travel infinitely far land inwards and will therefore overestimate the extent of the floodplain, in particular on coastlines with wide flat areas far land inwards. On the other hand, the spatial averaging of the DEM may result in smoothing of local depressions, causing underestimation of the flood extent in areas with relatively large upward gradients land inwards. Also, coastal cities are often in delta areas where the river may propagate the surge into the hinterland. In this sense, the approach may underestimate the flood extent.

References

1. Resio, D. T. & Westerink, J. J. Modeling the physics of storm surges. *Phys. Today* **61,** 33–38 (2008).
2. Needham, H. F., Keim, B. D. & Sathiaraj, D. A review of tropical cyclone-generated storm surges: Global data sources, observations, and impacts. *Rev. Geophys.* **53,** 545–591 (2015).
3. Pugh, D. T. *Tides, Surges and Mean Sea-Level* (John Wiley & Sons, 1987).
4. McGranahan, G., Balk, D. L. & Anderson, B. The rising tide: assessing the risks of climate change and human settlements in low elevation coastal zones. *Environ. Urban.* **19,** 17–37 (2007).
5. Hinkel, J. *et al.* Coastal flood damage and adaptation costs under 21st century sea-level rise. *Proc. Natl Acad. Sci. USA* **111,** 3292–3297 (2014).
6. Spencer, T., Brooks, S. M. & Möller, I. Storm-surge impact depends on setting. *Nature* **505,** 26 (2013).
7. Rappaport, E. N. Fatalities in the United States from Atlantic tropical cyclones: new data and interpretation. *Bull. Am. Meteorol. Soc.* **95,** 341–346 (2014).
8. Jongman, B., Ward, P. J. & Aerts, J. C. J. H. Global exposure to river and coastal flooding: long term trends and changes. *Glob. Environ. Chang.* **22,** 823–835 (2012).
9. Nicholls, R. J. & Cazenave, A. Sea-level rise and its impact on coastal zones. *Science* **328,** 1517–1520 (2010).
10. Syvitski, J. P. M. *et al.* Sinking deltas due to human activities. *Nat. Geosci.* **2,** 681–686 (2009).
11. Bouwer, L. M. Have disaster losses increased due to anthropogenic climate change? *Bull. Am. Meteorol. Soc.* **6,** 791–793 (2011).
12. Hallegatte, S., Green, C., Nicholls, R. J. & Corfee-Morlot, J. Future flood losses in major coastal cities. *Nat. Clim. Chang.* **3,** 802–806 (2013).
13. Kemp, A. C. & Horton, B. P. Contribution of relative sea-level rise to historical hurricane flooding in New York City. *J. Quat. Sci.* **28,** 537–541 (2013).
14. Reed, A. J. *et al.* Increased threat of tropical cyclones and coastal flooding to New York City during the anthropogenic era. *Proc. Natl Acad. Sci. USA* **112,** 12610–12615 (2015).
15. Brown, S., Nicholls, R. J., Lowe, J. A. & Hinkel, J. Spatial variations of sea-level rise and impacts: an application of DIVA. *Clim. Change* **134,** 403–416 (2013).
16. Hinkel, J. *et al.* Sea-level rise impacts on Africa and the effects of mitigation and adaptation: an application of DIVA. *Reg. Environ. Chang.* **12,** 207–224 (2011).
17. Hinkel, J., Nicholls, R. J., Vafeidis, A. T., Tol, R. S. J. & Avagianou, T. Assessing risk of and adaptation to sea-level rise in the European Union: an application of DIVA. *Mitig. Adapt. Strateg. Glob. Chang.* **15,** 703–719 (2010).
18. Hinkel, J. & Klein, R. J. T. Integrating knowledge to assess coastal vulnerability to sea-level rise: The development of the DIVA tool. *Glob. Environ. Chang.* **19,** 384–395 (2009).
19. Vafeidis, A. T. *et al.* A new global coastal database for impact and vulnerability analysis to sea-level rise. *J. Coast. Res.* **244,** 917–924 (2008).
20. Merrifield, M., Genz, A. S., Kontoes, C. P. & Marra, J. J. Annual maximum water levels from tide gauges: Contributing factors and geographic patterns. *J. Geophys. Res. Ocean* **118,** 2535–2546 (2013).
21. Marcos, M., Tsimplis, M. N. & Shaw, A. G. P. Sea level extremes in southern Europe. *J. Geophys. Res.* **114,** C01007 (2009).
22. Bernier, N. B. & Thompson, K. R. Predicting the frequency of storm surges and extreme sea levels in the northwest Atlantic. *J. Geophys. Res.* **111,** C10009 (2006).
23. Haigh, I. D. *et al.* Estimating present day extreme water level exceedance probabilities around the coastline of Australia: tides, extra-tropical storm surges and mean sea level. *Clim. Dyn.* **42,** 121–138 (2013).
24. Kernkamp, H. W. J., Van Dam, A., Stelling, G. S. & de Goede, E. D. Efficient scheme for the shallow water equations on unstructured grids with application to the Continental Shelf. *Ocean Dyn.* **61,** 1175–1188 (2011).
25. Dee, D. P. *et al.* The ERA-Interim reanalysis: configuration and performance of the data assimilation system. *Q. J. R. Meteorol. Soc.* **137,** 553–597 (2011).
26. Carrère, L. *et al.* in *Proc. 20YPRA Symp.*, 3–8 (ESA/CNES, Venice-Lido, Italy, 2012).
27. Rabus, B., Eineder, M., Roth, A. & Bamler, R. The shuttle radar topography mission - A new class of digital elevation models acquired by spaceborne radar. *ISPRS J. Photogramm. Remote Sens.* **57,** 241–262 (2003).
28. Center for International Earth Science Information Network (CIESIN) — Columbia University, International Food Policy Research Institute (IFPRI) — The World Bank, and Centro Internacional de Agricultura Tropical (CIAT). *Global Rural-Urban Mapping Project, Version 1 (GRUMPv1): Population Count Grid* (NASA Socioeconomic Data and Applications Center (SEDAC), Palisades, NY, 2011).
29. Balk, D. L. *et al.* Determining global population distribution: methods, applications and data. *Adv. Parasitol.* **62,** 119–156 (2011).
30. Muis, S., Verlaan, M., Winsemius, H., Aerts, J. C. J. H. & Ward, P. J. Global Tide and Surge Reanalysis (GTSR). *3TU. Datacentrum* https://dx.doi.org/10.4121/uuid:aa4a6ad5-e92c-468e (2016).
31. Menéndez, M. & Woodworth, P. L. Changes in extreme high water levels based on a quasi-global tide-gauge data set. *J. Geophys. Res.* **115,** C10011 (2010).
32. Cid, A., Castanedo, S., Abascal, A. J., Menéndez, M. & Medina, R. A high resolution hindcast of the meteorological sea level component for Southern Europe: the GOS dataset. *Clim. Dyn.* **43,** 2167–2184 (2014).
33. Haigh, I. D. *et al.* Estimating present day extreme water level exceedance probabilities around the coastline of Australia: tropical cyclone-induced storm surges. *Clim. Dyn.* **42,** 139–157 (2013).
34. Flather, R. A., Smith, J. A., Richards, J. D., Bell, C. & Blackman, D. L. Direct estimates of extreme storm surge elevations from a 40-year numerical model simulation and from observations. *Glob. Atmos. Ocean Syst.* **6,** 165–176 (1998).
35. Needham, H. F. & Keim, B. D. A storm surge database for the US Gulf Coast. *Int. J. Climatol.* **32,** 2108–2123 (2012).
36. Mahar, A. *et al.* Devastating storm surges of Typhoon Haiyan. *Int. J. Disaster Risk Reduct.* **11,** 1–12 (2014).
37. Mori, N. *et al.* Local amplification of storm surge by Super Typhoon Haiyan in Leyte Gulf. *Geophys. Res. Lett.* **41,** 1–8 (2014).
38. Holland, G. J. An analytic model of the wind and pressure profiles in hurricanes. *Mon. Weather Rev.* **108,** 1212–1218 (1980).
39. Lin, N., Emanuel, K. A., Oppenheimer, M. & Vanmarcke, E. Physically based assessment of hurricane surge threat under climate change. *Nat. Clim. Chang.* **2,** 462–467 (2012).
40. Emanuel, K. A., Ravela, S., Vivant, E. & Risi, C. A statistical deterministic approach to hurricane risk assessment. *Bull. Am. Meteorol. Soc.* **87,** 299–314 (2006).
41. Woodruff, J. D., Irish, J. L. & Camargo, S. J. Coastal flooding by tropical cyclones and sea-level rise. *Nature* **504,** 44–52 (2013).
42. Haigh, I. D., Nicholls, R. J. & Wells, N. C. A comparison of the main methods for estimating probabilities of extreme still water levels. *Coast. Eng.* **57,** 838–849 (2010).
43. Tsimplis, M. N. & Woodworth, P. L. The global distribution of the seasonal sea level cycle calculated from coastal tide gauge data. *J. Geophys. Res.* **99,** 16031 (1994).
44. Bernier, N. B. & Thompson, K. R. Tide-surge interaction off the east coast of Canada and northeastern United States. *J. Geophys. Res.* **112,** 1–12 (2007).
45. Horsburgh, K. J. & Wilson, C. Tide-surge interaction and its role in the distribution of surge residuals in the North Sea. *J. Geophys. Res.* **112,** C08003 (2007).
46. Wolf, J. Coastal flooding: impacts of coupled wave–surge–tide models. *Nat. Hazards* **49,** 241–260 (2008).
47. Zijl, F., Verlaan, M. & Gerritsen, H. Improved water-level forecasting for the Northwest European Shelf and North Sea through direct modelling of tide, surge and non-linear interaction. *Ocean Dyn.* **63,** 823–847 (2013).
48. Dietrich, J. C. *et al.* Modeling hurricane waves and storm surge using integrally-coupled, scalable computations. *Coast. Eng.* **58,** 45–65 (2011).
49. Bunya, S. *et al.* A high-resolution coupled riverine flow, tide, wind, wind wave, and storm surge model for southern louisiana and mississippi. Part I: model development and validation. *Mon. Weather Rev.* **138,** 345–377 (2010).

50. Wesselink, A., Warner, J. & Kok, M. You gain some funding, you lose some freedom: The ironies of flood protection in Limburg (The Netherlands). *Environ. Sci. Policy* **30,** 113–125 (2013).
51. Muis, S., Güneralp, B., Jongman, B., Aerts, J. C. J. H. & Ward, P. J. Flood risk and adaptation strategies under climate change and urban expansion: A probabilistic analysis using global data. *Sci. Total Environ.* **538,** 445–457 (2015).
52. IPCC. *Managing the Risks of Extreme Events and Disasters to Advance Climate Change Adaptation* (Cambridge University Press, 2012).
53. Zhang, K., Li, Y., Liu, H., Xu, H. & Shen, J. Comparison of three methods for estimating the sea level rise effect on storm surge flooding. *Clim. Change* **118,** 487–500 (2012).
54. Brown, J. M., Wolf, J. & Souza, A. J. Past to future extreme events in Liverpool Bay: model projections from 1960–2100. *Clim. Change* **111,** 365–391 (2011).
55. Lowe, J. A., Gregory, J. M. & Flather, R. A. Changes in the occurrence of storm surges around the United Kingdom under a future climate scenario using a dynamic storm surge model driven by the Hadley Centre climate models. *Clim. Dyn.* **18,** 179–188 (2001).
56. UNISDR. *Global Assessment Report on Disaster Risk Reduction: Revealing Risk, Redefining Development.* 178 (United Nations International Strategy for Disaster Reduction, Geneva, Switzerland, 2011).
57. Cid, A. *et al.* Long-term changes in the frequency, intensity and duration of extreme storm surge events in southern Europe. *Clim. Dyn.* **46,** 1503–1516 (2015).
58. Losada, I. J. *et al.* Long-term changes in sea-level components in Latin America and the Caribbean. *Glob. Planet. Chang.* **104,** 34–50 (2013).
59. Ward, P. J. *et al.* Strong influence of El Niño Southern Oscillation on flood risk around the world. *Proc. Natl Acad. Sci. USA* **111,** 15659–15664 (2014).
60. Wahl, T., Jain, S., Bender, J., Meyers, S. D. & Luther, M. E. Increasing risk of compound flooding from storm surge and rainfall for major US cities. *Nat. Clim. Chang.* **5,** 1093–1097 (2015).
61. Kew, S. F., Selten, F. M., Lengaigne, M. & Hazeleger, W. The simultaneous occurrence of surge and discharge extremes for the Rhine delta. *Nat. Hazards Earth Syst. Sci.* **13,** 2017–2029 (2013).
62. Zheng, F., Westra, S., Leonard, M. & Sisson, S. A. Modeling dependence between extreme rainfall and storm surge to estimate coastal flooding risk. *Water Resour. Res.* **50,** 2050–2071 (2014).
63. Keef, C., Svensson, C. & Tawn, J. A. Spatial dependence in extreme river flows and precipitation for Great Britain. *J. Hydrol.* **378,** 240–252 (2009).
64. van den Hurk, B. J. J. M., van Meijgaard, E., de Valk, P., van Heeringen, K.-J. & Gooijer, J. Analysis of a compounding surge and precipitation event in the Netherlands. *Environ. Res. Lett.* **10,** 035001 (2015).
65. Verlaan, M., De Kleermaeker, S. & Buckman, L. in *Australasian Coasts & Ports Conference 2015: 22nd Australasian Coastal and Ocean Engineering Conference and the 15th Australasian Port and Harbour Conference*, 229–234 (Auckland, New Zealand & Engineers Australia and IPENZ, 2015).
66. Stammer, D. *et al.* Accuracy assessment of global barotropic ocean tide models. *Rev. Geophys.* **52,** 1–40 (2014).
67. IOC, IHO, & BODC. *Centenary Edition of the GEBCO Digital Atlas, Published on CD-ROM on Behalf of the Intergovernmental Oceanographic Commission and the International Hydrographic Organization as Part of the General Bathymetric Chart of the Oceans.* (British Oceanographic Data Centre, Liverpool, UK, 2003).
68. Charnock, H. Wind stress on a water surface. *Q. J. R. Meteorol. Soc.* **81,** 639 (1955).
69. Rijkswaterstaat. TIDEMAT Tidal Analysis. User's Manual 1.01, 30 p. (The Hague, The Netherlands, 2004).
70. Gumbel, E. J. The return period of flood flows. *Ann. Math. Stat.* **12,** 163–190 (1941).
71. Jenkinson, A. F. The frequency distribution of maximum values. *Q. J. R. Meteorol. Soc.* **81,** 158–171 (1954).

Acknowledgements

The research leading to these results has received funding from the European research project RISES-AM- (grant agreement no. 603396). P.J.W. received additional funding from the Netherlands Organisation for Scientific Research (NWO) in the form of a VENI grant (grant no. 863-11-011). J.C.J.H.A. and S.M. received funding from the NWO in the form of a VICI Grant (grant no. 453-14-006). We thank SURFsara (www.surfsara.nl) for the support in using the Lisa Computer Cluster.

Author contributions

S.M. was responsible for carrying out the model runs and the validation of GTSR, including the tidal analysis of the UHCL data set. M.V. developed the GTSM model, provided a script to apply the FES2012 model, and assisted with carrying out the model runs. H.C.W. was responsible for the post-processing of the data. S.M., M.V., H.C.W. and P.J.W. were involved in the development of the data set and the data analysis. All authors discussed the results and implications and were involved in the writing of the paper.

Additional information

Competing financial interests: The authors declare no competing financial interests.

3

Microelectrode characterization of coral daytime interior pH and carbonate chemistry

Wei-Jun Cai[1,2], Yuening Ma[1,2], Brian M. Hopkinson[2], Andréa G. Grottoli[3], Mark E. Warner[1], Qian Ding[2,4], Xinping Hu[2,5], Xiangchen Yuan[2,6], Verena Schoepf[3,7], Hui Xu[2,4,8], Chenhua Han[2,4,9], Todd F. Melman[10], Kenneth D. Hoadley[1], D. Tye Pettay[1], Yohei Matsui[3], Justin H. Baumann[3], Stephen Levas[3], Ye Ying[4] & Yongchen Wang[2,‡]

Reliably predicting how coral calcification may respond to ocean acidification and warming depends on our understanding of coral calcification mechanisms. However, the concentration and speciation of dissolved inorganic carbon (DIC) inside corals remain unclear, as only pH has been measured while a necessary second parameter to constrain carbonate chemistry has been missing. Here we report the first carbonate ion concentration ($[CO_3^{2-}]$) measurements together with pH inside corals during the light period. We observe sharp increases in $[CO_3^{2-}]$ and pH from the gastric cavity to the calcifying fluid, confirming the existence of a proton (H^+) pumping mechanism. We also show that corals can achieve a high aragonite saturation state (Ω_{arag}) in the calcifying fluid by elevating pH while at the same time keeping [DIC] low. Such a mechanism may require less H^+-pumping and energy for upregulating pH compared with the high [DIC] scenario and thus may allow corals to be more resistant to climate change related stressors.

[1] School of Marine Science and Policy, University of Delaware, Newark, Delaware 19716, USA. [2] Department of Marine Sciences, University of Georgia, Athens, Georgia 30602, USA. [3] School of Earth Sciences, The Ohio State University, Columbus, Ohio 43210, USA. [4] Ocean College, Zhejiang University, Hangzhou 310058, China. [5] Department of Physical and Environmental Sciences, Texas A&M University - Corpus Christi, Corpus Christi, Texas 78412, USA. [6] Key Laboratory of Marine Bio-resources Sustainable Utilization, South China Sea Institute of Oceanology, Chinese Academy of Sciences, Guangzhou 510301, China. [7] ARC Centre of Excellence for Coral Reef Studies, School of Earth and Environment and UWA Oceans Institute, University of Western Australia, Crawley, Western Australia 6009, Australia. [8] School of Mechanical Engineering and Automation, Zhejiang Sci-Tech University, Hangzhou 310023, China. [9] Key Laboratory of Marine Ecosystem and Environment, State Oceanic Administration (SOA) Second Institute of Oceanography, Hangzhou 310012, China. [10] Reef Systems Coral Farm, New Albany, Ohio 43054, USA. Correspondence and requests for materials should be addressed to W.-J.C. (email: wcai@udel.edu).
‡Deceased

Atmospheric CO_2 has increased from 280 parts per million (ppm) during pre-industrial times to 400 ppm today[1]. This increase has led to invasion of CO_2 into the surface ocean, which has shifted the acid-base equilibrium and modified the dissolved inorganic carbon (DIC) species composition such that the proton concentration ($[H^+]$) has increased, and pH and $[CO_3^{2-}]$ have decreased; a process commonly known as ocean acidification (OA)[2,3]. Since OA reduces calcification rates in many, though not all, corals[4–8], it is particularly detrimental to the health of coral reefs and their associated ecosystem functions and biogeochemical processes[4,7,9]. The decline in coral calcification rates is thought to be linked to reductions in seawater $[CO_3^{2-}]$ and thus Ω_{arag}. However, because calcification occurs in a semi-isolated compartment where conditions are largely controlled by the coral[10], the exact mechanism by which OA reduces calcification is difficult to decipher[11]. To understand and predict how corals might respond to current and future CO_2 increases, climate warming and other stressors, we need a better understanding of coral calcification mechanisms, particularly how a high pH and Ω_{arag} state is achieved inside the calcifying fluid and the sources as well as speciation and concentration of DIC that sustain calcification[5,11–16].

In order to precipitate $CaCO_3$ in a place separated from seawater by multiple layers of tissue, corals ultimately must be able to import Ca^{2+} ions and DIC from seawater and transport them to the site of calcification. These processes remain poorly understood and controversial. Although proposed routes of ion and molecule import differ in their details, potential import processes can be broadly grouped into three categories: (1) passive CO_2 diffusion, (2) active bicarbonate and other ion transport and (3) whole seawater import though the time scale of each of these processes will differ[10,16–19]. Which of these three processes dominates the import pathway for DIC species remains unknown in part because a complete characterization of the internal carbonate chemistry of corals is lacking. In addition to the import of necessary ions, corals must raise the aragonite saturation state of the calcifying fluid to obtain rapid rates of calcification[18]. It is widely believed that this is achieved through a H^+ efflux pump that creates high pH and, via chemical equilibrium and kinetic limitations, high $[CO_3^{2-}]$ in the calcifying space[5,19]. The H^+ efflux pump is proposed to be coupled to Ca^{2+} influx via a Ca-H ATPase[10]. This theoretical model for coral calcification is considered the foundation for hypotheses explaining the response of calcification to future acidification[5], yet direct measurements of at least two of the several carbonate system parameters to fully substantiate this conceptual model for coral calcification are lacking. Based on either indirect geochemical tracer evidence[12] or sensor-based pH data combined with the assumption of either active HCO_3^- transport or whole seawater import[13,20], several studies have also suggested that corals concentrate DIC to roughly twice the seawater value at the site of calcification. Others, however, have assumed that coral internal [DIC] is similar to or lower than that in seawater[5,21]. Furthermore, while a linkage between H^+-pumping and carbon species transportation has recently been suggested[12], confirmation by direct measurements is required. In summary, despite decades of research, our knowledge of the internal carbonate chemistry of corals is still limited and direct evidence is needed to definitively constrain coral calcification mechanisms.

Microelectrodes are a powerful tool for the direct measurement of chemical concentrations within organisms. The reliability and accuracy of pH and other microelectrodes have been well-established in the literature and in our work (see Methods section below). Previous pH microelectrode studies have shown that pH increases near the coral surface and in the upper region of the polyp's gastric cavity due to photosynthesis, but decreases deeper in the cavity (coelenteron)[22]. Most importantly, a high pH has been measured in the calcifying fluid using both pH microelectrodes (pH: 9.3–10.1) and pH sensitive dyes (pH: 8.5–9.0) (refs 5,13,23). These experiments showed that corals increase the pH of the calcifying fluid to achieve a high Ω_{arag}, which facilitates rapid rates of calcification. However, to fully characterize internal coral carbonate chemistry, pH alone is not sufficient and another carbonate system parameter such as $[CO_3^{2-}]$, [DIC], or total alkalinity (TA) must be assumed[12,13] or preferably directly measured.

Here we use CO_3^{2-} microelectrodes together with pH microelectrodes to directly measure two necessary system parameters and calculate concentrations of all other carbonate species within corals. Our novel findings are that $[CO_3^{2-}]$ and carbonate saturation state inside the coral calcifying fluid are high and calcifying fluid DIC concentration is low.

Results

pH and $[CO_3^{2-}]$ depth profiles. pH and CO_3^{2-} microelectrodes were used to characterize the interior of coral polyps (see Methods section below) under ambient seawater pH and in the light (500 μmol photons $m^{-2}s^{-1}$), when photosynthesis stimulates high rates of calcification. We advanced a microelectrode inside the polyp's gastric cavity at a resolution of 50 μm per step (5 μm in one case) parallel to the polyp wall until it broke on the solid $CaCO_3$ base (Fig. 1a,b). Repeated micro-profile measurements from seawater into the lower part of the coelenteron revealed that the profiles were reproducible (Fig. 1b, Supplementary Figs 1a,b and 2) even though polyp contraction sometimes affected the observed depths (Supplementary Fig. 2) and introduced noise in the electrode data.

We obtained several full-depth profiles of pH and CO_3^{2-} through the polyp mouth, the coelenteron, and into the calcifying fluid for three tropical coral species: *Orbicella faveolata, Turbinaria reniformis* and *Acropora millepora*. For *O. faveolata*, pH (in National Bureau of Standards, NBS, scale) increased from 8.1 and $[CO_3^{2-}]$ from 180 μmol kg^{-1} in the overlying seawater to 8.3–8.5 and ~300 μmol kg^{-1}, respectively, in the upper part of the polyp due to photosynthesis (Fig. 2a,b). Deeper in the polyps the pH decreased to a value of 7.7, which was consistent with earlier studies[13,22,24]. Correspondingly, in the deeper portion of the polyp, $[CO_3^{2-}]$ decreased to 100 μmol kg^{-1}. Then, within a short distance, pH increased sharply to as high as 9.7, confirming previous work[5,13,23], while $[CO_3^{2-}]$ increased six fold to 600 μmol kg^{-1} before the electrode tip broke. We interpret this rapid increase in pH and $[CO_3^{2-}]$ as evidence that the microelectrodes obtained measurements within the calcifying fluid prior to breaking.

Similar profiles were measured inside coral polyps of *T. reniformis* with highest pH in the calcifying fluid ranging from 8.8 to 9.3 (Fig. 2c) and highest $[CO_3^{2-}]$ from 1000 to 1400 μmol kg^{-1} (Fig. 2d). In *A. millepora*, the electrode advancement step was reduced to 5 μm due to the smaller polyps compared with the other species, resulting in a very fine-scale pH profile, which demonstrated a sharp increase in pH from 8.15 to 8.65 within only a short distance of 10–60 μm (Fig. 2c). Similarly, we observed a 15-fold increase in $[CO_3^{2-}]$ from ~100 μmol kg^{-1} in the coelenteron to ~1500 μmol kg^{-1} in the calcifying fluid within a distance of 50 μm in this species (Fig. 2d).

Composite depth profiles of CO_2, DIC, TA and Ω_{arag}. Our methods did not allow simultaneous profiling of pH and $[CO_3^{2-}]$ at the same spots due to both technical challenges and high heterogeneity inside a coral polyp. Therefore, we constructed composite profiles of pH and $[CO_3^{2-}]$ based on the range of

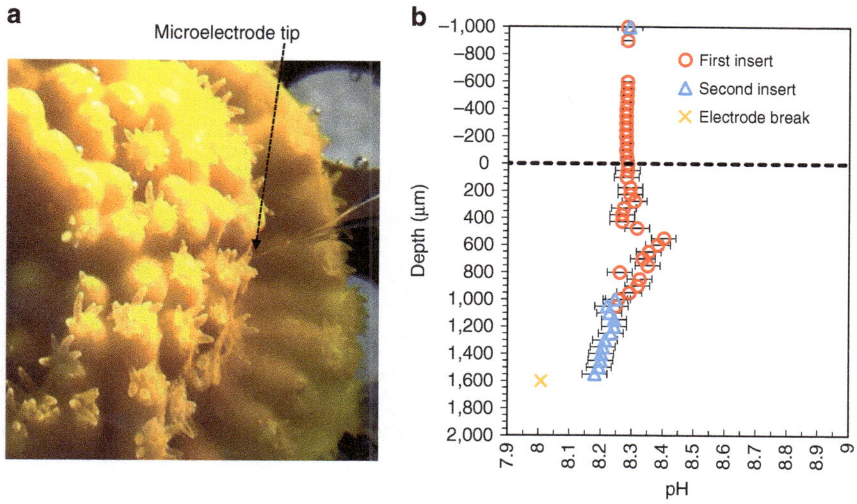

Figure 1 | Microelectrode profiling and electrode performance. (**a**) Close-up view of the coral *T. reniformis* showing the tip of a microelectrode as it enters the polyp. (**b**) Repeated pH microelectrode profile readings inside a *T. reniformis* coral polyp. Between the last point of the second insert and the next reading (which is a noise-free reading and is marked as ×), the microelectrode tip broke. pH microelectrodes were calibrated on the NBS scale, which is about 0.1-0.15 unit greater than that on the total scale. Error bars (s.d.) are defined in the Methods section. Positive depths are inside the coral polyp while negative depths are in the overlying seawater. Error bars are smaller than the symbols in the overlying water.

our data across all three species and calculated [CO_2], DIC, TA and Ω_{arag} profiles (Supplementary Table 1), assuming the DIC system was in chemical equilibrium (see Methods section below). These composite profiles were then used to generate a more complete picture and a conceptual model of coral calcification.

Overall, our study shows that DIC decreases slightly near the polyp surface and in the upper part of the coelenteron due to photosynthetic CO_2 consumption (Fig. 3a,b). Farther down in the polyp calculated DIC increases due to respiratory input of CO_2 and reaches a peak at the bottom of the coelenteron before rapidly declining again in the calcifying space (Fig. 3b). Calculated TA in the coelenteron is nearly constant throughout most of the gastric cavity space, but increases dramatically in the calcifying fluid (Fig. 3b). Nonetheless, in agreement with direct TA measurements of the coelenteron fluid[25] and some model assumptions[5,21], DIC and TA values in the coelenteron solution are not substantially different from those of the overlying seawater.

Notably, a sharp increase in [CO_2] to $\sim 30\,\mu mol\,kg^{-1}$ occurs toward the bottom of the coelenteron solution (Fig. 3b). This increase is not only due to an increase in DIC from respiration but is substantially enhanced by the nearly invariable TA (Fig. 3b, Supplementary Table 1). Thus, we attribute this [CO_2] increase to both respiration input and the conversion of CO_3^{2-} and HCO_3^- to CO_2 by H^+ exported from the calcifying fluid (Fig. 3a) - a mechanism suggested earlier as being important for *Symbiodinium* spp. photosynthetic needs[19]. Inside the calcifying fluid [CO_2] is nearly zero (Fig. 3b), primarily due to the fluid's very high pH (Fig. 2a,c). This very large [CO_2] difference ($\sim 30\,\mu mol\,kg^{-1}$, Supplementary Table 1), based on direct measurements for the first time, confirms recent indirect evidence from geochemical tracers[12] that a high rate of CO_2 molecular diffusion from the coelenteron through the tissue into the calcifying fluid provides an important source of DIC for $CaCO_3$ precipitation.

CO_2 flux into the calcifying fluid. We estimated the CO_2 influx to the calcifying fluid by both a 1-D diffusion model and an exchange model between the coelenteron and the calcifying fluid[21,26] (equations and parameters are given in the Methods section and Supplementary Table 1). The CO_2 diffusive flux was estimated to be as high as 217–433 $nmol\,cm^{-2}\,h^{-1}$. Alternatively, the CO_2 influx was estimated by the exchange model to be 82–301 $nmol\,cm^{-2}\,h^{-1}$ (see Methods section). This range of CO_2 influx and consumption rate compares favourably with coral calcification rates measured in the past (150–1000 $nmol\,cm^{-2}\,h^{-1}$) (refs 10,27). Based on the observations and model calculations presented here, we postulate that the coral's proton pumping ability is essential not only for creating a high aragonite saturation state in the calcifying fluid (favouring calcification), but also for creating a sharp CO_2 gradient facilitating DIC influx to support calcification. This CO_2 transport mechanism is consistent with dual isotopic evidence[18,28] and geochemical proxies[12] showing that respired CO_2 is an essential source of DIC for calcification. This passive CO_2 transport mechanism may reduce the need for the active transport of HCO_3^- by processes such as transmembrane or vacuolar transport though the ultimate source of DIC for calcification comes from external seawater via photosynthesis, diffusive flux and whole seawater import.

Low [DIC] in the calcifying fluid. Despite the great CO_2 flux into the coral calcifying fluid, our analysis reveals that the fluid [DIC] is not that different from seawater concentrations due to the consumption of DIC during calcification, which balances the CO_2 influx (Fig. 3a). Even though TA in the calcifying fluid is elevated (1.2–1.9 times that of overlying seawater) it is not as high as previously reported values[12]. Allison *et al.*[12] have recently used geochemical tracers to determine pH (from B isotope ratio) and estimate DIC (from B/Ca ratio) and concluded that in order to achieve a high aragonite saturation state, corals must internally concentrate DIC. Others also assumed that corals concentrate internal DIC to about twice, or more, the concentration of seawater[13,29]. Our direct pH and [CO_3^{2-}] measurements show that while Ω_{arag} inside the calcifying fluid is elevated, DIC is not elevated with respect to background seawater.

Discussion

Our microelectrode profiles of pH and CO_3^{2-} through coral polyps provide the first direct evidence for elevated [CO_3^{2-}] in the calcifying fluid. [CO_3^{2-}] (600–1500 $\mu mol\,kg^{-1}$) and corresponding aragonite mineral saturation states ($\Omega_{arag} = 8$–22) in the calcifying fluid are high and consistent with expectations based on inorganic $CaCO_3$ formation experiments[11], previous experiments[12] and model frameworks[5,13].

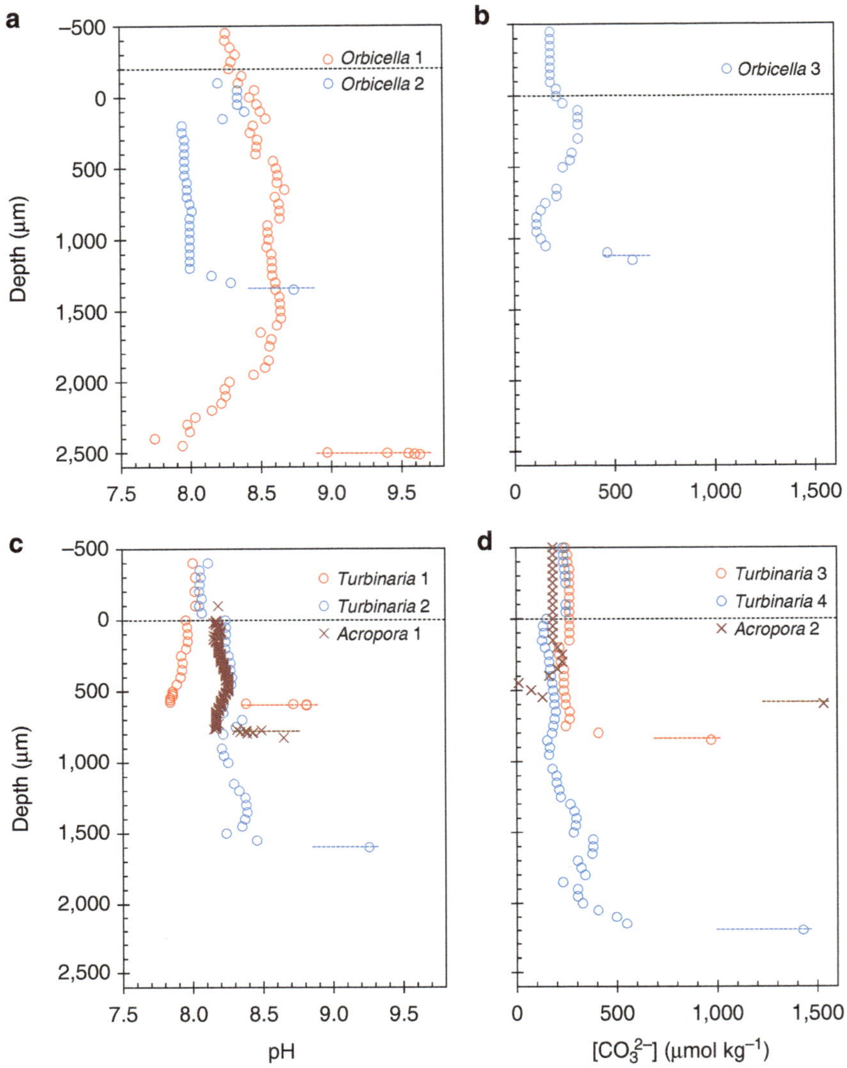

Figure 2 | Coral internal pH and $[CO_3^{2-}]$ profiles through the polyp mouths. (**a**) pH and (**b**) $[CO_3^{2-}]$ of *Orbicella faveolata* corals. (**c**) pH and (**d**) $[CO_3^{2-}]$ of *Turbinaria reniformis* and *Acropora millepora* corals. Dashed black lines indicate the top of the polyp mouths. Colored dashed lines roughly indicate the location of the calcifying fluid and differ in depth for each polyp.

Data from the calcifying fluid was only obtained in a small subset of the numerous profiles collected, most likely because the calcifying fluid layer is very thin (likely only a few micrometers) and existed in pockets[13,16,17]. In most cases, the electrode tips broke within one step (50 μm) between the bottom of the polyp gastric cavity space (coelenteron) and the hard $CaCO_3$ skeleton. In some cases, however, the data suggest the space between the coelenteron and the skeleton could be thicker than previously thought (up to 100 μm). Alternatively, it is possible that the newly formed $CaCO_3$ minerals are porous allowing the glass microelectrode tip to slightly penetrate the mineral surface, thus giving an inaccurate measure of the calcifying fluid thickness. It would be desirable to collect profiles with a better constraint of vertical and lateral heterogeneity of carbonate properties inside corals. Nonetheless a high pH, high $[CO_3^{2-}]$ layer was unequivocally detected right above the $CaCO_3$ skeleton, though further work is needed to better define its thickness and assess its spatial variability. While the pH upregulation in the calcifying fluid measured by microelectrodes does overlap with those by the dye-based method[13] and boron isotope method[12], the highest values have been obtained from microelectrodes (9.7 here, 9.3 and 10.1 by two others, see Table 1)[5,23]. Differences between the approaches may result from differences in coral species, colony size[30], measurement condition and location within the colony[13,14], and time-scales over which the approaches integrate[30].

If the very high pH of 9.0 or higher measured by microelectrodes reported here or elsewhere is typical of coral calcifying fluid, a high DIC internal solution (2 × of seawater DIC or higher) would indicate that internal Ω_{arag} could be as high as 30 to 55 (Fig. 4a), which is inconsistent with Ω_{arag} values inferred from abiotic experiments[18]. Therefore, it seems that the most reasonable internal [DIC] is about 1 × of seawater value, which is consistent with our results from the two direct carbonate parameter measurements. While further research is required to confirm our conclusion, our findings strongly shift the argument towards one of a lower calcifying fluid DIC concentration[5,21].

One important difference between low versus high DIC scenarios is the necessary amount of protons that a coral must pump out (which results in an increase in TA) to raise its internal pH by one unit (in mmol-H^+ per kg per pH unit). This buffer index for the high DIC scenario is at least twice that for the low DIC scenario, thus more protons must be removed in a high DIC scenario to achieve the observed calcifying fluid pH (Fig. 4b).

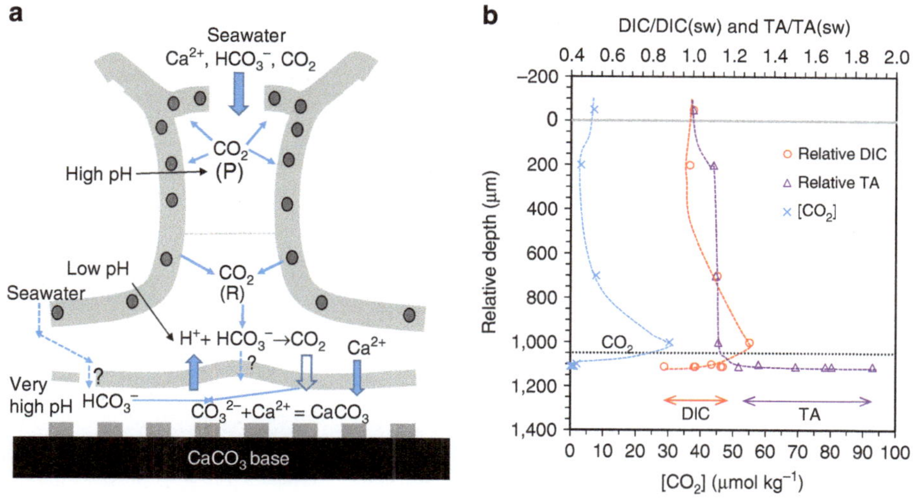

Figure 3 | Conceptual model of coral calcification and CO_2 transport mechanism and distributions of internal [CO_2] and relative DIC and TA to seawater values. (**a**) Illustrative model and (**b**) concentration distributions. Dark circles are photosynthetic endosymbionts. P stands for photosynthesis, R for respiration, and sw for seawater. In (**b**), for DIC (red line) and TA (purple line) the calculated ranges of possible concentrations inside the calcifying fluid are provided based on calculations of all species from Supplementary Table 1.

Table 1 | Comparison of pH values measured in coral calcifying fluid by microelectrodes and other methods.

Technique	Reported by	Coral species	Light condition	Location	pH_T	pH_{NBS}	Notes
Microelectrode	Al-Horani *et al.*[23]	*Galaxea fascicularis* (tropical coral)	Light	Apexes		9.28	Sensor buried under tissue
Microelectrode	Ries[5]	*Astrangia poculata* (temperate coral)	Light	Apexes, between septal ridges		10.1	Through a predrilled incision
Microelectrode	This work	*Orbicella faveolata* (tropical coral)	Light	Apexes, under polyp mouth		8.75-9.65	Sensor penetrated via the mouth
Microelectrode	This work	*Turbinaria reniformis* (tropical coral)	Light	Apexes, under polyp mouth		8.8-9.3	Sensor penetrated via the mouth
Microelectrode	This work	*Acropora millepora* (tropical coral)	Light	Apexes, under polyp mouth		8.65	Sensor penetrated via the mouth
pH sensitive dye	Venn *et al.*[13]	*Stylophora pistillata* (tropical coral; microcolonies)	Light	Distal margin (edge)	8.55-8.85	8.70-9.00	Laterally grown on slide
B-isotopes	Allison *et al.*[12]	*Porites* spp. (tropical coral)	Light/dark cycle	Average	~8.5	~8.65	See citations therein

Note that only one B-isotope based result is given for comparison but many more are cited in the reference. Note, we convert pH_T to pH_{NBS} by adding 0.15.

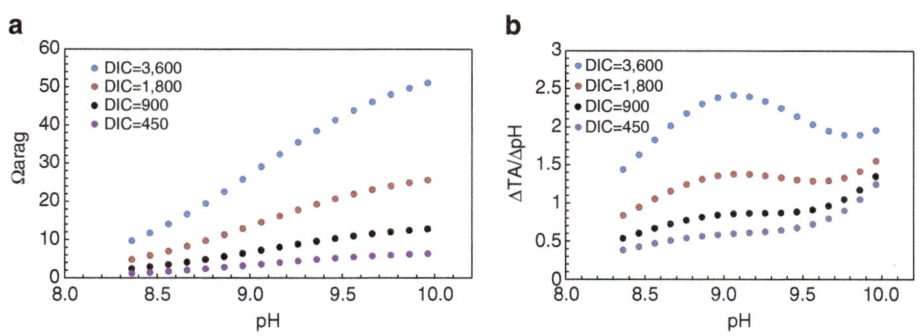

Figure 4 | Aragonite mineral saturation state (Ω_{arag}) and buffer index in the coral calcifying fluid as functions of pH and [DIC]. (**a**) Ω_{arag} and (**b**) buffer index. The buffer index is given as $mmol\,kg^{-1}$ of H^+ pumping (H^+ loss or TA gain) per unit of pH upregulation. This index is calculated numerically with CO2SYS (ref. 37) (see Methods section). Here [DIC] = 1800 $\mu mol\,kg^{-1}$ is roughly the same as seawater value. pH is on the NBS scale.

Therefore, we conclude that the low DIC scenario would require less energy than the high DIC scenario as proton-pumping requires ATP. Comparisons of a relative change in energy requirement to upregulate internal pH between high external CO_2 under future OA conditions and 400 ppm current conditions following the method of Ries[5] also supported our conclusion (Supplementary Note 1 and Supplementary Table 5). As a result, if proton removal from the coelenteron is increasingly difficult as a result of higher seawater H^+ concentrations under OA conditions[31], the low DIC scenario should be advantageous as less H^+-pumping will be required to further raise pH in the calcifying fluid.

Overall, this suggests that corals that use passive CO_2 diffusion as an important pathway for DIC supply, and maintain low DIC in their calcifying fluid, may require less energy to calcify at high rates, which could potentially make them more resistant to OA and other stressors. However, at the same time, the lower buffering capacity of low DIC and moderate TA calcifying fluid means that the ability to maintain high fluid pH will be vulnerable to changes in the physiological and biochemical conditions that control H^+-pumping. For example, a weakening of proton pumping during dark periods may lead a coral with low internal DIC towards decalcification more readily than otherwise expected in a high DIC scenario. In addition, a high internal DIC could allow corals to reach a high carbonate saturation state at lower internal pH (Fig. 4a)[13]. Therefore, additional studies using microelectrodes and other techniques in corals under OA conditions in combination with other stressors and environmental conditions (i.e., elevated temperature, nutrient levels and daytime versus nighttime) are a critical next step in elucidating coral calcification mechanisms under future ocean conditions.

Methods

Coral collection and preparation. The measurements focused on three species of corals: *O. faveolata*, *T. reniformis* and *A. millepora*. All species contain photosynthetic endosymbionts, often called zooxanthellae (*Symbiodinium* spp.), and form an aragonite skeleton. Small fragments of *O. faveolata* were sampled from the Florida Keys and transported to the laboratory in the University of Georgia (Florida Keys National Marine Sanctuary permit 2014-2015). *Turbinaria reniformis* and *Acropora millepora* coral fragments were provided by Reef Systems Coral Farm in New Albany, OH. The parent colonies of both species have been maintained in recirculating indoor aquaria with natural light (greenhouse, 700–1000 μmol photons $m^{-2} s^{-1}$) and commercially available artificial seawater (Instant Ocean Reef Crystals) for over 10 years. Several fragments of each colony were shipped to the Cai lab at the University of Georgia, where they were acclimated to a closed-system laboratory aquarium with artificial seawater of salinity 35 and TA of 2.4 mmol kg^{-1} at 26 °C with light levels of about 200 μmol photons $m^{-2} s^{-1}$ for at least two weeks prior to conducting pH and CO_3^{2-} microelectrode profile measurements. TA in the aquarium was depleted at the time of microelectrode measurements (Supplementary Tables 2, 3 and 4) but was amended weekly.

Microelectrode construction and properties. pH microelectrodes[32,33] and CO_3^{2-} microelectrodes[34,35] were constructed as previously described. We made our microelectrodes with a flat tip of diameter between 10 and 15 μm (Supplementary Fig. 3), which enabled us to make pH measurements along a profile into the coral polyp. pH microelectrodes were calibrated with three commercial pH NBS standards (pH = 4, 7 and 10) at 26 °C to determine their response slopes. CO_3^{2-} microelectrodes were calibrated in CO_3^{2-} standard solutions pumped sequentially into a sealed chamber where an electrode was hosted[35]. Carbonate concentrations were set at approximately 50, 150, 250, 350, 500 and 1,000 μmol kg^{-1} by adding controlled amounts of HCl and $NaHCO_3$ into filtered seawater collected from the Gulf of Mexico[35]. Standard solutions were then bubbled with room air for 24 h to minimize further exchange of CO_2 with air during the experiment. Finally, CO_3^{2-} concentrations were precisely calculated by CO2SYS with TA and pH measured by Gran titration and a commercial Ross glass electrode, respectively.

Data collection. In brief, a coral fragment was transferred from the incubation tank together with the tank water into a small measurement chamber that was kept at 26 °C and stirred with a magnetic stirrer (Supplementary Fig. 4). The coral fragment was allowed to acclimate in the small container for 1 h under 500 μmol photons $m^{-2} s^{-1}$ until the microelectrode reading was stable in the chamber water (Fig. 1b & Supplementary Fig. 1). The slopes of pH and CO_3^{2-} microelectrodes were calibrated immediately before use, and chamber water pH measured by a commercial electrode and measured TA values were used as the reference point before beginning the profiles. Measurement conditions are provided in Supplementary Tables 2,3 and 4. The point at which the microelectrode entered the polyp (defined as depth = 0) was observed under a microscope. Typically the coral reacted by closing its mouth making this point unambiguous. The electrode was then advanced through the polyp's gastric cavity in increments of 5–50 μm until the microelectrode broke on the skeleton. The microelectrode was held at each step until a stable reading was obtained. The pH reading was considered stable when drift was <0.2 mV or 0.003 pH min^{-1} (or <0.2 mV min^{-1} for CO_3^{2-}) or a total of 3 min had passed, whichever was first. A 3-min limit was established because our pH microelectrodes had a much shorter response time[32,35] but, in some cases, a stable reading was not realistic due to polyp movement.

Several replicate profiles were obtained to ensure that the microelectrodes were not fouled or otherwise damaged by profiling through the corals. In these tests a profile was obtained about half way through the coral coelenteron, at which point the electrode was returned to the overlying seawater, and the seawater readings were compared with values obtained before profiling was started. Then, the profile was resumed, repeating a portion of the earlier profile to ensure agreement between the two (Fig. 1b and Supplementary Figs 1 and 2). However, this practice cannot be extended into the calcifying fluid as an electrode is destined to be damaged and a repeating profile cannot be collected.

Finally, based on electrode calibration result, stability test and repeated measurements as shown above, we assign errors as ± 0.02 pH unit for the overlying seawater, ± 0.04 pH unit for the coelenteron and ± 0.1 pH unit for the calcifying fluid. For $[CO_3^{2-}]$, a maximum uncertainty of ± 10% is established based on our experiences and literature[35,36].

Carbonate chemistry calculation. While the CO2SYS program[37] is popularly used for carbonate equilibrium calculation based on two of the four parameters (TA, DIC, pH and pCO_2), it does not accept the pH and $[CO_3^{2-}]$ pair as input parameters. However, measured pH and $[CO_3^{2-}]$ data allowed us to calculate concentrations of other DIC species, [DIC] and the TA in a straightforward way. For example,

$$[\text{DIC}] = [CO_3^{2-}]([H^+]^2/(K_1 * K_2) + [H^+]/K_2 + 1) \quad (1)$$

where, K_1 and K_2 are the dissociation constants of the carbonic acid[38]. All equations are described in Supplementary Note 2. As in previous studies, we assumed that the coral's internal salinity was the same as that of the external seawater salinity (that is corals are osmoconformers) for our thermodynamic calculation[5,12,21]. While a lower internal salinity would lead to a higher calculated DIC from the same observed pH and $[CO_3^{2-}]$, for example, calculated [DIC] is 0.8–12% higher at salinity = 30 than that at 35, conclusion derived here holds well within a reasonable salinity range (see more details in Supplementary Note 3). In this work, pH in NBS scale is used. However, when needed, pH in total proton scale is converted to NBS scale roughly by adding 0.15 unit, $pH_{NBS} = pH_T + 0.15$, according to CO2SYS simulation[37] and Dickson[39].

Although internal $[Ca^{2+}]$ was reported to be about 10% higher than that in seawater[23], as it was the only reported value and as seawater $[Ca^{2+}]$ is very high (~10.3 mM), we assumed that $[Ca^{2+}]$ in the calcifying fluid is the same as in seawater in calculating the saturation state Ω. Since $[CO_3^{2-}]$ is directly measured in this work, the uncertainty in Ω calculation is about the same as that in $[Ca^{2+}]$, roughly within 10%, and it does not affect the conclusions and discussion of this work.

Calculation of CO_2 flux into the calcifying fluid. In a 1-D diffusion model, CO_2 molecular diffusional flux is defined as:

$$F_{CO_2} = D_{CO_2} \Delta[CO_2]/\Delta x, \quad (2)$$

where, D is the diffusion coefficient, and Δx is the distance corresponding to the concentration gradient. We assume D determined in seawater[40] is applicable to the coral interior space. We calculated CO_2 flux using the $\Delta[CO_2]$ (Supplementary Table 1) and the corresponding distance between the two data points in the coelenteron and the calcifying fluid, respectively, as observed with our microelectrode (50–100 μm).

Using the calculated $[CO_2]$ gradient (29.8 μmol kg^{-1}) from Supplementary Table 1, CO_2 flux (F_{CO_2}) from the coelenteron into the calcifying fluid was calculated. Note the $[CO_2]$ gradient determined by the microelectrode data is not affected much by the uncertainty caused by relatively large pH and $[CO_3^{2-}]$ ranges inside the calcifying fluid as $[CO_2]$ is nearly 0 (that is an uncertainty of $[CO_2]$ between 0.5 and 0.1 μmol kg^{-1} only changes $[CO_2]$ gradient from 29.9 to 30.3 μmol kg^{-1}) (Supplementary Table 1). In other words, a conservative internal pH value of 9.1 and a high value of 9.5 lead to similar estimates of CO_2 gradient and flux.

Flux can also be calculated as the exchange rate between two adjacent 'boxes', the coelenteron and the calcifying fluid:

$$F_{CO_2} = k_{CO_2} \Delta[CO_2], \quad (3)$$

where, k_{CO_2} is the exchange coefficient (0.76–2.8 cm s^{-1}) across biological tissues determined by an isotopic exchange method[41], which has been used in two recent coral CO_2 flux model simulations[21,26]. In principle, the two coefficients are linked by:

$$k_{CO_2} = D_{CO_2}/\Delta x. \quad (4)$$

A lower CO_2 flux calculated based on equation 3 than equation 2 may indicate that the CO_2 diffusion coefficient across the coral tissue (or the coral tissue's permeability) is lower than that of the seawater.

Amount of H^+-pumping per unit pH upregulation. We define this parameter based on the buffer factor concept, discussed in details by Frankignoulle[42], Hofmann *et al.*[43] and Egleston *et al.*[44]. We take the buffer factor β defined below as a measure of how strong the calcifying fluid can buffer its pH:

$$\beta = \left(\frac{\partial \ln[H^+]}{\partial \mathrm{TA}}\right)^{-1} = -\left(\frac{\partial \mathrm{TA}}{2.3\partial \mathrm{pH}}\right) \quad (5)$$

We rearranged the definition to $\Delta\mathrm{TA}/\Delta\mathrm{pH} = -2.3\beta$, which represents the amount of proton pumping (or TA gain) a coral must do to upregulate its calcifying fluid pH by one unit. This was calculated numerically by following the same approach as the Revelle Factor calculation in the CO2SYS program[37]. We first set four DIC values and for each DIC value we allowed a range of pH changes. Thus, we derived TA values for the various DIC and pH combinations. Then, a $\pm 1\ \mu\mathrm{mol\,kg}^{-1}$ disturbance is applied to TA values and the corresponding perturbation in pH is derived using the TA and DIC pairs as the input parameters in CO2SYS (Fig. 4b).

Energy consumption for pH upregulation. We calculated the relative energy changes of pH upregulation between different OA conditions as specified by Ries[5] using the Nernst equation. The equations and results are given in the Supplementary Note 1 and Supplementary Table 5.

References

1. IPCC. Climate Change 2013. The Physical Science Basis. Contribution of Working Group I to the Fifth Assessment Report of the Intergovernmental Panel on Climate Change (2014).
2. Caldeira, K. & Wickett, M. E. Anthropogenic carbon and ocean pH. *Nature* **425,** 365–365 (2003).
3. Orr, J. C. *et al.* Anthropogenic ocean acidification over the twenty-first century and its impact on calcifying organisms. *Nature* **437,** 681–686 (2005).
4. Hoegh-Guldberg, O. & Bruno, J. F. The impact of climate change on the world's marine ecosystems. *Science* **328,** 1523–1528 (2010).
5. Ries, J. B. A physicochemical framework for interpreting the biological calcification response to CO_2-induced ocean acidification. *Geochim. Cosmochim. Acta* **75,** 4053–4064 (2011).
6. Shamberger, K. E. F. *et al.* Diverse coral communities in naturally acidified waters of a Western Pacific reef. *Geophys. Res. Lett.* **41,** 499–504 (2014).
7. Langdon, C. *et al.* Effect of elevated CO_2 on the community metabolism of an experimental coral reef. *Global Biogeochem. Cycles* **17,** 1011 (2003).
8. Schoepf, V. *et al.* Coral energy reserves and calcification in a high-CO_2 world at two temperatures. *PLoS ONE* **8,** e75049 (2013).
9. Doney, S. C. The growing human footprint on coastal and open-ocean biogeochemistry. *Science* **328,** 1512–1516 (2010).
10. Allemand, D. *et al.* Biomineralisation in reef-building corals: from molecular mechanisms to environmental control. *C. R. Palevol* **3,** 453–467 (2004).
11. Cohen, A. L. & Holcomb, M. Why corals care about ocean acidification: Uncovering the mechanism. *Oceanography* **22,** 118–127 (2009).
12. Allison, N., Cohen, I., Finch, A. A., Erez, J. & Tudhope, A. W. Corals concentrate dissolved inorganic carbon to facilitate calcification. *Nat. Commun.* **5,** 5741 (2014).
13. Venn, A., Tambutté, E., Holcomb, M., Allemand, D. & Tambutté, S. Live tissue imaging shows reef corals elevate pH under their calcifying tissue relative to seawater. *PLoS ONE* **6,** e20013 (2011).
14. Holcomb, M. *et al.* Coral calcifying fluid pH dictates response to ocean acidification. *Sci. Rep* **4,** 5207 (2014).
15. Gagnon, A. C. Coral calcification feels the acid. *Proc. Natl Acad. Sci.* **110,** 1567–1568 (2013).
16. Tambutté, S. *et al.* Coral biomineralization: from the gene to the environment. *J. Exp. Mar. Biol. Ecol.* **408,** 58–78 (2011).
17. Gagnon, A. C., Adkins, J. F. & Erez, J. Seawater transport during coral biomineralization. *Earth Planet. Sci. Lett.* **329–330,** 150–161 (2012).
18. Cohen, A. L. & McConnaughey, T. A. Geochemical perspectives on coral mineralization. *Rev. Mineral. Geochem.* **54,** 151–187 (2003).
19. McConnaughey, T. Zooxanthellae that open calcium channels: implications for reef corals. *Mar. Ecol. Prog. Ser.* **460,** 277–287 (2012).
20. Zoccola, D. *et al.* Bicarbonate transporters in corals point towards a key step in the evolution of cnidarian calcification. *Sci. Rep.* **5,** 9983 (2015).
21. Hohn, S. & Merico, A. Modelling coral polyp calcification in relation to ocean acidification. *Biogeosciences* **9,** 4441–4454 (2012).
22. Kuhl, M., Cohen, Y., Dalsgaard, T., Jergensenl, B. & Revsbech, N. Microenvironment and photosynthesis of zooxanthellae in scleractinian corals studied with microsensors for O_2, pH and light. *Mar. Ecol. Prog. Ser.* **117,** 1995 (1995).
23. Al-Horani, F. A., Al-Moghrabi, S. M. & de Beer, D. The mechanism of calcification and its relation to photosynthesis and respiration in the scleractinian coral Galaxea fascicularis. *Mar. Biol.* **142,** 419–426 (2003).
24. Barott, K. L., Venn, A. A., Perez, S. O., Tambutté, S. & Tresguerres, M. Coral host cells acidify symbiotic algal microenvironment to promote photosynthesis. *Proc. Natl Acad. Sci.* **112,** 607–612 (2015).
25. Agostini, S. *et al.* Biological and chemical characteristics of the coral gastric cavity. *Coral Reefs* **31,** 147–156 (2012).
26. Nakamura, T., Nadaoka, K. & Watanabe, A. A coral polyp model of photosynthesis, respiration and calcification incorporating a transcellular ion transport mechanism. *Coral Reefs* **32,** 779–794 (2013).
27. Barnes, D. J. & Devereux, M. J. Productivity and calcification on a coral reef: A survey using pH and oxygen electrode techniques. *J. Exp. Mar. Biol. Ecol.* **79,** 213–231 (1984).
28. Furla, P., Galgani, I., Durand, I. & Allemand, D. Sources and mechanisms of inorganic carbon transport for coral calcification and photosynthesis. *J. Exp. Biol.* **203,** 3445–3457 (2000).
29. McCulloch, M., Falter, J., Trotter, J. & Montagna, P. Coral resilience to ocean acidification and global warming through pH up-regulation. *Nat. Clim. Change* **2,** 623–627 (2012).
30. Honisch, B. *et al.* Assessing scleractinian corals as recorders for paleo-pH: Empirical calibration and vital effects. *Geochim. Cosmochim. Acta* **68,** 3675–3685 (2004).
31. Jokiel, P. L. Ocean acidification and control of reef coral calcification by boundary layer limitation of proton flux. *Bull. Mar. Sci.* **87,** 639–657 (2011).
32. Zhao, P. & Cai, W.-J. pH polymeric membrane microelectrodes based on neutral carriers and their application in aquatic environments. *Anal. Chim Acta* **395,** 285–291 (1999).
33. Cai, W.-J., Zhao, P. & Wang, Y. pH and pCO_2 microelectrodes measurement and diffusive behavior of carbon dioxide species in coastal marine sediments. *Mar. Chem.* **70,** 133–148 (2000).
34. de Beer, D. *et al.* A microsensor for carbonate ions suitable for microprofiling in freshwater and saline environments. *Limnol. Oceanogr. Meth.* **6,** 532–541 (2008).
35. Han, C., Cai, W.-J., Wang, Y. & Ye, Y. Calibration and evaluation of a carbonate microsensor for studies of the marine inorganic carbon system. *J. Oceanogr.* **70,** 425–433 (2014).
36. Lee, B. H., Shim, Y.-B. & Park, S. B. A lipophilic sol – gel matrix for the development of a carbonate-selective electrode. *Anal. Chem.* **76,** 6150–6155 (2004).
37. Pierrot, D., Lewis, E. & Wallace, D. W. R. *MS Excel Program Developed for CO_2 System Calculations* (Oak Ridge National Laboratory, 2006).
38. Millero, F. J., Graham, T. B., Huang, F., Bustos-Serrano, H. & Pierrot, D. Dissociation constants of carbonic acid in seawater as a function of salinity and temperature. *Mar. Chem.* **100,** 80–94 (2006).
39. Dickson, A. G. pH buffers for sea water media based on the total hydrogen ion concentration scale. *Deep-Sea Res.* **40,** 107–118 (1993).
40. Li, Y. H. & Gregory, S. Diffusion of ions in sea water and in deep-sea sediments. *Geochim. Cosmochim. Acta* **38,** 703 (1974).
41. Sültemeyer, D. & Rinast, K.-A. The CO_2 permeability of the plasma membrane of Chlamydomonas reinhardtii: mass-spectrometric ^{18}O-exchange measurements from $^{13}C^{18}O_2$ in suspensions of carbonic anhydrase-loaded plasma-membrane vesicles. *Planta* **200,** 358–368 (1996).
42. Frankignoulle, M. A complete set of buffer factors for acid-base CO_2 system in seawater. *J. Mar. Syst.* **5,** 111–118 (1994).
43. Hofmann, A., Soetaert, K., Middelburg, J. & Meysman, F. R. AquaEnv: an aquatic acid–base modelling environment in R. *Aquat. Geochem.* **16,** 507–546 (2010).
44. Egleston, E. S., Sabine, C. L. & Morel, F. M. M. Revelle revisited: Buffer factors that quantify the response of ocean chemistry to changes in DIC and alkalinity. *Global Biogeochem. Cycles* **24,** GB1002 (2010).

Acknowledgements

The authors acknowledge financial supports from the NSF EF-1041124 to A.G., 1040940 to M.W., 1041070 to W-J.C., and EF-1315944 to B.H. & W-J.C. D.Q. and X.Y. acknowledge the supports of China Scholarship Council for supporting their visits to the Cai laboratory at the University of Georgia.

Author contributions

W-J.C., A.G., M.W., B.H. and Y.W. conceived and designed the experiments; Y.M., D.Q. and Y.W. performed the experiments; Y.Y., C.H., H.X., B.H. and Y.W. contributed materials/analysis tools; W-J.C. and Y.M. analysed the data, and H.X. and W-J.C. performed the buffer index calculation. W-J.C. prepared the manuscript. All authors contributed to discussion and writing of the paper.

Additional information

Competing financial interests: The authors declare no competing financial interests.

Western Pacific hydroclimate linked to global climate variability over the past two millennia

Michael L. Griffiths[1], Alena K. Kimbrough[2], Michael K. Gagan[2], Russell N. Drysdale[3,4], Julia E. Cole[5,6], Kathleen R. Johnson[7], Jian-Xin Zhao[8], Benjamin I. Cook[9], John C. Hellstrom[10] & Wahyoe S. Hantoro[11]

Interdecadal modes of tropical Pacific ocean-atmosphere circulation have a strong influence on global temperature, yet the extent to which these phenomena influence global climate on multicentury timescales is still poorly known. Here we present a 2,000-year, multiproxy reconstruction of western Pacific hydroclimate from two speleothem records for southeastern Indonesia. The composite record shows pronounced shifts in monsoon rainfall that are antiphased with precipitation records for East Asia and the central-eastern equatorial Pacific. These meridional and zonal patterns are best explained by a poleward expansion of the Australasian Intertropical Convergence Zone and weakening of the Pacific Walker circulation (PWC) between ~1000 and 1500 CE Conversely, an equatorward contraction of the Intertropical Convergence Zone and strengthened PWC occurred between ~1500 and 1900 CE. Our findings, together with climate model simulations, highlight the likelihood that century-scale variations in tropical Pacific climate modes can significantly modulate radiatively forced shifts in global temperature.

[1] Department of Environmental Science, William Paterson University, Wayne, New Jersey 07470, USA. [2] Research School of Earth Sciences, Australian National University, Canberra, Australian Capital Territory 0200, Australia. [3] School of Geography, University of Melbourne, Parkville, Victoria 3010, Australia. [4] EDYTEM, UMR CNRS 5204, Université de Savoie, 73376 Le Bourget du Lac, France. [5] Department of Geosciences, University of Arizona, Tucson, Arizona 85721, USA. [6] Department of Atmospheric Sciences, University of Arizona, Tucson, Arizona 85721, USA. [7] Department of Earth System Science, University of California, Irvine, California 92697-3100, USA. [8] School of Earth Sciences, University of Queensland, Brisbane, Queensland 4072, Australia. [9] NASA Goddard Institute for Space Studies, New York, New York 10025, USA. [10] School of Earth Sciences, The University of Melbourne, Parkville, Victoria 3010, Australia. [11] Research Center for Geotechnology, Indonesian Institute of Sciences, Bandung 40135, Indonesia. Correspondence and requests for materials should be addressed to M.L.G. (email: griffithsm@wpunj.edu).

Deep atmospheric convection over the Indo-Pacific Warm Pool (IPWP) redistributes heat and moisture throughout the global climate system. The strength and location of IPWP convection is highly variable and strongly coupled to the Pacific Walker Circulation (PWC) and the associated modes of tropical climate variability, the El Niño-Southern Oscillation (ENSO) and the Interdecadal Pacific Oscillation. Changes in IPWP convection impact extra-tropical climate via meridional shifts in the average position of the intertropical convergence zone (ITCZ), which influences the intensity of the Asian Summer Monsoon and the Australian-Indonesian Summer Monsoon (AISM; Supplementary Figs 1 and 2). Meridional and zonal variations of the Indo-Pacific climate system have very different impacts on Earth's climate. Meridional shifts in the position of the ITCZ rapidly transmit hydroclimate perturbations across the hemispheres in response to changes in interhemispheric temperature gradients[1]. By contrast, zonal shifts in the PWC, which alter the trade wind fields and the location and strength of deep convection (both zonally and meridionally), extend the geographic reach of ENSO and contribute to global temperature changes[2–4]. Therefore, unravelling the history of deep atmospheric convection in the western equatorial Pacific region before instrumental measurements is critical for understanding the role of the tropics in modulating both regional and global climate variability.

Paleoclimate records spanning the last ~2,000 years have highlighted the potential influence of Northern Hemisphere temperature changes in modulating Australasian atmospheric circulation via meridional shifts in the ITCZ[5–8]. However, the influence of zonal displacements in deep convection (associated with ENSO) in driving tropical climate is implicated by other studies[9,10]. As a result, there are different hypotheses regarding the precise drivers of tropical Pacific hydroclimate over the past millennium, and the relative importance of meridional forcing from the extratropics versus zonal ENSO-like shifts inherent to the tropics. To further examine these interactive climate systems, we have constructed a high-resolution multi-proxy hydroclimate record for southeast Indonesia spanning the last ~2,000 years based on stalagmite oxygen ($\delta^{18}O$) and carbon ($\delta^{13}C$) isotopes, elemental ratios (Mg/Ca, Sr/Ca), and initial $^{234}U/^{238}U$ activity ratios. These data reveal substantial changes in AISM rainfall and deep atmospheric convection at the ascending limb of the PWC over the past two millennia. Comparison of our results with coupled general circulation model (CGCM) simulations reveal century-scale periods of divergence between the paleodata and model simulations, highlighting a deficiency in the models to capture the lower-frequency variability of the PWC. As such, the influence of the PWC to amplify the radiatively forced warming and cooling trends of the past millennium is limited in the models. These results have important implications for future projections of global temperatures because they show that state-of-the-art last millennium (AD 850–1850) and historical (AD 1850–2005) CGCM experiments may be underestimating the potential for low-frequency shifts in the tropical Pacific mean state to modulate Earth's climate.

Results

Cave setting. The two stalagmites used in this study (LR06-B1 and LR06-B3) were collected from Liang Luar, an ~1.7-km-long cave situated on the east Indonesian island of Flores (8° 32'N, 120° 26'E; 550 m above sea level; Fig. 1 and Supplementary Figs 1 and 2). Both specimens were located in a large chamber ~600 m from the cave entrance and were actively growing at the time of collection in August 2006. Liang Luar resides close to the southerly limit of the austral summer ITCZ and most of the annual rainfall (~70%) occurs during the AISM season (December to March)[11].

Stable isotope and trace element records. The hydroclimate record derived from stalagmites LR06-B1 and LR06-B3 is fixed in time by 35 uranium-thorium (U-Th) ages (Supplementary Figs 4–6) and spans the period 3 to 2005 CE (Common Era, Supplementary Table 1). All ages lie in stratigraphic order (within 2σ uncertainty) and have typical uncertainties of 1–3%. The age models for both records were first calculated using a Bayesian–Monte Carlo approach[12] and then refined using the intra-site correlation age modelling programme (*iscam*)[13] written in MATLAB (see the Methods for details). The age models give an average growth rate of ~125–135 µm per year for both stalagmites.

The $\delta^{18}O$ and $\delta^{13}C$ records are based on 1,157 measurements performed on calcite powders micromilled along the central growth axes of LR06-B1 and LR06-B3. The Mg/Ca and Sr/Ca ratios in stalagmite LR06-B1, previously presented in Griffiths *et al.*[14], were analysed on H_3PO_4 residues following calcite dissolution for stable isotope analysis. The strong degree of replication between the two stable isotope profiles (Fig. 2, Supplementary Fig. 7 and Supplementary Table 2), along with 'Hendy' screening tests[15] conducted on these specimens[11], rules out significant disruption of the records by kinetic isotope fractionation effects.

The $\delta^{18}O$ values for LR06-B1 and LR06-B3 range from −5.6 to −6.9‰ and average −6.2‰ between 1 and 2005 CE, whereas the $\delta^{13}C$ values range from −7.8 to −13.4‰ and average −10.8‰. Broadly speaking, the $\delta^{18}O$ and $\delta^{13}C$ records display multidecadal- to centennial-scale oscillations with generally lower values from ~500 to 900 CE and ~1300 to 1600 CE, and generally higher values from ~1000 to 1300 CE The Mg/Ca and Sr/Ca profiles show patterns similar to the stable isotopes over the last millennium, with generally higher values from ~1000 to 1300 CE followed by a trend to lower values until ~1600 CE Between ~1600 CE and ~2004 values have generally increased, signifying an overall reduction in rainfall over this period with the trend being most significant during the twentieth century.

Interpretation of the stalagmite $\delta^{18}O$ record. The seasonal pattern of monsoon rainfall at Liang Luar is reflected in the $\delta^{18}O$ of precipitation ($\delta^{18}O_p$), with summer rainfall (DJFM) ~6–7‰ lighter than its winter monsoon (JJA) counterpart[11]. Hence, a change in the fraction of the year dominated by the summer monsoon (that is, change in seasonality), or a shift in monsoon rainfall amount in summer, should have a significant impact on the average annual $\delta^{18}O$ value. The strong relationship between Indonesian rainfall amount and $\delta^{18}O_p$ has been identified in modern isotope studies[11] and CGCM simulations of modern[16,17] and past[18] hydroclimates in Indonesia. Model simulations of precipitation $\delta^{18}O_p$ using the National Center for Environmental Prediction (NCEP)/National Center for Atmospheric Research (NCAR) nudged IsoGSM[17] for the grid point closest to Liang Luar reveal a significant correlation with sea-surface temperatures (SSTs) over the NINO3.4 region and, to a lesser extent, in the western Indian Ocean; this pattern is similar to the observed pattern of rainfall amount at Liang Luar (Supplementary Fig. 3). Moreover, comparison of Liang Luar IsoGSM $\delta^{18}O_p$ with observed rainfall amount, outgoing longwave radiation and zonal 850 mb winds reveals that $\delta^{18}O_p$ is strongly influenced by western Pacific rainfall variability, which is governed by the strength of the AISM, the position of the ITCZ and ENSO (Supplementary Fig. 3).

Based on observed and modelled precipitation $\delta^{18}O$ systematics, we interpret variations in the Flores speleothem $\delta^{18}O$ records to primarily reflect regional rainfall amount. However, on relatively short timescales (centuries), other factors may affect

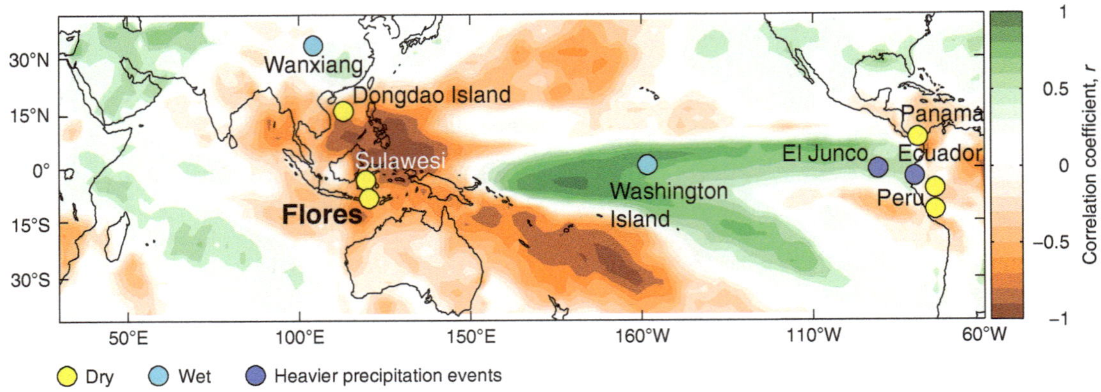

Figure 1 | ENSO influence on tropical hydroclimate. Field correlation map of annual rainfall and east equatorial Pacific sea surface temperatures. Correlation coefficients (r) were calculated between GPCP v2.2 precipitation (at each 1° × 1° grid point) and HadISST1 NINO3.4 SSTs (1979–2010). The coloured dots show the location of our study site and other paleoclimate records mentioned in the text. Yellow indicates relatively dry conditions during the Medieval Climate Anomaly[5,6,9,44,46,48] (950–1250 CE), whereas cyan/purple indicates relatively wet conditions[7,8]/intense precipitation events[10,47].

stalagmite $\delta^{18}O$, such as changes in cave temperature (due to shifts in regional SSTs), the $\delta^{18}O$ of surface seawater[5], the proportion of summer- and winter-sourced moisture[11] and multidecadal changes in the relationship between $\delta^{18}O_p$ and ocean-atmosphere climate modes, such as ENSO[19] and the Indian Ocean Zonal Mode[20]. In addition, mixing of meteoric water in the soil-karst overburden (~50–100 m) above Liang Luar reduces the amplitude of the highest frequencies[21] (for example, seasonal-annual variations), which likely varies with time. Therefore, some of the observed variability may be an artefact of filtering through the karst network. However, given the amount of cave recharge through the epikarst each monsoon season, it is likely that the $\delta^{18}O$ of the cave drip-water represents decadal (if not multi-year) shifts in precipitation.

A composite multi-proxy stalagmite record. To constrain the Liang Luar paleomonsoon record, we compare the stalagmite $\delta^{18}O$ with the contemporaneous carbon isotope ($\delta^{13}C$), Mg/Ca and Sr/Ca ratios, which primarily reflect Liang Luar karst hydrology[14,22]. Recent geochemical modelling of the soil-karst system above Liang Luar demonstrated that the $\delta^{13}C$ in the late Holocene section of stalagmite LR06-B1 was controlled by the degree of prior calcite precipitation (PCP)[22] in the epikarst 'upstream' of the stalagmite. Under this process, periods of reduced recharge result in partial dewatering of the karst fracture system overlying the cave, which causes a greater proportion of the CO_2 degassing to occur before the drips reach the growing stalagmite[23]. The waters therefore become supersaturated in $CaCO_3$ before their emergence in the cave, causing upstream (prior) calcite precipitation in fractures and on stalactite tips. In the case of $\delta^{13}C$, carbonate precipitation results in a higher stalagmite $\delta^{13}C$ owing to the greater removal of $^{12}CO_2$ during the degassing process[24,25]. The opposite occurs during wetter intervals[25] (Fig. 2). Additional climate-related factors may have influenced the stalagmite $\delta^{13}C$, such as changes in the ratio of C3:C4 plants, soil respiration rates, contributions from atmospheric CO_2 (refs 26,27) and the relative contribution of bedrock carbon due to changes in open- versus closed-system dissolution[22]. However, these factors are all susceptible to hydrological changes and fluctuate in the same sense as PCP (that is, higher $\delta^{13}C$ values indicate drier conditions). Therefore, regardless of their relative contributions to the carbon-isotope mass balance, shifts in stalagmite $\delta^{13}C$ are inextricably linked to hydroclimate variations at this site.

Additional support for a PCP signal is provided by the pattern of Sr/Ca and Mg/Ca changes over the last 2,000 years (Supplementary Fig. 8). Covariation between these two ratios is diagnostic of a hydrologic control[23,28]: increases in both indicate drier conditions because the preferential loss of Ca^{2+} from solution (relative to both Mg^{2+} and Sr^{2+}) in the percolation waters during PCP raises the Mg/Ca and Sr/Ca of the drip waters. This is transmitted to the stalagmite as higher Mg/Ca and Sr/Ca. During wetter intervals, PCP is either reduced or absent, leading to lower Mg/Ca and Sr/Ca in the stalagmite[23]. The significant correlation between $\delta^{13}C$ and both trace element ratios provides even more compelling evidence of hydroclimate-driven changes in the degree of PCP over the last 2,000 years (Supplementary Fig. 8). It not only reinforces the similar findings recorded in Liang Luar for the entire Holocene[14], but is in good agreement with previous studies of short-term (seasonal to annual)[26] and long-term (century to orbital)[29] hydroclimate change.

To effectively combine the climate proxies for stalagmites LR06-B1 and LR06-B3 into one composite hydroclimate record, we conducted an unrotated principal component (PC) analysis on the six individual stable isotope and trace element time series. The leading PC (from here referred to as 'LL_{PC1}') represents the dominant hydrologic signal embedded in the records[30] (Fig. 2). The stalagmite $\delta^{18}O$ does not significantly load on LL_{PC1} indicating that the processes important in altering precipitation $\delta^{18}O$ (vapour transport, oceanic source and regional temperatures)[31] may not strongly affect the amount of moisture at the cave site. These additional effects on $\delta^{18}O_p$ are best captured in the second PC (Fig. 2). Therefore, at century timescales, $\delta^{18}O_p$ probably reflects the over-riding influence of large-scale ocean-atmospheric processes that do not strongly affect the total amount of precipitation at the site[32]. These findings highlight the importance of utilizing multiple speleothem geochemical tracers to reconstruct tropical hydroclimate during the relatively stable background climate state of the last two millennia.

Additional evidence to support LL_{PC1} as a hydrologic proxy is provided by initial uranium isotope activity ratios ($[^{234}U/^{238}U]_i$) in the Liang Luar stalagmites (Supplementary Table 1 and Fig. 2d). Holocene $[^{234}U/^{238}U]_i$ ratios of LR06-B1 and LR06-B3 reflect the extent of dissolution related to the rate of groundwater movement in the epikarst of Liang Luar[14,30]. The $[^{234}U/^{238}U]_i$ values increase during drier periods when relatively slow-moving percolation waters (in approximately chemical equilibrium with the host rock) preferentially remove ^{234}U (produced by alpha recoil or preferential leaching and loosely bound in limestone bedrock)[33]. In contrast, during wetter periods, the more rapid traverse of the percolation water through the karst network can

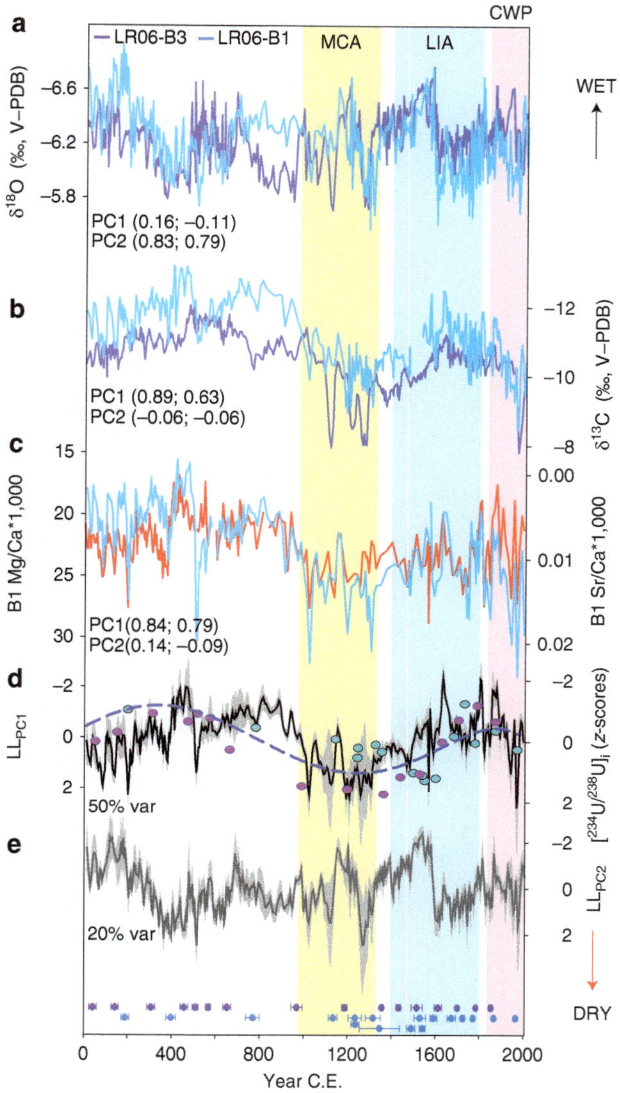

Figure 2 | Multi-proxy reconstruction of east Indonesian hydroclimate. Comparison of stalagmites LR06-B3 and LR06-B1 (**a**) $\delta^{18}O$, (**b**) $\delta^{13}C$, (**c**) Mg/Ca (cyan) and Sr/Ca (red), (**d**) LL_{PC1} (black line) and $[^{234}U/^{238}U]_i$ (purple dots: LR06-B3; blue dots: LR06-B1; dashed line: cubic spline fitted to the data) and (**e**) LL_{PC2}. We interpret higher (lower) values in LL_{PC1} and $[^{234}U/^{238}U]_i$ in **d** to reflect drier (wetter) conditions. Mg/Ca and Sr/Ca records were previously reported in Griffiths *et al.*[14]. The PC factor loadings for each time series of PC1 and PC2 are shown in parentheses. PC1 accounts for 50% of the variance and PC2 accounts for 20%. Colour-coded symbols along the bottom indicate U-Th ages (purple: LR06-B3; blue: LR06-B1) and their associated 2σ uncertainties. The grey shading in **d**,**e** indicates the 95% bootstrap confidence interval. Vertical bars indicate the approximate timing of the Medieval Climate Anomaly (MCA; yellow), Little Ice Age (LIA; blue) and Current Warm Period (CWP; pink) in Flores, defined by decreased AISM rainfall during the MCA and CWP, and increased AISM rainfall during the LIA.

result in greater dissolution rates causing both U isotopes to be leached approximately equally from the host rock. This can lead to relatively low $[^{234}U/^{238}U]_i$ values in a stalagmite. Comparison of the $[^{234}U/^{238}U]_i$ values with LL_{PC1} reveals similar multicentury trends over the past 2,000 years.

A final confirmation that LL_{PC1} reflects a regional hydroclimate signal over the western Pacific is provided by the significant correlation with other paleohydrologic records from the region, including a seawater $\delta^{18}O$ record[5] ($r = 0.5$, 50-year smoothed, significant at the 99% level) and a leaf wax δD record[6] ($r = 0.28$, 50-year smoothed, significant at the 95% level) from marine sediment cores drilled in the Makassar Strait immediately west of Sulawesi (Fig. 1).

Meridional shifts in the Australasian ITCZ. The LL_{PC1} record exhibits pronounced century-scale excursions, with positive values (drier climate) at ~0–400 CE, ~1000–1400 CE and ~1900–2004 CE, and negative excursions (wetter climate) at ~400–1000 CE and ~1400–1900 CE (Fig. 3). The dry period at around the turn of the last millennium coincides with the approximate timing of the Medieval Climate Anomaly (MCA; ~950–1250 CE)[34]. This ~400-year reduction in AISM rainfall begins to recover in strength at around 1300, and by 1500 is above average during the Little Ice Age (LIA). Maximum rainfall in Flores at ~1600 CE, among the wettest periods of the past 2,000 years, is synchronous (within dating uncertainty) with peak cooling in the Northern Hemisphere[34] and maximum ice discharge in the North Atlantic[35] (Fig. 3a).

Comparison of the Flores LL_{PC1} record with subtropical paleomonsoon records from China[7] (Fig. 3b) reveals synchronous (antiphased) multicentury fluctuations in monsoon rainfall across the hemispheres. The cross-equatorial antiphase relationship is consistent with both CGCMs[36] and glacial paleoclimate records[37,38], showing that cooling of the high northern latitudes increases the interhemispheric thermal gradient, displacing the Australasian ITCZ southward. However, zonal changes in the PWC are also known to strongly influence western Pacific hydroclimate[39], and are thought to be a significant driver of low-latitude climate change on century scales[40]. Indeed, an in-phase relationship between LL_{PC1} and paleohydrologic records from the South China Sea[9] (Fig. 3c) challenges the paradigm of regional north–south ITCZ migrations as the primary control of lower-frequency monsoon shifts in the western Pacific over the past millennium. Critically, a recent synthesis of paleohydrologic records for the Australasian monsoon region[41], including a new northwest Australian flood record[42] (Fig. 3d), demonstrated that, rather than moving southward during the LIA, the latitudinal range of monsoon-ITCZ migration probably contracted equatorward, perhaps in response to lower solar irradiance[43] (Fig. 3e).

Zonal teleconnections via the PWC. Evidence for a potential link between the PWC and multi-century changes in western Pacific rainfall is provided by the inverse relationship between the AISM[5,6] and rainfall in the central[8]-eastern[10,44–48] equatorial Pacific (Fig. 4) over the past millennium. Specifically, the western Pacific and Peruvian Andes (Fig. 4a–e) were generally wetter during the LIA, whereas the central and eastern equatorial Pacific (EEP) experienced drier conditions (Fig. 4f) or reduced heavy precipitation events (Fig. 4g,h), similar to La Niña events today when the PWC strengthens (Fig. 1). Conversely, drier conditions in Indonesia and Peru (along with Panama[48]) during the MCA were matched by wetter conditions in the central and EEP, signifying a more 'El Niño-like' mean state. Wavelet-transform analysis of the Flores LL_{PC1} shows significant multidecadal to centennial periodicities in the AISM record (Supplementary Fig. 8) that have also been identified in reconstructions of ENSO amplitude from North American tree rings[49] and NINO3.4 SSTs[40] (Supplementary Fig. 9).

The multi-proxy record of eastern Indonesian hydroclimate presented here (LL_{PC1}) lends support to a LIA characterized by a stronger PWC, whereas the LL_{PC1} shows drier conditions during the MCA, indicating a weaker PWC. These findings are compatible with the ITCZ expansion (LIA)/contraction (MCA)

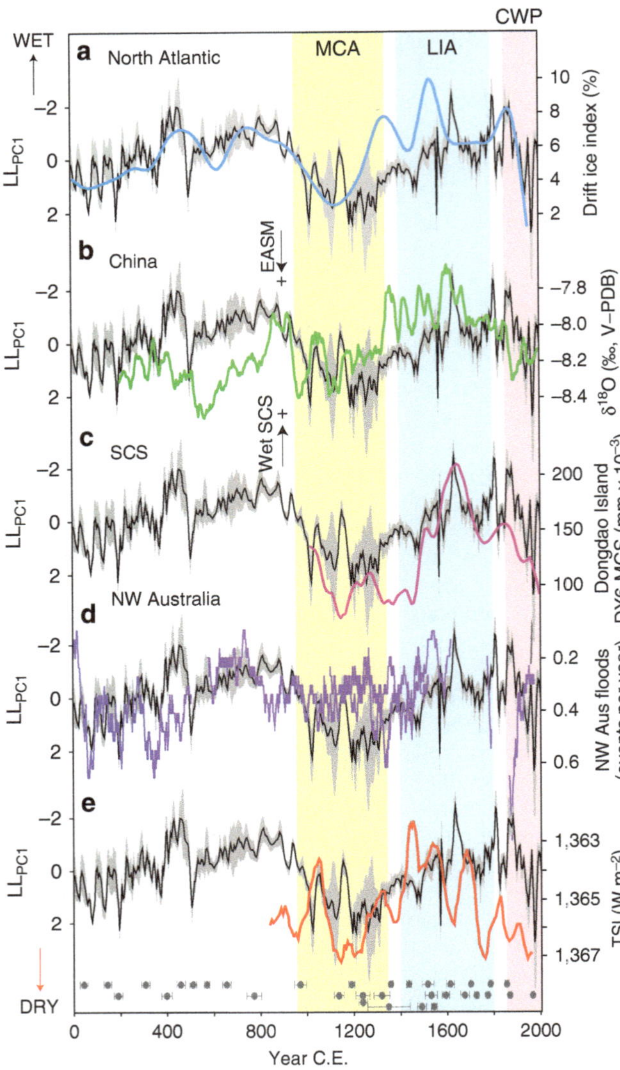

Figure 3 | Tropical and high northern latitude teleconnections. (**a**) North Atlantic drift ice index[35]. (**b**) Stalagmite $\delta^{18}O$ record from Wanxiang Cave, China[7]. (**c**) Mean grain size (MGS) of lake core DY6 from Dongdao Island in the South China Sea (SCS)[9]. (**d**) Northwest Australian flood events as recorded in a stalagmite from KNI-51 cave[42]. (**e**) Total solar irradiance reconstructed from cosmogenic nuclides[43]. The black line and grey shading in all panels shows the LL_{PC1} record and corresponding 95% bootstrap confidence interval. The *y* axis for the LL_{PC1} record is oriented to show a weaker AISM with warmer North Atlantic temperatures. Positions of the U-Th dates (with 2σ uncertainties) for both stalagmites are shown along the bottom. Vertical bars indicate the approximate timing of the MCA (yellow), LIA (blue) and CWP (pink) in Flores, defined by decreased AISM rainfall during the MCA and CWP, and increased AISM rainfall during the LIA.

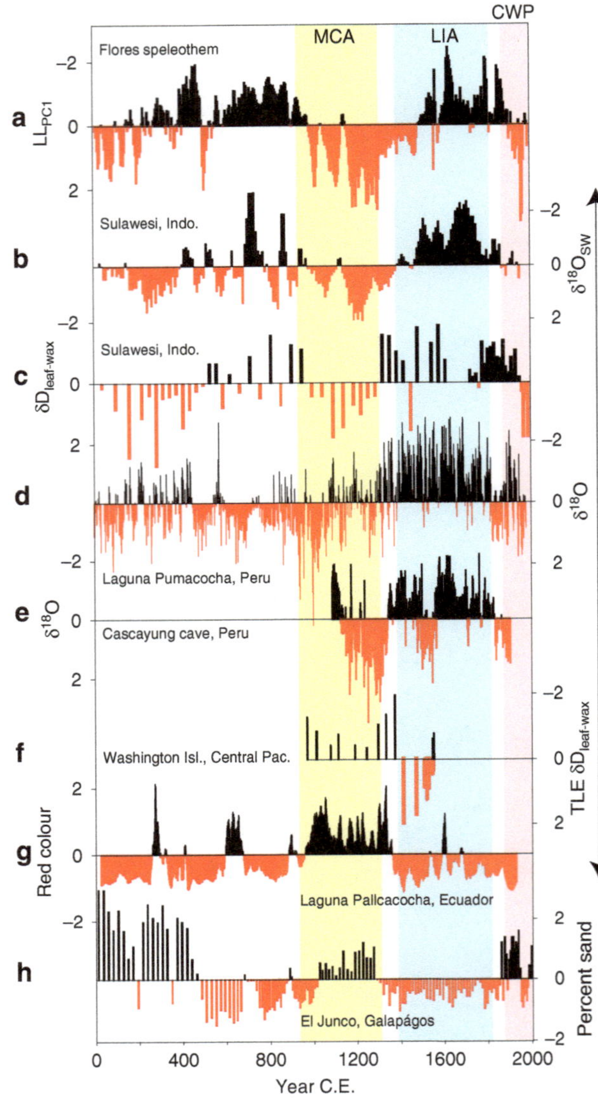

Figure 4 | Hydroclimate records for the tropical western and eastern Pacific. (**a**) Flores LL_{PC1} record. (**b**) Marine foraminifera $\delta^{18}O_{sw}$ (ref. 5) and (**c**) terrestrial $\delta D_{leaf\text{-}wax}$ (ref. 6) records recovered from marine sediment cores located in the Makassar Strait on the Sulawesi margin. (**d**) $\delta^{18}O$ of lake sediment calcite in Laguna Pumacocha in the central Peruvian Andes (proxy for the strength of the South American summer monsoon)[44]. (**e**) Speleothem $\delta^{18}O$ record from Cascayunga cave in northeast Peru[46]. (**f**) $\delta D_{leaf\text{-}wax}$ record from Washington Island in the central equatorial Pacific[8]. (**g**) Red-colour intensity from Laguna Pallcacocha, southern Ecuador[47]. (**h**) Percent sand in El Junco lake, Galápagos Islands[10]. For clarity, all records have been converted to standard (*z*) scores with blue indicating wetter conditions (**a–f**) or heavier precipitation events (**g–h**) and vice versa for red. Vertical bars indicate the approximate timing of the MCA (yellow), LIA (blue) and CWP (pink) in Flores.

hypothesis of Yan *et al.*[41] The temporal changes in zonal precipitation across the equatorial Pacific[5,6,8–10,40] are also consistent with some terrestrial reconstructions of ENSO[40,45,50] activity (Fig. 5), which show strengthening of the PWC during the LIA, and weakening during the MCA. These findings run counter to inferences from some SST-based ENSO reconstructions from the western[5,51], central[52] and EEP[53], which suggest that the tropical Pacific was characterized by 'La Niña-like' conditions during the MCA compared with the LIA. However, there is still no consensus on the marine paleo-ENSO signals; a recent reconstruction of EEP SSTs[54] indicates that the MCA was marked by relatively warm SSTs in the EEP, contradicting the findings of earlier studies.

Theoretical and physical models of the PWC are also not entirely consistent within the context of Indo-Pacific climate change during the past millennium. Modern dynamical studies suggest that under warmer conditions, the PWC should weaken[55,56], and at least one paleoclimate study supports the inference of a stronger PWC during the LIA[41]. On the other hand, the 'ocean dynamical thermostat' mechanism[57] supports a potential weakening of the PWC during the LIA. Modern observations do not yet distinguish the PWC's response to global

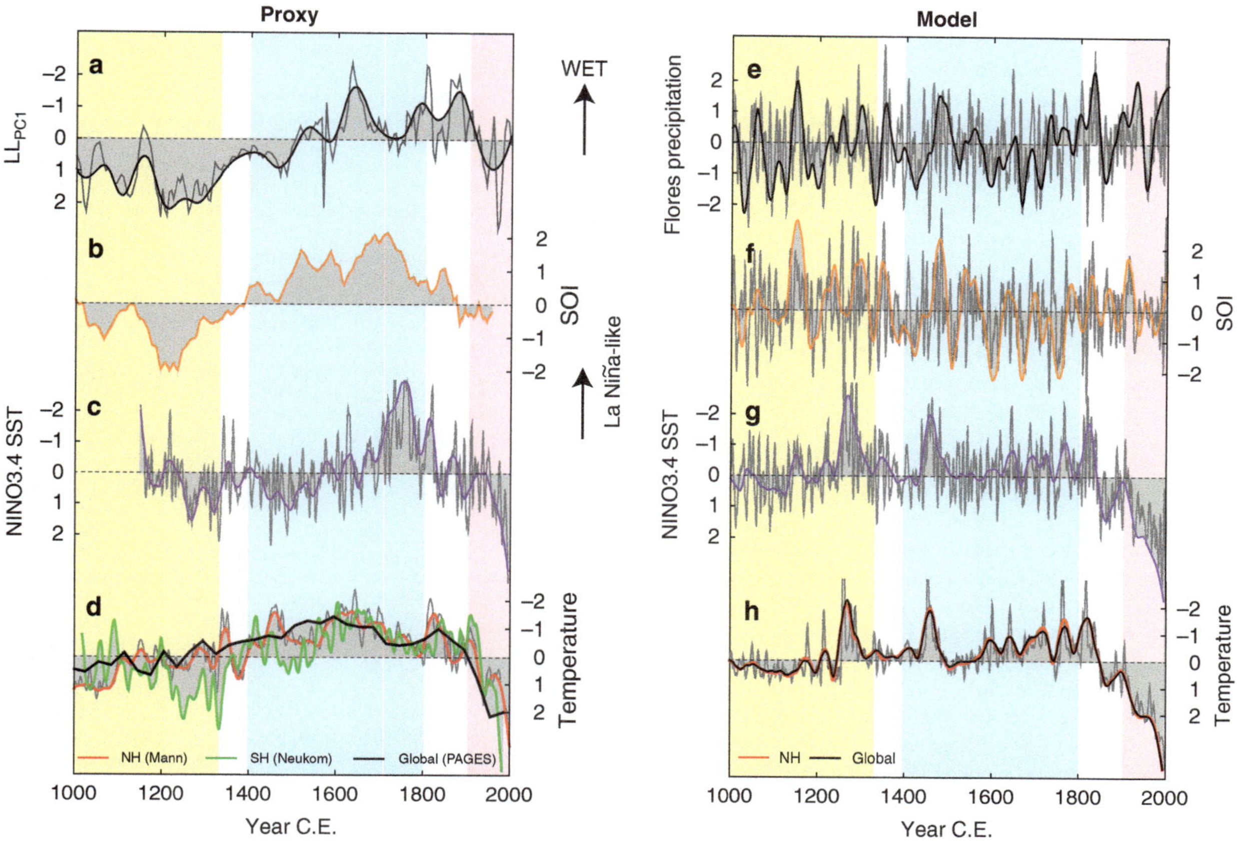

Figure 5 | Western Pacific rainfall and ocean-atmosphere climate modes from paleoclimate records and climate model simulations. Panels **a-d** show paleoclimate records[34,40,50,59,60] and **e-h** show CCSM4 LM + hist CGCM simulations. Comparison of Flores precipitation (**a**,**e**), SOI (**b**,**f**), NINO3.4 SSTs (**c**,**g**) and land temperatures (red: NH; blue: global; **d**,**h**). The records have been smoothed with a 50-year (thick coloured lines) and 7-year (thin grey lines) loess filter. All time series were converted to standard (*z*) scores to facilitate comparison of proxy reconstructions with model simulations. Vertical bars indicate the approximate timing of the MCA (yellow), LIA (blue) and CWP (pink) in Flores.

warming. Yan *et al.*[41] recently noted that there are numerous interacting components that may contribute to the discrepancies among paleoclimate records: (i) misinterpretation of SST or hydrological signals with respect to meridional versus zonal changes in ocean-atmosphere circulation; (ii) decoupling of ocean and atmospheric processes in response to different forcings over century time scales and (iii) the interacting influence of multidecadal climate modes (for example, Atlantic Multidecadal Oscillation) with ENSO, which are difficult to identify in the proxy records that originate from outside these system' centres of action.

Palaeoclimate proxy-model comparisons. To explore the potential role of multidecadal- to century-scale changes in the PWC on global climate variability, we examined the relationship between Flores rainfall, NINO3.4 SSTs and the Southern Oscillation Index (SOI) with similar indices in CGCM simulations (Fig. 5). To do so, we employed the Paleoclimate Modelling Intercomparison Project Phase 3 (PMIP3) last millennium (LM; 850–1849 CE) and historical (1850–2005 CE) CGCM runs from the US NCAR CCSM4 (ref. 58) model. The CCSM4 model was employed because its simulation of the modern hydroclimate dynamics of east Indonesia compares well with observations. Importantly, model preindustrial control runs (500-year-long) accurately simulate the influence of ENSO on rainfall variability around Flores (Supplementary Fig. 10 and Supplementary Table 3). Although the association of SST with rainfall is somewhat stronger and more widespread in models compared with observations, the regional pattern of SST is consistent. The dominant influence of ENSO on east Indonesian rainfall is also apparent for the LM and historical (LM + hist; 850–2005 CE) simulations (Supplementary Table 3), although the correlation is not as strong as for the preindustrial control runs.

If variations in global temperatures over the LM were amplified by decadal to multidecadal variations in the tropical Pacific, as is the case during the instrumental period[2–4], then model skill in the tropical Indo-Pacific region is clearly important for projecting future climate change. Therefore, we compared Flores precipitation, the SOI and NINO3.4 SSTs captured in the proxy records[40,50] (Fig. 5a–c) with the same indices in the LM + hist CCSM4 simulations (Fig. 5e–g). In the LM + hist CGCM experiments, the variance in Pacific SST and Indonesian rainfall is concentrated at interannual/decadal time scales, whereas the paleoclimate record for Indonesia exhibits strongest variance at multidecadal to centennial periodicities (Supplementary Fig. 8). However, the coherent patterns of lower-frequency variability evident in the network of western Pacific hydroclimate records suggest that the data-model mismatch likely stems from model deficiencies with respect to century-scale variations in the PWC (Fig. 5b,c,f,g), particularly during the MCA and LIA. Consequently, the models may be underestimating the magnitude and duration of reconstructed global[59] and hemispheric[34,60] Medieval warming, and the persistent LIA cooling between ~1400 and ~1850 CE (Fig. 5d,h).

Our results highlight significant discrepancies between the proxy records and model simulations for the past millennium. Critically, these discrepancies coincide with century-scale anomalies in the strength of the PWC. We cannot rule out the possibility that some of the low-frequency Pacific variability was a forced response to variable solar intensity[36] and changing teleconnections to higher latitudes[61] that are not simulated by the models, or that non-climatic processes have influenced the proxies[62]. However, of particular importance is that the paleodata-model mismatch supports the possibility that unforced, low-frequency internal climate variability (that is difficult for models to simulate) was responsible for at least some of the global temperature change of the past millennium[62,63].

Discussion

The broad east–west antiphasing of Pacific hydroclimate during the past millennium, and clear connections with higher-latitude climates, suggests the deep tropics may have played a more active role in global climate change than previously assumed. Indeed, Northern and Southern Hemisphere temperature reconstructions[34,60] (Fig. 5d) reveal symmetrical shifts in air-temperature over the past millennium (particularly during the LIA), suggesting that changes in the interhemispheric temperature gradient, and north–south shifts in the ITCZ, cannot explain the tropical precipitation variability evident in the proxy records. Instead, it is more likely that the LIA global cooling was initiated by a decline in solar irradiance (and lower atmospheric CO_2 and potentially increased volcanic activity)[60] and amplified by the strengthened PWC that occurred in parallel with a latitudinal contraction of the ITCZ[41]. Although it would be premature to suggest that LIA cooling was triggered by century-scale shifts in the tropical mean state, a stronger PWC during the LIA may have intensified broad-scale cooling already underway, and a weaker PWC during the MCA could have acted to warm the planet.

Our findings have significant implications for projections of decadal-scale changes in tropical atmospheric convection and global temperatures. For example, from the beginning of this century until recently, the tropical Pacific was locked into a negative Interdecadal Pacific Oscillation phase (that is, low-frequency La Niña-like pattern) in association with increased Walker and Hadley circulation winds and eastern Pacific cooling[2,3]. The La Niña-like pattern is thought to be a factor contributing to the recent so-called 'warming hiatus'[2,3] and earlier twentieth century cool and warm decades[4]. Therefore, our analysis of multicentury hydroclimate variability suggests that projections of tropical rainfall patterns, and global temperature extremes, will remain uncertain until paleoclimate records and models consistently capture the lower-frequency variability, and associated feedbacks, in the tropical Pacific.

Methods

^{230}Th dating and age models. A total of 35 pieces of calcite, each weighing ~50–300 mg, were extracted from the central growth axis of stalagmites LR06-B1 and LR06-B3. Samples were removed using a carbide dental bur fitted to an air drill. $^{230}Th/^{234}U$ age determinations were conducted on both a thermal ionisation mass spectrometer and a multi-collector inductively coupled plasma mass spectrometer (MC-ICPMS; Supplementary Table 1 and Supplementary Figs 4 and 5). Thermal ionisation mass spectrometer analyses were carried out at the University of Queensland using a Fisons VG Sector 54-30 mass spectrometer equipped with a WARP filter and an ion counting Daly detector, as described by Yu *et al.*[64] The MC-ICPMS U-Th analyses were conducted on two separate Nu-Instruments Nu Plasma mass spectrometers housed at the University of Melbourne and the University of Queensland. For a description of the MC-ICPMS methods employed at the University of Melbourne, refer to study by Hellstrom[65]. The methods employed at the University of Queensland are outlined in Zhou *et al.*[66]. All samples were corrected for initial ^{230}Th using an initial $^{230}Th/^{232}Th$ value of 3.4 ± 1.7. The initial $^{230}Th/^{232}Th$ value was determined by the methods described in the study by Hellstrom[65]. Corrected ages were calculated using half-lives specified in Cheng *et al.*[67]

The age models for LR06-B1 and LR06-B3 were first calculated using a Bayesian–Monte Carlo approach[12], producing two $\delta^{18}O$ time series that display a good level of coherence (Fig. 2 and Supplementary 7), despite the minor discrepancy between the two isotopic profiles during the period ~700–1000 CE, which we attribute to slight differences in the resolution of U-Th dates through this interval (see Supplementary Figs 4 and 7 for details). Given this degree of replication, we refined the age models using *iscam*[13] (that is, Intra-Site Correlation Age Modelling) with a prescribed point-wise linear interpolation scheme between adjacent U-Th ages. This novel age-model algorithm finds the best correlation between the proxy records for two (or more) neighbouring dated stalagmites (highest *r* value of 0.73 between LR06-B1 and LR06-B3) using a Monte Carlo approach (100,000 simulations using a 100-year smoothing), and then calculates the most probable age model for both records based on the calculated correlation coefficients. Because the two stalagmites were collected from the same cave chamber, we assume that the geochemical excursions preserved in the overlapping sections of both records reflect a real climate signal, and that minor discrepancies between the two are likely due to the uncertainties in the U-Th dates and slight differences in calcite stratigraphy. Significance levels were calculated against a red-noise background from 2,000 pairs of artificially simulated first-order autoregressive time series (AR1). The *iscam* age model provides a number of advantages to simple linear interpolation, which include: (i) quantitative assessment of whether the isotopic signals from multiple stalagmites have a common signal within age error; (ii) linking climatic proxies from multiple samples to construct a composite record and (iii) significant reduction of age uncertainty within overlapping time intervals through the enlargement of the signal-to-noise ratio.

Stable isotope analysis. The oxygen ($\delta^{18}O$) and carbon ($\delta^{13}C$) isotope profiles for stalagmite LR06-B1 were established from 632 calcite powders micromilled along the central growth axis. The top 74 mm (that is, top of stalagmite down to a depth of 74 mm) of LR06-B1, which represents the ~1172–2004 CE period of growth, was micromilled at 0.15 mm increments (~1- to 4-year resolution). 30–70 μg of sample powders were analysed on a Thermo Finnigan Kiel IV carbonate device coupled with a Delta V Plus isotope ratio mass spectrometer at the University of California, Irvine ($n = 458$ samples). Measurement precision (1σ) for the NBS-19 standard material was ± 0.06‰ for $\delta^{18}O$ and 0.03‰ for $\delta^{13}C$. The lower section of LR06-B1, representing ~1–1172 CE period of growth (that is, depth between 74 and 246 mm from the top), was drilled at 1 mm increments (~7- to 10-year resolution). The $\delta^{18}O$ and $\delta^{13}C$ for these lower-resolution samples was measured on ~1 mg powders using a GV Instruments GV2003 continuous-flow isotope ratio mass spectrometer at the University of Newcastle, Australia ($n = 174$ samples). The precision (1σ) of in-run measurements of the standard material (a Carrara Marble called New1) was ± 0.08‰. These lower-resolution samples were previously reported in Griffiths *et al.*[11]

The $\delta^{18}O$ and $\delta^{13}C$ profiles for LR06-B3 were established from 525 calcite powders micromilled at 0.5 mm increments along the growth axis. The isotope measurements were performed at the University of Arizona on a Micromass Optima dual-inlet stable isotope ratio mass spectrometer with an automated carbonate preparation system. The precision (1σ) for in-run measurements ($n = 113$) of laboratory standard material Luxor was ± 0.07‰ for $\delta^{18}O$ and $\delta^{13}C$. All results are expressed as the deviation in per mil (‰) between the sample and Vienna Peedee Belemnite standard using the delta notation.

Trace element analysis. The Mg/Ca and Sr/Ca ratios in stalagmite LR06-B1 were analysed on H_3PO_4 residues after calcite dissolution for stable isotope analysis. The samples were analysed for Mg, Sr and Ca on a Varian Liberty 4,000 inductively coupled plasma atomic emission spectrometer. Molar concentrations of the elements were calculated from inductively coupled plasma atomic emission spectrometer intensities using four internal working standards of known concentration and a blank, with concentrations expressed as ratios to Ca. Acid blank corrections were applied to both sample and standard data before conversion of the data to ratios. The relative standard deviation of replicate standards of Ca, Mg and Sr was 1.5%, 1.6% and 2.4%, respectively. These data were previously published in Griffiths *et al.*[14,22,30]

PC analysis. The speleothem hydroclimate record was constructed using the $\delta^{18}O$ and $\delta^{13}C$ for stalagmites LR06-B1 and LR06-B3, and trace elements (Mg/Ca, Sr/Ca) for LR06-B1. Due to the average resolution among all records being ~5 years, we linearly interpolated each proxy record to a common 5-year temporal resolution before analysis. To extract the dominant climate signal embedded within the stable isotope and trace element records, we performed an unrotated PC analysis on the six profiles, where the leading PC (explains 50% of the variance) represents the dominant hydrologic signal common to all individual time series. Because other factors besides rainfall can affect the $\delta^{18}O$ (for example, cave temperature, changes in ratio of summer versus winter moisture delivery) and $\delta^{13}C$ (for example, above-cave vegetation, soil productivity, cave ventilation) and also the trace elements (for example, cave temperature, speleothem growth rate), this method has proven useful in isolating the dominant hydrologic signal from other potential influences[30]. However, the $\delta^{18}O$ time series load weakly on PC1, and load much more strongly on PC2. This suggests that factors in addition to local rainfall amount likely influenced the speleothem $\delta^{18}O$. Confidence intervals (95%) for the PCs were calculated using bias-corrected and accelerated nonparametric bootstrap technique performed in MATLAB.

General circulation modelling. To assess the long-term dynamical influence of the Walker circulation on IPWP convection, and NH and global temperatures, we used the PMIP3 LM (850–1849 CE) and historical (1850–2005 CE) CGCM runs from the US NCAR CCSM4 (ref. 58) model. The skill of the CGCMs in simulating the observed (that is, instrumental) ENSO influence on western Pacific hydroclimate was demonstrated by comparing field correlation maps between east Indonesian rainfall (annual precipitation averaged over a gridbox encompassing the area 12°–6° S and 110°–125° E) and SSTs in the historical simulations (1850–2005), with the same variables in the instrumental (1950–2010) records (Supplementary Fig. 10). In addition, comparison of east Indonesian rainfall with both NINO3.4 SSTs and the SOI in model preindustrial control runs (500-year-long) and the LM + hist (850–2005 CE) simulations reveals statistically significant r values (Supplementary Table 3), which again demonstrates the strength of the model in simulating the dominant climate mode in the western Pacific. Details of the model characteristics (that is, forcings, atmospheric levels, atmospheric resolution) for the LM and historical simulations, and preindustrial control runs, can be found in ref. 58.

References

1. Denton, G. H. *et al.* The last glacial termination. *Science* **328,** 1652–1656 (2010).
2. England, M. H. *et al.* Recent intensification of wind-driven circulation in the Pacific and the ongoing warming hiatus. *Nat. Clim. Change* **4,** 222–227 (2014).
3. Kosaka, Y. & Xie, S.-P. Recent global-warming hiatus tied to equatorial Pacific surface cooling. *Nature* **501,** 403–407 (2013).
4. Meehl, G. A., Hu, A. & Santer, B. D. The mid-1970s climate shift in the Pacific and the relative roles of forced versus inherent decadal variability. *J. Clim* **22,** 780–792 (2009).
5. Oppo, D. W., Rosenthal, Y. & Linsley, B. K. 2,000-year-long temperature and hydrology reconstructions from the Indo-Pacific warm pool. *Nature* **460,** 1113–1116 (2009).
6. Tierney, J. E., Oppo, D. W., Rosenthal, Y., Russell, J. M. & Linsley, B. K. Coordinated hydrological regimes in the Indo-Pacific region during the past two millennia. *Paleocean* **25 (2010).**
7. Zhang, P. Z. *et al.* A Test of climate, sun, and culture relationships from an 1810-year Chinese cave record. *Science* **322,** 940–942 (2008).
8. Sachs, J. P. *et al.* Southward movement of the Pacific intertropical convergence zone AD 1400–1850. *Nat. Geosci* **2,** 519–525 (2009).
9. Yan, H. *et al.* South China Sea hydrological changes and Pacific Walker Circulation variations over the last millennium. *Nat. Commun* **2,** 293 (2011).
10. Conroy, J. L., Overpeck, J. T., Cole, J. E., Shanahan, T. M. & Steinitz-Kannan, M. Holocene changes in eastern tropical Pacific climate inferred from a Galapagos lake sediment record. *Quat. Sci. Rev.* **27,** 1166–1180 (2008).
11. Griffiths, M. L. *et al.* Increasing Australian-Indonesian monsoon rainfall linked to early Holocene sea-level rise. *Nat. Geosci* **2,** 636–639 (2009).
12. Drysdale, R. N., Zanchetta, G., Hellstrom, J. C., Fallick, A. E. & Zhao, J.-X. Stalagmite evidence for the onset of the Last Interglacial in southern Europe at 129 ± 1 ka. *Geophys. Res. Lett.* **32,** L24708 (2005).
13. Fohlmeister, J. A statistical approach to construct composite climate records of dated archives. *Quat. Geochron* **14,** 48–56 (2012).
14. Griffiths, M. L. *et al.* Evidence for Holocene changes in Australian-Indonesian monsoon rainfall from stalagmite trace element and stable isotope ratios. *Earth Planet. Sci. Lett.* **292,** 27–38 (2010).
15. Hendy, C. H. The isotopic geochemistry of speleothems: The calculations of the effects of different modes of Formation on the isotopic composition of speleothems and their applicability as palaeclimate indicators. *Geochim. Cosmochim. Acta* **35,** 801–824 (1971).
16. Vuille, M., Werner, M., Bradley, R. S., Chan, R. Y. & Keimig, F. Stable isotopes in east African precipitation record Indian ocean zonal mode. *Geophys. Res. Lett.* **32,** L21705 (2005).
17. Yoshimura, K., Kanamitsu, M., Noone, D. & Oki, T. Historical isotope simulation using reanalysis atmospheric data. *J. Geophys. Res. Atmos* **113,** D19108 (2008).
18. Lewis, S. C., LeGrande, A. N., Kelley, M. & Schmidt, G. A. Water vapour source impacts on oxygen isotope variability in tropical precipitation during Heinrich events. *Clim. Past* **6,** 325–343 (2010).
19. Lewis, S. C., LeGrande, A. N., Schmidt, G. A. & Kelley, M. Comparison of forced ENSO-like hydrological expressions in simulations of the preindustrial and mid-Holocene. *J. Geophys. Res. Atmos* **119,** 7064–7082 (2014).
20. Konecky, B., Russell, J., Vuille, M. & Rehfeld, K. The Indian Ocean zonal mode over the past millennium in observed and modeled precipitation isotopes. *Quat. Sci. Rev.* **103,** 1–18 (2014).
21. Fairchild, I. J. & Baker, A. *Speleoethem Science* (John Wiley & Sons, 2012).
22. Griffiths, M. L. *et al.* Hydrological control of the dead-carbon fraction in a Holocene tropical speleothem. *Quat. Geochron* **14,** 81–93 (2012).
23. Fairchild, I. J. *et al.* Controls on trace element (Sr-Mg) compositions of carbonate cave waters: implications for speleothem climatic records. *Chem. Geol.* **166,** 255–269 (2000).
24. Dulinski, M. & Rozanski, R. Formation of $^{13}C/^{12}C$ isotope ratios in speleothems: a semi-dynamic model. *Radiocarbon* **32,** 7–16 (1990).
25. Drysdale, R. *et al.* Late Holocene drought responsible for the collapse of Old World civilizations is recorded in an Italian cave flowstone. *Geology* **34,** 101–104 (2006).
26. Johnson, K. R., Hu, C., Belshaw, N. S. & Henderson, G. M. Seasonal trace-element and stable-isotope variations in a Chinese speleothem: The potential for high-resolution paleomonsoon reconstruction. *Earth Planet. Sci. Lett.* **244,** 394–407 (2006).
27. Genty, D. *et al.* Precise dating of Dansgaard–Oeschger climate oscillations in western Europe from stalagmite data. *Nature* **421,** 833–837 (2003).
28. Fairchild, I. J. & Treble, P. C. Trace elements in speleothems as recorders of environmental change. *Quat. Sci. Rev.* **28,** 449–468 (2009).
29. Cruz, F. W. *et al.* Evidence of rainfall variations in Southern Brazil from trace element ratios (Mg/Ca and Sr/Ca) in a Late Pleistocene stalagmite. *Geochim. Cosmochim. Acta* **71,** 2250–2263 (2007).
30. Griffiths, M. L. *et al.* Abrupt increase in east Indonesian rainfall from flooding of the Sunda Shelf ~9500 years ago. *Quat. Sci. Rev.* **74,** 273–279 (2013).
31. Lachniet, M. S. Climatic and environmental controls on speleothem oxygen-isotope values. *Quat. Sci. Rev.* **28,** 412–432 (2009).
32. Konecky, B., Russell, J. & Bijaksana, S. Glacial aridity in central Indonesia coeval with intensified monsoon circulation. *Earth Planet. Sci. Lett.* **437,** 15–24 (2016).
33. Hellstrom, J. C. & McCulloch, M. T. Multi-proxy constraints on the climatic significance of trace element records from a New Zealand speleothem. *Earth Planet. Sci. Lett.* **179,** 287–297 (2000).
34. Mann, M. E. *et al.* Global signatures and dynamical origins of the Little Ice Age and Medieval Climate Anomaly. *Science* **326,** 1256–1260 (2009).
35. Bond, G. *et al.* Persistent solar influence on north Atlantic climate during the Holocene. *Science* **294,** 2130–2136 (2001).
36. Broccoli, A. J., Dahl, K. A. & Stouffer, R. J. Response of the ITCZ to Northern Hemisphere cooling. *Geophys. Res. Lett.* **33,** L01702 (2006).
37. Wang, Y. J. *et al.* A high-resolution absolute-dated Late Pleistocene monsoon record from Hulu Cave, China. *Science* **294,** 2345–2348 (2001).
38. Ayliffe, L. K. *et al.* Rapid interhemispheric climate links via the Australasian monsoon during the last deglaciation. *Nat. Commun* **4,** 2908 (2013).
39. Tokinaga, H., Xie, S.-P., Deser, C., Kosaka, Y. & Okumura, Y. M. Slowdown of the Walker circulation driven by tropical Indo-Pacific warming. *Nature* **491,** 439–443 (2012).
40. Emile-Geay, J., Cobb, K. M., Mann, M. E. & Wittenberg, A. T. Estimating central equatorial Pacific SST variability over the past millennium. Part 2: Reconstructions and implications. *J. Clim* **26,** 2329–2352 (2012).
41. Yan, H. *et al.* Dynamics of the intertropical convergence zone over the western Pacific during the Little Ice Age. *Nat. Geosci* **8,** 315–320 (2015).
42. Denniston, R. F. *et al.* Extreme rainfall activity in the Australian tropics reflects changes in the El Niño/Southern Oscillation over the last two millennia. *Proc. Natl Acad. Sci. USA* **112,** 4576–4581 (2015).
43. Bard, E., Raisbeck, G., Yiou, F. & Jouzel, J. Solar irradiance during the last 1200 years based on cosmogenic nuclides. *Tellus B* **52,** 985–992 (2000).
44. Bird, B. W. *et al.* A 2,300-year-long annually resolved record of the South American summer monsoon from the Peruvian Andes. *Proc. Natl Acad. Sci. USA* **108,** 8583–8588 (2011).
45. Thompson, L. *et al.* Annually resolved ice core records of tropical climate variability over the past ~1800 years. *Science* **340,** 945–950 (2013).
46. Reuter, J. *et al.* A new perspective on the hydroclimate variability in northern South America during the Little Ice Age. *Geophys. Res. Lett.* **36,** L21706 (2009).
47. Moy, C. M., Seltzer, G. O., Rodbell, D. T. & Anderson, D. M. Variability of El Niño/Southern Oscillation activity at millennial timescales during the Holocene epoch. *Nature* **420,** 162–165 (2002).
48. Lachniet, M. S. *et al.* A 1500-year El Niño/Southern Oscillation and rainfall history for the isthmus of Panama from speleothem calcite. *J. Geophys. Res. Atmos* **109,** D20117 (2004).
49. Li, J. B. *et al.* Interdecadal modulation of El Niño amplitude during the past millennium. *Nat. Clim. Change* **1,** 114–118 (2011).
50. Yan, H. *et al.* A record of the Southern Oscillation Index for the past 2000 years from precipitation proxies. *Nat. Geosci* **4,** 611–614 (2011).
51. Newton, A., Thunell, R. & Stott, L. Climate and hydrographic variability in the Indo-Pacific Warm Pool during the last millennium. *Geophys. Res. Lett.* **33** (2006).
52. Cobb, K. M., Charles, C. D., Cheng, H. & Edwards, R. L. El Niño/Southern Oscillation and tropical Pacific climate during the last millennium. *Nature* **424,** 271–276 (2003).
53. Conroy, J. L., Overpeck, J. & Cole, J. El Niño/Southern Oscillation and changes in the zonal gradient of tropical Pacific sea surface temperature over the last 1.2 ka. *PAGES News* **18,** 32–34 (2010).

54. Rustic, G. T., Koutavas, A., Marchitto, T. M. & Linsley, B. K. Dynamical excitation of the tropical Pacific Ocean and ENSO variabilty by Little Ice Age cooling. *Science* **350,** 1537–1541 (2015).
55. Vecchi, G. A. *et al.* Weakening of tropical Pacific atmospheric circulation due to anthropogenic forcing. *Nature* **441,** 73–76 (2006).
56. Held, I. M. & Soden, B. J. Robust responses of the hydrological cycle to global warming. *J. Clim* **19,** 5686–5699 (2006).
57. Clement, A. C., Seager, R., Cane, M. A. & Zebiak, S. E. An ocean dynamical thermostat. *J. Clim* **9,** 2190–2196 (1996).
58. Landrum, L. *et al.* Last millennium climate and its variability in CCSM4. *J. Clim* **26,** 1085–1111 (2011).
59. Ahmed, M. *et al.* Continental-scale temperature variability during the past two millennia. *Nat. Geosci* **6,** 339–346 (2013).
60. Neukom, R. *et al.* Inter-hemispheric temperature variability over the past millennium. *Nat. Clim. Change* **4,** 362–367 (2014).
61. Timmermann, A., An, S.-I., Krebs, U. & Goosse, H. ENSO suppression due to weakening of the North Atlantic thermohaline circulation. *J. Clim* **18,** 3122–3139 (2005).
62. Ault, T., Deser, C., Newman, M. & Emile-Geay, J. Characterizing decadal to centennial variability in the equatorial Pacific during the last millennium. *Geophys. Res. Lett.* **6,** 3450–3456 (2013).
63. Laepple, T. & Huybers, P. Ocean surface temperature variability: large model–data differences at decadal and longer periods. *Proc. Natl Acad. Sci. USA* **111,** 16682–16687 (2014).
64. Yu, K. F. *et al.* U-series dating of dead *Porites* corals in the South China Sea: Evidence for episodic coral mortality over the past two centuries. *Quat. Geochron* **1,** 129–141 (2006).
65. Hellstrom, J. C. U–Th dating of speleothems with high initial ^{230}Th using stratigraphical constraint. *Quat. Geochron* **1,** 289–295 (2006).
66. Zhou, H. Y., Zhao, J. X., Qing, W., Feng, Y. X. & Tang, J. Speleothem-derived Asian summer monsoon variations in central China, 54–46 ka. *J. Quat. Sci* **26,** 781–790 (2011).
67. Cheng, H. *et al.* Improvements in ^{230}Th dating, ^{230}Th and ^{234}U half-life values, and U–Th isotopic measurements by multi-collector inductively coupled plasma mass spectrometry. *Earth Planet. Sci. Lett.* **371,** 82–91 (2013).

Acknowledgements

We thank the Indonesian Institute of Sciences (LIPI) for operational support and are particularly grateful to Neil Anderson, Garry Smith, Bambang Suwargadi and Linda Ayliffe, who provided expertise in cave conservation and technical assistance. Heather Scott-Gagan provided technical assistance at ANU. The work in Indonesia was carried out under LIPI research permit number 04057/SU/KS/2006. Financial support for the research was provided by Australian Research Council *Discovery* grants DP0663274 and DP1095673 to M.K.G., R.N.D., J.-X.Z., J.C.H. and W.S.H. This research was also supported by a 2010–2012 NOAA/UCAR Climate and Global Change Postdoctoral Fellowship to M.L.G. and an internal summer 2013 research grant through William Paterson University. Support for A.K.K. was provided by a 2007–2010 Lewis and Clark College Mellon Research Initiative Fellowship and support for J.E.C. was provided by the US NSF.

Author contributions

M.K.G., R.N.D. and W.S.H collected the samples from Liang Luar cave in August 2006. M.L.G. sampled stalagmite LR06-B1, carried out the data analysis and interpreted the results. A.K.K. and J.E.C. undertook the sample preparation and isotope analyses for LR06-B3. K.R.J. carried out the isotope measurements on LR06-B1. J.C.H. and J.-X. Z. performed the ^{230}Th/^{234}U dating and age-modelling. B.I.C. carried out the model analysis and interpretation. M.L.G., M.K.G, A.K.K., R.N.D. and J.E.C. wrote the manuscript with contributions from the other authors.

Additional information

Competing financial interests: The authors declare no competing financial interests.

5

Arctic cut-off high drives the poleward shift of a new Greenland melting record

M. Tedesco[1,2], T. Mote[3], X. Fettweis[4], E. Hanna[5], J. Jeyaratnam[6], J.F. Booth[6], R. Datta[1,6,7] & K. Briggs[8]

Large-scale atmospheric circulation controls the mass and energy balance of the Greenland ice sheet through its impact on radiative budget, runoff and accumulation. Here, using reanalysis data and the outputs of a regional climate model, we show that the persistence of an exceptional atmospheric ridge, centred over the Arctic Ocean, was responsible for a poleward shift of runoff, albedo and surface temperature records over the Greenland during the summer of 2015. New records of monthly mean zonal winds at 500 hPa and of the maximum latitude of ridge peaks of the 5,700 ± 50 m isohypse over the Arctic were associated with the formation and persistency of a cutoff high. The unprecedented (1948-2015) and sustained atmospheric conditions promoted enhanced runoff, increased the surface temperatures and decreased the albedo in northern Greenland, while inhibiting melting in the south, where new melting records were set over the past decade.

[1] Lamont-Doherty Earth Observatory, Columbia University, New York, New York 10964, USA. [2] NASA Goddard Institute of Space Studies, New York, New York 10025, USA. [3] University of Georgia, Athens, Georgia 30602-2502, USA. [4] University of Liege, Liege 4000, Belgium. [5] University of Sheffield, Sheffield S10 2TN, UK. [6] The City College of New York, New York, New York 10031, USA. [7] The Graduate Center of the City University of New York, New York, New York 10016, USA. [8] University of Leeds, Leeds LS2 9JT, UK. Correspondence and requests for materials should be addressed to M.T. (email: mtedesco@ldeo.columbia.edu or cryocity@gmail.com).

Atmospheric circulation affects the energy and mass budgets of the Greenland ice sheet[1–5] by controlling cloud coverage and optical depth[6], and by driving the spatial distribution, and amount of surface melting and accumulation[7,8]. Improving our understanding of the impact of atmospheric circulation on the Greenland's surface mass balance is, therefore, crucial for the refinement of climate and ice-sheet models, and will ultimately enable improved estimates of current and future contributions to sea level, by the largest ice body in the Northern Hemisphere.

Here we show that a poleward shift of melting record over the Greenland ice sheet in 2015 was driven by the exceptional atmospheric conditions characterized by new records in mean zonal winds and jet stream wave amplitude associated with the formation and evolution of a Arctic cutoff high.

Results

Atmospheric conditions and indicators. Our analysis of the geopotential height at 500 hPa (Methods) shows that during July 2015 a persistent atmospheric ridge was centred over the Arctic Ocean (Lincoln Sea, north of Greenland), with geopotential height anomalies being up to 3.7 s.d.'s (σ, ~ 150 m) above the 1981–2010 long-term mean (Fig. 1a). The North Atlantic Oscillation (NAO; Methods) and Greenland Blocking Index (GBI, defined as the 500 hPa geopotential height area averaged between 60–80° N and 20–80° W (ref. 9); Methods) have been associated with extreme melting events over the Greenland[9–12]. The summer average (June-July-August) value for NAO in 2015 of − 1.61 was close to the summer value in 2012 of − 1.59. Differently from 2015, however, the atmospheric ridge in 2012 was centred over the Greenland ice sheet[2,10] (Supplementary Figs 1 and 2). The July monthly averaged NAO value set a new record low of − 1.23 (since 1899), being 3.2σ below the 1981–2010 mean (Fig. 1c). Concurrently, the GBI also set a new record for the month of July (Fig. 1c; Supplementary Fig. 3b), being 2.8σ above the 1981–2010 mean. The June and August conditions in 2015 were not as exceptional in 2012, with mean June and August NAO values in 2015 being higher than the same quantities in 2012 (Supplementary Fig. 3a).

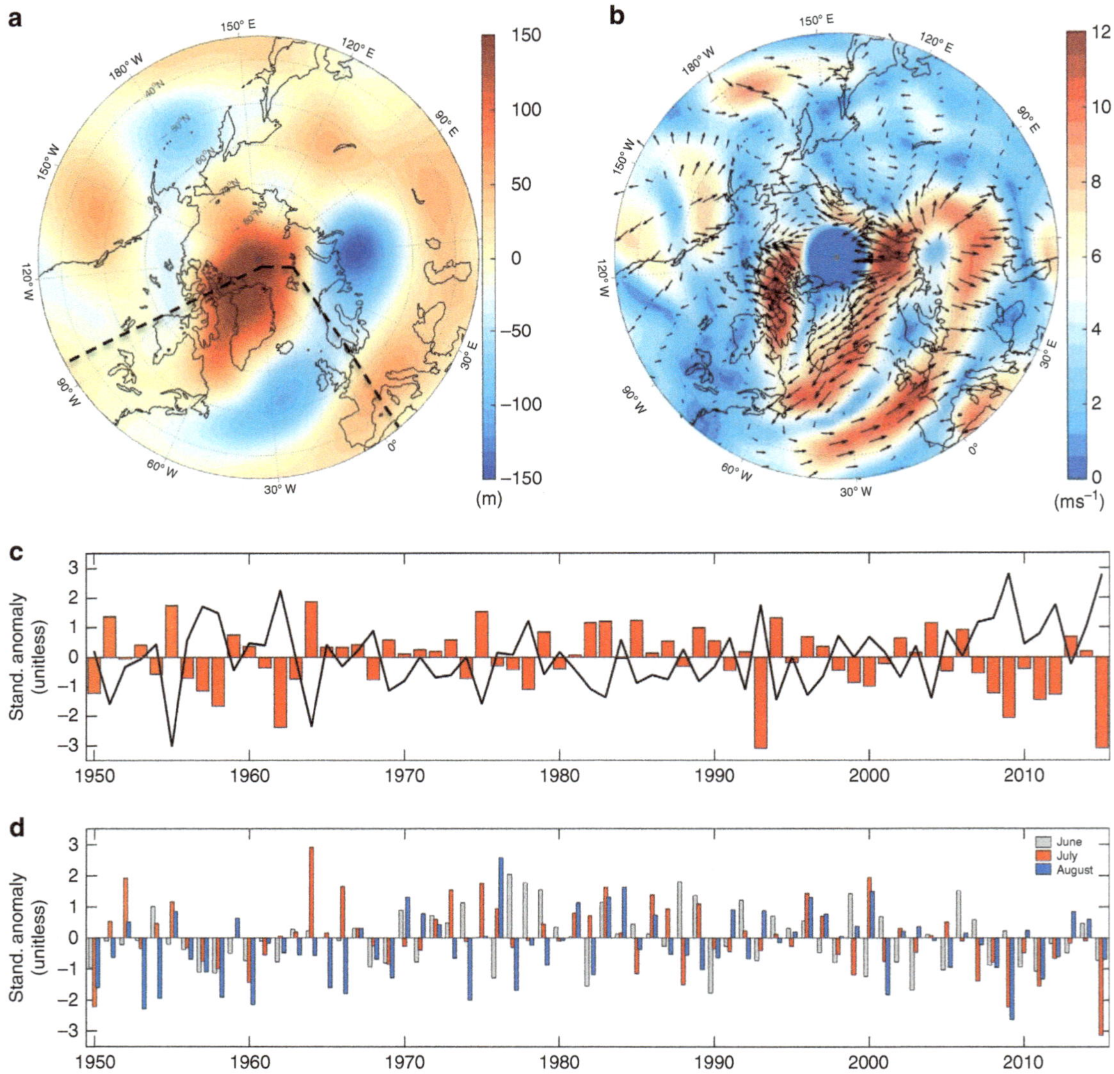

Figure 1 | Atmospheric conditions and selected quantities over the Arctic and over the Greenland ice sheet. (**a**) 500 hPa geopotential height composite anomaly (m) for the month of July 2015, with respect to the 1981-2010 baseline period (using NCEP-NCARv1 reanalysis); (**b**) same as **a**, but for the vector winds (ms^{-1}). (**c**) Time series of monthly averaged July NAO (red bars) and GBI (black line) indices (unitless) for the period 1950-2015. (**d**) Time series of standardized anomalies for the zonal winds at 500 hPa (unitless) averaged over the months of June (light gray), July (red) and August (pale blue) over the region bounded between 45-85° N and 100° W-0° E (included in the area marked by the dashed lines in **a**).

Arctic cutoff high and new surface Greenland records. The July 2015 high-pressure ridge over the Greenland evolved from a cutoff high that formed along the eastern coast of Greenland at the end of June (white circle in Fig. 2b). Over the same period, the jet stream, here characterized through the 5,700 ± 500 m 500 hPa isoheights[13], broke into three positive heights around the Northern Hemisphere (Fig. 2c). These conditions reinforced the atmospheric ridge over the Greenland ice sheet, which moved westward and persisted until mid-July (Fig. 2b–e). This promoted new records for meltwater production, runoff, albedo and surface temperature over northwest Greenland (Fig. 3; Supplementary Fig. 4), as simulated by the Modèle Atmosphérique Régionale[1,3,7] (MAR; Methods). The monthly averaged record setting values for simulated albedo and surface temperature in northwest Greenland for July 2015 were, respectively, ~2.5σ below and above the 1981–2010 mean, while runoff was up to ~3σ above the mean (Fig. 3). The spatial distribution of the 2015 surface albedo anomaly simulated by MAR (Supplementary Fig. 5a–c) indicates that the July 2015 negative anomaly was driven by an albedo decrease at relatively high elevations, promoted by the reduced summer snowfall (associated with anticyclonic conditions) and by increased surface melting and runoff. The same atmospheric conditions that promoted these new records over the northern Greenland also inhibited melting in the south, where enhanced melting and new records have been occurring over the past recent years[1,10]. This had implications for the surface mass balance of Greenland at both regional and ice-sheet scales (Supplementary Fig. 6). The summer exposure of bare ice and the presence of surface impurities have been suggested to be driving the enhanced melting observed over the past ~20 years[1]. However, the 2015 summer atmospheric conditions promoted the flow of cold air from the Arctic Ocean (Fig. 1b), favouring the accumulation of fresh new snow with a high albedo along western Greenland, hence offsetting the effects of bare ice exposure (Supplementary Figs 5–7).

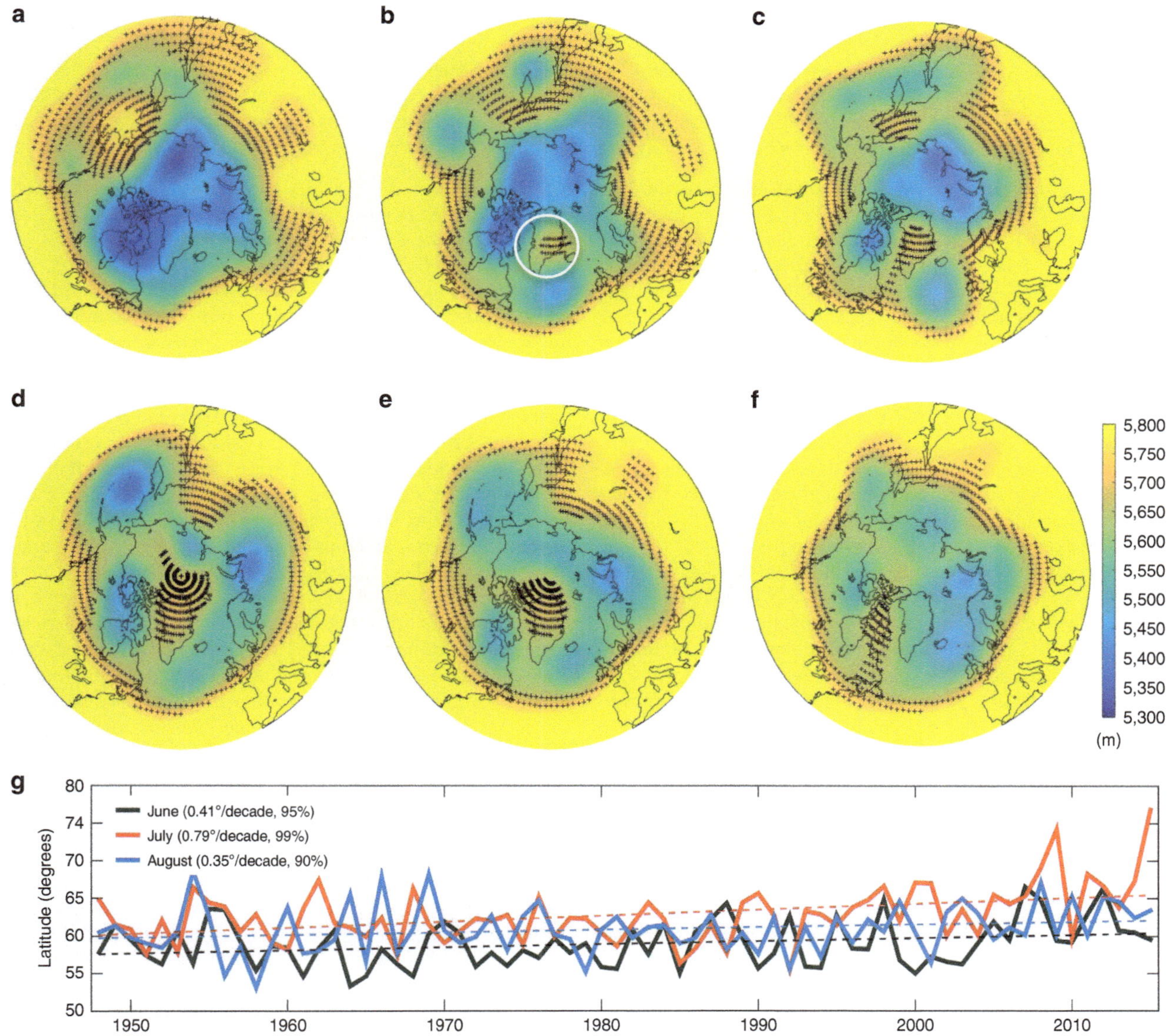

Figure 2 | Spatial and temporal evolution of the jet stream conditions and wave amplitude. Jet stream is here characterized through the 5,700 ± 50 m 500 hPa isoheights[13]. **(a–f)** Five day average geopotential height (m) at 500 hPa for the period 18th June to 22nd July 2015. For each day, the daily averaged values for the 2 days before and after were averaged with the daily average value of that day. Crosses show the locations where geopotential height values are 5,700 ± 50 m. **(g)** Maximum latitude of ridge peaks computed from the 500 hPa 5,700 ± 50 m isoheight for the period 1948-2015 over the region bounded between 45-85° N and 100° W-0° E (included in the area marked by the dashed lines in Fig. 1b), averaged over the months of June (black), July (red) and August (blue). Linear trends and significance levels are reported in the figure.

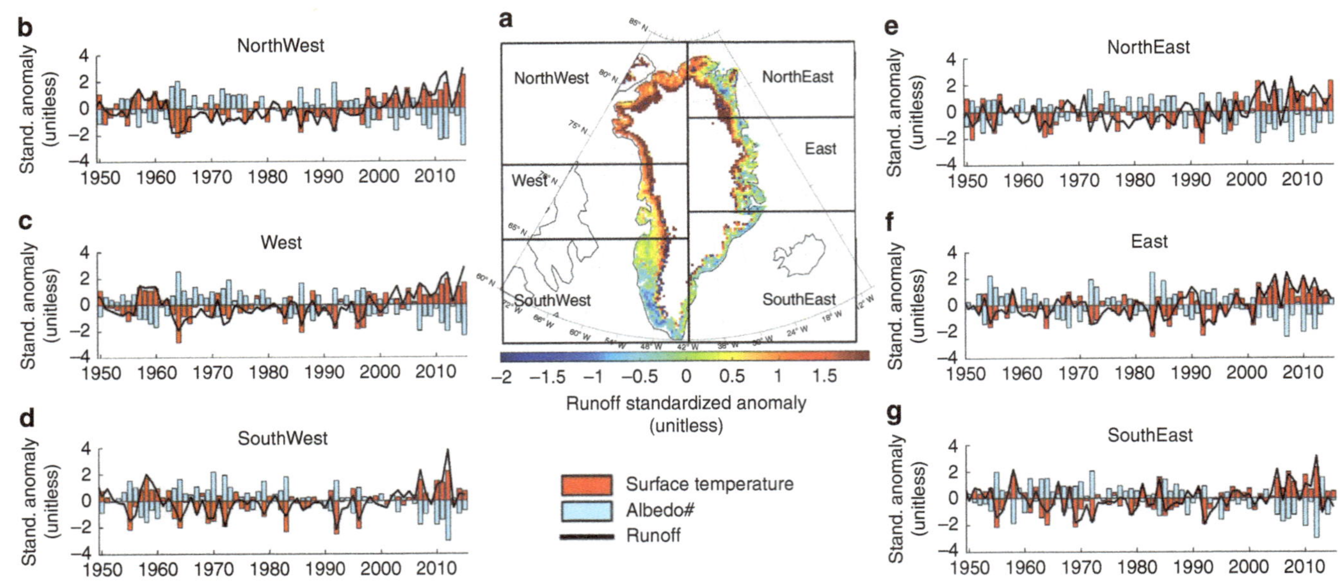

Figure 3 | Surface mass balance and energy balance quantities over the Greenland ice sheet. (**a**) Spatial distribution of the July 2015 MAR-simulated runoff anomaly (unitless) over the Greenland ice sheet (1981-2010). The boxes in the map display the boundaries of the different drainage regions. (**b-g**) Time series of standardized anomalies (unitless) for mean July runoff (black line), surface temperature (red bars) and surface broadband albedo (cyan bars) for the period 1950-2015 over the different drainage basin regions identified in **a**, as simulated by the MAR model.

Mean zonal winds and jet stream wave amplitude records. The westward shift of the cutoff high between the end of June and the beginning of July 2015 is associated with new records for both the mean zonal winds at 500 hPa (Fig. 1d) and the maximum latitude of ridging (Fig. 2g) over the region bounded between 45–85° N and 100° W–0° E. The July-averaged value of the 500 hPa zonal winds speed over the Greenland for latitudes between 60° and 80° N was $-1.5\,\mathrm{m\,s^{-1}}$ (here, the negative sign indicates easterly flow), $\sim 3\sigma$ below the mean. The two other occurrences of easterly flow for the same quantity during the 1948–2015 period happened in 1950 and 2009. However, the mean wind speed values were only $\sim -0.15\,\mathrm{m\,s^{-1}}$ in both cases, hence much smaller in magnitude than the 2015 record value. The monthly mean maximum latitude of ridge peaks of the $5{,}700 \pm 50$ m isohypse over the Atlantic sector (45–85° N and 100° W–0° E) also set the new record of 76.61° N (3.4σ) in July 2015, exceeding the previous record of 74.92° N (2.9σ) set in 2009 (Fig. 2g). The trend for the maximum latitude of the ridge peaks for the period 1948–2015 for the month of July is $0.79 \pm 0.09°$ per decade. The same trends for the months of June and August are, respectively, $0.41 \pm 0.14°$ per decade and $0.35 \pm 0.19°$ per decade. Such trends are even larger when considering only the satellite era (1979–2015), being up to $\sim 2.4°$ per decade for the month of July and are confirmed by the analysis of different global atmospheric reanalysis data sets (Supplementary Fig. 8).

Discussion

The mechanisms that created and maintained the 2015 observed ridge may be linked with forcing from very strong extratropical cyclones[14], to forcings from southern regions[15] or to latent heat release[16]. Another possibility is the local forcing related to Arctic amplification[13,17]. Although recent melt records over the Greenland have been linked to exceptional mid-tropospheric atmospheric conditions, with episodes of atmospheric blocking ridges being associated with Greenland's melting extremes[9,12], little or no attention has been given to the impact of the anticipated effects of Arctic amplification on the surface mass balance of the Greenland ice sheet. In this regard, the 2015 records for both the 500 hPa zonal winds and the maximum ridging latitude are consistent with the proposed effects on upper level atmosphere characteristics associated with Arctic amplification[13,17].

The 2015 poleward shift of the surface melting record in 2015, clearly indicates that improving our understanding of the impact of exceptional atmospheric conditions on the spatial distribution of extreme melting is crucial. Besides modulating the contribution of Greenland to sea level through the volume of meltwater production, the location of enhanced melting can influence ocean/ice interaction processes and ocean circulation[18], and bio-productivity, by altering salinity and temperature profiles of the surrounding ocean. Furthermore, the evolution of surface melting strongly impacts the Greenland's hydrological system, with implications for the englacial and subglacial systems, as well as ice discharge and dynamics[19]. Currently, several general circulation global climate models and Earth System models do not properly capture summer Arctic atmospheric forcing[8], limiting our capability to properly project the evolution of the surface mass balance and melting under future warming scenarios. Our work presented here demonstrates a strong need to identify the mechanisms that create and maintain strong cutoff highs. The new atmospheric records, and the trends of mean zonal winds and wave amplitude of the jet stream are consistent with the suggested effects of Arctic amplification[13,17]. Recent studies provide theoretical arguments that slowing zonal winds might be associated with larger planetary wave amplitudes[20] and that Arctic amplification and/or sea-ice loss do intensify existing ridges, thereby contributing to their persistence[21,22]. In the event studied here, however, the exceptional melting followed the ridging, rather than preceding it in alignment with other studies, indicating that observations and models results do not support the above mentioned expected effects of Arctic amplification[23–27]. Be that as it may, understanding the impact of cutoff highs on the Greenland's surface mass balance, and studying the mechanisms driving the trends and extremes of the anticipated effects of Arctic amplification are crucial tasks in view of the potential regional and global impacts long-time effects of enhanced melting over Greenland.

Methods

NAO data set. For the NAO index, we used the Hurrell[28] NAO values distributed by the Climate Research Unit of the University of Anglia (http://www.cru.uea.ac.uk/cru/data/nao/). The Hurrell NAO index is computed from the difference of normalized sea-level pressure between Lisbon, Portugal and Stykkisholmur/Reykjavik, and Iceland. More detail on the Hurrell NAO values can be found at https://climatedataguide.ucar.edu/climate-data/hurrell-north-atlantic-oscillation-nao-index-pc-based (ref. 28).

Reanalysis data. Geopotential heights and zonal winds at 500 hPa are obtained from the National Centers for Environmental Prediction (NCEP)/National Center for Atmospheric Research (NCEP/NCAR) reanalysis data set. The NCEP/NCAR data set consists of globally, gridded data sets, incorporating observations and outputs from a numerical weather prediction model from 1948 to present[29]. Values of the GBI are computed from the daily and monthly values of the NCEP/NCAR 500 hPa geopotential height values averaged over the area between 60–80° N and 280–340° E (ref. 9). ERA-Interim reanalysis data is used to complement the results obtained with NCEP/NCAR and was downloaded from http://apps.ecmwf.int/datasets/data/interim-full-daily/levtype=sfc/.

The MAR regional climate model. Simulations of surface quantities over the Greenland ice sheet are performed using the MAR[1,7,8] MAR is a modular, hydrostatic, and compressible atmospheric model that uses the sigma-vertical coordinate to better simulate airflow over complex terrain and the Soil Ice Snow Vegetation Atmosphere Transfer scheme surface model. The snow model in MAR is the CROCUS model[30]. The MAR model configuration used here has 25 terrain-following sigma layers between the Earth's surface and the top. The horizontal resolution of the outputs used here is 20 km. The lateral boundary conditions are prescribed every 6 h from NCEP/NCAR meteorological fields. Here we use daily outputs obtained from the average of 120 s outputs produced by the model. The temporal configuration for the runs is from 1948 to the present. The sea surface temperature and the sea-ice cover are also prescribed every 6 h in the model, using NCEP/NCAR reanalysis data[29]. No nudging or interactive nesting is used in any of the experiments, with the atmospheric fields over the Greenland ice sheet computed by the atmospheric module of MAR. The atmospheric model, in turn, interacts with the CROCUS model, which provides the state of the snowpack, and associated surface mass balance and energy balance quantities (for example, albedo and runoff).

Code availability. MAR is an open-source code available to the scientific community. The source code for the MAR version used in this study is available at ftp://tedesco-dell.ldeo.columbia.edu/cryoftp/MARv3.5.2src_2015-03-18.tgz. The codes used for analysing the reanalysis data are available upon request from the authors.

References

1. Tedesco, M. *et al.* The role of albedo and accumulation in the 2010 melting record in Greenland. *Environm. Res. Lett.* **6,** 014005 (2011).
2. Tedesco, M. *et al.* Evidence and analysis of 2012 Greenland records from spaceborne observations, a regional climate model and reanalysis data. *The Cryosphere* **7,** 615–630 (2013).
3. Fettweis, X. *et al.* Brief communication 'Important role of the mid-tropospheric atmospheric circulation in the recent surface melt increase over the Greenland ice sheet'. *The Cryosphere* **7,** 241–248 (2013).
4. Hanna, E. *et al.* The influence of North Atlantic atmospheric and oceanic forcing effects on 1900–2010 Greenland summer climate and ice melt/runoff. *Int. J. Climatol.* **33,** 862–880 (2013).
5. Hanna, E. *et al.* Atmospheric and oceanic climate forcing of the exceptional Greenland ice sheet surface melt in summer 2012. *Int. J. Climatol.* **34,** 1022–1037 (2014).
6. Van Tricht *et al.* Clouds enhance Greenland ice sheet meltwater runoff. *Nat. Commun.* **7,** 10266 (2016).
7. Fettweis, X., Tedesco, M., Broeke, M. & Ettema, J. Melting trends over the Greenland ice sheet (1958–2009) from spaceborne microwave data and regional climate models. *The Cryosphere* **5,** 359–375 (2011).
8. Tedesco, M. & Fettweis, X. 21st century projections of surface mass balance changes for major drainage systems of the Greenland ice sheet. *Environ. Res. Lett.* **7,** 045405 (2012).
9. Hanna, E., Cropper, T., Hall, R. & Cappelen, J. Greenland Blocking Index 1851-2015: a regional climate change signal. *Int. J. Climatol.* doi:10.1002/joc.4673 (2016).
10. Tedesco, M. *et al.* Greenland, in states of the climate in 2013. *Bull. Amer. Meteor. Soc.* **95,** S1–S279 (2014).
11. Nghiem, S. V. *et al.* The extreme melt across the Greenland ice sheet in 2012. *Geophys. Res. Lett.* **39,** L20502 (2012).
12. McLeod, J. T. & Mote, T. L. Linking interannual variability in extreme Greenland blocking episodes to the recent increase in summer melting across the Greenland ice sheet. *Int. J. Climatol* **36,** 1484–1499 (2016).
13. Francis, J. A. & Vavrus, S. J. Evidence linking Arctic amplification to extreme weather in mid-latitudes: Arctic links to mid-latitude weather. *Geophys. Res. Lett.* **39,** L06801 (2012).
14. Archambault, H. M., Bosart, L. F., Keyser, D & Cordeira, J. M. A climatological analysis of the extratropical flow response to recurving western north pacific tropical cyclones. *Mon. Wea. Rev.* **141,** 2325–2346 (2013).
15. Krishnamurti, T. N. *et al.* A Pathway connecting the monsoonal heating to the rapid Arctic ice Melt. *J. Atmos. Sci.* **72,** 5–34 (2015).
16. Pfahl, S., Schwierz, C., Croci-Maspoli, M., Grams, C. M. & Wernli, H. Importance of latent heat release in ascending air streams for atmospheric blocking. *Nat. Geosci.* **8,** 610–614 (2015).
17. Francis, J. & Vavrus, S. J. Evidence for a wavier jet stream in response to rapid Arctic warming. *Environ. Res. Lett.* **10,** 014005 (2015).
18. Rahmstorf, S. *et al.* Exceptional twentieth-century slowdown in Atlantic Ocean overturning circulation. *Nat. Clim. Change* **5,** 475–480 (2015).
19. Tedesco, M. *et al.* Ice dynamic response to two modes of surface lake drainage on the Greenland ice sheet. *Environ. Res. Lett.* **8,** 034007 (2013).
20. Huang, C. S. & Nakamura, N. Local finite-amplitude wave activity as a diagnostic of anomalous weather events. *J. Atmos. Sci.* **73,** 211–229 doi:http://10.1175/JAS-D-15-0194.1 (2016).
21. Lee, M.-Y., Hong, C.-C. & Hsu, H.-H. Compounding effects of warm sea surface temperature and reduced sea ice on the extreme circulation over the extratropical North Pacific and North America during the 2013–2014 boreal winter. *Geophys. Res. Lett.* **42,** 1612–1618 (2015).
22. Kug, J. S. *et al.* Two distinct influences of Arctic warming on cold winters over North America and East Asia. *Nat. Geosci.* **8,** 759–762 (2015).
23. Hoskins, B. & Woollings, T. in *Current Climate Change Reports.* (eds Sobel, A. & Camargo, S. J) **1,** 115–124 (2015).
24. Barnes, E. A. & Screen, J. A. The impact of Arctic warming on the midlatitude jet-stream: Can it? Has it? Will it?: Impact of Arctic warming on the midlatitude jet-stream. *Wiley Interdiscip. Rev. Clim. Change* **6,** 277–286 (2015).
25. Barnes, E. A. Revisiting the evidence linking Arctic amplification to extreme weather in midlatitudes: Arctic Amplification and weather. *Geophys. Res. Lett.* **40,** 4734–4739 (2013).
26. Screen, J. & Simmonds., I. The central role of diminishing sea ice in recent Arctic temperature amplification. *Nature* **464,** 1334–1337 (2010).
27. Hassanzadeh, P. & Kuang, Z. Blocking variability: Arctic amplification versus Arctic Oscillation. *Geophys. Res. Lett.* **42,** 8586–8589 (2015).
28. Hurrell, J. *The Climate Data Guide: Hurrell North Atlantic Oscillation (NAO) Index (station-based).* Available at https://climatedataguide.ucar.edu/climate-data/hurrell-north-atlantic-oscillation-nao-index-station-based.
29. Kalnay, E. *et al.* The NCEP/NCAR 40-year reanalysis project. *Bull. Amer. Meteor. Soc.* **77,** 437–471 (1996).
30. Brun, E., David, P., Sudul, M. & Brunot, G. A numerical model to simulate snow-cover stratigraphy for operational avalanche forecasting. *J. Glaciol.* **38,** 13–22 (1992).

Acknowledgements

M.T. and J.J. were supported by the National Science Foundation grants PLR1304807 and ANS 0909388, and NASA grant NNX1498G. T.M. was supported by the NASA grant NNX1498G. M.T. thanks Adriana Basagni for reviewing the draft of the manuscript and for the discussions concerning the Arctic atmospheric variability. We also wish to thank the three anonymous reviewers for their suggestions that led to an improved version of this manuscript.

Author contributions

M.T. conceived the study, analysed the data and prepared the main body of the manuscript. M.T., T.M., E.H., R.D., X.F. and J.F.B. contributed to the analysis of the atmospheric data. X.F. and M.T. generated MAR outputs. J.J. contributed to the extraction and processing of MAR outputs. K.B. and M.T. contributed to the analysis of MAR data. All authors contributed to the editing of the final version of the manuscript.

Additional information

Competing financial interests: The authors declare no competing financial interests.

6

Unexpected seasonality in quantity and composition of Amazon rainforest air reactivity

A.C. Nölscher[1], A.M. Yañez-Serrano[1,2], S. Wolff[1,2], A. Carioca de Araujo[3], J.V. Lavrič[4], J. Kesselmeier[1] & J. Williams[1]

The hydroxyl radical (OH) removes most atmospheric pollutants from air. The loss frequency of OH radicals due to the combined effect of all gas-phase OH reactive species is a measureable quantity termed total OH reactivity. Here we present total OH reactivity observations in pristine Amazon rainforest air, as a function of season, time-of-day and height (0-80 m). Total OH reactivity is low during wet ($10\,s^{-1}$) and high during dry season ($62\,s^{-1}$). Comparison to individually measured trace gases reveals strong variation in unaccounted for OH reactivity, from 5 to 15% missing in wet-season afternoons to mostly unknown (average 79%) during dry season. During dry-season afternoons isoprene, considered the dominant reagent with OH in rainforests, only accounts for ~20% of the total OH reactivity. Vertical profiles of OH reactivity are shaped by biogenic emissions, photochemistry and turbulent mixing. The rainforest floor was identified as a significant but poorly characterized source of OH reactivity.

[1] Air Chemistry and Biogeochemistry Department, Max Planck Institute for Chemistry, Mainz 55128, Germany. [2] Clima e Ambiente (CLIAMB), Instituto Nacional de Pesquisas da Amazônia (INPA), Avenue André Araújo 2936, Manaus, Amazonas CEP 69083-000, Brazil. [3] Embrapa Amazônia Oriental, Empresa Brasileira de Pesquisa Agropecuaria, Belem, Pará CEP 66095-100, Brazil. [4] Biogeochemical Systems Department, Max Planck Institute for Biogeochemistry, Jena 07745, Germany. † Present address: Division of Geological and Planetary Sciences, California Institute of Technology, Pasadena, California 91125, USA. Correspondence and requests for materials should be addressed to A.C.N. (email: a.noelscher@mpic.de).

The Amazon rainforest is the largest, most productive, biodiverse, contiguous terrestrial ecosystem on Earth (about 6 million km^2)[1] and of global importance in terms of carbon, water and energy fluxes[2–4]. Its canopy delineates one of the most important interfaces between the terrestrial biosphere and atmosphere. This is where a complex mixture of highly reactive biogenic volatile organic compounds (BVOC) is released to the atmosphere[5], among them isoprene, monoterpenes, other isoprenoids and various oxygenated compounds. Plant metabolic processes induce BVOC emissions that depend on light, temperature, stress, ambient concentration levels, water, nutrient availability, soil type, plant species and phenology[6–10]. After entering the atmosphere, BVOC are oxidized primarily by the hydroxyl (OH) radical—typically within seconds to hours. Therefore BVOC have a direct impact on ambient OH radical concentrations, or in other words the regional oxidation capacity of the atmosphere[11]. Since many greenhouse gases (for example, methane, CH_4) and pollutants (for example, carbon monoxide, CO) are predominately oxidized in the Tropics by OH, variations of OH levels in these regions may have global scale consequences. The oxidation of BVOC by OH leads to a multitude of photochemical products that often remain undetected[12,13]. Recently, it has been found that some of these compounds partition to and from surfaces[9,14,15] or participate in aerosol formation and growth[16–18]. Currently there are large uncertainties in how tropical aerosol influences the Earth's energy balance by directly effecting radiation or by modifying cloud characteristics[19].

The combined effect of all OH reactive species on the instantaneous atmospheric OH loss frequency (in s^{-1}) is defined as total OH reactivity, the inverse of the chemical lifetime of OH. It is dependent on the atmospheric concentration of all OH reactive trace gases $[X_i]$ and their reaction rate coefficient with OH (k_{OH+Xi}).

$$R_{total} = \sum_{i=1}^{n} k_{OH+X_i} \times [X_i] \quad (1)$$

When total OH reactivity measurements are compared with individually detected BVOC levels, the relative contribution of each BVOC to the total OH reactivity can be determined. Moreover, the extent of missing or unaccounted for OH sinks can be evaluated. Previous studies in forested regions revealed a significant gap between the budget of individually detected known OH sinks and the directly measured total OH reactivity[13,20,21]. The missing reactivity has been shown to be dependent on the ecosystem, season, time of the day and environmental conditions. These unknown and undetected atmospheric OH sinks are likely to be either forest emissions[20,22], the oxidation products of such[13,23], or a complex mixture of both[21,24]. High missing OH reactivity implies an incomplete understanding of the processes occurring at the forest canopy–atmosphere interface such as biogenic emissions, gas-to-particle partitioning of oxidation products, atmospheric ozone formation potential, trace gas deposition inside a forest canopy, canopy dynamics, and the photooxidation of global pollutants and greenhouse gases.

Here we present the first detailed characterization of total OH reactivity seasonality, diel variation and vertical distribution throughout the canopy of a pristine Amazon rainforest. The results reveal how total OH reactivity is strongly influenced by biogenic emissions or uptake, atmospheric chemistry and physical mixing processes. Finally, we assess the composition and chemical nature of the total atmospheric OH reactivity through comparison with co-measured volatile organic compounds (VOC).

Results

Observations at the Amazonian tall tower observatory. Measurements were conducted in Central Amazonia within the framework of the Brazilian-German ATTO (Amazonian Tall Tower Observatory) project[25]. The measurement site is located about 150 km north-east of Manaus (120 m above sea level) in a dense, non-flooded upland forest. The immediate surroundings as well as large areas to the north-east, upwind of the site, are covered by an old-growth undisturbed forest, which permits natural process studies with minimal anthropogenic impact. Meteorological conditions at the ATTO site vary through the year according to the position of the inter-tropical convergence zone, with a cooler rainy season (February–May) and a hotter drier season (June–October)[26].

Observations of greenhouse gases, reactive trace gases and aerosol within and above the forest canopy were conducted from an 80-m walk-up (INSTANT)[25] tower (S02°08′38.8″, W58°59′59.5″). Micrometeorological conditions were continuously monitored and phenological observations were performed regularly. Since early 2012, this comprehensive set of measurements has served to characterize the tropical biosphere–atmosphere interface of the ATTO site[25]. Within this framework, total OH reactivity measurements were conducted by the comparative reactivity method (CRM)[27,28], which is based on the competitive reaction of OH with a non-atmospheric molecule and all atmospheric OH sink compounds. In parallel, a quadrupole proton transfer reaction-mass spectrometer (PTR-MS) provided on-line measurements of various VOC[29]. Occasionally, air samples were collected on absorbent tubes for offline analysis of selected VOC in the laboratory using gas chromatographic methods.

Air for analysis was drawn down at ca. 16 l min^{-1} through eight continuously flushed, heated and insulated PTFE inlet lines (3/8″ = 0.95 cm outer diameter) from the forest floor (0.05, 0.5 m), throughout understory and canopy (4, 12, 24, 38.3 m ≈ average top of canopy), up to far above the canopy (53, 79.3 m). A valve system switched sequentially between the eight levels, completing one entire profile from 0.05 to 79.3 m in 16 min (Supplementary Fig. 1). Parallel to the total OH reactivity measurements, VOC (methanol, acetonitrile, acetaldehyde, acetone, isoprene, isoprene oxidation products, methyl ethyl ketone (MEK), benzene, toluene and the sum of monoterpenes), reactive trace gases (ozone (O_3), nitric oxide (NO), nitrogen dioxide (NO_2)) and carbon dioxide (CO_2)) were sampled via the same inlet system. The inlet system was tested with a gas standard for all detected VOC and no significant differences between the various heights were observed. The line losses (difference between expected and detected VOC mixing ratios) were small, on average 9% (see Methods).

Seasonal variation. Four intensive field measurement campaigns were performed over a 1-year period from October 2012 to September 2013 at the ATTO site (Supplementary Fig. 2).

The first observations were made in the transition period between dry and wet season (31 October 2012–29 November 2012). Typically, the days were hot, reaching maximum values inside the forest canopy of up to 35 °C. Photosynthetically active radiation (PAR) was highly variable due to isolated convective clouds moving overhead and occasional rain events (Fig. 1, Table 1). Significant diel variability of total OH reactivity was found ranging from the instrument's detection limit (3–4 s^{-1}) to 135 s^{-1} (observed at the 79.3 m level in the afternoon). Inside the canopy (24 m) and above the canopy (79.3 m) comparable averages of 20.7 ± 16.8 and 19.6 ± 14.2 s^{-1} (campaign mean ± s.d.) were determined, respectively.

During the colder, generally overcast and rainier wet season measurement campaign (27 February 2013–6 March 2013) total OH reactivity decreased to about half of what was observed during the dry-to-wet-season transition period, to on average $9.9 \pm 5.2\,s^{-1}$ inside the canopy at 24 m. These levels of total OH reactivity are, for example, comparable to the typical summertime boreal forest (no prolonged heat, no anthropogenic or biomass burning influence)[21,30].

In the wet-to-dry-season transition period (8–14 June 2013), average temperature and radiation increased relative to the previous campaign. The total OH reactivity rose inside the forest canopy to on average $26.1 \pm 13.8\,s^{-1}$. This was higher than the average canopy total OH reactivity observed during the dry-to-wet-season transition. Comparable total OH reactivity levels and variability were recently observed in a Mediterranean Oak forest which copiously emitted isoprene[31].

At the midpoint of the dry season (20–29 September 2013), days were typically dry and hot with overall highest annual midday averages for PAR (Table 1). At that time of the year, total OH reactivity reached maximum values up to $241\,s^{-1}$ inside the Amazon rainforest's canopy in the afternoon of 23 September 2013 (Supplementary Fig. 2). Similar high total OH reactivity has been reported previously only in the extremely polluted environment of Mexico City at rush hour[32]. Overall, this was the season with the largest average total OH reactivity of $62.4 \pm 41.3\,s^{-1}$ within the canopy.

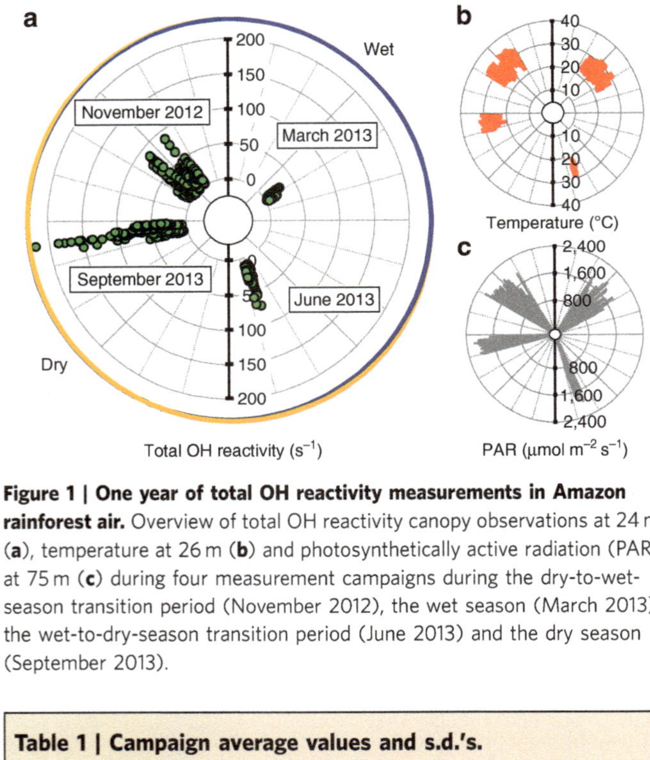

Figure 1 | One year of total OH reactivity measurements in Amazon rainforest air. Overview of total OH reactivity canopy observations at 24 m (**a**), temperature at 26 m (**b**) and photosynthetically active radiation (PAR) at 75 m (**c**) during four measurement campaigns during the dry-to-wet-season transition period (November 2012), the wet season (March 2013), the wet-to-dry-season transition period (June 2013) and the dry season (September 2013).

Diel variation. Figure 2 presents diel hourly medians of the total OH reactivity and its inverse, the hydroxyl radical chemical lifetime. Independent of the season, sunrise and sunset were at approximately 06:00 and 18:00 l.t. (local time) and the peak of radiation was at noon. The diurnal peak in ambient temperature was shifted compared with PAR, reaching its maximum in the afternoon at 13:00 l.t. for the wet season and 15:00 l.t. for the dry season.

The scales for total OH reactivity in Fig. 2 differ by about a factor of 7 with significantly higher median values in the dry season. Total OH reactivity ranged between 6 and $13\,s^{-1}$ within the canopy at 24 m for the wet season. Clear diel trends were observed for the dry season with median total OH reactivity inside the canopy varying between 31 and $85\,s^{-1}$. Maximum values were found during daytime, decreasing throughout the night. For both seasons the weakest diel variation and the lowest absolute values of total OH reactivity were found at the lowest level, 0.05 m from the soil. During the dry season, the median total OH reactivity inside the canopy (24 m) reached peak values with maximum PAR at noon and maximum temperature in the afternoon. This is in close accordance with previous studies[5,7,33] that identified light and temperature as the major driving forces for biogenic emissions.

The inverse of the total OH reactivity measurements directly provides the OH chemical lifetime. It is determined by all instantaneous OH loss reactions at each observed level within and above the Amazon rainforest. The dry-season median OH chemical lifetime inside the canopy in the afternoon was extremely short and reached minimum values of < 10 ms, while during the wet season it became an order of magnitude longer (100–200 ms).

Vertical variation. Before sunrise (03:00–06:00 l.t., September 2013) the dry-season median total OH reactivity (35.8–$42.2\,s^{-1}$) stayed rather constant throughout understory, canopy and atmosphere above the canopy (12–79.3 m, Fig. 3). At that time the ambient temperature (23.0–25.1 °C) did not vary significantly from soil to uppermost level. At sunrise vertical gradients were smallest and total OH reactivity at its diel minimum (Fig. 2). This was when the shallow nocturnal boundary layer started to break up. Aged and less-reactive air from higher altitudes was likely mixed down and diluted the rainforest reactive conglomerate[34].

With the onset of solar irradiation, the forest canopy began to heat up. Already by 09:00–12:00 l.t. the canopy temperature was

Table 1 | Campaign average values and s.d.'s.

	November 2012	March 2013	June 2013	September 2013
Total OH reactivity				
0.05 m	$(11.0 \pm 11.7)\,s^{-1}$	$(7.4 \pm 4.7)\,s^{-1}$	$(21.2 \pm 9.3)\,s^{-1}$	$(38.4 \pm 28.3)\,s^{-1}$
24 m	$(20.7 \pm 16.8)\,s^{-1}$	$(9.9 \pm 5.2)\,s^{-1}$	$(26.1 \pm 13.8)\,s^{-1}$	$(62.4 \pm 41.3)\,s^{-1}$
79.3 m	$(19.6 \pm 14.2)\,s^{-1}$	$(11.3 \pm 6.5)\,s^{-1}$	$(31.4 \pm 15.4)\,s^{-1}$	$(60.2 \pm 37.8)\,s^{-1}$
Temperature (26 m)	(27.3 ± 3.2) °C	(24.6 ± 2.4) °C*	(25.4 ± 2.5) °C	(26.8 ± 3.4) °C
PAR (at noon)†	$(1{,}433 \pm 744)\,\mu mol\,m^{-2}s^{-1}$	$(929 \pm 620)\,\mu mol\,m^{-2}s^{-1}$	$(1{,}170 \pm 690)\,\mu mol\,m^{-2}s^{-1}$	$(1{,}775 \pm 670)\,\mu mol\,m^{-2}s^{-1}$
Rainday frequency	7.3	25.3	18.8	15.0

Calculated for the total OH reactivity, canopy temperatures and noon-time photosynthetically active radiation (PAR, 11:00-13:00 l.t.). In addition, the rainday frequency, the number of days with rain in the month of measurements are included. Details about the precipitation data for the four measurement intensives described here are given in Yañez-Serrano *et al.*[29].
*No gradient measurements for temperature were available in March 2013. Instead the temperature monitored by a radiation sensor is given.
†Noon time (11:00-13:00 l.t.) PAR data were averaged over the course of the campaign.

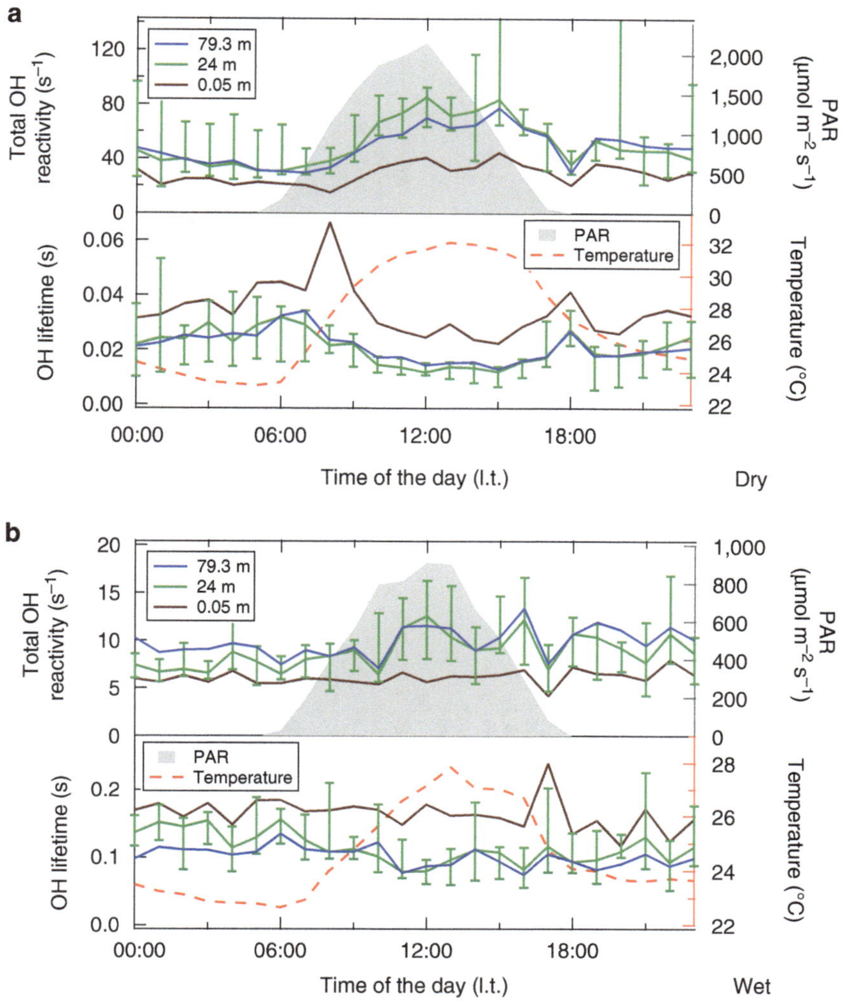

Figure 2 | Dry and wet season total OH reactivity and OH chemical lifetime. Figure panels (**a**) and (**b**) show soil (0.05 m), canopy (24 m) and atmospheric (79.3 m) total OH reactivity and OH chemical lifetime during dry (September 2013) and wet season (March 2013), respectively. Diel hourly medians are plotted as function of the time of the day, with ambient temperature (at 26 m) and the photosynthetically active radiation (PAR). Exemplarily the ±75/25 percentiles are pictured as error bars for the 24 m total OH reactivity (OH lifetime) medians.

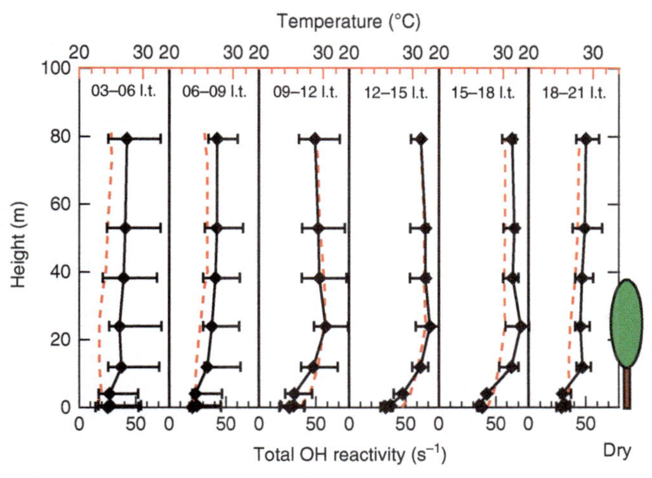

Figure 3 | Dry-season vertical profiles of total OH reactivity and temperature between forest floor and 79.3 m. Six different profiles present the temporal variation throughout the day for 3 h-binned median total OH reactivity and ambient temperature. Positive error bars include the 75 percentiles of the dry-season campaign (September 2013) total OH reactivity measurements, negative error bars the 25 percentiles.

significantly enhanced. With rising irradiation and temperature, the total OH reactivity increased strongly to maximum levels inside the canopy. A strong temperature inversion at 4 m isolated the lower cooler layer dynamically throughout the entire day which led to minimum total OH reactivities between 20 and 40 s^{-1} (Fig. 3). Interestingly, surface soil temperature (23.7–26.2 °C) and water content (0.16–0.27) did not vary significantly, neither during the day nor throughout the tropical seasons (Supplementary Fig. 3). During the dry-season day (12:00–15:00 l.t.) the total OH reactivity developed a strong gradient, decreasing rapidly from 71.5 s^{-1} (3 h median, 24 m) towards the soil level to 32.3 s^{-1} (3 h median, 0.05 m) and decreasing more moderately above the canopy to 63.7 s^{-1} (3 h median, 79.3 m). This development reflects the emission source location (canopy), the rate of oxidation of reactive primary emissions to less-reactive products (above canopy) and the turbulent exchange rates at each height (mixing above the canopy, stratification below the canopy).

Figure 4 compares the relative direction of vertical gradients above the canopy as function of season and time of the day for the total OH reactivity and two prominent atmospheric reactants with OH. Isoprene is copiously emitted by tropical vegetation and is highly reactive ($k_{OH+ISO}(298\,K) = 1 \times 10^{-10}\,cm^3\,molecule^{-1}\,s^{-1}$, IUPAC)[16,35]. Prominent oxidation products from the reaction of

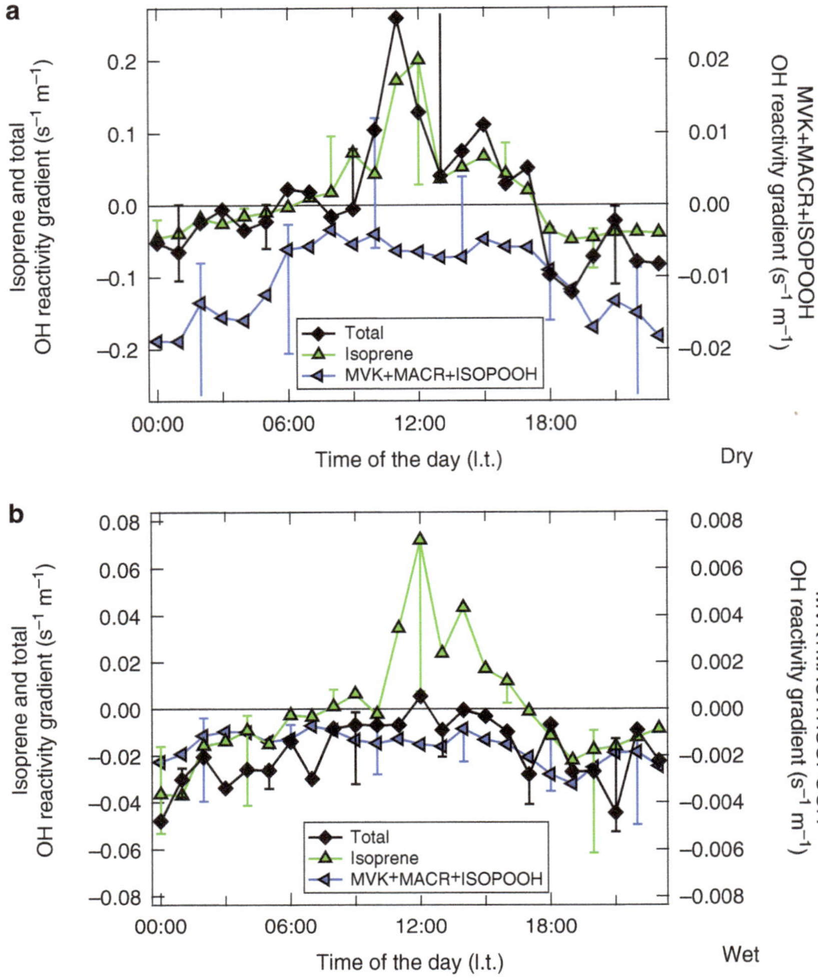

Figure 4 | Dry and wet season diurnal variation of OH reactivity median gradients above the canopy. The gradients were calculated as linear regression slope for the total OH reactivity, isoprene OH reactivity and isoprene products (MVK + MACR + ISOPOOH) OH reactivity above the canopy. A gradient equal to zero corresponds to constant OH reactivity with height. A negative gradient goes along with higher values in 79.3 m, and a positive gradient with higher values inside the canopy (24 m). Median OH reactivity gradients above the canopy were calculated for dry (**a**) and wet (**b**) season measurement campaigns in September 2013 and March 2013. Error bars picture the variability of the median OH reactivity gradients above the canopy as the 75-(positive) and 25-(negative) percentiles.

isoprene and OH[12] are methyl vinyl ketone (MVK), methacrolein (MACR) and isoprene hydroxy hydroperoxides (ISOPOOH) whose summed signal is detected by the PTR-MS (*m/z* 71)[36,37]. Detected atmospheric isoprene and isoprene product (MVK + MACR + ISOPOOH) mixing ratios were used to determine the OH reactivity of isoprene (representing light and temperature-dependent primary emissions) and of isoprene products (representing photochemistry products). Total, isoprene and MVK + MACR + ISOPOOH OH reactivities were plotted as a function of height in and above the canopy (24, 38.3, 53, 79.3 m) and a linear regression slope was calculated (Supplementary Fig. 4). Gradients above the canopy were positive when OH reactivity was greater inside the canopy (24 m), and negative when OH reactivity was greater at the top levels (79.3 m).

Wet-season vertical gradients relative to the canopy were less pronounced than in the dry season. However, the night-to-daytime variability of isoprene and MVK + MACR + ISOPOOH OH reactivity gradients was consistent throughout the seasons. Both compounds had negative gradients during the night (higher levels at 79.3 m). After sunrise isoprene was released from the forest canopy (positive gradients), whereas MVK + MACR + ISOPOOH were photochemically produced above the canopy (negative gradients). Unlike these two classes of important OH sinks (emissions and oxidation products), the total OH reactivity vertical distribution unexpectedly changed its diurnal pattern with the season. During the wet season, total OH reactivity gradients were mostly negative and close to zero, similar to the observed isoprene products. Due to the relatively slow reaction with OH ($k_{OH+MVK+MACR+ISOPOOH}(298\,K) = 5.62 \times 10^{-11}$ $cm^3\,molecule^{-1}\,s^{-1}$, average MVK + MACR + ISOPOOH rate coefficient, IUPAC, MCM v3.3), the absolute gradient of MVK + MACR + ISOPOOH OH reactivity was about a factor of 10 smaller than the total OH reactivity. During the dry season, total OH reactivity gradients matched isoprene, being positive by day and negative at night. This shows that the vertical distribution of predominant OH sinks strongly depends on the tropical season. Photooxidation products, such as MVK + MACR + ISOPOOH-like compounds, shape the vertical gradient of the wet season total OH reactivity. These species persist and can be transported further above the canopy due to their longer lifetimes. Within the canopy they can be efficiently uptaken by the wet surfaces decreasing their OH sink potential. In the dry season, isoprene-like compounds dominate the total OH reactivity vertical distribution. The significantly enhanced total OH reactivity in

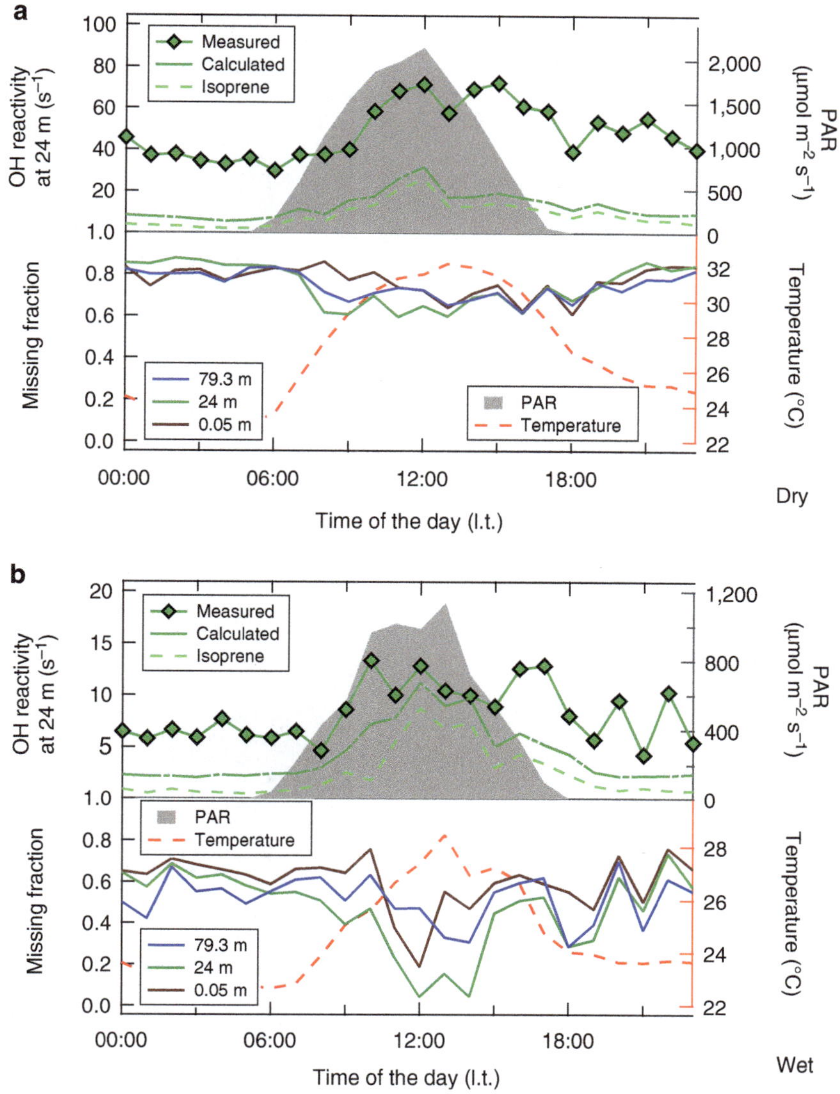

Figure 5 | Comparison of measured and calculated OH reactivity for dry and wet season. Dry (**a**) and wet (**b**) season OH reactivity was measured, and calculated due to isoprene alone and all co-detected reactive compounds. The relative difference between measured and calculated OH reactivity is the missing fraction, the ratio of unmeasured OH sinks to the total measured OH reactivity. The missing fraction is presented for three levels: soil (0.05 m), canopy (24 m) and top of the tower (79.3 m). Total OH reactivity hourly medians slightly differ from Fig. 2, because only data with full coverage of all measured compounds have been considered in the calculation. For comparison, PAR and temperature diel hourly medians were included in this figure.

the dry-season canopy layer was potentially caused by strong biogenic emissions of reactive compounds due to increased PAR and leaf temperatures, *in situ* oxidation by strong irradiation reaching inside the canopy, and the reduced deposition due to drier surfaces, through less-opened stomata[38], and therefore a smaller leaf area[39]. Another possibility is that microbial communities on leaf surfaces modulate ambient BVOC levels[40,41] due to seasonal variations in number and species.

Missing reactivity. The atmospheric OH reactivity was calculated as the sum of all individually detected OH sinks: methanol, acetonitrile, acetaldehyde, acetone, isoprene, MVK + MACR + ISOPOOH, MEK, monoterpenes, benzene, toluene, CO, NO, NO_2, O_3 and CH_4 (Supplementary Note 1; Supplementary Table 1; Supplementary Fig. 5). Figure 5 presents a comparison of the diurnal variation for wet- and dry-season OH reactivity—calculated, measured and missing. The gap between calculated and measured total OH reactivity was significant and variable as a function of season, time of the day and height. On average, $49 \pm 24\%$ ($4.1 \pm 3.0\,s^{-1}$) of the measured total OH reactivity inside the canopy remained unaccounted for during the wet-season measurement campaign (March 2013). At midday most of the total OH reactivity within the forest canopy could be explained with only 5–15% missing, whereas below and above the forest canopy this fraction was greater. The dry season missing reactivity was greater with on average $79 \pm 11\%$ ($51.5 \pm 29.4\,s^{-1}$) in 24 m. Even though isoprene was the strongest OH sink among the detected VOC, it only accounted for about one-fifth of the total OH reactivity inside the canopy during the dry-season afternoon. For both seasons the measured total OH reactivity could be better explained during the daytime inside the canopy when isoprene emissions were high.

High missing OH reactivity has been previously observed in forested environments and explained as either forest primary emissions[20,22] or photooxidation products of such[13,23] or a mixture of both[21,24]. In the Amazon rainforest the most prominent OH sinks are isoprene, emitted by trees with high light and temperature, and its oxidation products, formed above the canopy in the sunlit atmosphere. Total and missing OH reactivity are weakly, but positively correlated to the observed isoprene and MVK + MACR + ISOPOOH OH reactivities (Supplementary Fig. 6). Figure 6 and Table 2 show that this correlation can be partitioned according to the level of missing OH reactivity: Extremely high missing OH reactivity ($>60\,s^{-1}$) that occurred during dry-season nights (no daytime observations available); medium missing OH reactivity ($<60\,s^{-1}$) during several days at the end of the September 2013 dry-season campaign; and low missing OH reactivity ($<20\,s^{-1}$) during the wet season (March 2013).

Dry season total OH reactivity correlated best with isoprene and its oxidation products (0.74 < Pearson correlation coefficient (Pearson correlation coefficient (PCC)) < 0.82) for both high and medium missing reactivity, whereas correlations were worse for the wet-season observations in times of low missing reactivity (0.49 < PCC < 0.50) (Table 2). For all periods, missing OH reactivity correlated not as well as the total OH reactivity with isoprene or its measured oxidation products (MVK + MACR + ISOPOOH). Missing OH reactivity during the wet season (low missing reactivity) was even anti-correlated to isoprene and its oxidation products (PCC = − 0.27).

The slope between total OH reactivity and isoprene OH reactivity equals 2.2 ± 0.1 for medium missing reactivity, which is similar to the value reported by Edwards *et al.*[13] for the Borneo tropical rainforest (measured at a height of ~5 m). In Amazon rainforest air, the slope is higher (7.5 ± 0.5) for high missing OH reactivity and lower (0.8 ± 0.04) for low missing OH reactivity during the wet season. Edwards *et al.*[13] contend that higher slopes (up to 10) indicate low physical losses or higher OH, whereas smaller slopes are caused by enhanced physical loss.

Whether the behaviour of the missing OH reactivity is more like biogenic forest emissions or atmospheric photooxidation

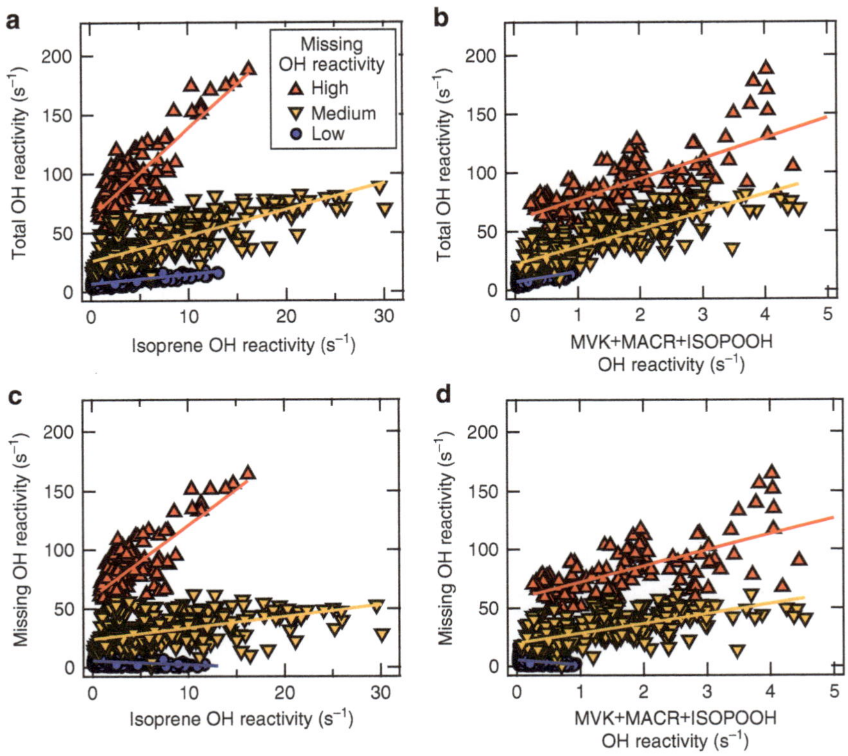

Figure 6 | Total or missing OH reactivity correlations with isoprene or MVK + MACR + ISOPOOH OH reactivity. For low (wet season), medium (dry season, $<60\,s^{-1}$) and high (dry season, $>60\,s^{-1}$) missing OH reactivity, total OH reactivity was correlated with isoprene and its oxidation product's MVK + MACR + ISOPOOH OH reactivity in (**a**) and (**b**), and missing OH reactivity was correlated with isoprene and MVK + MACR + ISOPOOH OH reactivity in (**c**) and (**d**).

Table 2 | Correlations between total or missing OH reactivity and isoprene or MVK + MACR + ISOPOOH OH reactivity

Total OH reactivity		Missing OH reactivity		Regimes of missing reactivity
Isoprene OH reactivity				
y = (7.5 ± 0.5)x + 63.4	PCC = 0.82	y = (6.1 ± 0.4)x + 60.2	PCC = 0.76	High
y = (2.2 ± 0.1)x + 26.0	PCC = 0.75	y = (1.0 ± 0.1)x + 23.6	PCC = 0.46	Medium
y = (0.8 ± 0.04)x + 6.8	PCC = 0.56	y = (− 0.3 ± 0.04)x + 5.3	PCC = − 0.27	Low
MVK + MACR + ISOPOOH OH reactivity				
y = (17.2 ± 1.4)x + 60.2	PCC = 0.74	y = (13.7 ± 1.4)x + 58.4	PCC = 0.69	High
y = (14.8 ± 0.6)x + 22.1	PCC = 0.79	y = (8.5 ± 0.6)x + 19.6	PCC = 0.62	Medium
y = (8.7 ± 0.6)x + 6.9	PCC = 0.49	y = (− 4.2 ± 0.6)x + 5.3	PCC = − 0.27	Low

Linear fit parameters and PCC for correlations between total or missing OH reactivity to isoprene or isoprene products (MVK + MACR + ISOPOOH) OH reactivity as pictured in were calculated for high (dry season, $>60\,s^{-1}$), medium (dry season, $<60\,s^{-1}$) and low (wet season) missing OH reactivity regimes.

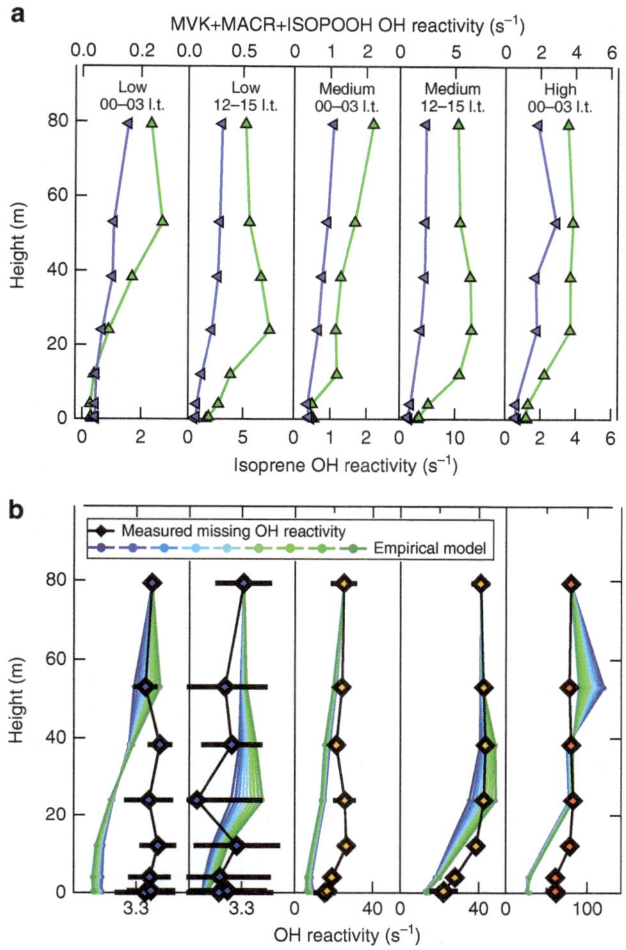

Figure 7 | Median vertical profiles for night and day during three regimes of missing OH reactivity. (**a**) Isoprene and isoprene products (MVK + MACR + ISOPOOH) OH reactivity. (**b**) Missing OH reactivity (markers with uncertainties ± 95 percentiles) and empirical model that is calculated as the sum of isoprene and isoprene products OH reactivity multitudes. To account for the missing OH reactivity in 79.3 m the mixture of typical biogenic emissions (isoprene) and typical photooxidation products (MVK + MACR + ISOPOOH) was varied in ratios from 90:10 to 10:90, and the resulting typical vertical profile (empirical model) was coloured green for an emission dominance and blue for a products dominance.

products is examined in Fig. 7. Median vertical profiles for isoprene, detected isoprene oxidation products and missing OH reactivity were calculated for night (00:00–03:00 l.t.) and day (12:00–15:00 l.t.) during low, medium and high missing reactivity periods. Similar to Fig. 4, it can be seen that MVK + MACR + ISOPOOH typically reached highest values at the top level (79.3 m) where oxidation, mixing and transport dominate the atmospheric composition, and that isoprene typical profiles build up maximum levels during daytime inside the canopy (24 m) following the diurnal cycle of light- and temperature-driven forest emissions (Fig. 7, top panel). In Fig. 7 (bottom panel), we apply an empirical model to elucidate missing OH reactivity. The entire missing OH reactivity determined at 79.3 m was assumed to consist of a mixture of unknown biogenic tree emissions and undetected oxidation products. For example, in the case of the dry-season daytime medium missing reactivity, the missing OH reactivity at the top level was 41 s^{-1}. If this missing OH reactivity is attributed to 50% emissions and 50% oxidation products, then the mixture would have had to be as large as two times the observed isoprene and almost nine times the MVK + MACR + ISOPOOH reactivities. Within the framework of this empirical model, the ratio of isoprene to isoprene oxidation products was varied (from 90:10 to 10:90) to account for the 79.3 m missing OH reactivity. This way, the typical vertical profiles of isoprene and MVK + MACR + ISOPOOH reactivities were scaled and summed, while varying between plausible extremes in primary versus products mixtures (Supplementary Note 2; Supplementary Table 2). Therefore, we can map out the span of possible vertical profiles: from being shaped like biogenic tree emissions (isoprene-like) to photooxidation products (MVK + MACR + ISOPOOH-like; Fig. 7, bottom panel).

Daytime medium missing OH reactivity vertical gradients were well reproduced in the empirical model by the mixture of biogenic emissions and photooxidation products inside and above the forest canopy. Below the canopy the model decreased faster towards the soil than the measured missing OH reactivity profile. This difference is depicted as a residual greater than zero in Supplementary Fig. 7. Close to the ground this gap persists for all analysed median profiles—independent of the season. Especially during nighttime, the typical profiles of tree emissions (isoprene) and (isoprene) photooxidation products fail to explain the vertical distribution of missing OH reactivity. In addition, during wet-season days, observed missing OH reactivity and model profiles show opposite tendencies, which is consistent with the wet-season anti-correlation between missing OH reactivity and isoprene or isoprene products (Fig. 6).

The assumption that the Amazon rainforest missing OH reactivity consists of isoprene-like and isoprene oxidation products-like compounds simplifies the question about the source of unknown and unmeasured OH sinks. Applying this assumption within a simple empirical model has revealed that the vertical distribution of missing OH reactivity cannot be explained by the sum of biogenic tree emissions peaking inside the canopy and photooxidation products formed above the canopy alone. Additional sources of OH reactive compounds need to be considered from lower down, potentially in the understory vegetation, the forest floor or soil.

Potential candidates for missing reactivity. If the average dry-season missing OH reactivity inside the canopy ($51.5 \pm 29.4\,s^{-1}$) was due to a single compound ($k_{OH+X} = 5 \times 10^{-11}\,cm^3\,molecule^{-1}\,s^{-1}$), then this compounds mixing ratio would need to be over 40 p.p.b.V. (parts per billion by volume). This mixing ratio is greater than that of isoprene and would not likely have remained undetected by our measurement systems. More likely is that a myriad of primary and secondary VOC are present at substantially lower mixing ratios.

Decaying plant material, soil or litter emissions can provide additional sources of VOC and thereby OH reactivity. Isidorov *et al.*[42] reported 77 species emitted from microbiologically decomposed leaf litter of 5 different deciduous trees. For 59 of those compounds a rate coefficient with OH is reported in the literature. Their average rate coefficient ($k_{OH+LITTER} = 3.22 \times 10^{-11}\,cm^3\,molecule^{-1}\,s^{-1}$) was applied for compounds with no literature value available. Excluding the compounds that were determined during our measurements and hence are already included in the OH reactivity budget, the calculated OH reactivity of typical litter emissions is 2.8 s^{-1}. This assumes that all of these compounds were present in the atmosphere with levels of 100 p.p.t.V. Clearly higher mixing ratios could be present for many of these species. Some samples of Amazon rainforest air that were analysed offline with a gas chromatograph-flame ionization detector (GC-FID) had hexane in mixing ratios up to 700 p.p.t.V. Hexane was part of the Isidorov *et al.*[42] list along with methylfuran, 2-butene and ethyl mercaptan, all of which are very reactive to OH. Likely, the tropical soil and leaf litter emitted a broader variety of compounds than the leaf litter of the five different deciduous trees examined in Isidorov *et al.*[42]. Overall, the tropical forest floor seems to be a potentially important but as yet insufficiently characterized source of OH reactivity.

Physical damage and herbivory elicits specific stress emissions from vegetation that can also contribute to reactivity, especially in the lower canopy and understory where light- and temperature-driven emissions decrease. These include benzenoid compounds that have been recently reported by Misztal *et al.*[43]. However, even assuming 100 p.p.t.V. of the 27 compounds by Misztal *et al.*[43] leads to only about $2\,s^{-1}$ OH reactivity. This calculation includes as an approximation for missing rate coefficients the average rate coefficient of the 14 known rate constants ($k_{OH+BENZENOID} = 2.67 \times 10^{-11}\,cm^3\,molecule^{-1}\,s^{-1}$). Benzene and toluene have been excluded as OH sinks because these two compounds were already part of the calculated OH reactivity. It has to be noted that the benzenoid compounds may represent only a fraction of stress-induced biogenic emissions, and that the photochemical products of these species are often more reactive than the emissions themselves[44].

Courtois *et al.*[45] measured 264 and identified 206 stress-related VOC from tropical forest in French Guiana using gas chromatographic-mass spectrometry (GC-MS). Methyl salicylate, hexanal, linalool and many highly reactive sesquiterpenes are part of this budget. The 33 known literature reaction rate coefficients with OH were averaged ($k_{OH+TROPICAL} = 8.5 \times 10^{-11}\,cm^3\,molecule^{-1}\,s^{-1}$) and used for all VOC without known rates. As monoterpenes were measured and already were part of the calculated OH reactivity, they were excluded at this point. For atmospheric levels of 100 p.p.t.V., we estimate the OH reactivity of $36.1\,s^{-1}$ that can be produced from these stress-related species. This is in the order of missing OH reactivity observed in the dry-season Amazon rainforest.

In fact, there are thousands, if not tens of thousands of such reactive organic compounds in the atmosphere[46] emitted from the biosphere and oxidized to numerous products in the air. Many of these compounds are present in relatively low concentrations, below the detection limit of many instruments, or not detected by currently available field equipment. However, their summed effect is visible in the total OH reactivity, as ozone production potential, or as a trigger for aerosol formation and growth.

Discussion

Important driving forces of total OH reactivity at the ATTO site were identified to be of biological (BVOC emissions), chemical (oxidation of BVOC) and physical (deposition, turbulence and thermal stratification) nature.

The OH chemical lifetime inside the canopy varied enormously (by a factor of 10), between wet and dry season (Fig. 2). At the time of the shortest local, chemical OH lifetime, during dry season, a large fraction of the total OH reactivity remained unaccounted for (on average 79% at 24 m) despite isoprene and its main photooxidation products (MVK + MACR + ISOPOOH) being measured. In that season a good match between isoprene and the total OH reactivity vertical profiles above the canopy was observed (Fig. 4). The missing OH reactivity correlated well with both isoprene and its oxidation products in September 2013 (Fig. 6). Accordingly, the daytime missing OH reactivity gradient could be mimicked well by combining the typical biogenic emission profile of isoprene and the typical atmospheric oxidation product of isoprene photooxidation products (Fig. 7, medium missing reactivity). At those times, the missing OH reactivity was likely dominated by an as yet undefined mixture of biogenic emissions and photooxidation products. Potential candidates for these unmeasured species are forest emissions triggered by temperature and light or by mechanical and herbivoral stress, such as green leaf volatiles (C6 aldehydes and ketones), benzenoid compounds[43], multifunctional entities containing an alkene bond (for example, hexenal, hexenol)[5,6] or sesquiterpenes in very low levels. In addition, several product families of VOC photooxidation[13,47,48], such as peroxides, aldehydes (for example, formaldehyde) and epoxides, that are challenging to determine during field campaigns, were not measured at the ATTO site.

For the time of the longest local, chemical OH lifetime, during wet season, missing OH reactivity was smallest. Vertical gradients of total OH reactivity matched those of photooxidation products (Fig. 4). The daytime total OH reacitivity within the canopy was mostly explained by co-measured OH sinks. However, the remaining missing OH reactivity gradient could not be accounted for by any combination of biogenic emissions and oxidation products (Figs 6 and 7), especially during nighttime and below the canopy. Overall, the less oxidative atmosphere, the relatively high deposition of oxidation products on wet surfaces[49] and possible enhanced leaf surface microbial activity[40,41] helped to close the total OH reactivity budget in the wet season.

Throughout all seasons the total OH reactivity was lowest at the forest floor, but interestingly the average missing OH reactivity was always high there (Fig. 5). The vertical gradients of missing OH reactivity in the lowermost layers of the rainforest could not be explained by the vertical distribution of biogenic tree emissions (isoprene-like), or by the profile of oxidation products (isoprene products-like), or by a combination of both (Fig. 7). A thermal inversion below the forest canopy hindered mixing between canopy and soil level (Fig. 3). Hence, soil, litter and understory vegetation likely affected the budget of reactive compounds in this lowermost layer through emission and deposition[5,50]. Compounds such as nitrous acid (HONO)[51] and some VOC (for example, organic acids and multifunctional carbonyls)[5,50] emitted as a result of microbial activities in soils and leaf litter were not measured in this study, but likely also contributed to the Amazon rainforest total OH reactivity near the forest floor.

The seasonal, diel and vertical variations of total and unaccounted for air reactivity have important implications and research challenges for several adjacent fields such as plant–plant and plant–insect interactions, microbiology, soil science, biogeochemistry and carbon cycling. High reactivity and therefore short OH lifetime can impact the effectiveness of insect pollination or the timing of windborne pollen dispersal. Microbial communities in soils and on leaves may play a very active, but as yet poorly understood role in regulating total OH reactivity at the biosphere–atmosphere interface. Their effect on atmospheric composition and oxidation potential was exposed in this study as a missing OH reactivity at soil level and postulated on leaf surfaces based on the wet-season data (negative total OH reactivity vertical gradients and minimum missing reactivity). Long-term observations of Amazon rainforest OH reactivity of the kind shown here also have the potential to reveal how the natural forest–atmosphere interface changes when under the influence of biomass burning, anthropogenic pollution or even El Niño events. Total OH reactivity can also be considered as a measure for carbon released by the tropical vegetation or soils weighted by its reaction potential with OH. Especially during the dry season the amount of OH reactive carbon in the Amazon rainforest air was substantial presenting a new challenge to carbon cycling models.

This comprehensive characterization of tropical rainforest total OH reactivity was achieved by the high temporal and vertical resolution of the total OH reactivity observations. The vertical gradients spanning eight measurement levels from the forest floor up to almost 80 m into the atmosphere have revealed unexpected variations in seasonal, diel and vertical total OH reactivity distribution, composition and strength, as well as a hitherto overlooked influence of the forest floor on OH reactivity.

Methods

Details about the ATTO measurement site. ATTO[25] is embedded in undisturbed old-grown rainforest. The rainforest upwind of the site as well as in close distance is a species rich 'terra firme' forest. Highest tree species richness was observed at the plateau adjacent to the 80 m tower. Here, over 400 species have been identified[25]. The average tree height was about 38 m. Observations from the 80 m walk-up tower of the leaf area index (LAI) showed a high variability with height and time. On average the LAI was 5.5 with timely variable biomass at different levels inside the canopy or in the understory (monitored with LICOR 2200, personal communication Giordane Martins). The understory surrounding the measurement tower is rich in small, young, growing trees, palms and ferns. Mosses and lichen frequently cover stems and leaves of old-grown trees. Leaf litter and other decaying plant material such as branches or fine roots cover densely the clumpy, red soil (Supplementary Fig. 1). Detailed information about forest phenology and soil characteristics at the ATTO site can be found in Andreae *et al.*[25].

Since 2012, continuous meteorological observations are available for the ATTO site. The near-equatorial location defines two characteristic seasons depending on the inter-tropical convergence zone (ITCZ) and its associated wind systems[26]. As the ITCZ moves to the north, the main wind direction is east and southeast during the dry season (July–November). During this time potential influence from biomass burning or human activities from eastern Brazil may be observed at the site. As the ITCZ moves to the South, wet-season winds are directed from northeast (January–March). At that time, air masses pass over long distances of pristine and undisturbed rainforest before arriving at the ATTO site.

Continuously monitored are also aerosol parameters (that is, black carbon, aerosol scattering, particle number concentration, aerosol number size distribution, particle mass and chemical speciation), greenhouse gases (CO_2, CH_4) and other trace gases (CO, NO, NO_2, O_3). The VOC and total OH reactivity observations have been performed during intensive field measurement campaigns.

Vertical gradient set-up. At the ATTO site an 80-m walk-up (INSTANT)[25] tower was installed hosting an inlet system that allowed sampling air from eight different heights (0.05, 0.5, 4, 12, 24, 38.3, 53 and 79.3 m). The three lowermost inlets (0.05, 0.5 and 4 m) were mounted on a tripod in close proximity (12 m) to the 80-m walk-up tower but at a more undisturbed location inside the forest. The inlet lines were made of 3/8″ (0.95 cm) diameter non-transparent PTFE tubing, they were heated, insolated and continuously flushed with a high flow. At the top of each line, the sample was passed through a regularly renewed PTFE filter (5 μm) to avoid impact of aerosol and insects. For sampling, a Teflon pump drew down air from all inlets with a flow rate of $\sim 16\,\mathrm{l\,min^{-1}}$.

Every 2 min a Teflon valve system switched to the next level, completing one entire profile from 0.05 to 79.3 m in 16 min. With this inlet system, vertical gradients of ozone (O_3), nitric oxide (NO), nitrogen dioxide (NO_2), carbon dioxide (CO_2), VOC and total OH reactivity were monitored simultaneously. The equipment was hosted in a temperature-controlled laboratory container adjacent to the 80-m walk-up tower. The instruments were subsampling from the gradient system via short and insolated Teflon lines.

Inlet line tests. The inlet lines of the gradient system have been tested for losses and variable response that could result from the different length of the tubing. The VOC calibration standard gas was used to spike into the lines while sampling ambient air at ground level (0.5 m), just below the canopy (12 m) and the top level (79.3 m). The typical 2 min switch cycle of the gradient measurements was kept and the 'spiked' level was compared with the previous and next measurement point at the same height. The concentration levels of the spikes were chosen high enough to stay above ambient levels (12–16 p.p.b.V.). On average the line losses were about 9% varying between the different VOC. Most importantly we observed no significant differences between the three tested inlet lines.

Since the losses were minor, the inlet line response independent of the length and both instruments sampled in parallel from the same lines, no corrections were applied and the original data are presented.

Total OH reactivity measurements. The total OH reactivity was directly measured using the CRM[21,27]. This method is based on a competitive reaction of OH with a reagent which is not usually present in ambient air, and OH with the atmospheric OH reactive compounds. We used an automated set-up for the CRM and a PTR-MS (operated in standard conditions: 2.2 mbar drift pressure, 600 V drift voltage and 127 Td) as detector for the reagent pyrrole (C_4H_5N) as described in Nölscher *et al.*[21] Pyrrole was mixed inside a Teflon-coated glass reaction cell with zero air (dry) and nitrogen (dry) to obtain the initial concentration level of pyrrole [C1]. Then, nitrogen was humidified and flushed over a ultraviolet lamp to generate OH molecules that react inside the glass reactor with pyrrole. The detected pyrrole level [C2] dropped according to the generated amount of OH inside the reactor. During this step catalytically scrubbed air (Platinum catalytic converter, heated >500 °C) was used instead of dry synthetic air for pyrrole dilution to match and follow ambient humidity. Next, the catalytically scrubbed air was exchanged with ambient air, atmospheric OH reactive compounds entered the reactor, and competed with pyrrole for the available OH molecules. The resulting increase in pyrrole levels was detected via the PTR-MS as concentration level [C3]. These three monitored pyrrole levels were used to calculate the atmospheric total OH reactivity via the following equation.

$$R_{\mathrm{air}} = \mathrm{C1} \times k_{\mathrm{OH+PYR}} \frac{\mathrm{C3}-\mathrm{C2}}{\mathrm{C1}-\mathrm{C3}} \quad (2)$$

The reaction rate coefficient for pyrrole with OH has recently been determined in dependency of temperature and pressure by Dillon *et al.*[52] According to the typical temperatures and pressure inside the glass reactor a value of $k_{\mathrm{OH+PYR}} = 1.2 \times 10^{-10}\,\mathrm{cm^3\,molecule^{-1}\,s^{-1}}$ was applied for analysis.

Typically the instrument was operated with initial pyrrole mixing ratios about 50–60 p.p.b.V. [C1] and OH mixing ratios of about 35–45 p.p.b.V. Errors in the detector (5%), rate coefficient (14%), gas standard (5%) and dilution (2%) lead to an overall uncertainty of 16%.

For the measurements from the vertical gradient system, the CRM was programmed to switch automatically between [C2] and [C3] in an hourly cycle (48 + 12 min). This allowed conducting total OH reactivity measurements for the first three gradients (each 16 min) and a background [C2] measurement for the last 12 min of the hour. During the [C2] measurements the gradient system switched between the different levels with a higher rate (1:30 min per height). The [C1] parts were measured regularly every other day.

For each measurement height 2 min averages were calculated from the 9 s PTR-MS raw data and the s.d. was given as uncertainty. The detection limit during the four Amazon rainforest measurement campaigns was about $3\text{–}4\,\mathrm{s^{-1}}$ (2σ of background [C2] noise). Throughout all campaigns the linear response of the CRM has been tested using an isoprene gas standard. The instrument showed excellent linearity over the wide range of tested OH reactivities and its response was close to 1 (Supplementary Fig. 8).

Total OH reactivity data analysis. The procedures of data evaluation, corrections and analysis for the total OH reactivity measurements using CRM were previously described in detail[21,27,28]. Here we provide an overview of the total OH reactivity analysis for the presented Amazon rainforest observations.

Humidity dependence of PTR-MS sensitivity. The instrumental response of the PTR-MS to pyrrole is dependent on humidity[53]. Therefore separate calibrations for pyrrole in dry and humid conditions were performed. The dry calibration factor was applied to dry measurements, that is, during level [C1]. The wet calibration factor was applied to all wet measurements, that is, the [C2] and the [C3] parts. It has been observed that for relative humidity >30% the sensitivity of the PTR-MS to pyrrole did not vary. Since nitrogen is humidified for OH generation and the humid ambient sample is added to the mixture inside the glass reactor, our measurements during [C2] and [C3] always stay above this threshold, which allows working with one calibration factor for all humid parts.

Humidity dependence of OH generation. Although catalytically scrubbed air was used for the [C2] parts, humidity potentially varied during and between the [C2] and [C3] measurements. As the OH concentration inside the reactor strongly depends on the given humidity, a correction for these changes in the humidity was applied. For example, sections with higher humidity would have higher OH concentrations, and a smaller step between [C2] and [C3], hence the true total OH reactivity would be underestimated. A calibration with different known humidity levels in accordance to the PTR-MS water cluster (*m*/*z* 37) was performed and the data (during [C3]) accordingly corrected. This correction was especially critical, because the different levels of the 80 m profile experienced variable relative humidity.

The background [C2] was measured every hour, while the gradient system switched between the eight different levels. During that time, differences of the humidity at individual levels could be used as calibration since OH concentrations inside the CRM reactor varied accordingly. In extreme cases the humidity-dependent variability of OH concentration inside the CRM reactor caused a change in [C2] pyrrole levels of about 2 p.p.b.V. This was also the range of maximum [C2] drift over the time of the day that was mainly due to the ambient humidity variations.

Accounting for not operating in pseudo-first-order conditions. The calculation using the differences between the three pyrrole levels is based on the assumption of pseudo-first-order reaction conditions. This applies if pyrrole molecules are in excess in comparison to the generated OH. However, to provide detectable differences inside the reactor pyrrole and OH have been adjusted to the same order of magnitude (Pyr/OH ≈ 1.5–2.5). A chemical model (Facsimile) helped to calculate theoretically expected pyrrole variations for the given experimental conditions. A comparison between the true total OH reactivity and the OH reactivity calculated using the model results within the above-mentioned equation provided a correction which was for the given conditions about 4–7% of the total measured OH reactivity.

Dilution. Finally, we had to account for the dilution of the sampled air. The addition of humidified nitrogen and pyrrole diluted the sample. The exact flow rates of all introduced gases have been monitored regularly and served to calculate a dilution factor.

Possible interferences. A known interference for the CRM is the recycling of OH within the reactor when high NO concentrations are apparent. Our instrument has shown that the reaction of ambient NO with the HO_2 produced inside the reactor significantly impacts the results for total OH reactivity measurements when ambient NO levels are higher than 10 p.p.b.V. (ref. 27). However, during the Amazon rainforest measurement campaigns highest mixing ratios of NO reached about 1 p.p.b.V. at the soil level, hence no such interference was expected.

Detection of VOC. The quadrupole PTR-MS(Ionicon, Austria) is capable of detecting gas-phase species with a higher proton affinity than water. This is the case for many VOC but not for molecular nitrogen, oxygen and argon. The molecules of interest are protonated through the reaction with H_3O^+, accelerated through a drift tube (2.2 mbar, 600 V, 127 Td) and their mass to charge ratio determined using a quadrupole mass spectrometer. Details about the measurement principal can be found elsewhere[29,54].

The PTR-MS was calibrated for different humidity levels with a standard gas mixture (Apel-Riemer Environmental, 2013) for the following VOC: methanol (detected as protonated mass-to-charge ratio *m*/*z* 33), acetonitrile (*m*/*z* 42), acetaldehyde (*m*/*z* 45), acetone (*m*/*z* 59), dimethyl sulphide (DMS, *m*/*z* 63), isoprene (*m*/*z* 69), methyl vinyl ketone (MVK, *m*/*z* 71), methacrolein (MACR, *m*/*z* 71), methyl ethyl ketone (*m*/*z* 73), benzene (*m*/*z* 79), toluene (*m*/*z* 93), *o*-xylene (*m*/*z* 107), α-pinene (*m*/*z* 81 and *m*/*z* 137). For regular background measurements with VOC free air a catalytic converter (Platinum pellets, 400 °C) was automatically switched into the sampling line every hour.

The overall uncertainty of the VOC measurements was about 6–30%, the detection limit about 0.01–1 p.p.b.V., both strongly depending on the detected compound[29].

References

1. Food and Agriculture Organization. of the United Nations and International Tropical Timber Organization. *The State of Forests in the Amazon Basin, Congo Basin and Southeast Asia.* ISBN: 9789251068885 (2011).
2. Davidson, E. A. *et al.* The Amazon basin in transition. *Nature* **481**, 321–328 (2012).
3. Ter Steege, H. *et al.* Hyperdominance in the Amazonian Tree Flora. *Science* **342**, 6156 (2013).
4. Mahli, Y. *et al.* The regional variation of aboveground live biomass in old-growth Amazonian forests. *Global Change Biol.* **12**, 1107–1138 (2006).
5. Kesselmeier, J. & Staudt, M. Biogenic volatile organic compounds (VOC): an overview on emission, physiology and ecology. *J. Atmos. Chem.* **33**, 23–88 (1999).
6. Laothawornkitkul, J., Taylor, J. E., Paul, N. D. & Hewitt, C. N. Biogenic volatile organic compounds in the Earth system. *New Phytol.* **183**, 27–51 (2009).
7. Rinne, H. J. I., Guenther, A. B., Greenberg, J. P. & Harley, P. C. Isoprene and monoterpene fluxes measured above Amazonian rainforest and their dependence on light and temperature. *Atmos. Environ.* **36**, 2421–2426 (2002).
8. Jardine, K. J. *et al.* Within-plant isoprene oxidation confirmed by direct emissions of oxidation products methyl vinyl ketone and methacrolein. *Global Change Biol.* **18**, D19301 (2011).
9. Rottenberger, S. *et al.* Exchange of short-chain aldehydes between Amazonian vegetation and the atmosphere. *Ecol. Appl.* **14**, S247–S262 (2004).
10. Kesselmeier, J. *et al.* Concentrations and species composition of atmospheric volatile organic compounds (VOCs) as observed during the wet and dry season in Rondonia (Amazonia). *J. Geophys. Res.* **107**, D20 (2002).
11. Lelieveld, J. *et al.* Atmospheric oxidation capacity sustained by a tropical forest. *Nature* **452**, 737–740 (2008).
12. Karl, T. *et al.* Rapid formation of isoprene photo-oxidation products observed in Amazonia. *Atmos. Chem. Phys.* **9**, 7753–7767 (2009).
13. Edwards, P. M. *et al.* OH reactivity in a South East Asian tropical rainforest during the Oxidant and Particle Photochemical Processes (OP3) project. *Atmos. Chem. Phys.* **13**, 9497–9514 (2013).
14. Karl, T. *et al.* Efficient atmospheric cleansing of oxidized organic trace gases by vegetation. *Science* **330**, 816 (2010).
15. Park, J.-H. *et al.* Active atmosphere-ecosystem exchange of the vast majority of detected volatile organic compounds. *Science* **341**, 643–647 (2013).
16. Kuhn, U. *et al.* Isoprene and monoterpene fluxes from Central Amazonian rainforest inferred from tower-based and airborne measurements, and implications on the atmospheric chemistry and the local carbon budget. *Atmos. Chem. Phys.* **7**, 2855–2879 (2007).
17. Pöschl, U. *et al.* Rainforest aerosols as biogenic nuclei of clouds and precipitation in the Amazon. *Science* **329**, 1513–1516 (2010).
18. Ehn, M. *et al.* A large source of low-volatility secondary organic aerosol. *Nature* **506**, 476–479 (2014).
19. Working Group I Contribution to the Fifth Assessment Report of the Intergovernmental Panel on Climate Change. *Climate Change 2013: The Physical Science Basis.* ISBN: 9781107661820 (Cambridge University Press, 2013).
20. Di Carlo, P. *et al.* Missing OH reactivity in a forest: evidence for unknown reactive biogenic VOCs. *Science* **304**, 722–725 (2010).
21. Nölscher, A. C. *et al.* Summertime total OH reactivity measurements from boreal forest during HUMPPA-COPEC 2010. *Atmos. Chem. Phys.* **12**, 8257–8270 (2012).
22. Nölscher, A. C. *et al.* Seasonal measurements of total OH reactivity emission rates from Norway spruce in 2011. *Biogeosciences* **10**, 4241–4257 (2013).
23. Kim, S., Guenther, A. B., Karl, T. & Greenberg, J. P. Contributions of primary and secondary biogenic VOC to total OH reactivity during the CABINEX (Community Atmosphere-Biosphere INteractions Experiments)-09 field campaign. *Atmos. Chem. Phys.* **11**, 8613–8623 (2011).
24. Hansen, R. F. *et al.* Measurements of total hydroxyl radical reactivity during CABINEX 2009—Part 1: field measurements. *Atmos. Chem. Phys.* **14**, 2923–2937 (2014).
25. Andreae, M. O. *et al.* The Amazon Tall Tower Observatory (ATTO) in the remote Amazon Basin: overview of first results from ecosystem ecology, meteorology, trace gas, and aerosol measurements. *Atmos. Chem. Phys.* **15**, 10723–10776 (2015).
26. McGregor, G. R. & Nieuwolt, S. *Tropical Climatology* (John Wiley & Sons Ltd, 1977).
27. Sinha, V., Williams, J., Crowley, J. N. & Lelieveld, J. The comparative reactivity method—a new tool to measure total OH reactivity in ambient air. *Atmos. Chem. Phys.* **8**, 2213–2227 (2008).
28. Nölscher, A. C., Sinha, V., Bockisch, S., Klüpfel, T. & Williams, J. Total OH reactivity measurements using a new fast Gas Chromatographic Photo-Ionization Detector (GC-PID). *Atmos. Meas. Tech.* **5**, 2981–2992 (2012).
29. Yañez-Serrano, A. M. *et al.* Diel and seasonal changes of biogenic volatile organic compounds concentrations in and above the Amazon rainforest. *Atmos. Chem. Phys.* **15**, 3359–3378 (2015).
30. Sinha, V. *et al.* OH reactivity measurements within a boreal forest: evidence for unknown reactive emissions. *Environ. Sci. Technol.* **44**, 6614–6620 (2010).
31. Zannoni, N. *et al.* OH reactivity and concentrations of biogenic volatile organic compounds in a Mediterranean forest of downy oak trees. *Atmos. Chem. Phys. Discuss* **15**, 22047–22095 (2015).
32. Shirley, T. R. *et al.* Atmospheric oxidation in the Mexico City Metropolitan Area (MCMA) during April 2003. *Atmos. Chem. Phys.* **6**, 2753–2765 (2006).
33. Guenther, A. B., Monson, R. K. & Fall, R. Isoprene and monoterpene emission rate variability: model evaluations and sensitivity analyses. *J. Geophys. Res.* **98**, 12609–12617 (1993).
34. Vilà-Guerau de Arellano, J. *et al.* The role of boundary layer dynamics on the diurnal evolution of isoprene and the hydroxyl radical over tropical forests. *J. Geophys. Res.* **116**, D07304 (2011).
35. Zimmermann, P. R. *et al.* Measurements of atmospheric hydrocarbons and biogenic emission fluxes in the amazon boundary layer. *J. Geophys. Res.* **93**, 1407–1416 (1988).
36. Liu, Y. J., Herdlinger-Blatt, I., McKinney, K. A. & Martin, S. T. Production of methyl vinyl ketone and methacrolein via the hydroperoxyl pathway of isoprene photooxidation. *Atmos. Chem. Phys.* **13**, 5715–5730 (2013).
37. Rivera-Rios, J. C. *et al.* Conversion of hydroperoxides to carbonyls in field and laboratory instrumentation: observational bias in diagnosing pristine versus anthropogenically controlled atmospheric chemistry. *Geophys. Res. Lett.* **41** (2014).
38. Tezara, W. *et al.* Water relations and photosynthetic capacity of two species of Clotropis in a tropical semi-arid ecosystem. *Ann. Bot.* **107**, 397–405 (2011).
39. Myneni, R. B. *et al.* Large seasonal swings in leaf area of Amazon rainforests. *PNAS* **104**, 4820–4823 (2007).
40. Peñuelas, J. *et al.* Removal of floral microbiota reduces floral terpene emissions. *Sci. Rep* **4**, 6727 (2014).
41. Lindow, S. E. & Brandl, M. T. Microbiology of the Phyllosphere. *Appl. Environ. Microbiol.* **69**, 1875–1883 (2003).
42. Isidorov, V. & Jdanova, M. Volatile organic compounds from leaves litter. *Chemosphere* **48**, 975–979 (2002).
43. Misztal, P. K. *et al.* Atmospheric benzenoid emissions from plants rival those from fossil fuels. *Sci. Rep.* **5**, 12064 (2015).
44. Atkinson, R. & Arey, J. Atmospheric degradation of volatile organic compounds. *Chem. Rev.* **103**, 4605–4638 (2003).
45. Courtois, E. A. *et al.* Diversity of volatile organic compounds emitted by 55 species of tropical trees: a survey in French Guiana. *J. Chem. Ecol.* **35**, 1349–1362 (2009).
46. Goldstein, A. H. & Galbally, I. E. Known and unexplored organic constituents in the Earth's atmosphere. *Environ. Sci. Technol.* **41**, 1514–1521 (2007).
47. Paulot, F. *et al.* Unexpected epoxide formation in the gas-phase photooxidation of isoprene. *Science* **325**, 730–733 (2009).

48. Wolfe, G. M. *et al.* Photolysis, OH reactivity and ozone reactivity of a proxy for isoprene-derived hydroperoxynals (HPALDs). *Phys. Chem. Chem. Phys.* **14,** 7276–7286 (2012).
49. Langford, B. *et al.* Fluxes and concentrations of volatile organic compounds from a Southeast Asian tropical rainforest. *Atmos. Chem. Phys.* **10,** 8391–8412 (2010).
50. Veres, P. R., Behrendt, T., Klapthor, A., Meixner, F. X. & Williams, J. Volatile organic compound emissions from soil: using Proton-Transfer-Reaction Time-of-Flight Mass Spectrometry (PTR-TOF-MS) for the real time observation of microbial processes. *Biogeosci. Discuss.* **11,** 12009–12038 (2014).
51. Oswald, R. *et al.* HONO emissions from soil bacteria as a major source of atmospheric reactive nitrogen. *Science* **341,** 1233–1235 (2013).
52. Dillon, T. *et al.* Reaction of hydroxyl radicals with C4H5N (pyrrole): temperature and pressure dependent rate coefficients. *J. Phys. Chem. A* **116,** 6051–6058 (2012).
53. Sinha, V., Custer, T. G., Klüpfel, T. & Williams, J. The effect of relative humidity on the detection of pyrrole by PTR-MS for OH reactivity measurements. *Intern. J. Mass. Spectr.* **282,** 108–111 (2010).
54. Lindinger, W., Hansel, A. & Jordan, A. On-line monitoring of volatile organic compounds at pptv levels by means of proton-transfer-reaction mass spectrometry (PTR-MS): Medical applications, food control and environmental research. *Int. J. Mass. Spectrom. Ion Processes* **173,** 191–241 (1998).

Acknowledgements

We thank the Max Planck Society and the Instituto Nacional de Pesquisas da Amazonia for continuous support. We acknowledge the support by the German Federal Ministry of Education and Research (BMBF contract 01LB1001A) and the Brazilian Ministério da Ciência, Tecnologia e Inovação (MCTI/FINEP contract 01.11.01248.00) as well as the Amazon State University (UEA), FAPEAM, LBA/INPA and SDS/CEUC/RDS-Uatumã. Particularly acknowledged are the contributions for technical and logistical support (Reiner Ditz, Hermes Braga Xavier, Nina Maria Knothe). Thanks to Guenther Schebeske for the GC-FID analysis, to Thomas Klüpfel for PTR-MS related help, and to Jan Winderlich and Leonardo Ramos de Oliveira for contributing to CO measurements. We greatly appreciate helpful discussions with J.N. Crowley.

Author contributions

A.C.N. conducted the total OH reactivity measurements, analysis and wrote the manuscript. A.M.Y.S. was in charge of the VOC PTR-MS observations and analysis. All the work was achieved under the supervision of J.W. and J.K. who helped in the scientific analysis and the manuscript writing process. S.W. provided the gradient system and NOx and Ozone data. J.V.L. provided CO data and A.C.A. was in charge of the micrometeorological observations.

Additional information

Competing financial interests: The authors declare no competing financial interests.

Mid Pleistocene foraminiferal mass extinction coupled with phytoplankton evolution

Sev Kender[1,2], Erin L. McClymont[3], Aurora C. Elmore[3,4], Dario Emanuele[5], Melanie J. Leng[1,2] & Henry Elderfield[4,‡]

Understanding the interaction between climate and biotic evolution is crucial for deciphering the sensitivity of life. An enigmatic mass extinction occurred in the deep oceans during the Mid Pleistocene, with a loss of over 100 species (20%) of sea floor calcareous foraminifera. An evolutionarily conservative group, benthic foraminifera often comprise >50% of eukaryote biomass on the deep-ocean floor. Here we test extinction hypotheses (temperature, corrosiveness and productivity) in the Tasman Sea, using geochemistry and micropalaeontology, and find evidence from several globally distributed sites that the extinction was caused by a change in phytoplankton food source. Coccolithophore evolution may have enhanced the seasonal 'bloom' nature of primary productivity and fundamentally shifted it towards a more intra-annually variable state at ~0.8 Ma. Our results highlight intra-annual variability as a potential new consideration for Mid Pleistocene global biogeochemical climate models, and imply that deep-sea biota may be sensitive to future changes in productivity.

[1]Centre for Environmental Geochemistry, School of Geography, University of Nottingham, Nottingham NG7 2RD, UK. [2]British Geological Survey, Keyworth, Nottingham NG12 5GG, UK. [3]Department of Geography, Durham University, Durham DH1 3LE, UK. [4]Department of Earth Sciences, University of Cambridge, Cambridge CB2 3EQ, UK. [5]DST—Università del Sannio, Via dei Mulini 59a, Benevento 82100, Italy. Correspondence and requests for materials should be addressed to S.K. (email: sev.kender@nottingham.ac.uk) or to (sesev@bgs.ac.uk).
‡Deceased

High-resolution continuous records, such as those found in the marine realm, provide an outstanding opportunity to pair fossil occurrences with geochemical environmental proxies to examine the possible causes of mass extinctions. The Mid Pleistocene transition (MPT, ~1.2–0.6 Ma) was characterized by global cooling, glacial stage lengthening, changing ocean circulation and evolution of terrestrial and marine biota[1–3]. During the MPT, a stepwise extinction of over 100 species of deep-water benthic foraminifera occurred, targeted to a specific morphological group (largely the elongated 'stilostomellids' with ornamented apertures, referred to here as the 'extinction group') within just three families[4–7]. Benthic foraminifera are an important group often comprising >50% of the total benthic eukaryote biomass on the deep-ocean floor[8]. The cause of this last mass extinction, their only significant extinction in the last 15 Myr, is still unknown[6,7,9–11]. The deep ocean is the largest and most stable habitat on Earth[10], and the only other two mass extinctions of benthic foraminifera in the Cenozoic occurred over the Palaeocene–Eocene and Eocene–Oligocene boundaries[10]. Understanding the ecological sensitivity of the deep sea is important as it is a habitat increasingly under stress[12], and may be sensitive to current and future changes in productivity[13]. This last mass extinction in benthic foraminifera, completed by ~0.8–0.6 Ma, has been extensively studied and shown to be global in extent, although slightly diachronous and apparently stepwise during glacials[4,7,11,14]. Higher latitude localities (above ~60°) have not been studied in detail[7], but Bering Sea[15] and Kerguelen Plateau Southern Ocean[7] foraminiferal assemblages show very low abundances of the extinction group before the MPT. A lack of living representatives has hampered our understanding of their ecological preferences and therefore the cause of the extinction. Leading hypotheses include the following: changing deep-water physical properties, such as cooling and increased oxygenation[6,7,10,14]; a change in productivity or extinction of a particular phytoplankton food source[7,11]; and increased 'seasonality' or annual irregularity of organic carbon export[9–11]. Although bottom-water temperature, oxygenation and corrosiveness have been challenged as the cause, partly because of a global uniform change being unlikely[7,10], these properties are yet to be directly tested with environmental parameters reconstructed alongside extinction group data[7].

In this study, we address these hypotheses by examining the first records of extinction group abundance generated together with proxies for bottom-water temperature, carbonate saturation ($\Delta[CO_3^{2-}]$, related to corrosiveness), sea-surface temperature (SST) and productivity. We also generated records of phytoplankton assemblages from two sites to compare with a published South Pacific record[11]. Here we provide direct evidence from benthic foraminiferal Mg/Ca which disproves the hypothesis that changing bottom-water temperature controlled foraminiferal decline[6]. However, we do find compelling micropalaeontological evidence for a change in phytoplankton food source as the cause of the extinction. We show that a global dominance of the nannoplankton 'small' *Gephyrocapsa* occurred at ~0.8 Ma, coinciding with the largest decline in benthic foraminifera. We conclude that small *Gephyrocapsa* evolution may have enhanced intra-annual variability of primary productivity and carbon flux reaching much of the mid-low latitude ocean floor where the extinction was centred, and that the extinction group may have been unable to cope with long periods of seasonally reduced organic carbon flux during parts of the year.

Results

Benthic extinction in the Tasman Sea. We reconstructed environmental parameters using a wide range of proxies at Deep Sea Drilling Project (DSDP) Site 593 over the period ~0.75–1 Ma, covering the benthic extinction (Fig. 1). Site 593 is situated in 1,063 m of water on the Challenger Plateau of the Tasman Sea (SW Pacific Ocean, Supplementary Fig. 1), lying to the north of the modern Subtropical Front, which is a complex zone delineated by large gradients in SST and salinity[16]. SSTs in the Tasman Sea are considered to be sensitive to glacial–interglacial displacement of the Subtropical Front[7,17]. Site 593 is bathed by Antarctic Intermediate Water, which is broadly characterized by low salinity (34.3–34.5 PSU), low temperatures (3.5–10 °C) and high dissolved oxygen (200–250 μmoles kg^{-1}; refs 18,19).

Previous low-resolution (~22 kyr) benthic foraminiferal data from Site 593 (ref. 5) indicated that the extinction occurred between 0.8 and 0.9 Ma. Our higher-resolution (~3.5 kyr) results indicate the extinction group declined in the Tasman Sea, irrespective of size fraction, throughout the study interval with an initial overall reduction in abundance at ~0.95 Ma, and a major decline towards very low abundance at ~0.83 Ma (Fig. 1g). Our data are consistent with other studies that conclude that the architecture of the extinction is captured in all size fractions[7], and suggest that it was not associated with an early shortening of the life cycle, which might be apparent with an increased proportion of small specimens. Sites below ~1 km water depth typically have the highest extinction group diversity[7], and species richness at Site 593 is relatively low at ~10–12 species per ~10-cc sample (Fig. 1h). Site 593 is dominated by *Strictocostella scharbergana* and *Siphonodosaria lepidula*, which decline abruptly at 0.85 and 0.8 Ma, respectively (Fig. 1i,j). Different species are dominant in extinction assemblages in the Pacific, Atlantic and Mediterranean, although they are morphologically (and thus possibly ecologically) related and also became extinct during the MPT[7].

Bottom-water temperature and corrosiveness. Decreased intermediate/deep-water temperature is a hypothesized cause of the extinction[6], possibly due to increased oxygenation and its impact on an inferred microbial food source[6,10,14]. Bottom-water temperature at intermediate-depth Site 593, reconstructed from Mg/Ca of infaunal *Uvigerina peregrina* (see Methods), ranges from ~3 to 8 °C (modern temperature is 4.5 °C), with warmer interglacials (Fig. 1e). Seawater absolute magnesium (17 Ma residence time) and calcium (1 Ma residence time) concentrations would have been slightly different during the Mid Pleistocene[20], thus having an impact on the accuracy of the temperature estimates based on modern calibration (possibly by up to ~1 °C), although the overall trends will be unchanged. There is no apparent correlation between bottom-water temperature and faunal abundances during the pre-extinction period before ~0.83 Ma (Fig. 2a), nor any secular change over the extinction itself, and we conclude that the benthic extinction at Site 593 was not caused by temperature changes. Increased bottom-water corrosiveness is another physical property that has been proposed to have an impact on benthic foraminifera[21,22]. The living position of the extinction group is unknown, but has been inferred as infaunal[7,9,11]. If this were the case, pore water $[CO_3^{2-}]$ might be more relevant regarding the extinction, although pore waters have lower $[CO_3^{2-}]$ and changes in bottom-water $[CO_3^{2-}]$ would have influenced pore water values (in addition to other factors such as organic carbon flux and sedimentation rate). Bottom-water $\Delta[CO_3^{2-}]$ at Site 593, reconstructed from B/Ca of epifaunal *Cibicidoides wuellerstorfi* (see Methods), shows relatively low values ranging from 10 to 25 μmol kg^{-1}, with highest values recorded during cooler glacials and some potential negative outliers before 0.9 Ma (Fig. 1f). Similar to Mg/Ca, the absolute B/Ca of seawater would have been slightly different from

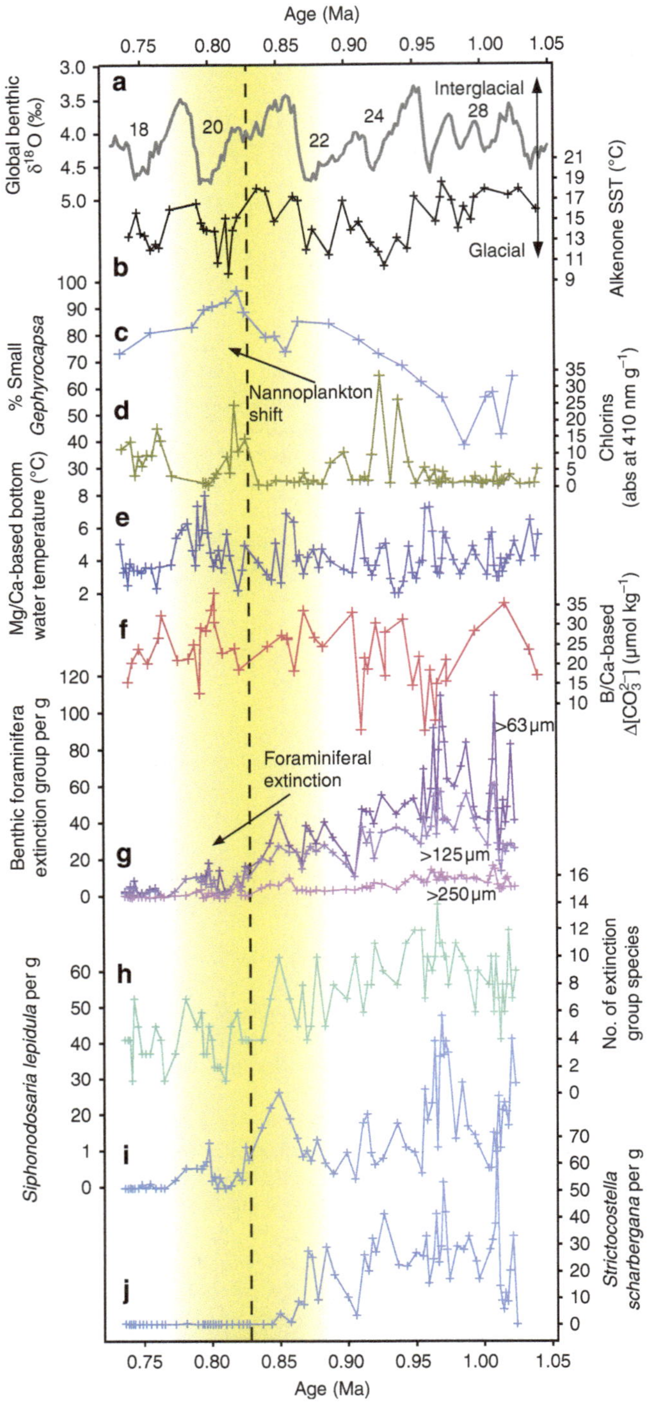

Figure 1 | Reconstructed environmental proxies for Tasman Sea DSDP Site 593. (**a**) Global benthic foraminifera $\delta^{18}O$ composite[52] showing colder glacial (positive) and interglacial cycles. (**b**) Sea-surface temperature reconstructed from alkenones. (**c**) Small *Gephyrocapsa* as a % of total nannofossil assemblage. (**d**) Concentration of chlorin pigments used here as a proxy for photosynthetic material related to primary productivity[24]. (**e**) Bottom-water temperature reconstructed from benthic foraminiferal *U. peregrina* Mg/Ca ratios. (**f**) Bottom-water $\Delta[CO_3^{2-}]$ reconstructed from benthic foraminiferal *C. wuellerstorfi* B/Ca ratios. (**g-j**) Abundance per g and number of species of benthic foraminifera from the extinction group. Note that none of the environmental proxies (**d-f**) follows abundance of the foraminiferal extinction group (**g-j**). The vertical yellow bar indicates the interval over which the foraminiferal extinction occurs, and the vertical dashed line indicates where small *Gephyrocapsa* dominates the assemblage.

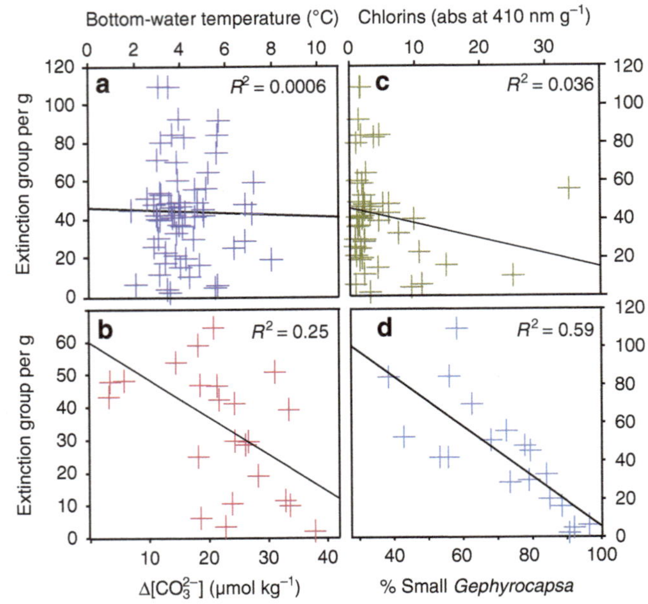

Figure 2 | Foraminiferal extinction group against other parameters for the pre-extinction interval ~0.8-1 Ma. (**a**) Bottom-water temperature derived from foraminiferal Mg/Ca, (**b**) bottom-water $\Delta[CO_3^{2-}]$ derived from foraminiferal B/Ca, (**c**) chlorin P410 from bulk sediment and (**d**) % small *Gephyrocapsa*.

modern values, thus potentially having an impact on the absolute values of our calculated $\Delta[CO_3^{2-}]$ even though the overall trends should be considered accurate[23]. Since all the values are oversaturated with respect to *in situ* $[CO_3^{2-}]$ (that is, all have positive $\Delta[CO_3^{2-}]$ values), and the extinction group abundance does not co-vary with $\Delta[CO_3^{2-}]$ during the pre-extinction interval (Fig. 2b), we conclude that the extinction group was tolerant to values in this range and could not have become globally extinct because of increased bottom-water corrosiveness.

Carbon flux. Benthic foraminifera are influenced by several ecological factors, which include bottom-water oxygenation, bottom-water sediment heterogeneity and hydrodynamics, temperature, corrosiveness and organic carbon type and flux[21,22]. However, above the lysocline (below which calcite dissolution occurs), typical open-ocean benthic ecology is primarily affected by the organic carbon flux—which is related to primary productivity (quantity, type and duration) and remineralization of particulate organic carbon as it is transported to depth[10]. A change in organic carbon export, linked to primary productivity, has been hypothesized as an alternative cause of the extinction[7,9,10]. SST at Site 593 (Fig. 1b), reconstructed using the alkenone proxy $U^{K}_{37}{}'$ (see Methods), shows significant variability over the study interval, ranging from 10 to 18 °C, with cooler temperatures during glacials after ~0.95 Ma. Since modern SSTs in the Tasman Sea are tightly coupled to the position of the productive Subtropical Front, we anticipate that shifts to Subtropical Front position might have influenced the organic matter flux. Sedimentary chlorin (Fig. 1d) is derived from photosynthetic material and specifically from chlorophyll pigments[24]. At this distal oceanic location they are likely to have originated from a proximal phytoplankton source, although a terrestrial contribution cannot be discounted by this study. Chlorin concentrations are typically higher during colder glacials at 0.75, 0.82 and 0.93 Ma, possibly indicating enhanced productivity[24] at times of Subtropical Front northward migration (cooler SSTs). However, despite significant variations, there is no prior correlation (Fig. 2c) nor secular change in chlorin concentration (and by extension phytoplankton carbon flux), which could account for the extinction in benthic foraminifera focused at ~0.83 Ma (Fig. 1).

Phytoplankton food source. The extinction group therefore appears to have been relatively tolerant to variations in physical water properties and potentially overall organic carbon flux changes at Site 593 before the ~0.83 Ma decline. Changing organic carbon *source* is another possible extinction mechanism[7,9–11]. The extinction group probably lived infaunally, according to their elongated morphology and lower shell $\delta^{13}C$ than epifaunal foraminifera[7,9,11], and preferred relatively high organic carbon flux at upper-abyssal to mid-bathyal depths; their global abundance reflects this distribution[7]. The specialized architectural function of extinction group apertures has been discussed at length[11] but remains unknown, and may have helped direct pseudopodial flow for detritus feeding[10,11], perhaps leaving them sensitive to a change in organic carbon supply.

Considering the strong benthic–pelagic coupling of benthic foraminifera[10], we compiled published high-resolution calcareous nannoplankton assemblage records to assess the potential for a global changing source of organic carbon causing the extinction. Coccolithophores are one of the major mid-low latitude phytoplankton groups contributing to the organic carbon pump, known to undergo rapid evolution in the Pleistocene[25]. From published records, we identified that a significant peak in the morphological genus 'small *Gephyrocapsa*' (<3 μm) occurred in the SE Atlantic[26], NW Pacific[27] and SW Pacific[11] centred at ~0.8 Ma (Fig. 3b). This peak in abundance is consistent with records from the Indian Ocean[25], and other nannoplankton records that do not differentiate this particular species from other small placoliths in the North Atlantic, South Atlantic and Mediterranean Sea (Supplementary Fig. 2). Although most of these sites do not have benthic foraminiferal data with which to directly compare the nannoplankton assemblages, Site MD97-2114 in the SW Pacific indicates that the benthic extinction coincided with this peak[11], and other sites[7] indicate that at a global scale the extinction had largely taken place by 0.8 Ma (Fig. 3c). To test the hypothesis that dominant small *Gephyrocapsa* could have been implicated in the benthic extinction, we paired records of nannoplankton and foraminiferal assemblage data in the Tasman Sea (Fig. 1c) and the North Atlantic (Supplementary Fig. 3). Both records show a consistently high % small *Gephyrocapsa* (>90%) during the extinction interval at ~0.83 Ma. Interestingly, the first step in the extinction at ~0.96 Ma at Site 593 (Fig. 1g) coincides with an initial increase in the % small *Gephyrocapsa* from ~50 to 65% (Fig. 1c), showing the possibility that % small *Gephyrocapsa* was already exerting some control over extinction group abundance. From our research, the abundance of small *Gephyrocapsa* is the only oceanic parameter that shows a correlation with the extinction group leading up to the extinction in the Tasman Sea (Fig. 2d). To address the possible significance of this correlation, we explore the various possible environmental changes controlling the two groups and describe a new conceptual model that can account for both data sets.

Discussion

Our new nannoplankton data (Fig. 1c and Supplementary Fig. 3), together with other compiled records (Fig. 3b and Supplementary Fig. 2), indicate that the global oceans became dominated by small *Gephyrocapsa* around ~0.8 Ma, a finding that has not previously been highlighted, perhaps because of the relatively short duration of this event compared with typical biostratigraphic sampling resolution. In some of the records, the dominance may have occurred sooner (for example, Site 980, N Atlantic) than others (for example, Site 593, Tasman Sea), such that there is diachroneity in the onset of this event; however, the crucial point is that all global records with coccolith data show high abundances of small *Gephyrocapsa* at ~0.8 Ma. In summary, evidence for a global peak in small *Gephyrocapsa* at ~0.8 Ma comes from sites in the SE Atlantic[26], SW Pacific[11], NW Pacific[27], Indian Ocean[25], N Atlantic and Tasman Sea (Supplementary Figs 2 and 3). The cause of the small *Gephyrocapsa* event remains unknown; however, it may have been an evolutionary adaptation[25,28] as the *Gephyrocapsa* lineage is thought to be relatively independent of temperature and nutrients, but dependent on light intensity and day length[25]. The well-known abundance peak of *G. caribbeanica* at ~0.6 Ma has been linked to increased blooms because of a sustained reduction in orbital eccentricity affecting day length and light intensity[25]; a sustained eccentricity minima is also observed during the benthic extinction event at ~0.8 Ma (Supplementary Fig. 2).

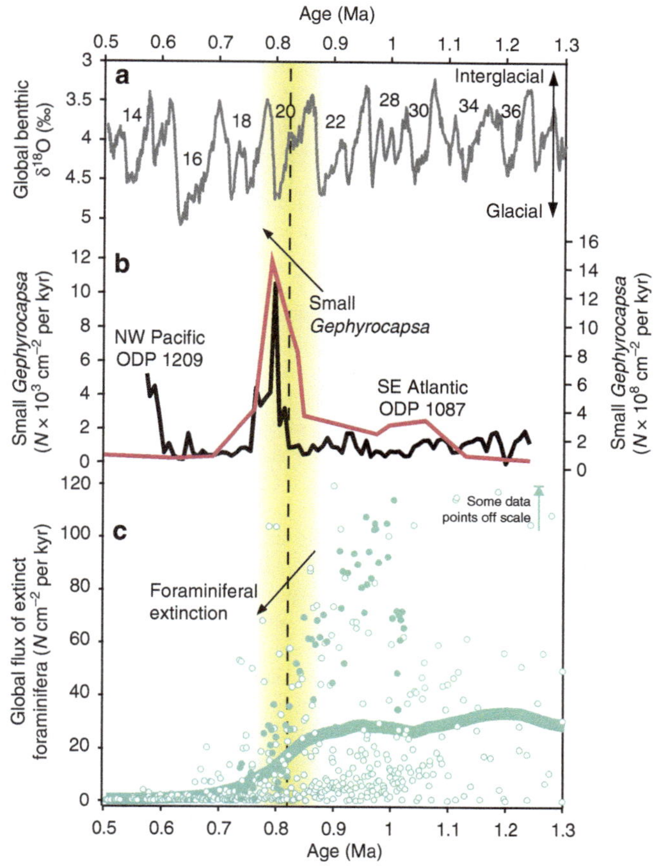

Figure 3 | Nannoplankton assemblages compared with extinct benthic foraminifera over the Mid Pleistocene. (**a**) Global deep-sea $\delta^{18}O$ composite[52]. (**b**) Accumulation rate of small *Gephyrocapsa* at Ocean Drilling Program (ODP) Site 1087 (ref. 26) and ODP Site 1209 (ref. 27). (**c**) Flux (accumulation rate) of extinct foraminifera from 15 global sites compiled by ref. 7 (open symbols), and including new data from this study (solid symbols), with a 0.2-pt LOESS smoothing spline (bold). Some data points are off the scale; smoothed line takes into account all data. Note how the peak in small *Gephyrocapsa* dominance at ~0.8 Ma occurs in the NW Pacific and SE Atlantic (and SW Pacific and N Atlantic, Supplementary Fig. 3), and coincides with persistently low abundance of the extinction group thereafter. Vertical yellow and dashed lines as in Fig. 1.

Regardless of the cause of the small *Gephyrocapsa* event, its correlation with the extinction at ~0.8 Ma provides a new piece of evidence when evaluating the cause of the benthic extinction. Correlation itself does not prove causation; however, as phytoplankton provide a food supply for marine benthos, how might this change in phytoplankton assemblages have had an impact on benthic foraminifera? We propose that the most plausible extinction mechanism would be a change towards a more enhanced variation in annual export production and delivery of organic carbon to the deep ocean[9–11]. There is micropalaeontological evidence for an increase in intra-annual variability of phytodetrital pulses during the extinction interval, with increased seasonal phytodetrital benthic foraminifera in the Indian Ocean[9] and SW Pacific[11] from ~0.8 Ma. Previous studies have suggested that this could not have caused the benthic extinction because physical processes that enhance seasonality could not have occurred throughout the global ocean at the same time, as the benthic extinction did[7,11]. However, if the ecology of small *Gephyrocapsa* was similar to its modern descendent, *Emiliania huxleyi*, a seasonal bloom species that dominates mid-low latitude nannoplankton assemblages during the Holocene[25,28], the increased global presence of small *Gephyrocapsa* may represent an early adaptation to more bloom-type ecology within nannoplankton, where intra-annual export production changed. Modern (interglacial) global productivity is dominantly annually uneven outside the oligotrophic ocean gyres (Fig. 4). Modelling studies indicate that mid-low latitude oceanic intra-annual variability of production may decrease with future projected warming[29], potentially implying that intra-annual production variability could have been higher (compared with modern variability) during colder glacials. Any increased intra-annual variability during MPT glacials would have reduced organic carbon flux during some seasons, while increasing flux at other times. In addition to preferring relatively high organic carbon flux[7,9], the unique apertural ornamentation of the extinction group may denote a different feeding strategy. We speculate that they required a more uniform supply of carbon throughout the year. This is in contrast to extant species, typically able to survive for years without fresh phytodetritus[30]. The phytoplankton change may have primed the group for complete extinction at ~0.8–0.6 Ma. There remains the possibility that both the small *Gephyrocapsa* event and the benthic extinction reflect an alternative, common cause. Although we consider it unlikely, as we have discounted various water property changes (Figs 1 and 2), perhaps an unconstrained factor such as a reduction in surface and deep-ocean CO_2 played a role.

Our proposed mechanism is consistent with the timing, bathymetric and geographic distribution of the extinction event. To summarize, our proposed ecology for the extinction group is that they preferred a relatively high flux of organic carbon *regularly* delivered to the sea floor such that there were no long periods within the year where flux was very low. This explains their distribution before the extinction[7], which was higher in mid-low latitude regions as opposed to high latitudes, as high latitudes would have had a more uneven supply of export production throughout the year (Fig. 4). It was also lower in oligotrophic regions such as the South Pacific Gyre, South China Sea and Mediterranean Sea, and higher in modern eutrophic regions[7]. Finally, it was higher at mid-bathyal to upper-abyssal depths compared with the oligotrophic lower abyssal ocean, where the carbon flux is very low. This proposed ecology also explains their disappearance first at deeper abyssal depths (oligotrophic), the Mediterranean (oligotrophic) and during glacials[7], which may have experienced more variability of seasonal export production, according to modelling studies[29]. Our proposed mechanism can explain the extinction because we invoke evolution within a phytoplankton group prevalent in the global oceans where the extinction group was most abundant before the extinction. Thus, although small *Gephyrocapsa* may not have been abundant at higher latitudes where siliceous phytoplankton predominate today, this was a region with low extinction group abundance before 0.8 Ma and thus not a suitable refuge for the group when mid-low latitude small *Gephyrocapsa* began to dominate nannoplankton assemblages.

The long-term decline of the extinction group during the Eocene–Oligocene, and increase in phytodetritus-exploiting species, has been suggested to be because of enhanced seasonality in primary productivity forced by cooling[10,31]. In this sense, the Cenozoic decline of the extinction group could be categorized as a 'slow' mass extinction[32], as export production may have exhibited gradually more intra-annual variability. However, the abrupt decline by ~0.8 Ma coinciding with the small *Gephyrocapsa* event would indicate a 'rapid' mass extinction with biological causes, perhaps with similarities to planktonic foraminiferal mass extinctions during the Palaeocene–Eocene and Eocene–Oligocene transitions[32]. The timing of our proposed increase in this mid-low latitude intra-annual variability, inferred from the global small *Gephyrocapsa* dominance and benthic extinction, may be important for the development of climate during the MPT as variability of primary production could have had an impact on carbon export and storage[33]. Further modelling studies are needed to ascertain the impact of an enhanced intra-annual variability of production on climate (during MPT cooling), forced by evolutionary changes within the biosphere rather than physical processes alone.

Methods

DSDP Site 593 stratigraphy and age model. DSDP Site 593 (40°30.47′S, 167°40.47′E, 1,063 m water depth) was cored on the Challenger Plateau of the Tasman Sea in the SW Pacific Ocean. The upper 393 m of recovered sediment is foraminiferal-bearing nannofossil ooze, hosting very abundant and visually well-preserved benthic foraminifera[34]. A low-resolution, orbitally tuned stratigraphy was available to guide sampling based on a shipboard bio- and magnetostratigraphy and coarse-resolution benthic $\delta^{18}O$ analyses on infaunal *Uvigerina* spp.[35]. The updated age model here has been generated after re-assigning this to the GTS2012 timescale[36]. Samples for the new isotope stratigraphy were analysed at 10–20-cm resolution in cores 593Z-1H-1 through 593Z-5H-2 (~0–36.3 m depth) to yield a mean sample resolution of ~5 kyr for the time period 0–1.1 Ma. The revised glacial–interglacial stratigraphy for the interval 0–0.4 Ma, based on *C. wuellerstorfi* oxygen isotopes and the Brunhes/Matuyama palaeomagnetic reversal, is presented in ref. 37. Here we created a revised isotope stratigraphy for the period 0.4–1.1 Ma using new $\delta^{18}O$ analyses of the epifaunal species *C. wuellerstorfi* (Supplementary Fig. 4 and Supplementary Data 1). The tuning targets are shown in Supplementary Table 1. The Potaka tephra (1 Ma (ref. 38)) is clearly identified at 21.50 mbsf, and lies above a distinct benthic $\delta^{18}O$ minimum, which is thus aligned to MIS 31. The top of the Olduvai magnetochron is not well represented, but the base of the Olduvai and the Gauss/Matuyama boundary were used to guide identification of the key isotope stages. A linear sedimentation rate is assumed between all tie points.

Micropalaeontology analysis. Benthic foraminiferal analyses were carried out at the British Geological Survey, Keyworth, UK, and at the University of Leicester, UK, using an Olympus SZX10 binocular microscope. Seventy-eight sediment samples of ~10 cc were oven dried at 40 °C and washed with deionized water over a 63-µm sieve to remove clays (at Durham University). The >63-µm fraction was then dried, and individual species of foraminifera were counted under a binocular microscope and transferred into cardboard reference slides. Residues were sieved into size fractions, and picked separately, according to a procedure outline in ref. 7, so that the >250-µm fraction had 100% picked, the >125-µm fraction had an average of 37% picked and the >63-µm fraction had an average of 21% picked (Supplementary Data 2). Fraction splits were used in order to increase the number of samples analysed because of the high volume of sediment. Abundance in splits was then normalized for dry sediment weight to estimate a total number of extinction group specimens per g of dried sediment for each sample[7]. A total of 31 species and taxonomic groups were identified using the taxonomic work of Hayward *et al.*[7], and photographed using a Hitachi S-3600N scanning electron microscope in the Geology Department at the University of Leicester (Supplementary Fig. 5). The BFAR (benthic foraminiferal accumulation rate,

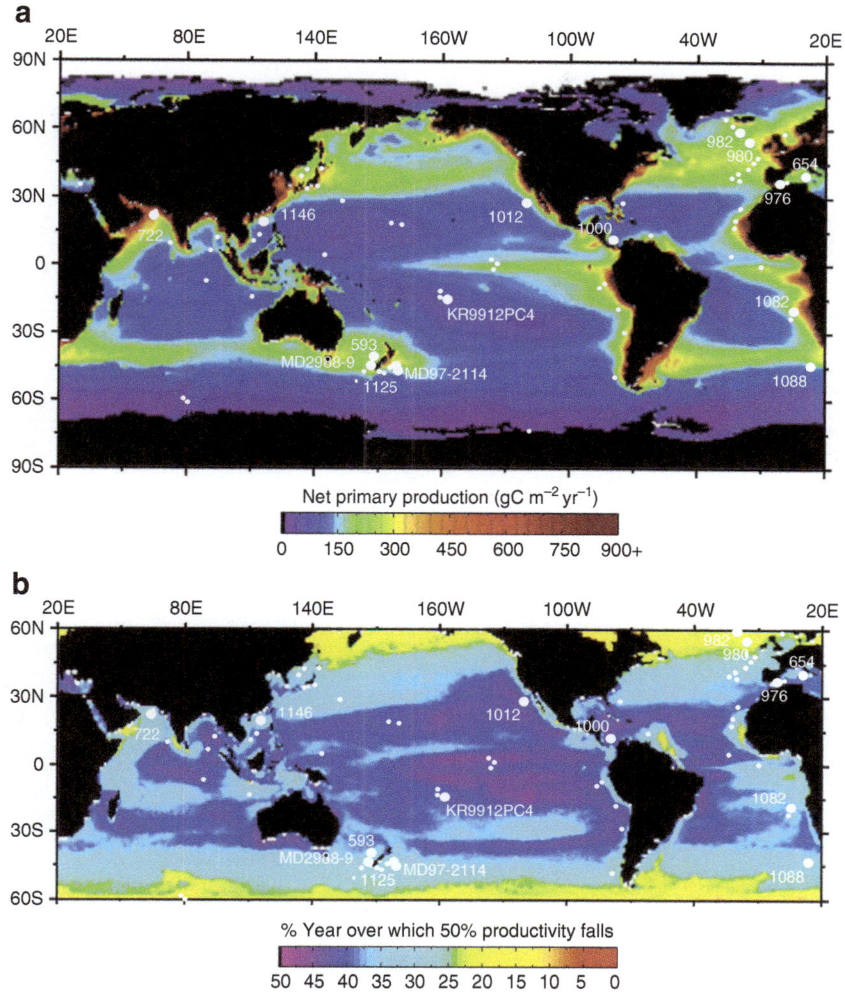

Figure 4 | Distribution of extinction group study sites and modern productivity. Location of core sites on maps showing the mathematical estimates of modern net primary production, and modern unevenness of annual primary production (figure adapted from ref. 53). Sites shown indicate where the Mid Pleistocene benthic foraminiferal extinction group has been studied; the DSDP/ODP site numbers are given for higher-resolution studies[7] that correspond to the sites used to construct Fig. 3. (**a**) Global average net primary productivity[53]. (**b**) The geographic distribution of the seasonal variation of net primary production (seasonality) for the years 1998-2007. The colours refer to the % of the year over which 50% productivity takes place, such that purple indicates no seasonality (colours modified from ref. 53). We propose that glacial MPT seasonality may have been higher than modern seasonality. Note that seasonality is generally lowest in regions where primary productivity is low.

specimens $cm^2 kyr^{-1}$) was calculated by using sediment dry bulk density (ρ, in $g cm^{-3}$), sedimentation rate (v, in $cm kyr^{-1}$) and number of specimens per g (x, specimens g^{-1}) as follows:

$$\mathrm{BFAR} = \rho v x \tag{1}$$

Calcareous nannofossil species were analysed at the British Geological Survey, Keyworth, UK. A total of 33 smear slides were prepared from the central part of fresh core sediment samples and were analysed using a Zeiss cross-polarizing light microscope at × 1,000 magnification. A counting phase was performed to obtain a qualitative and quantitative description of the assemblages based on percentage abundances of each recognized taxon (Supplementary Data 3). Nannofossil counts were performed in random visual fields on slides where the nannofossils were homogeneously distributed. At least 300 specimens > 3 µm were counted per slide in a varying number of fields of view. Specimens < 3 µm were included in a separate set of counts to quantify their dominance. As a result, two counts were obtained: the total assemblage and the subdominant one excluding specimens < 3 µm.

Most of the species identified in this study belong to the family Noelaerhabdaceae and are recognized at generic and specific levels: *G. oceanica*, *G. caribbeanica*, *G. muellerae* recognized during this interval as *G. margereli*[39], *G. omega* and small *Gephyrocapsa* (specimens < 3 µm, mainly constituted of *G. aperta* and *G. ericsonii*). Other taxa are present within the assemblages recognized at generic, specific and subspecific levels. At the generic level, minor taxa are represented by the genres *Syracosphaera*, *Pontosphaera*, *Reticulofenestra* and *Rhabdosphaera*. At the specific level, minor taxa are represented by the species *Pseudoemiliania lacunosa*, *R. asanoi*, *Helicosphaera carteri*, *H. hyalina* and *Umbilicoshpaera sibogae*. Finally, *Coccolithus pelagicus* and *Calcidiscus leptoporus* were distinguished at subspecific and submorphotype levels on the basis of their coccolith length. Refs 40,41 have documented the existence of two *C. pelagicus* subspecies: *C. pelagicus pelagicus*, the cold form, and *C. pelagicus braarudii*, the temperate and upwelling waters form[42]. Three *C. leptoporus* morphotypes have been distinguished: *C. leptoporus* type small (< 5.5 µm (refs 41,43)); *C. leptoporus* type medium (5.5–8 µm; (refs 41,43)); and *C. leptoporus* type large (> 8 µm (refs 41,43)).

Chlorin and alkenone analyses. Separate original samples were freeze-dried, homogenized, and alkenones and chlorins (diagenetic transformation products of chlorophyll)[24,44] were extracted with an organic solvent mixture of dichloromethane/methanol (Supplementary Data 4), following the microwave-assisted protocol of ref. 45. The microwave temperature was increased from room temperature to 70 °C over 2 min, held at this temperature for 5 min and then allowed to cool. Extracts were subsequently centrifuged in test tubes, and the supernatant dried by rotary evaporation before being divided into two aliquots for analysis. All analyses were carried out in the Geography Department at Durham University. Chlorins were analysed by ultraviolet-visible spectrophotometry, quantified at the 410 and 665 nm wavelengths, and normalized for extracted sample weight in g (ref. 45). One aliquot was dissolved again in a known volume of acetone, and analysed using a dionex photodiode array detector coupled to a quaternary pump[45]. Absorbance across the 350–850-nm wavelength was recorded and quantified at 410 and 665 nm, which corresponds to the diagenetic transformation products of chlorophyll[24,44]. Samples were analysed in triplicate, and the means are reported here (Supplementary Data 4). The average s.d. within samples was 0.44 units (410 nm) and 0.08 units (665 nm).

Alkenones were isolated from the second aliquot of the lipid extract using silica column chromatography, eluting with *n*-hexane (for hydrocarbons), dichloromethane (for ketones) and methanol (for polar compounds). Alkenones were quantified by Thermo Scientific Trace 1310 gas chromatograph fitted with a flame ionization detector. Separation was achieved with a fused silica column (30 m × 0.25 mm inner diameter) coated with 0.25 μm of 5% phenyl methyl siloxane (HP-5MS) and using He as the carrier gas. After injection, the following oven temperature programme was used: 60–200 °C at 20 °C min^{-1}, 200–320 °C at 6 °C min^{-1}, and then held at 320 °C for 35 min. SST was calculated using the $U^{K}_{37}{}'$ index[46] and the global mean annual SST calibration[47].

Oxygen isotope analysis. Approximately four individual tests of *C. wuellerstorfi* from the >250-μm fraction of each sample were analysed using an IsoPrime dual inlet mass spectrometer plus Multiprep device at the NERC Isotope Geosciences Laboratory. $\delta^{18}O$ is reported as per mille (‰) deviations of the isotopic ratios ($^{18}O/^{16}O$), and is calculated on the V-PDB scale using an internal laboratory standard that is routinely calibrated against NBS-19 standards (Supplementary Data 1). Average analytical reproducibility for $\delta^{18}O$ of the calcite standard is <0.1‰.

Trace elemental analysis. Approximately 10 pristine individuals (with white calcite, no visible clay and no visible recrystallization) per sample of benthic foraminifera *U. peregrina* and *C. wuellerstorfi* were selected from the >250-μm fraction. Particular care was taken to select the correct *U. peregrina* and *C. wuellerstorfi* morphotypes, as *Planulina/Cibicidoides* have been shown to have significant species-specific elemental fractionation[48,49]. Benthic foraminifera were cleaned following the oxidative procedure of ref. 50, and analysed first for calcium concentration using inductively coupled plasma optical emission spectrometer (ICP-OES) (Varian Vista). *U. peregrina* were then re-analysed at 100 p.p.m. [Ca] for Mg/Ca by ICP-OES and the *C. wuellerstorfi* samples were re-analysed at 10 p.p.m. [Ca] by ICP-MS (Element) for B/Ca. All trace element sample preparation and analysis occurred at the Godwin Laboratory for Palaeoclimate Research, Cambridge University (Supplementary Data 5). All analysed tests appeared visually well preserved, and low Fe/Ca and Mn/Ca values recorded simultaneously by ICP-OES indicate that samples were effectively cleaned and were devoid of diagenetic effects. Estimations of past IWT (intermediate water temperature) were then generated using the most recently published calibration from the Mg/Ca ratio of *Uvigerina*[2,51] (in mmol mol^{-1}), which has been shown to be suitable for this site for the interval from 0.4–0 Ma (ref. 37).

$$Mg/Ca_{U.peregrina} = 0.1IWT + 1.0 \quad (2)$$

Bottom-water $\Delta[CO_3^{2-}]$ (defined as the difference between $[CO_3^{2-}]_{in\ situ}$ and $[CO_3^{2-}]_{saturation}$) was calculated from the B/Ca ratio of epifaunal foraminifera *C. wuellerstorfi* (in μmol mol^{-1}) using the following calibration[49]. This proxy has also previously been used at this site for the interval from 0 to 0.4 Ma (ref. 37).

$$B/Ca_{C.wuellerstrorfi} = 1.14 \pm 0.048\Delta[CO_3^{2-}] + 177.1 \pm 1.41 \quad (3)$$

References

1. Clark, P. U. *et al.* The middle Pleistocene transition: characteristics, mechanisms, and implications for long-term changes in atmospheric pCO_2. *Quat. Sci. Rev.* **25,** 3150–3184 (2006).
2. Elderfield, H. *et al.* Evolution of ocean temperature and ice volume through the mid-Pleistocene climate transition. *Science* **337,** 704–709 (2012).
3. McClymont, E. L., Sosdian, S. M., Rosell-Melé, A. & Rosenthal, Y. Pleistocene sea-surface temperature evolution: early cooling, delayed glacial intensification, and implications for the mid-Pleistocene climate transition. *Earth-Sci. Rev.* **123,** 173–193 (2013).
4. Weinholz, P. & Lutze, G. F. in *Proceedings of the Ocean Drilling Program, Scientific Results* (eds Ruddiman, W. *et al.*) **108,** 113–117 (the Ocean Drilling Program, 1989).
5. Hayward, B. W. Late Pliocene to middle Pleistocene extinctions of deep-sea benthic foraminifera ('Stilostomella extinction') in the south-west Pacific. *J. Foraminiferal Res.* **32,** 274–306 (2002).
6. Hayward, B. W. *et al.* Last global extinction in the deep sea during the mid-Pleistocene climate transition. *Paleoceanography* **22,** PA3103 (2007).
7. Hayward, B. W. *et al. The last global extinction (Mid-Pleistocene) of deep sea benthic foraminifera (Chrysalogoniidae, Ellipsoidinidae, Glandulonodosariidae, Plectofrondiculariidae, Pleurostomellidae, Stilostomellidae), their Late Cretaceous-Cenozoic history and taxonomy* (Cushman Foundation for Foraminiferal Research Special Publication 43 408, 2012).
8. Gooday, A. J. *et al.* in *Deep-Sea Food Chains and the Global Carbon Cycle* (eds Rowe, G. T. & Pariente, V.) 63–91 (NATO ASI Series, Springer, Dordrecht, The Netherlands, 1992).
9. Hess, S. & Kuhnt, W. Neogene and quaternary paleoceanographic changes in the southern South China Sea (Site 1143): the benthic foraminiferal record. *Mar. Micropaleontol.* **54,** 63–87 (2005).
10. Thomas, E. Cenozoic mass extinctions in the deep sea: what perturbs the largest habitat on Earth? *Geol. Soc. Am.* **424,** 1–23 (2007).
11. Mancin, N., Hayward, B. W., Trattenero, I., Cobianchi, M. & Lupi, C. Can the morphology of deep-sea benthic foraminifera reveal what caused their extinction during the mid-Pleistocene climate transition? *Mar. Micropaleontol.* **104,** 53–70 (2013).
12. Mengerink, K. J. *et al.* A call for deep-ocean stewardship. *Science* **344,** 696–698 (2014).
13. Behrenfeld, M. J. *et al.* Climate-driven trends in contemporary ocean productivity. *Nature* **444,** 752–755 (2006).
14. Hayward, B. W., Grenfell, H. R., Carter, R. & Hayward, J. J. Depth distribution of Recent deep-sea benthic foraminifera east of New Zealand, and their potential for improving palaeobathymetric assessments of Neogene microfaunas. *N. Z. J. Geol. Geophys.* **44,** 555–587 (2001).
15. Takahashi, K., Ravelo, A. C. & Alvarez Zarikian, C. A. & the Expedition 323 Scientists. in *Proceedings of the International Ocean Drilling Program* **323** (2011).
16. Hamilton, L. J. Structure of the subtropical front in the Tasman Sea. *Deep Sea Res. Part I Oceanogr. Res. Pap.* **53,** 1989–2009 (2006).
17. Carter, R. M., Gammon, P. R. & Millwood, L. Glacial-interglacial (MIS 1–10) migrations of the subtropical front across ODP Site 1119, Canterbury Bight, Southwest Pacific Ocean. *Mar. Geol.* **205,** 29–58 (2004).
18. Bostock, H. C., Sutton, P. J., Williams, M. J. M. & Opdyke, B. N. Reviewing the circulation and mixing of Antarctic intermediate water in the South Pacific using evidence from geochemical tracers and Argo float trajectories. *Deep Sea Res. Part I Oceanogr. Res. Pap.* **73,** 84–98 (2013).
19. Talley, L. D. in *Mechanisms of Global Climate Change at Millennial Time Scales.* (eds Clark, P. U., Webb, R. S. & Keigwin, L. D.) 1–22 (AGU, Geophys. Mono. Series., 1999).
20. Evans, D. & Müller, W. Deep time foraminifera Mg/Ca paleothermometry: nonlinear correction for secular change in seawater Mg/Ca. *Paleoceanography* **27,** PA4205 (2012).
21. Levin, L. A. *et al.* Environmental influences on regional deep-sea species diversity. *Annu. Rev. Ecol. Syst.* **32,** 51–93 (2001).
22. Jorissen, F. J., Fontanier, C. & Thomas, E. in *Proxies in Late Cenozoic Palaeoceanography: Pt. 2: Biological tracers and biomarkers* (eds Hillaire-Marcel, C. & de Vernal, A.) 263–326 (Elsevier, 2007).
23. Kender, S., Yu, J. & Peck, V. L. Deep ocean carbonate ion increase linked to Miocene CO_2 decline. *Sci. Rep.* **4,** 4187 (2014).
24. Harris, P. G. *et al.* Chlorin accumulation rate as a proxy for quaternary marine primary productivity proxy. *Nature* **383,** 63–65 (1996).
25. Rickaby, R. E. M. *et al.* Coccolith chemistry reveals secular variations in the global ocean carbon cycle? *Earth Planet. Sci. Lett.* **253,** 83–95 (2007).
26. McClymont, E. L. *et al.* Alkenone and coccolith records of the mid-Pleistocene in the south-east Atlantic: implications for the $U^{K}_{37}{}'$ index and South African climate. *Quat. Sci. Rev.* **24,** 1559–1572 (2005).
27. Lupi, C., Bordiga, M. & Cobianchi, M. *Gephyrocapsa* occurrence during the Middle Pleistocene transition in the Northern Pacific Ocean (Shatsky Rise). *Geobios* **45,** 209–217 (2012).
28. Bollmann, J., Baumann, K. H. & Thierstein, H. R. Global dominance of *Gephyrocapsa coccoliths* in the late Pleistocene: selective dissolution, evolution, or global environmental change? *Paleoceanography* **13,** 517–529 (1998).
29. Henson, S., Cole, H., Beaulieu, C. & Yool, A. The impact of global warming on seasonality of ocean primary productivity. *Biogeosciences* **10,** 4357–4369 (2013).
30. Alve, E. Benthic foraminiferal responses to absence of fresh phytodetritus: a two-year experiment. *Mar. Micropaleontol.* **76,** 67–65 (2010).
31. Thomas, E. & Gooday, A. J. Deep-sea benthic foraminifera: tracers for Cenozoic changes in oceanic productivity? *Geology* **24,** 355–358 (1996).
32. Molina, E. Evidence and causes of the main extinction events in the Paleogene based on extinction and survival patterns of foraminifera. *Earth-Sci. Rev.* **140,** 166–181 (2015).
33. Lutz, M. J., Caldeira, K., Dunbar, R. B. & Behrenfeld, M. J. Seasonal rhythms of net primary production and particulate organic carbon flux to depth describe the efficiency of biological pump in the global ocean. *J. Geophys. Res.* **112,** C10011 (2007).
34. Shipboard Scientific Party in *Initial Reports of the Deep Sea Drilling Project* (eds Kennett, J. P., von der Borch, C. C. *et al.*) **90,** 551–651 (Deep Sea Drilling Project, 1996).
35. Head, P. S. & Nelson, C. S. in *Evolution of the Tasman Sea Basin* (eds van der Lingen, G. J., Swanson, K. M. & Muir, R. J.) 159–179 (A.A. Balkema, 1994).
36. Gradstein, F. M., Ogg, J. G., Schmitz, M. D. & Ogg, G. M. *The Geologic Time Scale 2012* (Elsevier, 2012).

37. Elmore, A. C. *et al.* Antarctic intermediate water properties since 400ka recorded in infaunal (*Uvigerina peregrina*) and epifaunal (*Planulina wuellerstorfi*) benthic foraminifera. *Earth Planet. Sci. Lett.* **428,** 193–203 (2015).
38. Shane, P. A. R. A widespread, early Pleistocene tephra (Potaka tephra, 1 Ma) in New Zealand: character, distribution, and implications. *N. Z. J. Geol. Geophys.* **37,** 25–35 (1994).
39. Bréhéret, J. Formes nouvelles quaternaires et actuelles de la famille des Gephyrocapsaceae (Coccolithophorides). *C. R. Acad. Sci.* **287,** 447–449 (1978).
40. Baumann, K. H., Böckel, B. & Frenz, M. in *Coccolithophores—from Molecular Processes to Global Impact* (eds Thierstein, H. R. & Young, J. R.) 367–402 (Springer, 2004).
41. Geisen, M. *et al.* Lifecycle associations involving pairs of holococcolithophorid species- intraspecific variation or cryptic speciation? *Eur. J. Phycol.* **37,** 531–550 (2002).
42. Marino, M., Maiorano, P. & Lirer, F. Changes in calcareous nannofossil assemblages during the Mid-Pleistocene revolution. *Mar. Micropaleontol.* **69,** 70–90 (2008).
43. Sáez, A. G., Probert, I., Young, J. R. & Medlin, L. K. in *Coccolithphores - from Molecular Processes to Global Impact* (eds Thierstein, H. R. & Young, J. R.) 251–270 (Springer, 2004).
44. Baker, E. W. & Louda, J. W. in *Biological Markers in the Sedimentary Record* (ed. Johns, R. B.) 125–226 (Elsevier, 1986).
45. Kornilova, O. & Rosell-Melé, A. Application of microwave-assisted extraction to the analysis of biomarker climate proxies in marine sediments. *Org. Geochem.* **34,** 1517–1523 (2003).
46. Prahl, F. G. & Wakeham, S. G. Calibration of unsaturation patterns in long-chain ketone compositions for palaeotemperature assessment. *Nature* **320,** 367–369 (1987).
47. Müller, P. J., Kirst, G., Ruhland, G., Storch, I. V. & Rosell-Melé, A. Calibration of the alkenone paleotemperature index $U^{k}_{37}{}'$ based on core-tops from the eastern South Atlantic and the global ocean (60°N–60°S). *Geochim. Cosmochim. Acta* **62,** 1757–1772 (1998).
48. Yu, J. & Elderfield, H. Benthic foraminiferal B/Ca ratios reflect deep water carbonate saturation state. *Earth Planet. Sci. Lett.* **258,** 73–86 (2007).
49. Rae, J. W. B., Foster, G. L., Schmidt, D. N. & Elliot, T. Boron isotopes and B/Ca in benthic foraminifera: proxies for the deep ocean carbonate system. *Earth Planet. Sci. Lett.* **302,** 403–413 (2011).
50. Barker, S., Greaves, M. & Elderfield, H. A study of cleaning procedures used for foraminiferal Mg/Ca paleothermometry. *Geochem. Geophys. Geosyst.* **4,** 1–20 (2003).
51. Elderfield, H. *et al.* A record of bottom water temperature and seawater d18O for the Southern Ocean over the past 440kyr based on Mg/Ca of benthic foraminiferal *Uvigerina spp. Quat. Sci. Rev.* **29,** 160–169 (2010).
52. Lisiecki, L. E. & Raymo, M. E. A Pliocene-Pleistocene stack of 57 globally distributed benthic $\delta^{18}O$ records. *Paleoceanography* **20,** PA1003 (2005).
53. Brown, C. W., Uz, S. S. & Corliss, B. H. Seasonality of oceanic primary production and its interannual variability from 1998 to 2007. *Deep Sea Res. Part I Oceanogr. Res. Pap.* **90,** 166–175 (2014).

Acknowledgements

We thank Mervyn Greaves for analytical assistance, Adeyinka Aturamu for SEM assistance, and the International Ocean Discovery Program for providing samples. This work was supported by the UK Natural Environment Research Council (awards NE/I024372/1 to S.K., NE/I027703/1 to E.L.M. and IP-1339-1112 to E.L.M.), a Philip Leverhulme Prize (awarded to E.L.M.) and The European Research Council (award 2010-NEWLOG ADG-267931 to H.E.).

Author contributions

S.K. and E.L.M. devised the study. A.C.E. produced the Mg/Ca and B/Ca data and constructed the age model, S.K. produced the foraminiferal assemblage data, E.L.M. produced the organic geochemistry data, D.E. and E.L.M. produced the nannofossil assemblage data, A.C.E. and M.J.L. produced the oxygen isotope data. All authors contributed to the ideas and writing of the manuscript.

Additional information

Competing financial interests: The authors declare no competing financial interests.

8

Shifting El Niño inhibits summer Arctic warming and Arctic sea-ice melting over the Canada Basin

Chundi Hu[1,2], Song Yang[1,3,4], Qigang Wu[2], Zhenning Li[1,3], Junwen Chen[3], Kaiqiang Deng[1,3], Tuantuan Zhang[1,3] & Chengyang Zhang[1,5]

Arctic climate changes include not only changes in trends and mean states but also strong interannual variations in various fields. Although it is known that tropical-extratropical teleconnection is sensitive to changes in flavours of El Niño, whether Arctic climate variability is linked to El Niño, in particular on interannual timescale, remains unclear. Here we demonstrate for the first time a long-range linkage between central Pacific (CP) El Niño and summer Arctic climate. Observations show that the CP warming related to CP El Niño events deepens the tropospheric Arctic polar vortex and strengthens the circumpolar westerly wind, thereby contributing to inhibiting summer Arctic warming and sea-ice melting. Atmospheric model experiments can generally capture the observed responses of Arctic circulation and robust surface cooling to CP El Niño forcing. We suggest that identification of the equator-Arctic teleconnection, via the 'atmospheric bridge', can potentially contribute to improving the skill of predicting Arctic climate.

[1]Institute of Earth Climate and Environment System, Sun Yat-sen University, 135 West Xingang Road, Guangzhou, Guangdong 510275, China. [2]School of Atmospheric Sciences, Nanjing University, Nanjing, Jiangsu 210023, China. [3]School of Atmospheric Sciences, Sun Yat-sen University, Guangzhou, Guangdong 510275, China. [4]Guangdong Province Key Laboratory for Climate Change and Natural Disaster Studies, Sun Yat-sen University, Guangzhou, Guangdong 510275, China. [5]Climate Center, Guangxi Meteorological Bureau, Nanning, Guangxi 530022, China. Correspondence and requests for materials should be addressed to S.Y. (email: yangsong3@mail.sysu.edu.cn).

Since the late 1970s, the Arctic Ocean has experienced considerable changes[1,2], such as rapid warming and contracting sea ice. Growing evidence has attributed these changes to the changes in Arctic climate caused by external anthropogenic forcing[3–5] (that is, increasing aerosol and greenhouse gas emission) and internal natural variability, such as the Atlantic Multidecadal Oscillation[6], tropical sea surface temperature (SST) trend[7], as well as anomalous tropical convective heating forcing associated with the El Niño-Southern Oscillation (ENSO)[8–10] and the Madden–Julian oscillation[11–13]. The effect of Arctic climate change on society is becoming a hot research topic, given the more frequent mid-latitude extreme events because of the increasing Arctic sea-ice loss in the recent decades[14–16] and the anticipated global changes in future climate scenarios. It is estimated that, by the end of the twenty-first century, the Arctic will probably become an ice-free blue ocean[17]. Many studies have examined Arctic sea-ice responses to externally forced climate changes and found that the region has been warming more quickly than the rest of the world[18,19]. However, the ratio of trend to standard deviation of the interannual variability of detrended surface temperature and atmospheric circulation anomalies (Supplementary Fig. 1) is only 10% or so, suggesting that natural interannual variability is dominant and should not be neglected in the context of rapid warming.

Although attributions of the Arctic climate trends to human effects and tropical Pacific SST forcing are now well established, interannual variation of Arctic climate may respond more strongly to the changes in the two prominent types of interannual SST variability in the tropical Pacific, namely, the central Pacific (CP) El Niño and the eastern Pacific (EP) El Niño, both of which are linked to global climate teleconnections[20,21]. In the context of global warming, El Niño properties have exhibited considerable changes since the late 1970s, including the changes in frequency, location, intensity and meridional scale[20,22–24]. The CP El Niño is occurring more frequently than its counterpart, and tends to become more intense[25,26]. On the other hand, the Arctic warming is not spatially uniform but varies significantly with locations[7,27]. In addition, the Arctic sea-ice shrinks most strongly during summer and reaches a minimum in September[4], whereas there is no significant warming trend in the Arctic Ocean in the summer melting season[7] (see the Extended Data Figure 2 of ref. 7). These seemingly distinct phenomena bring up an intriguing question: Does the almost concurrent regime shift of El Niño play a role in the variability of summer Arctic climate on interannual timescale?

The influence of shifting El Niño on the Arctic climate, compared with the extra-polar climate[21,28–30], however, has received less attention. Projections of the effects of El Niño change on the concurrent Arctic climate change are inherently uncertain. Thus, understanding the impact of shifting El Niño on the Arctic climate is of profound interest.

In this study, we demonstrate that the CP El Niño contributes to Arctic cooling and sea-ice increase during the boreal summer (June–August, JJA) on interannual timescale. Using observational satellite data of summer Arctic sea-ice concentration (SIC) and oceanic/atmospheric reanalysis, together with numerical model simulations, we confirm a teleconnection that exists between the tropical Pacific Ocean and the Arctic Ocean; we also unravel a potential physical mechanism for the teleconnection.

Results

Teleconnection between shifting El Niño and Arctic climate. The upper panels of Fig. 1a,b show the EP and CP El Niño patterns, respectively, which are derived from a rotated empirical orthogonal function (REOF, see the Methods and Supplementary Note 1 for details and reasons) decomposition of normalized and detrended SST anomalies in a key region (120°E–80°W, 20°S–20°N) of the tropical Pacific. The spatial patterns of REOF1 and REOF2 highlight the positions and magnitudes of SST warming related to the two types of El Niño. As expected, the associated principal components (namely, RPC1 and RPC2) plotted in Fig. 1c,d (black lines) are directly related to detrended EP index and CP index (red lines, see the Methods for details) with large correlation coefficients up to 0.963 and 0.943, respectively. The SST changes related to shifting El Niño lead to changes in convective activity pattern (Fig. 1a,b and Supplementary Fig. 2a,b), which in turn stimulate large-scale Rossby waves via perturbation of upper-level divergence[31] (Supplementary Fig. 2c). As shown in Fig. 1e,f, the regressions of boreal summer 200-hPa geopotential height (H200) against RPC1 and RPC2 are characterized by planetary-scale atmospheric waves, which result in distant climate anomalies—teleconnections. In comparison with Fig. 1e, Fig. 1f reveals a significant atmospheric teleconnection between CP El Niño and the polar upper troposphere, that is, the significant deepening of tropospheric Arctic polar vortex (APV), in particular over the Canada Basin region, is closely associated with the CP warming (CPW). Similar REOF-related results can be reproduced by a more straightforward way of box-based regression and correlation analysis (Supplementary Figs 3 and 4). The corresponding quasi-geostrophic stream function and Rossby wave activity fluxes (Supplementary Fig. 5) exhibit two distinctive trajectories, which are consistent with the H200 teleconnection patterns (Fig. 1e,f). It is worth noting that the method is independent from the REOF analysis except for the choice of box location, suggesting that the above results are robust and reliable.

Deepening of the APV will inevitably lead to changes in the underlying Arctic climate at surface. As shown in Fig. 2a–c (Supplementary Figs 6 and 7), the regional changes in air temperature at 925-hPa (T925), land/sea surface temperature (LST/SST) and SIC over the Arctic are all significantly correlated with RPC2. The strongest correlation occurs in the Arctic Canada Basin where lower-level northwesterly anomalies (Fig. 2d, vectors; Supplementary Figs 6 and 7) and the resultant cold advection contribute to Arctic cooling and sea-ice increase. Moreover, such response may be amplified by local positive feedbacks (that is, thermal advections and feedbacks of ice-temperature, cloud-radiation and surface sensible/latent heat flux exchanges and so on) among atmosphere, ocean and sea ice[4,19,32,33], as shown in Supplementary Fig. 8. The 500-hPa geopotential height (H500) anomalies (Fig. 2d, shading; Supplementary Figs 6 and 7) also exhibit a similar wave-train-like feature as that shown in Fig. 1f, accompanied by deepening of the APV and strengthening of the circumpolar westerly wind (CWW) at both lower and upper levels (Fig. 2d,e). In addition, the coherence of lower- and upper-level winds (Fig. 2d,e, vectors) indicates that the circulation anomalies forced by CP El Niño warming are quasi-barotropic. The features linked to RPC1 (Supplementary Fig. 9) are distinct from those associated with RPC2. Motivated by previous studies[7,34,35], we thus hypothesize that the CP El Niño plays a key role in forcing the teleconnection pattern that induces the strengthening of CWW and deepening of APV, both of which are conductive to retaining cold air mass in the polar region, favouring regional Arctic cooling and inhibiting Arctic sea-ice melting.

Identification of key forcing region. To identify the key forcing region in the tropics that determines the wave trains to the Arctic, the combined regression and correlation maps of the tropical Pacific SST, precipitation and Outgoing Longwave Radiation

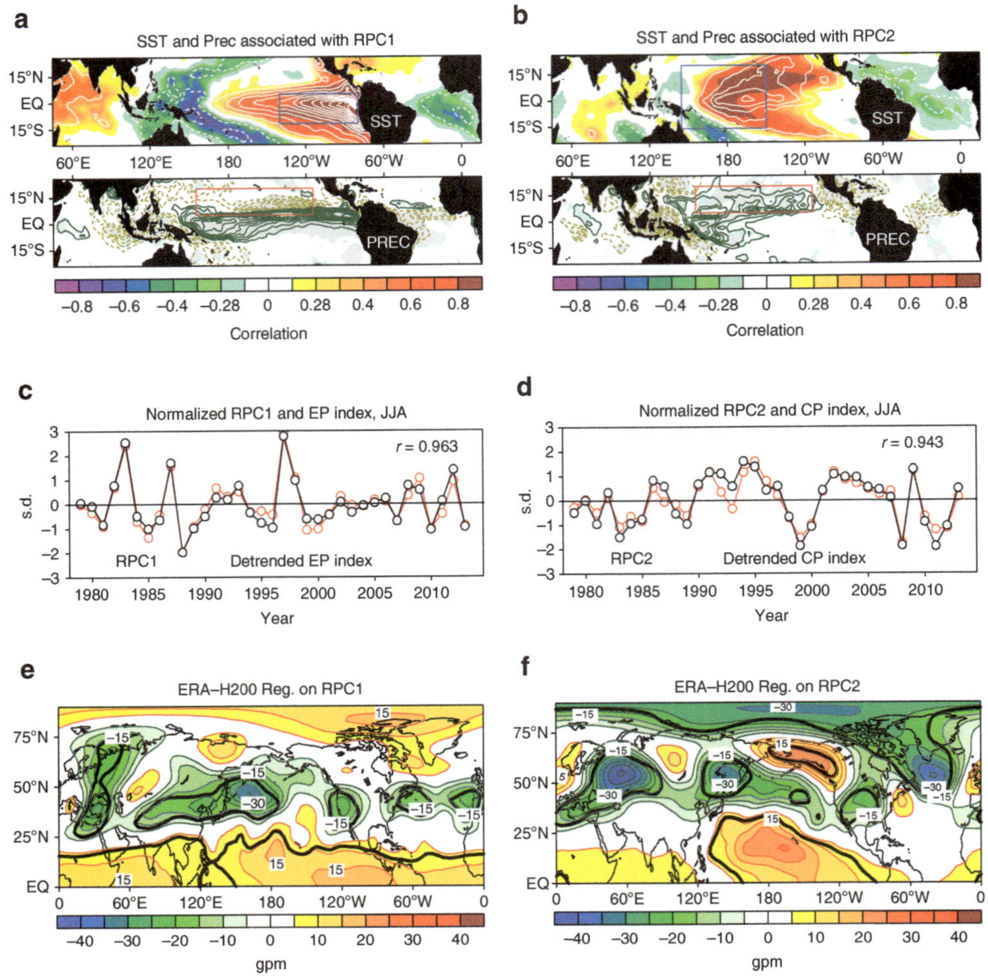

Figure 1 | Statistics of EP El Niño and CP El Niño. Spatial patterns of tropical SST: ((**a**) top: colour shadings are used for correlation, and contours are regression with an interval of 0.2 °C) marine precipitation ((**a**) bottom: regression with an interval of 0.4 mm per day) and ERA-H200 ((**e**), regression with an interval of 5 gpm) associated with normalized RPC1 (**c**, in black). Grey shadings and thick black lines indicate the correlation at the estimated 90% confidence level. Panels **b**,**d**,**f** are the same as panels **a**,**c**,**e**, except for RPC2. In **c**,**d**, the time series in red are the detrended and normalized EP and CP indices (see the Methods). Correlation (RPC1, detrended EP index) = 0.963, and correlation (RPC2, detrended CP index) = 0.943. Blue boxes outline the CP region (140°W-80°W, 12°S-5°N) and the EP region (145°E-150°W, 16°S-20°N), respectively. The two red boxes highlight the same region (155°E-115°W, 5°N-20°N) as a reference system.

(OLR) anomalies associated with detrended SIC-index and T925-index are shown in Supplementary Fig. 10. The two indices are defined as the area-weighted mean of SIC and T925 anomalies over the region of 160°W–90°W, 72°N–82°N (see the black boxes in Fig. 2a,c, respectively). We can see from the significant areas identified cooperatively by the black boxes in each physical field of SST, precipitation and OLR (Supplementary Fig. 10) that the key forcing region is located in the northern subtropical CP (155°E–115°W, 5°N–20°N). This also indirectly confirms that the interannual variability of Arctic climate anomalies over the Canada Basin is indeed closely linked to the circulation anomalies induced by the convective heating associated with the CP El Niño during summer.

Convective regimes associated with shifting El Niño. To further elucidate the regime differences in teleconnection patterns associated with the CP and EP El Niño episodes, we present the spatial distributions of total SST (that is, JJA SST climatology plus JJA SST anomalies associated with the CP or EP El Niño) and the composite pattern of precipitation (CP minus EP) in Fig. 3. The isotherms in Fig. 3a indicate the total SST above the warm pool threshold of 28 °C. Note that the black dashed lines (for EP El Niño case) are always surrounded by the red solid lines (for CP El Niño case) over the key forcing region (155°E–115°W, 5°N–20°N; see box in Fig. 3). In other words, during the CP El Niño, the 28 °C isotherm (north of the equator) extends into the white-shading areas where the JJA SST climatology is below 28 °C, which reinforces the climatological SST over the key forcing region and favours the poleward shift and strengthening of the inter-tropical convergence zone (ITCZ). Correspondingly, the enhanced convective precipitation along the ITCZ and its northern side over the key forcing region (Fig. 3b) confirms the northward shift of the intensifying ITCZ. Thus, the potential physical mechanism for the CPW-Arctic teleconnection could be understood as follows. A poleward shift of enhanced ITCZ, which indicates a poleward shift of enhanced tropical convection forcing, will inevitably lead to poleward shifts of upper-level divergence[31], northern flank of the Hadley cell, and the subtropical jet[36,37], reinforcing the poleward propagating Rossby waves[36,38] and favouring a further northward location of large-scale teleconnection linked more easily to the polar region.

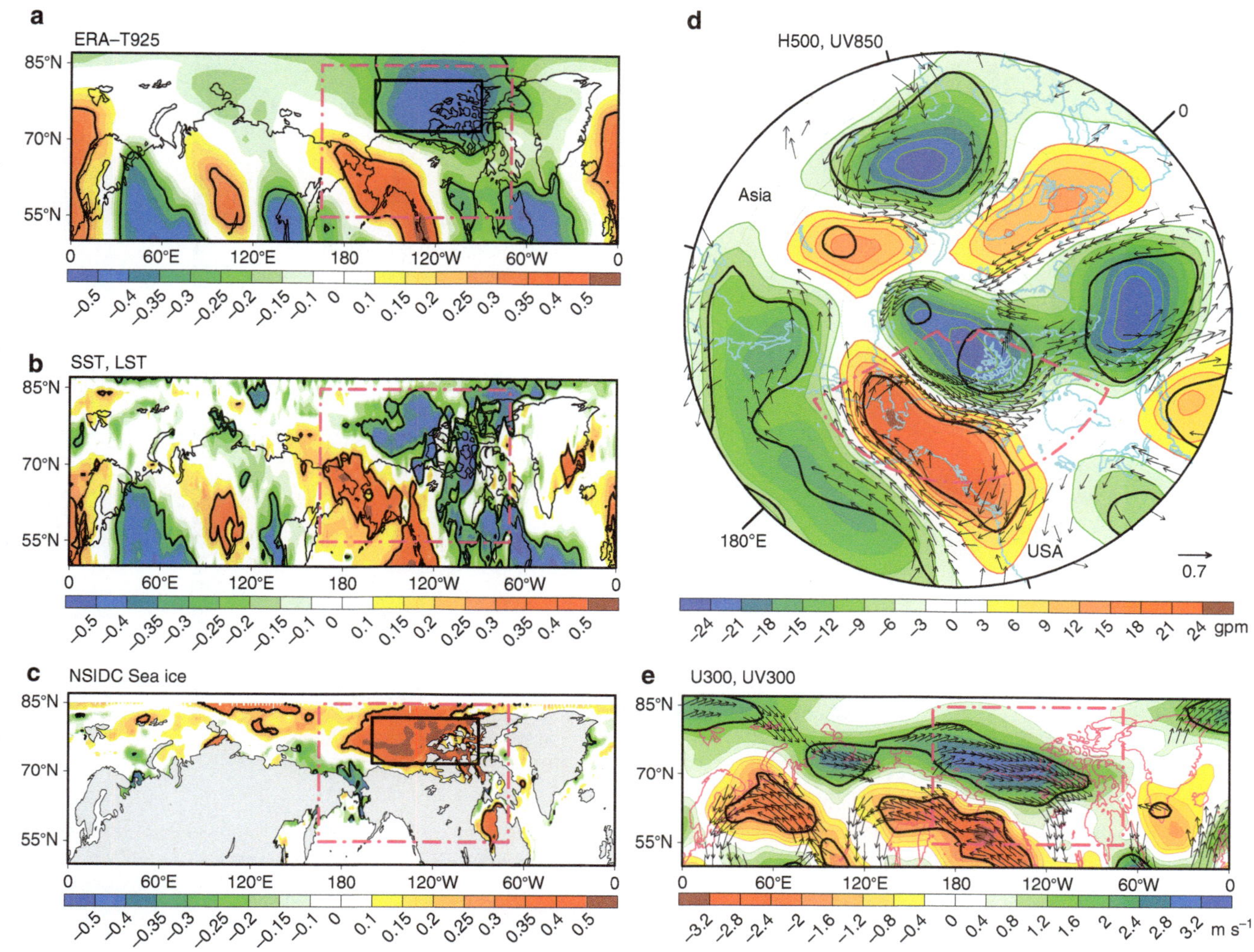

Figure 2 | Teleconnection patterns related to central Pacific SST warming. (**a**) Correlations of T925 with RPC2. (**b**,**c**) are the same as **a**, except for SST/LST and SIC, respectively. Shadings in **d**,**e** denote H500 and U300 regressed, respectively, against normalized RPC2. Black arrows in **d**,**e** show the correlations (vectors less than 0.28 omitted) of UV850 and UV300 with RPC2, respectively. Thick black lines indicate correlation at the estimated 90% confidence level. Area-weighted means of T925 and SIC anomalies over the black boxes (160°W–90°W, 72°N–82°N) in **a**,**c** are used to define the T925-index and the SIC-index. The area outlined by the purple dashed box is the same in all five panels. Here, the atmospheric data sets are from the ERA-Interim reanalysis.

Modelling support from NCAR CAM4. To further assess the causal relationship between CPW forcing and Arctic circulation response, numerical simulations are performed using the US National Center for Atmospheric Research (NCAR) global Community Atmosphere Model version 4 (CAM4; see the Methods for details). The simulated Arctic H200 response (Supplementary Fig. 11a) to CPW is significantly negative and comparable to the observed H200 pattern over the Arctic Ocean (Fig. 1f), suggesting that the model is capable of reproducing the deepening of the APV, in spite of some excursions in its position and magnitude. The CAM4 also well reproduces the observed strengthening of CWW and Arctic cooling (Supplementary Fig. 11b,c), implying that the observed Arctic cooling and SIC increase are the consequence of changes in upper-level circulation driven by tropical CPW to a certain extent. Thus, the declining H200 and widespread cooling over the Arctic Ocean can be viewed as an integral of linear responses to the imposed tropical SST anomalies related to the CP El Niño.

Although the simulated Arctic cooling is robust, the simulated position and magnitude of Arctic climate responses are not stationary, leading to some excursions in various composite patterns of the CAM4 results (Supplementary Fig. 11). Moreover, the simulated responses do not match the REOF results completely and therefore should be interpreted carefully. The difference is possibly because (i) atmospheric circulation and temperature are also affected by the changes in SIC, snow cover/depth and others; and (ii) inevitable difference exists between the observed and simulated climatologies of the background flow and temperature and so on. Therefore, the total response induced by the changes in atmospheric circulation may depend on snow–atmosphere–SIC–SST interactions, which are not fully included in the model. Nonetheless, the simulated results suggest that the deepening APV associated with CP El Niño warming contributes to the regional Arctic cooling.

Case study. The above empirical findings are directly derived from the data sets for 1979–2013. If shifting El Niño really contributes to Arctic cooling, there should be less Arctic warming and less sea-ice melting in JJA of 2014 (CP El Niño) and more Arctic warming and more sea-ice melting in JJA of 2015 (EP El Niño; see http://www.emc.ncep.noaa.gov/research/cmb/sst_analysis/images/archive/monthly_anomaly/). For verification, we show these two latest cases in Fig. 4 (and Supplementary Fig. 12).

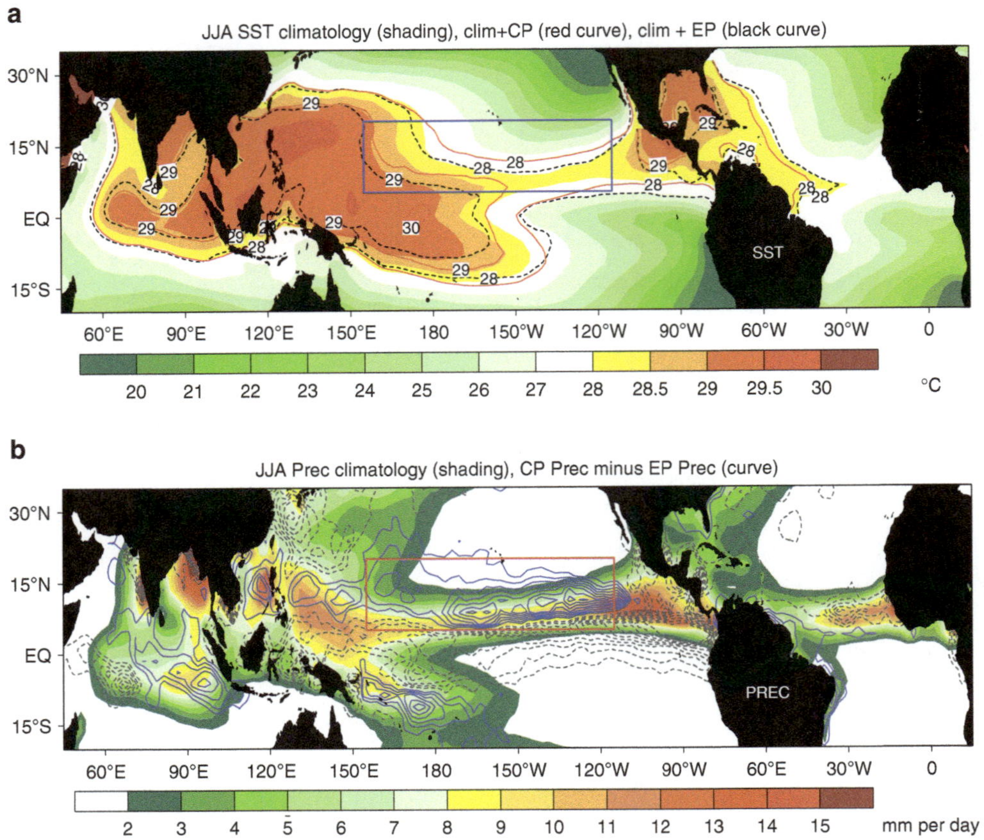

Figure 3 | Regime differences between CP El Niño and EP El Niño. (**a**) Composite patterns of total SST for CP El Niño (red solid lines) and EP El Niño (black dashed lines). Total SST is defined as the sum of SST anomalies shown in Fig. 1 plus the SST climatology in the boreal summer (only the total SST above 28 °C is shown, with an interval of 1 °C). (**b**) Composite marine precipitation differences ('CP minus EP', 0.4 mm per day interval) between CP El Niño and EP El Niño. Shadings in **a**,**b** show the climatology of SST and precipitation, respectively. Blues/red boxes outline the key forcing region (155°E–115°W, 5°N–20°N).

The tropical Pacific SST anomalies during JJA in 2014 and in 2015 mirror an episode of robust CP El Niño and a typical event of EP El Niño (Fig. 4a,b), respectively. As shown in Fig. 4c–f, the corresponding phases of low-level air temperature and SIC anomalies over the Arctic Canada Basin also exhibit highly consistent one-to-one relationship to the two types of El Niño in the two summers, respectively. Thus, the two contrasting cases provide further support that shifting El Niño contributes to inhibiting summer Arctic warming and Arctic sea-ice melting over the Canada Basin.

Discussion

This study demonstrates a linkage between CPW and summer Arctic climate from the perspective of detrended interannual variability. Our straightforward analyses of the tropical Pacific SST and corresponding precipitation anomalies reveal the following features. First, the linear response of upper-level circulation to CP El Niño warming is robust over the Arctic polar region, whereas the response to EP El Niño warming is not apparent although the intensity of EP El Niño is approximately twice of that of CP El Niño. Second, the discrepancy in the effects of two types of El Niño on the Arctic circulation can be explained by the background of higher SST over the CP than over the EP, favouring deeper convection along the ITCZ and to its north. Poleward shift of the enhanced ITCZ favours a further northward location of large-scale teleconnection to the Arctic region. Third, the teleconnection induced by the CP El Niño can cause deepening and strengthening APV and CWW. Finally, the changes in the Arctic circulation in turn provide favourable large-scale dynamic conditions for cold air mass to be retained in the polar region, leading to Arctic cooling and sea-ice increase, especially over the Canada Basin. Two more points should be emphasized: (i) observed results are consistent with those reproduced based on different EP and CP El Niño definitions; (ii) the results derived from the detrended CP index are most robust as the CP index contains more signals, such as the warming trend that is not well captured by other CP El Niño definitions (see Supplementary Note 2, Supplementary Tables 1 and 2 and Supplementary Figs 13–16 for more details).

Hence, the CP El Niño contributes to inhibiting summertime Arctic warming and sea-ice loss on interannual timescale to a certain extent. In other words, the summer Arctic sea-ice loss would be more intense without the effect of more frequent CP El Niño events over the past few decades (Fig. 5). It is clear that the regression coefficients of '$T^{\#}$' and '$CP^{\#}$' in the two equations shown in Fig. 5 are comparable but opposite with each other, suggesting that both the linear trend and the interannual CP El Niño affect the climate variability over the Arctic Canada Basin, at least for the recent decades. However, it may be just a matter of time before external forcing dominates the Arctic warming because of the foreseeable sustainable increase in greenhouse gases and aerosols[27]. Our study highlights one of the often-overlooked aspects of Arctic climate responses to the anticipated change in El Niño events on interannual timescale in the context of global warming. On the other hand, Kim *et al.*[28] has demonstrated that the predictability of CPW (for example, CP El Niño) has no obvious spring barrier that exists in the case of

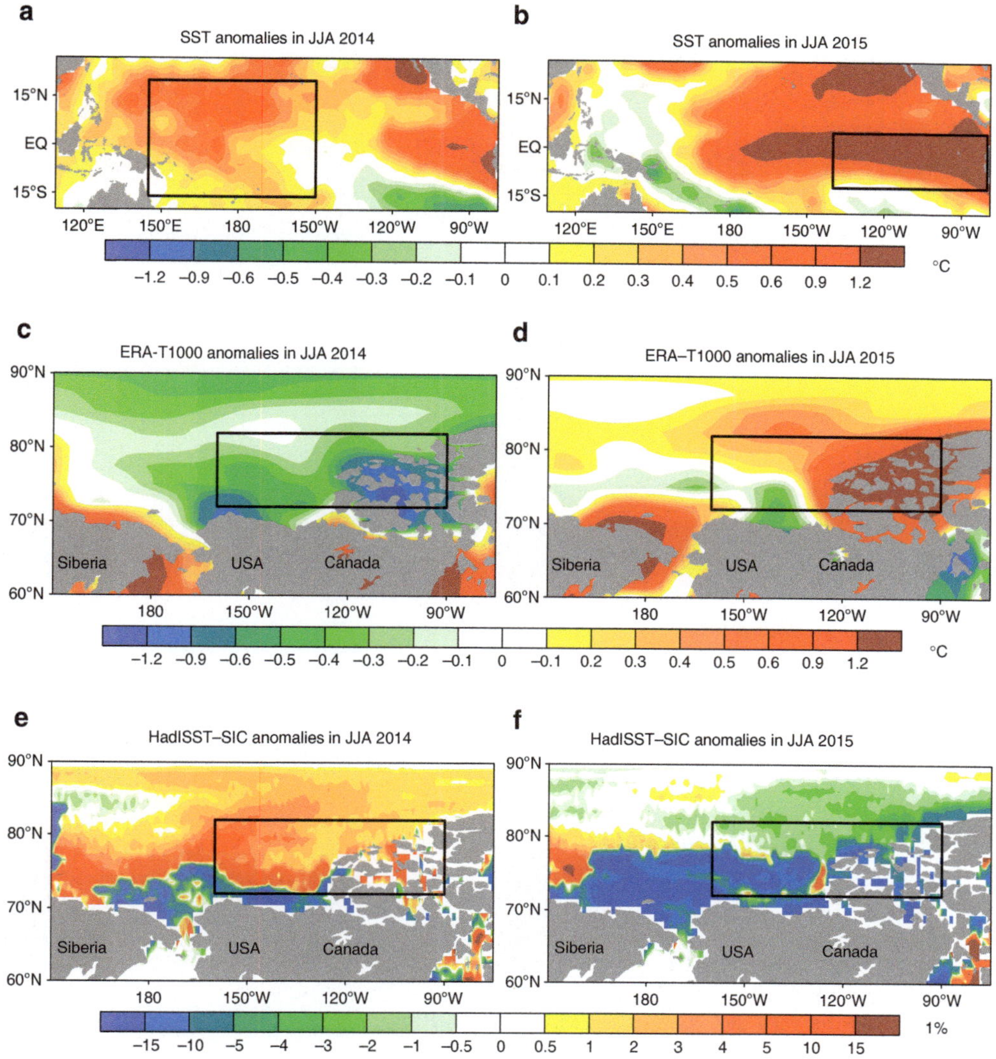

Figure 4 | Case verifications in JJA of 2014 and 2015. Shown in (**a**,**c**,**e**) are the SST, 1,000-hPa air temperature (ERA-Interim) and HadISST SIC anomalies in JJA of 2014, respectively. (**b**,**d**,**f**) are the same as in **a**,**c**,**e**, except for JJA of 2015. The boxes in **a**,**b** outline the CP region (140°W–80°W, 12°S–5°N) and the EP region (145°E–150°W, 16°S–20°N), respectively. The boxes in **c**–**f** mark the same region (160°W–90°W, 72°N–82°N).

EP El Niño[39]. Therefore, as shown in Supplementary Fig. 17, the CP El Niño-Arctic teleconnection potentially enhances our understanding of the predictability of the changes in summer Arctic climate under the greenhouse warming background.

We note that our argument for summer is different from that in the winter situation—EP El Niño can lead to winter Arctic cooling via a tropically excited Arctic warming mechanism[8]. However, there is no evidence indicating that the EP El Niño can also lead to summer Arctic cooling via the tropically excited Arctic warming mechanism. It is seen that the seasonal evolutions of Canada Basin surface temperature associated with the CP and EP indices are different from each other (Supplementary Fig. 18), although the features are similar among various seasons for both CP El Niño patterns and EP El Niño patterns. Different teleconnection mechanisms between summer and winter seem to be due mainly to seasonal reversal of climatological background state and to different air–sea interactions within and/or outside the Pacific Ocean[40], that is, different amplitudes and locations of the warm pool[41], ITCZ/heating and the basic flow[37] (Supplementary Fig. 19). For example, evidence has indicated that the ENSO modulates the polar vortex mainly by the Aleutian low via the Pacific-North American pattern during the boreal winter[37,42,43], but this mechanism seems less relevant in summer. This argument is consistent with that of Sung *et al.*[44], which suggested that the differences in polar responses between CP El Niño and EP El Niño arose from variation of seasonal evolution of extratropical waves, which depends strongly on the source location of tropical heating. Accordingly, here we emphasize that the summer teleconnections associated with CP and EP El Niño events are substantially distinct from the winter teleconnections[8,45] because of the differences in background climate.

We also note that the response of winter stratospheric polar vortex to the ENSO is sensitive to the phase of Quasi-biannual Oscillation[43,46,47] and to the stratospheric sudden warming[48]. Other factors such as the Arctic sea ice[49], Eurasian snow[50], extra-polar SST trend[51] and volcanic eruption[52] can also strongly affect the winter stratospheric polar vortex. Whether the response of boreal summer tropospheric APV to CP El Niño also depends on a third factor (for example, sea ice, snow, Madden–Julian oscillation, Atlantic Multidecadal Oscillation and so on) is a subject of further investigation. Nonetheless, the respective role of each factor has not been systematically examined, albeit these factors may be interconnected. In short, the changes in Arctic climate (that is, air temperature, sea ice, APV, CWW and so on) are influenced by many factors, including atmospheric

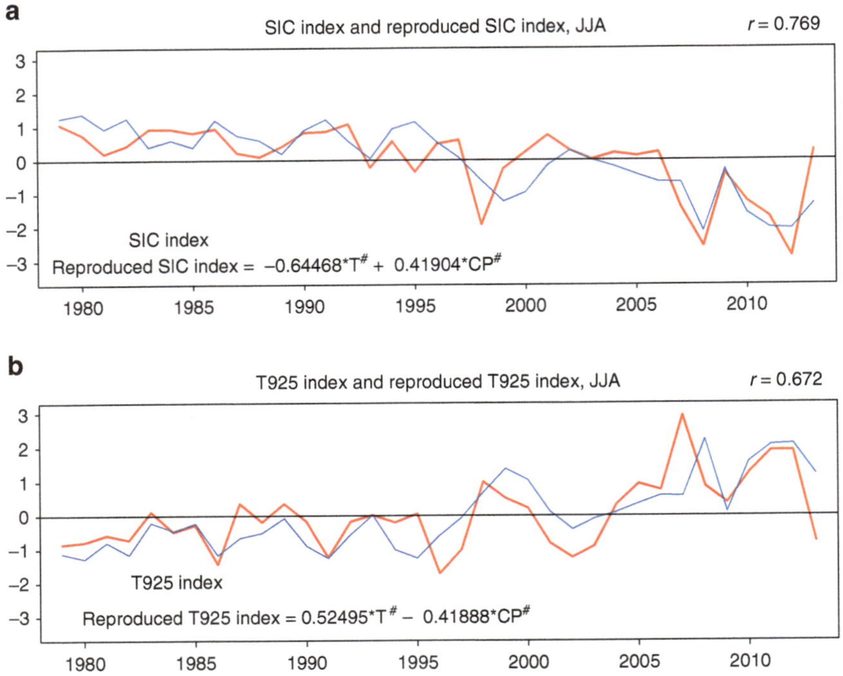

Figure 5 | Original and reproduced time series of SIC-index and T925-index. (**a**) The normalized SIC-index and the reproduced SIC-index, as well as their correlation. (**b**) The normalized T925-index and the reproduced T925-index, as well as their correlation. The reproduced indices are derived from the multiple regression equations as shown in the figure (in the two equations, '$T^{\#}$' and '$CP^{\#}$' indicate the normalized trend time series and the normalized and detrended CP index, respectively).

circulation, surface heat flux, ocean heat transport, ocean temperature, ice-albedo feedback, atmospheric CO_2 and surface air temperature and so on. The respective roles and their sensitivities to tropical SST forcing of these factors may be different between winter and summer. The CPW-Arctic teleconnection proposed here provides an addition to the possible factors that affect tropospheric APV and Arctic climate during the boreal summer.

Methods

Observational data sets. The averages of two SST data sets obtained from the NOAA Extended Reconstructed SST version 3b[53] (ERSSTv3b; http://www.esrl.noaa.gov/psd/data/gridded/data.noaa.ersst.html) and the Hadley Centre Global Sea Ice and Sea Surface Temperature[54] (HadISST; http://www.metoffice.gov.uk/hadobs/hadisst/data/download.html) are used. The HadISST SST is interpolated onto the same 2° × 2° longitude–latitude grid as the NOAA SST. Also used are the averages of two precipitation data sets, namely, the Global Precipitation Climatology Project version 2.2[55] (http://www.esrl.noaa.gov/psd/data/gridded/data.gpcp.html) and the Climate Prediction Center Merged Analysis of Precipitation[56] (http://www.esrl.noaa.gov/psd/data/gridded/data.cmap.html). The NOAA Interpolated OLR[57] (http://www.esrl.noaa.gov/psd/data/gridded/data.interp_OLR.html), the NSIDC SIC[58] (http://nsidc.org/data/docs/daac/nsidc0079_bootstrap_seaice.gd.html) from the Nimbus-7 SMMR and DMSP SSM/I-SSMIS Version 2, the HadISST SIC (http://www.metoffice.gov.uk/hadobs/hadisst/data/download.html) and the COBE[59] SIC (http://www.esrl.noaa.gov/psd/data/gridded/data.cobe2.html) are also used. The ERA-Interim reanalysis[60] (http://apps.ecmwf.int/datasets/) is provided by the European Centre for Medium-Range Weather Forecasts, and the MERRA reanalysis[61] (http://disc.sci.gsfc.nasa.gov/daac-bin/DataHoldings.pl) by the Global Modeling and Assimilation Office at NASA Goddard Space Flight Center. The JRA-55 reanalysis[62] (http://jra.kishou.go.jp/JRA-55/index_en.html) is provided by the Climate Prediction Division, Global Environment and Marine Department, of the Japan Meteorological Agency. In addition to the reanalysis temperatures, combined observed air/marine temperature anomalies are obtained from (i) the NASA Goddard Institute for Space Studies surface temperature analysis[63] (http://www.esrl.noaa.gov/psd/data/gridded/data.gistemp.html) and (ii) the HadCRUT4 data set[64] (http://www.esrl.noaa.gov/psd/data/gridded/data.crutem4.html). Also used is the NOAA GHCN_CAMS 2 m LST[65] (http://www.esrl.noaa.gov/psd/data/gridded/data.ghcncams.html). For detection of interannual variation, all data sets are first normalized and detrended, unless noted otherwise. The base period used in this study is 1979–2013.

Analysis methodology and significance test. As suggested by Garfinkel *et al.*[45], attentions are needed when selecting indices/methods to identify the CP El Niño. Accordingly, following Lian and Chen[66], in this study we apply the method of REOF decomposition[67] (see Supplementary Note 1 for details and reasons) to identify the dominant modes (namely, the two types of El Niño) of tropical Pacific SST anomalies (120°E–80°W, 20°S–20°N). All regression results are shown by the values of regression coefficient scaled by a factor of 2. Correspondingly, a value of correlation (r) above 0.28 is used to estimate the 90% confidence level for a 35-year length of 1979–2013. Here, r is defined as

$$r = \frac{\sum_{i=1}^{n}(x_i - \bar{x})(y_i - \bar{y})}{\sqrt{\sum_{i=1}^{n}(x_i - \bar{x})^2}\sqrt{\sum_{i=1}^{n}(y_i - \bar{y})^2}}, \quad (1)$$

where $\bar{x}$ and $\bar{y}$ are the sample mean values of x_i and y_i from $i = 1$ to $i = n$ (n is the sample size). For composite analysis, we determine the statistical significance level based on the Student's t-test. Here, t is defined as

$$t = \frac{\bar{x} - \bar{y}}{\sqrt{\frac{(m-1)s_1^2 + (n-1)s_2^2}{m+n-2}}\sqrt{\frac{1}{m} + \frac{1}{n}}} \quad (2)$$

where $\bar{x}$ and $\bar{y}$ are the sample mean values of x_i from $i = 1$ to $i = m$ and y_i from $i = 1$ to $i = n$ (m and n are the sample sizes); s_1^2 and s_2^2 are the sample variances; $m + n - 2$ is the degrees of freedom.

EP index and CP index. The detrended EP and CP indices are individually defined as the area mean of normalized and detrended SST anomalies at each grid in the regions marked by the blue boxes shown in Fig. 1a (140°W–80°W, 12°S–5°N) and Fig. 1b (145°E–150°W, 16°S–20°N), respectively; and the resulted indices are further normalized to obtain a unit s.d.

Rossby wave activity flux. We use the Rossby wave activity analysis[68,69] to reveal quasi-stationary Rossby wave propagations. The analysis is suitable for detecting the propagating characteristics of large-scale quasi-stationary planetary waves.

NCAR climate model. To further confirm the CPW-Arctic teleconnection and establish the causal relationship between CP El Niño SST forcing and Arctic regional climate anomalies, experiments are performed with the global CAM4 (ref. 70) from the US NCAR. The CAM4 is the atmospheric component of the Community Earth System Model (CESM 1.2.0), with a 1.9° × 2.5° finite volume grid at 26 hybrid sigma levels. The atmospheric composition is kept at the values of year 2000 (for example, the CO_2 concentration is set as a constant 367.0 p.p.m. in the simulations). A 40-year control run is forced with annually cycled climatological monthly mean SST (1979–2013). The forced run is similar to the control run, but is forced by the total SST constructed by adding the climatological SST and the SST anomalies associated with the CP El Niño in each calendar month. The SST anomalies are constructed by regressing SST anomalies over the tropical region (40°E–80°W, 25°S–25°N) to the CP index and then scaling it by a factor of 2. The impacts produced by CP El Niño warming in the forced run are identified by subtracting the results from the control run. For a more comprehensive comparison, the Arctic responses during the corresponding successive 10 years (that is, the 31–40 years), 20 years (that is, the 21–40 years) and 30 years (that is, the 11–40 years) are used to investigate the model responses to specified SST forcing.

CAM4 code availability. The CAM4 code is available via the URL: https://svn-ccsm-models.cgd.ucar.edu/cesm1/release_tags/cesm1_2_2/.

References

1. Duarte, C., Lenton, T., Wadhams, P. & Wassmann, P. Abrupt climate change in the Arctic. *Nat. Clim. Change* **2,** 60–62 (2012).
2. Jeffries, M., Overland, J. & Perovich, D. The Arctic shifts to a new normal. *Phys. Today* **66,** 35–40 (2013).
3. Gillett, N. P. *et al.* Attribution of polar warming to human influence. *Nat. Geosci* **1,** 750–754 (2008).
4. Serreze, M. C., Barrett, A. P., Stroeve, J. C., Kindig, D. N. & Holland, M. M. The emergence of surface-based Arctic amplification. *Cryosphere* **3,** 11–19 (2009).
5. Polyakov, I., Walsh, J. E. & Kwok, R. Recent changes of Arctic multiyear sea ice coverage and the likely causes. *Bull. Am. Meteorol. Soc* **93,** 145–151 (2012).
6. Chylek, P., Folland, C. K., Lesins, G., Dubey, M. K. & Wang, M. Y. Arctic air temperature change amplification and the Atlantic multidecadal oscillation. *Geophys. Res. Lett.* **36,** L14801 (2009).
7. Ding, Q. *et al.* Tropical forcing of the recent rapid Arctic warming in northeastern Canada and Greenland. *Nature* **509,** 209–212 (2014).
8. Lee, S. Testing of the tropically excited Arctic warming mechanism (TEAM) with traditional El Niño and La Niña. *J. Clim* **25,** 4015–4022 (2012).
9. Lee, S., Feldstein, S. B., Pollard, D. & White, T. S. Do planetary wave dynamics contribute to equable climates? *J. Clim* **24,** 2391–2404 (2011).
10. Lee, S., Gong, T., Johnson, N., Feldstein, S. & Pollard, D. On the possible link between tropical convection and the Northern Hemisphere Arctic surface air temperature change between 1958 and 2001. *J. Clim* **24,** 4350–4367 (2011).
11. Yoo, C., Feldstein, S. & Lee, S. The impact of the Madden–Julian oscillation trend on the Arctic amplification of surface air temperature during the 1979–2008 boreal winter. *Geophys. Res. Lett.* **38,** L24804 (2011).
12. Yoo, C., Lee, S. & Feldstein, S. B. Arctic response to an MJO-like tropical heating in an idealized GCM. *J. Atmos. Sci.* **69,** 2379–2393 (2012).
13. Yoo, C., Lee, S. & Feldstein, S. B. Mechanisms of Arctic surface air temperature change in response to the Madden–Julian oscillation. *J. Clim* **25,** 5777–5790 (2012).
14. Screen, J. A. Influence of Arctic sea ice on European summer precipitation. *Environ. Res. Lett.* **8,** 044015 (2013).
15. Tang, Q., Zhang, X., Yang, X. & Francis, J. A. Cold winter extremes in northern continents linked to Arctic sea ice loss. *Environ. Res. Lett.* **8,** 014036 (2013).
16. Tang, Q., Zhang, X., Yang, X. & Francis, J. A. Extreme summer weather in northern mid-latitudes linked to a vanishing cryosphere. *Nat. Clim. Change* **4,** 45–50 (2014).
17. Boé, J., Hall, A. & Qu, X. September sea ice cover in the Arctic Ocean projected to vanish by 2100. *Nat. Geosci* **2,** 341–343 (2009).
18. Graversen, R. G., Mauritsen, T., Tjernström, M., Källén, E. & Svensson, G. Vertical structure of recent Arctic warming. *Nature* **451,** 53–56 (2008).
19. Screen, J. A. & Simmonds, I. The central role of diminishing sea ice in recent Arctic temperature amplification. *Nature* **464,** 1334–1337 (2010).
20. Yeh, S.-W., Kug, J.-S., Dewitte, B., Kwon, M.-H., Kirtman, B. P. & Jin, F.-F. El Niño in a changing climate. *Nature* **461,** 511–514 (2009).
21. Ashok, K., Behera, S. K., Rao, S. A., Weng, H. & Yamagata, T. El Niño Modoki and its possible teleconnection. *J. Geophys. Res.* **112,** C11007 (2007).
22. An, S.-I. & Wang, B. Interdecadal change of the structure of the ENSO mode and its impact on the ENSO frequency. *J. Clim* **13,** 2044–2055 (2000).
23. McPhaden, M. J., Zebiak, S. E. & Glantz, M. H. ENSO as an integrating concept in Earth science. *Science* **314,** 1740–1745 (2006).
24. Zhang, W., Li, J. & Jin, F.-F. Spatial and temporal features of ENSO meridional scales. *Geophys. Res. Lett.* **36,** L15605 (2009).
25. Lee, T. & McPhaden, M. J. Increasing intensity of El Niño in the central-equatorial Pacific. *Geophys. Res. Lett.* **37,** L14603 (2010).
26. Yu, J.-Y. & Kim, S. T. Identifying the types of major El Niño events since 1870. *Int. J. Climatol.* **33,** 2105–2113 (2013).
27. Bader, J. Climate science: the origin of regional Arctic warming. *Nature* **509,** 167–168 (2014).
28. Kim, H.-M., Webster, P. J. & Curry, J. A. Impact of shifting patterns of Pacific Ocean warming on North Atlantic tropical cyclones. *Science* **325,** 77–80 (2009).
29. Yu, J.-Y. & Zou, Y. The enhanced drying effect of Central-Pacific El Niño on US winter. *Environ. Res. Lett.* **8,** 014019 (2013).
30. Karori, M. A., Li, J. & Jin, F.-F. The asymmetric influence of the two types of El Niño and La Niña on summer rainfall over southeast China. *J. Clim* **26,** 4567–4582 (2013).
31. Sardeshmukh, P. D. & Hoskins, B. J. The generation of global rotational flow by steady idealized tropical divergence. *J. Atmos. Sci.* **45,** 1228–1251 (1988).
32. Kay, J. E., L'Ecuyer, T., Gettelman, A., Stephens, G. & O'Dell, C. The contribution of cloud and radiation anomalies to the 2007 Arctic sea ice extent minimum. *Geophys. Res. Lett.* **35,** L08503 (2008).
33. Ogi, M. & Wallace, J. M. The role of summer surface wind anomalies in the summer Arctic sea ice extent in 2010 and 2011. *Geophys. Res. Lett.* **39,** L09704 (2012).
34. Li, X., Holland, D. M., Gerber, E. P. & Yoo, C. Impacts of the north and tropical Atlantic Ocean on the Antarctic Peninsula and sea ice. *Nature* **505,** 538–542 (2014).
35. Graf, H.-F. & Zanchettin, D. Central Pacific El Niño, the "subtropical bridge," and Eurasian climate. *J. Geophys. Res.* **117,** D01102 (2012).
36. Hoskins, B. J. & Karoly, D. J. The steady linear response of a spherical atmosphere to thermal and orographic forcing. *J. Atmos. Sci.* **38,** 1179–1196 (1981).
37. Jin, F. & Hoskins, B. J. The direct response to tropical heating in a baroclinic atmosphere. *J. Atmos. Sci.* **52,** 307–319 (1995).
38. Lee, S. & Yoo, C. On the causal relationship between poleward heat flux and the Equator-to-Pole temperature gradient: a cautionary tale. *J. Clim.* **27,** 6519–6525 (2014).
39. Webster, P. J. & Yang, S. Monsoon and ENSO: selectively interactive systems. *Quart. J. Roy. Meteor. Soc.* **118,** 877–926 (1992).
40. Alexander, M. A. *et al.* The atmospheric bridge: the influence of ENSO teleconnections on air–sea interaction over the global oceans. *J. Clim.* **15,** 2205–2231 (2002).
41. Wang, B., Wu, R. & Fu, X. Pacific–East Asian teleconnection: How does ENSO affect East Asian climate? *J. Clim.* **13,** 1517–1536 (2000).
42. Horel, J. D. & Wallace, J. M. Planetary-scale atmospheric phenomena associated with the Southern Oscillation. *Mon. Wea. Rev.* **109,** 813–829 (1981).
43. Garfinkel, C. I. & Hartmann, D. L. Different ENSO teleconnections and their effects on the stratospheric polar vortex. *J. Geophys. Res.* **113,** D18114 (2008).
44. Sung, M.-K., Kim, B.-M. & An, S.-I. Altered atmospheric responses to eastern Pacific and central Pacific El Niños over the North Atlantic region due to stratospheric interference. *Clim. Dyn.* **42,** 159–170 (2014).
45. Garfinkel, C. I., Hurwitz, M. M., Waugh, D. W. & Butler, A. H. Are the teleconnections of central Pacific and eastern Pacific El Niño distinct in boreal wintertime? *Clim. Dyn.* **41,** 1835–1852 (2013).
46. Garfinkel, C. I. & Hartmann, D. L. Effects of the El Niño southern oscillation and the Quasi-Biennial oscillation on polar temperatures in the stratosphere. *J. Geophys. Res.* **112,** D19112 (2007).
47. Garfinkel, C. I., Hartmann, D. L. & Sassi, F. Tropospheric precursors of anomalous Northern Hemisphere stratospheric polar vortices. *J. Clim.* **23,** 3282–3299 (2010).
48. Iza, M. & Calvo, N. Role of stratospheric sudden warmings on the response to Central Pacific El Niño. *Geophys. Res. Lett.* **42,** 2482–2489 (2015).
49. Kim, B.-M. *et al.* Weakening of the stratospheric polar vortex by Arctic sea-ice loss. *Nat. Commun.* **5,** 4646 (2014).
50. Cohen, J., Barlow, M., Kushner, P. J. & Saito, K. Stratosphere–troposphere coupling and links with Eurasian land surface variability. *J. Clim.* **20,** 5335–5343 (2007).
51. Garfinkel, C. I., Hurwitz, M. M. & Oman, L. D. Effect of recent sea surface temperature trends on the Arctic stratospheric vortex. *J. Geophys. Res. Atmos.* **120,** 5404–5416 (2015).
52. Graf, H.-F., Zanchettin, D., Timmreck, C. & Bittner, M. Observational constraints on the tropospheric and near-surface winter signature of the Northern Hemisphere stratospheric polar vortex. *Clim. Dyn.* **43,** 3245–3266 (2014).
53. Smith, T. M., Reynolds, R. W., Peterson, T. Ç. & Lawrimore, J. Improvements to NOAA's historical merged land–ocean surface temperature analysis (1880–2006). *J. Clim.* **21,** 2283–2296 (2008).
54. Rayner, N. A. *et al.* Global analyses of sea surface temperature, sea ice, and night marine air temperature since the late nineteenth century. *J. Geophys. Res.* **108,** 4407 (2003).
55. Adler, R. F. *et al.* The Version-2 Global Precipitation Climatology Project (GPCP) monthly precipitation analysis (1979–present). *J. Hydrometeor* **4,** 1147–1167 (2003).
56. Xie, P. & Arkin, P. A. Global precipitation: A 17-year monthly analysis based on gauge observations, satellite estimates, and numerical model outputs. *Bull. Amer. Meteor. Soc.* **78,** 2539–2558 (1997).

57. Liebmann, B. & Smith, C. A. Description of a complete (Interpolated) Outgoing Longwave Radiation dataset. *Bull. Am. Meteorol. Soc.* **77,** 1275–1277 (1996).
58. Comiso, J. C. Updated 2015. Bootstrap Sea Ice Concentrations from Nimbus-7 SMMR and DMSP SSM/I-SSMIS Version 2. Boulder, Colorado USA: NASA National Snow and Ice Data Center Distributed Active Archive Center http://dx.doi.org/10.5067/J6JQLS9EJ5HU (2000).
59. Hirahara, S., Ishii, M. & Fukuda, Y. Centennial-scale sea surface temperature analysis and its uncertainty. *J. Clim.* **27,** 57–75 (2014).
60. Dee, D. P. *et al.* The ERA-Interim reanalysis: configuration and performance of the data assimilation system. *Quat. J. R. Meteorol. Soc.* **137,** 553–597 (2011).
61. Rienecker, M. M. *et al.* MERRA: NASA's modern-era retrospective analysis for research and applications. *J. Clim.* **24,** 3624–3648 (2011).
62. Kobayashi, S. *et al.* The JRA-55 Reanalysis: general specifications and basic characteristics. *J. Meteor. Soc. Jpn.* **93,** 5–48 (2015).
63. Hansen, J., Ruedy, R., Sato, M. & Lo, K. Global surface temperature change. *Rev. Geophys.* **48,** RG4004 (2010).
64. Morice, C. P., Kennedy, J. J., Rayner, N. A. & Jones, P. D. Quantifying uncertainties in global and regional temperature change using an ensemble of observational estimates: The HadCRUT4 dataset. *J. Geophys. Res.* **117,** D08101 (2012).
65. Fan, Y. & van den Dool, H. A global monthly land surface air temperature analysis for 1948–present. *J. Geophys. Res.* **113,** D01103 (2008).
66. Lian, T. & Chen, D. An evaluation of rotated EOF analysis and its application to tropical pacific SST variability. *J. Clim* **25,** 5361–5373 (2012).
67. Horel, J. D. A rotated principal component analysis of the interannual variability of the Northern Hemisphere 500-mb height field. *Mon. Wea. Rev.* **109,** 2080–2092 (1981).
68. Plumb, R. A. On the three-dimensional propagation of stationary waves. *J. Atmos. Sci.* **42,** 217–229 (1985).
69. Takaya, K. & Nakamura, H. A formulation of a phase-independent wave-activity flux for stationary and migratory quasigeostrophic eddies on a zonally varying basic flow. *J. Atmos. Sci.* **58,** 608–627 (2001).
70. Neale, R. B. *et al.* The mean climate of the Community Atmosphere Model (CAM4) in forced SST and fully coupled experiments. *J. Clim.* **26,** 5150–5168 (2013).

Acknowledgements

We thank Professor Jianping Li at the Beijing Normal University for providing the programme code of REOF, Dr Qinghua Ding at the University of Washington and Dr Ming Bao at the Nanjing University for helpful discussion. We also acknowledge the three anonymous reviewers, whose comments are helpful for improving the overall quality of the paper. This study was supported by the National Key Scientific Research Plan of China (grants 2012CB956002 and 2014CB953900), the National Natural Science Foundation of China (grants 41375076 and 41375081) and the China Special Fund for Meteorological Research in the Public Interest (grant 201206038). Computing resources for this study were provided by the China National Supercomputer Center in Guangzhou.

Author contributions

C.H., Z.L. and J.C. designed the CAM4 experiments. C.H. conceived the idea, conducted the data analysis and prepared the figures. S.Y. and Q.W. supervised the study. All authors discussed the results. C.H. and S.Y. wrote the paper.

Additional information

Competing financial interests: The authors declare no competing financial interests.

Radiocarbon evidence for enhanced respired carbon storage in the Atlantic at the Last Glacial Maximum

E. Freeman[1], L.C. Skinner[1], C. Waelbroeck[2] & D. Hodell[1]

The influence of ocean circulation changes on atmospheric CO_2 hinges primarily on the ability to alter the ocean interior's respired nutrient inventory. Here we investigate the Atlantic overturning circulation at the Last Glacial Maximum and its impact on respired carbon storage using radiocarbon and stable carbon isotope data from the Brazil and Iberian Margins. The data demonstrate the existence of a shallow well-ventilated northern-sourced cell overlying a poorly ventilated, predominantly southern-sourced cell at the Last Glacial Maximum. We also find that organic carbon remineralization rates in the deep Atlantic remained broadly similar to modern, but that ventilation ages in the southern-sourced overturning cell were significantly increased. Respired carbon storage in the deep Atlantic was therefore enhanced during the last glacial period, primarily due to an increase in the residence time of carbon in the deep ocean, rather than an increase in biological carbon export.

[1]Godwin Laboratory for Palaeoclimate Research, Department of Earth Sciences, University of Cambridge, Cambridge CB2 3EQ, UK. [2]LSCE/IPSL, Laboratoire CNRS-CEA-UVSQ, Domaine du CNRS, bât 12 91198 Gif-sur-Yvette, France. Correspondence and requests for materials should be addressed to E.F. (email: ef276@cam.ac.uk).

Numerous proxies suggest that ocean circulation at the Last Glacial Maximum (LGM) was different from modern, especially in the Atlantic Ocean[1–4]. Despite this, a consensus has yet to emerge on the LGM ocean circulation[5], primarily because of limitations associated with the various proxies, as well as a paucity of data. In agreement with stable oxygen isotope data[6], $\delta^{13}C$ (stable carbon isotopic ratio) and Cd/Ca data suggest that North Atlantic Deep Water (NADW) shoaled by up to 1,000 m during the LGM relative to today[3,4,7]. However, the combination of conservative and non-conservative impacts on nutrient proxies has permitted alternative interpretations, such as greater nutrient accumulation along the flow path[8] (because of greater export production or longer transit times), or variable contributions from northern- and southern-sourced waters between 2 and 4 km water depth[9] (without a complete elimination of NADW in the deep Atlantic). Although $^{231}Pa/^{230}Th$ has been used to reconstruct past rates of overturning circulation[10,11], its interpretation presents difficulties, as it represents an integrated flow speed across the water column and may be affected by scavenging[12].

Additional proxy data are therefore needed to reconstruct the pattern and rate of ocean circulation at the LGM, and to assess its impact on the marine carbon cycle. Radiocarbon-based ventilation ages could prove particularly useful in this regard as they provide an indication of the mean timescale for carbon exchange between the ocean and the atmosphere, including the effects of air–sea gas (and isotope) exchange in the regions of origin of a given parcel of water, and the mean time elapsed since that water was in the mixed layer. Furthermore, by combining ventilation ages with indicators of respired carbon accumulation, such as stable carbon isotopes, changes in the efficiency of the biological carbon pump can be estimated. This is not possible using reconstructions of flow rates or respired/total nutrient levels alone, as the efficiency of the biological carbon pump is controlled by both the export productivity (that is, the flux of carbon into the deep ocean) and the 'leakage' of this carbon out of the deep ocean (via all the processes that contribute to mass transfer from the deep ocean interior to the surface ocean mixed layer where air–sea exchange occurs).

Here we investigate the Atlantic Ocean overturning circulation at the LGM using both radiocarbon and stable carbon isotopes. We find a shallow well-ventilated northern-sourced cell overlying a poorly ventilated, predominantly southern-sourced cell. Furthermore, we find that the storage of respired carbon in the deep Atlantic Ocean was enhanced because of a long residence time, with a constant flux of respired carbon to the deep Atlantic.

Results

Radiocarbon and stable carbon isotopes. Here we present new radiocarbon-based ventilation ages along with stable carbon isotopes from 1,000 to 3,500 m on the Brazil Margin and from 1,100 to 4,700 m on the Iberian Margin (Supplementary Table 1). Ventilation ages are based on benthic-planktonic (B-P) radiocarbon age offsets, to which the estimated surface reservoir age is added to provide an estimate of the deep ocean versus atmospheric radiocarbon age offset (B-Atm). Given the proximity of cores in each of our study regions, surface reservoir ages are expected to be very similar for all cores in each transect. Therefore, despite some uncertainty in the surface reservoir ages, vertical gradients in ventilation ages within each transect will remain accurate.

Radiocarbon-based ventilation ages provide an indication of the extent of isotopic disequilibrium between the marine- and atmospheric carbon reservoirs. More radiocarbon-depleted waters (indicated by higher radiocarbon ventilation ages) will reflect less efficient radiocarbon (and therefore CO_2) exchange between the ocean interior and the atmosphere on average, and therefore a longer carbon sequestration time in the ocean interior. For continuous export of biologically fixed carbon to the ocean interior, a greater radiocarbon ventilation age and a longer carbon sequestration time (that is, relative to the atmosphere) should in turn reflect enhanced accumulation of respired carbon in the ocean interior and a more efficient (since less 'leaky') soft-tissue carbon pump[13].

Ventilation of the Brazil Margin. On the Brazil Margin, the radiocarbon ventilation age-depth profile for the LGM is significantly different from that of the modern ocean (Fig. 1). Although the shallowest core (1,000 m), has a similar ventilation age to that of the modern ocean, the deeper cores have considerably higher ventilation ages, with the deepest core being offset from the surface ocean by up to 1,700 years, compared with just 400 years today.

Stable carbon isotopes also had a distinctly different profile at the LGM compared with the modern. $\delta^{13}C$ shows a significant decrease with depth, ranging from ~ 1.1‰ at 1,000 m to ~ 0.2‰ in the deep ocean. The steepest gradient (~ -0.5‰/1,000 m) is seen between 2,300 and 3,550 m, whereas in the modern ocean the gradient is <0.02‰/1,000 m between these depths.

Ventilation of the Iberian Margin. The Iberian Margin also shows a very different ventilation age depth-profile at the LGM compared with the modern ocean (Fig. 1). Although the intermediate ocean ($<2{,}100$ m) was well ventilated, the deep ocean was poorly ventilated, with B-P age offsets more than 1,000 years greater than in the modern ocean. The shallowest core (1,127 m) has a negative B-P age offset suggesting that the surface was more poorly ventilated than the intermediate ocean (likely influenced by Mediterranean Outflow Water[14]). Beneath this, the B-P age offset increases with depth in the water column. A relatively steep gradient (~ 950 years/1,000 m) is seen for the upper 2,600 m, with a smaller increase in B-P age offset with depth below (~ 200 years/1,000 m). A similar profile is also seen in $\delta^{13}C$ with a steep gradient (~ -0.4‰/1,000 m) between 2 and 3.7 km but with relatively little change (< -0.1‰/1,000 m) with depth below $\sim 3{,}700$ m. The range of $\delta^{13}C$ (0–0.8‰) at the LGM is significantly greater than the range in the modern North Atlantic (1.0–1.1‰) between 2 and 4.7 km. $\delta^{13}C$ at 4.7 km are more than 0.9‰ lower at the LGM than in the modern ocean. However, extremely low $\delta^{13}C$, previously observed in both the deep South Atlantic (< -0.8‰)[15,16] and in the abyssal western North Atlantic (-0.3 to -0.5‰)[17,18], are not observed in any of the Iberian Margin cores.

Discussion

We combined our new ventilation ages with available published data in order to tentatively reconstruct the distribution of radiocarbon in the Atlantic Ocean during the LGM[18–25] (Fig. 2 and Supplementary Table 2). These radiocarbon-based ventilation ages show that, regardless of the uncertainties in LGM surface reservoir ages, the entire deep Atlantic Ocean (>2 km) was less ventilated during the LGM than the modern ocean (Fig. 3). Today, the deep North Atlantic is well ventilated by newly formed NADW, whereas the deep South Atlantic is primarily influenced by southern-sourced waters (Lower Circumpolar Deep Water, LCDW, and Antarctic Bottom Water, AABW), which have greater ventilation ages because of reduced air–sea gas exchange in the AABW formation regions[26] and because of mixing and entrainment of 'pre-aged' circumpolar deep water into AABW[27]. During the LGM this North–South

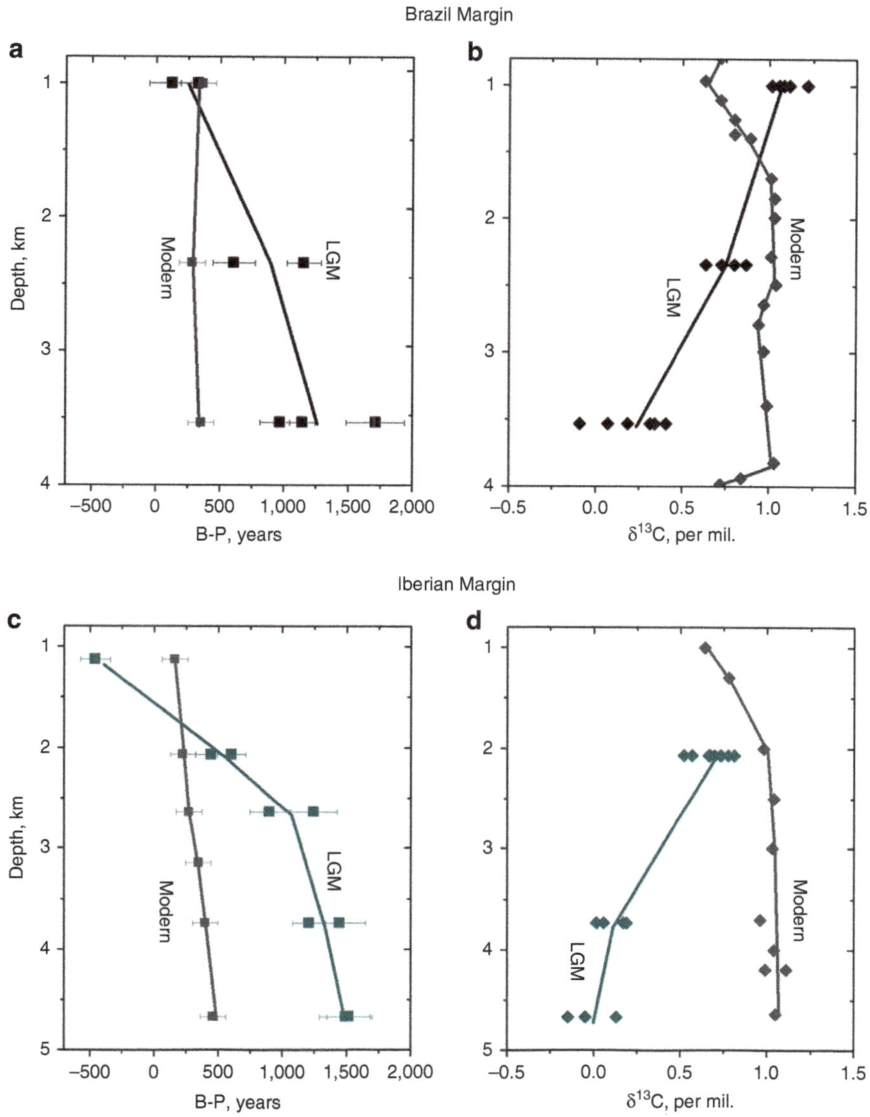

Figure 1 | LGM depth profiles for the Brazil Margin and the Iberian Margin. The benthic-planktonic radiocarbon age offset (B-P; **a**, **c**) and the stable carbon isotopic ratio, $\delta^{13}C$ (**b**, **d**). Modern B-P values (calculated by subtracting the modern surface reservoir age from the bottom water ventilation age) are show in grey (GLODAP[38]). Modern $\delta^{13}C$ of DIC is also shown in grey (GEOSECS[38]). Data are given in Supplementary Table 3.

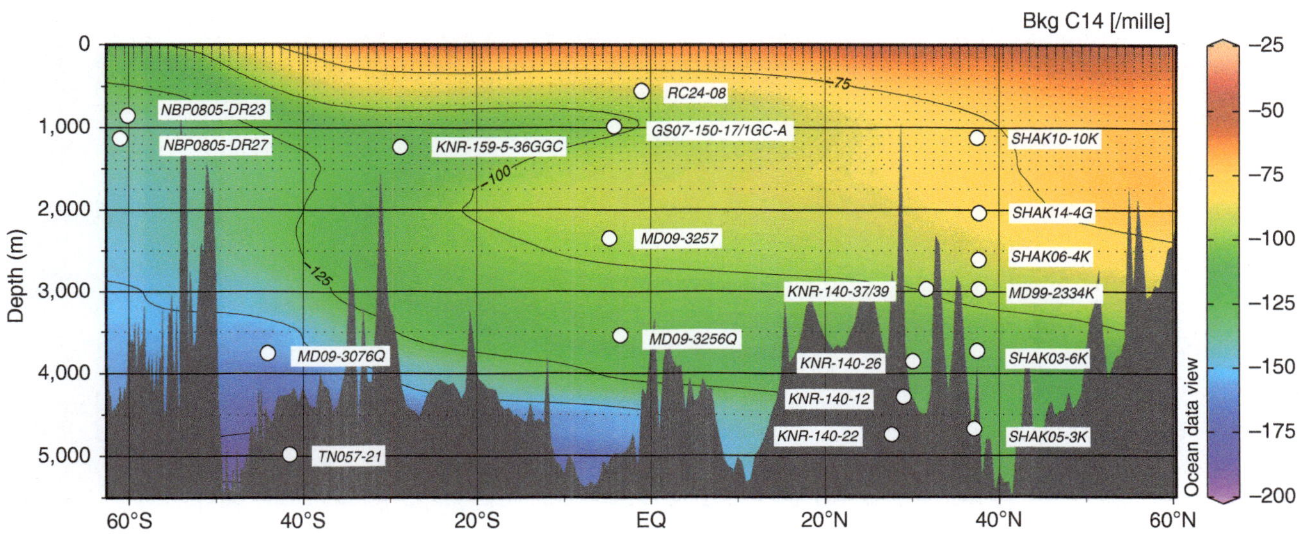

Figure 2 | Map of core locations. Map showing core/dredge locations with modern background $\Delta^{14}C$ (GLODAP[38]), using Ocean Data View[71].

ventilation age gradient in the deep Atlantic was significantly reduced or even eradicated (in the abyss), suggesting a distinctly different circulation must have prevailed (Figs 3 and 4). Based on the apparent northward expansion of poorly ventilated deep waters from the glacial South Atlantic to the deep equatorial and North Atlantic, the LGM circulation appears to have been characterized primarily by a greatly reduced influence of NADW in the deepest Atlantic (>2,000 m). It is notable that abyssal waters in the western North Atlantic and on the northern Brazil Margin had a similar ventilation age to that of the abyssal South Atlantic (Fig. 3). The lack of an increase in ventilation age along the flow path between the South Atlantic and the equatorial and North Atlantic suggests that neither the deep eastern Atlantic nor the deep western Atlantic contained pure southern-sourced waters. A persistent, albeit greatly reduced, northern source of radiocarbon was therefore provided to depths of at least 4.7 km, either via mixing with an overlying well-ventilated water mass or via direct episodic input of dense well-ventilated (for example, overflow) waters.

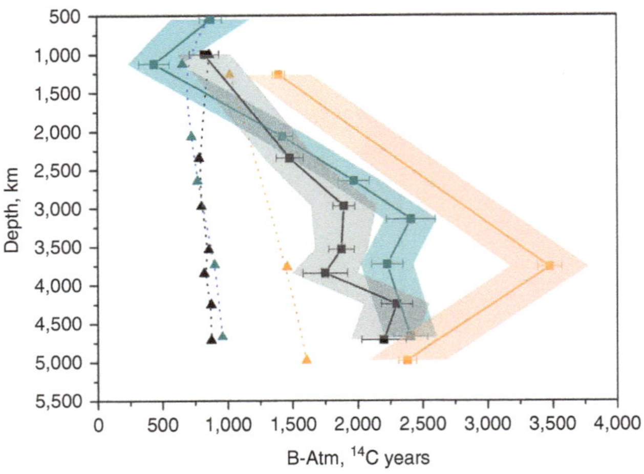

Figure 3 | Atlantic radiocarbon-based ventilation age depth profiles. Triangles: modern ventilation ages (dotted lines). Squares: LGM ventilation ages (solid lines)[18-24,34]. Error bars indicate the analytical error on the radiocarbon measurements. The shaded area represents the uncertainty of the surface reservoir age. Blue, East Atlantic; black, West Atlantic; orange, South Atlantic.

In the modern ocean, radiocarbon-based ventilation ages are well correlated with DIC and anti-correlated with $\delta^{13}C$ owing to the remineralization of organic carbon as waters flow from the North Atlantic to the North Pacific, via the Southern Ocean (Fig. 5). DIC thus increases in the modern ocean by $\sim 115\ \mu mol\ kg^{-1}$ (ref. 13), whereas the $\delta^{13}C$ of DIC decreases by ~0.8‰ (Fig. 5), for every 10% decrease in the deep ocean's radiocarbon content relative to the atmosphere (that is, for every ~850 ^{14}Cyr ventilation age increase). A similar gradient in $\delta^{13}C$ versus ventilation age is seen in the North and equatorial Atlantic

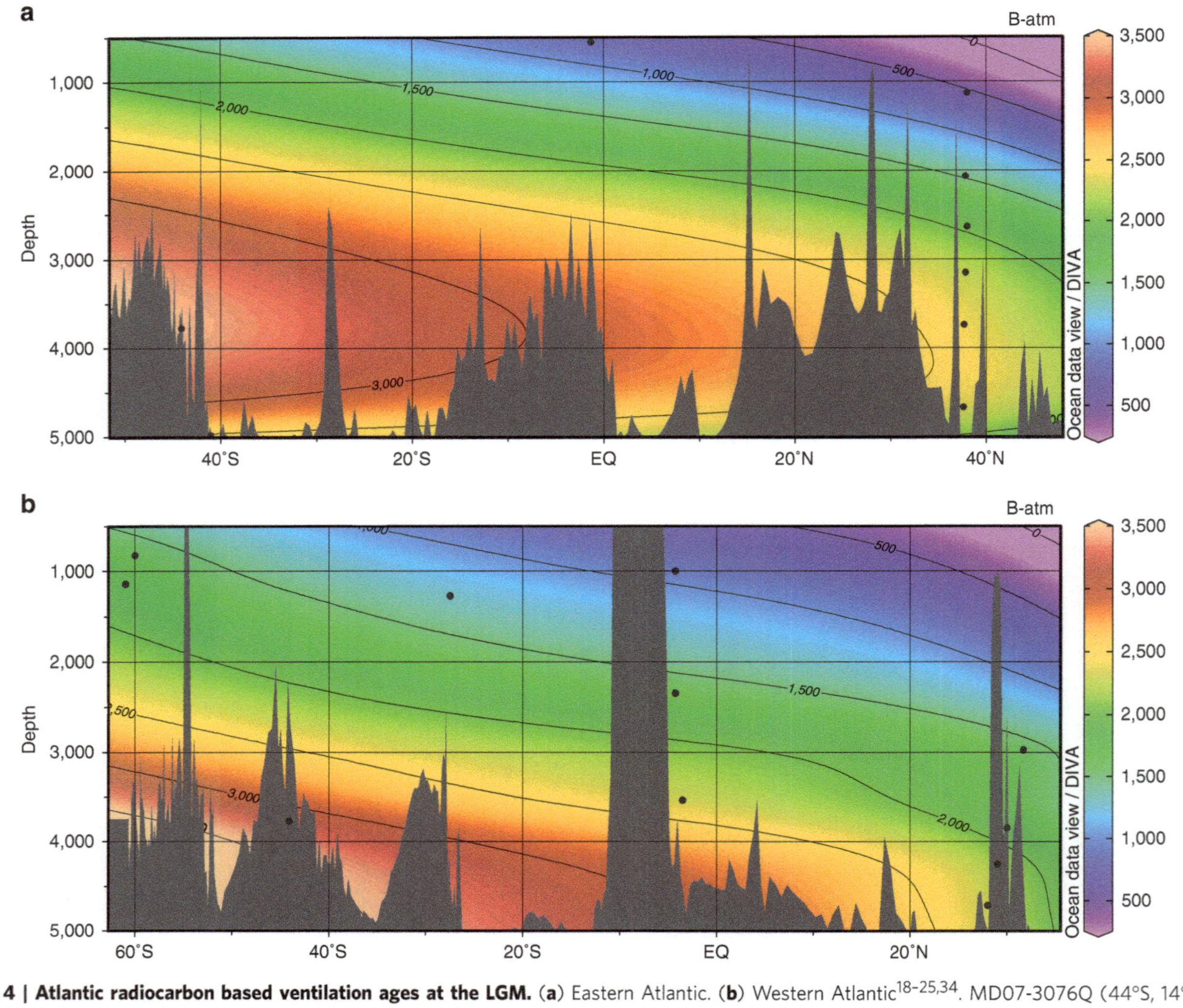

Figure 4 | Atlantic radiocarbon based ventilation ages at the LGM. (**a**) Eastern Atlantic. (**b**) Western Atlantic[18-25,34]. MD07-3076Q (44°S, 14°W, 3770 m) is included in both the eastern and the western profiles as it is located on the Mid-Atlantic Ridge. The dots indicate the position of the cores where ventilation ages have been determined. Gridded in ODV[71] using DIVA gridding with a signal-to-noise ratio of 72. Data are given in Supplementary Table 4.

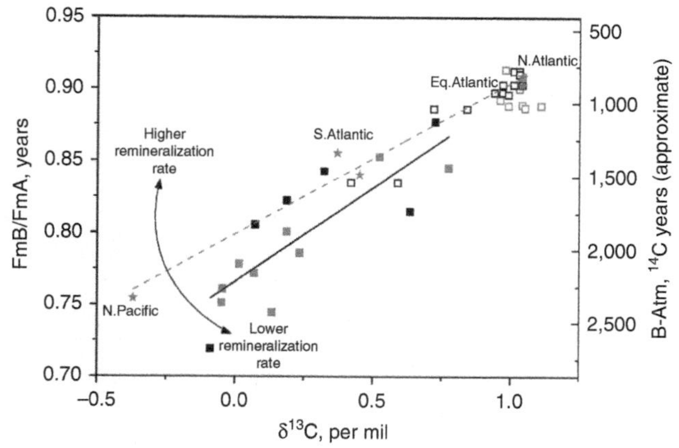

Figure 5 | Crossplot of ventilation age and stable carbon isotopic ratio. Ventilation ages are plotted as FmB/FmA, where FmB/FmA = exp(B-Atm/-8033), and Fm = fraction modern, because B-Atm ages do not mix linearly due to the exponential decay of radiocarbon. Approximate B-Atm values are shown as a guideline only. Modern whole ocean data (~2.5 km water depth; grey stars) is show along with modern data for the entire water column >2 km (open squares) on the Iberian Margin (blue) and Brazil Margin (black). LGM data are shown (filled squares) for the Iberian Margin (blue) and the Brazil Margin (black; Modern data are from GEOSECS and GLODAP[38]). The modern gradient is 0.11 ± 0.01 and the LGM gradient is 0.13 ± 0.02, equivalent to 1.1 ± 0.1‰/kyr and 0.9 ± 0.2‰/kyr in the modern and in the LGM, respectively. For a given rate of remineralization, an increase in residence time (that is, lower FmB/FmA) results in a greater amount of carbon storage.

at the LGM, suggesting that the remineralization rate of organic carbon in the deep Atlantic did not change significantly. On this basis, the LGM rate of DIC increase per year of radiocarbon decay (because of the soft-tissue pump alone) would also be expected to have remained roughly constant, as hypothesized in ref. 13. All else being equal, the ventilation age increase that we observe across the deep Atlantic would therefore imply a significant increase in the respired carbon inventory of the deep ocean at the LGM. This increase was proposed in ref. 8 but later questioned in ref. 9 because of an updated LGM state estimate. Here we not only propose an increase in the remineralized carbon inventory of the deep ocean (based on stable carbon isotope evidence, similar to that advanced previously), but also attribute it to an increase in residence time rather than an increase in biological carbon export rates.

Our results demonstrate a distinct difference in the distribution of radiocarbon in the Atlantic at the LGM relative to today. The deep ocean was more poorly ventilated at all latitudes, whereas the intermediate ocean remained as well ventilated as in the modern ocean. This strongly suggests a shoaling of northern-sourced waters, without a major reduction in their overturning strength, and an expansion of southern-sourced waters at depth. This is in agreement with other proxy reconstructions, which our radiocarbon data help to link more directly to carbon cycle impacts. The concentration of dissolved cadmium inferred from Cd/Ca measured on benthic foraminifera show a clear decrease in the upper ~2.2 km of the water column at the LGM relative to today, with the opposite sense of change in the deep ocean[4] (Supplementary Fig. 1a). This would suggest that high-nutrient waters expanded at depth while low-nutrient northern-sourced waters dominated in the intermediate ocean, consistent with $\delta^{13}C$ data that also show the opposite sense of change above and below ~2.2 km (ref. 4; Supplementary Fig. 1b). In principle, this change in nutrient distributions could have been driven by enhanced export productivity and/or longer ocean interior residence times.

As indicated above, our combined radiocarbon and stable isotope data indicate that the latter mechanism was dominant. Our interpretation that this change in circulation was primarily associated with an expansion of southern-sourced waters in the deepest Atlantic is further supported by εNd data, which indicate an increase in southern-sourced waters at depth with an increase in northern-sourced waters above[28] (Supplementary Fig. 1c). $^{231}Pa/^{230}Th$ ratios also display a distinct change of the opposite sign above and below ~2.2 km water depth[10] (Supplementary Fig. 1d). These $^{231}Pa/^{230}Th$ ratios suggest that the North Atlantic overturning circulation cell was just as strong during the LGM as it is today, but that it penetrated to shallower depths. Although this contrasts with many numerical modelling studies that tend to find a slower overturning when NADW shoals[29,30], our ventilation ages also suggest that the upper North Atlantic overturning rate was not reduced so much that it significantly decreased the supply radiocarbon to the intermediate ocean. The radiocarbon data are therefore consistent with the proposal that a significant shoaling of NADW may not necessarily imply a major reduction in the NADW formation/transport rate at the LGM.

However, while our results support the existence at the LGM of a shallow NADW cell overlying an expanded, radiocarbon-depleted AABW below, they also suggest that the deep southern-sourced cell was not completely isolated. In both the eastern and the western North Atlantic, a northern source of relatively radiocarbon-enriched and high $\delta^{13}C$ waters was mixed into the deep cell. This could imply a more complex Atlantic circulation than one characterized by two completely separate overturning cells at the LGM[31], where the northern overturning cell is either strong and deep or weak and shallow.

As in the landmark study of ref. 3, our stable isotope data demonstrate the existence of enhanced gradients in $\delta^{13}C$ between ~2,000 and 3,500 m in the equatorial and North Atlantic. This pattern of $\delta^{13}C$ in the Atlantic has been interpreted primarily to demonstrate a reduction in the proportion of northern-sourced waters with depth, and an increase in the total nutrient content of the deep Atlantic. Our new radiocarbon data demonstrate that this did not primarily reflect an increase in 'preformed' nutrients (that is, nutrients that were advected to the ocean interior, and that would be associated with carbon that was well-equilibrated with the atmosphere[32]), but rather an increase in respired nutrients (and therefore respired carbon). Alternatively, and equivalently for ocean versus atmosphere carbon inventory changes, the radiocarbon data might also indicate an increase in the 'disequilibrium' carbon inventory of the deep Atlantic[33]. However, in this case, a very large decrease in air–sea CO_2 exchange efficiency would need to be invoked in parallel with an increase in preformed nutrients, in order to account for the paired stable carbon isotope (and nutrient) reconstructions. The observed expansion of southern-sourced deep waters at the LGM was therefore very likely associated with an increase in the efficiency of the soft-tissue carbon pump, primarily due to an increase in the residence time of carbon in the deep Atlantic, rather than an increase in biological export productivity (that is, at low/mid-latitudes). Although global data coverage and better constraints on surface reservoir ages are required to assess the precise magnitude of the atmospheric CO_2 drawdown that is implied by the observed ocean circulation changes, our findings support a direct contribution from the Atlantic overturning circulation via its impact on the deep ocean's respired carbon inventory.

Methods

Radiocarbon and stable carbon isotope measurements. Radiocarbon ages were measured on benthic and planktonic foraminifera from a series of cores recovered from the Brazil Margin by the R/V *Marion Dufresne* (MD09-3256Q, MD09-3257) and the Iberian Margin by the R/V *James Cook* (SHAK03-6K, SHAK05-3K, SHAK06-4K, SHAK10-10K and SHAK14-4G; Fig. 2). Results from these cores are combined with previously published data from the same transect regions (MD99-2334K and GS07-150-17/1GC-A)[22,34]. Core locations are given in Supplementary Table 1. Benthic stable carbon isotopes were also measured in these cores on *Cibicidoides wuellerstorfi*. No stable carbon isotopes were measured for SHAK10-10K because of the lack of any specimens of *Cibicidoides*.

Foraminifera were picked from the > 212 μm size fraction and where necessary from the 150–212 μm fraction. Samples of *Globigerinoides ruber* (Brazil Margin) or *Globigerina bulloides* (Iberian Margin) and samples of mixed benthic foraminifera (excluding agglutinated species) were picked and graphitized in the Godwin Radiocarbon Laboratory at the University of Cambridge. Stable carbon isotope measurements were conducted on *Cibicidoides wuellerstorfi* and on DIC at the Godwin Laboratory, Cambridge.

Radiocarbon measurements. Samples were cleaned on a glass plate with deionized water to remove any loose material. Samples were then dried and acidified. The CO_2 produced was converted to graphite using a standard hydrogen/iron catalyst reduction method[35]. AMS-^{14}C dates were obtained for the graphite samples at the ^{14}Chrono Centre, Queens University Belfast. All dates are reported as conventional radiocarbon ages following[36].

Stable carbon isotope measurements on foraminifera. Stable carbon isotope measurements were conducted on *Cibicidoides wuellerstorfi* in the Godwin Laboratory. Each measurement was run on 1–3 individuals with a combined mass of 50–180 μg. The foraminifera were transferred into sample vials, crushed and then dried in an oven at 50 °C. Samples were reacted with orthophosphoric acid (100%) and the CO_2 produced was cryogenically dried and then admitted to the dual inlet mass spectrometer for isotopic analysis by comparison with a reference gas. Each run of 30 samples was accompanied by 10 reference carbonates and 2 control samples. The results are reported with reference to the international standard VPDB and the precision is better than ± 0.06‰ for $^{12}C/^{13}C$ and ± 0.08‰ for $^{16}O/^{18}O$.

Stable carbon isotope DIC measurements. DIC measurements were made using a Thermo Gas Bench attached to a Delta V Mass Spectrometer. Three or four drops of orthophosphoric acid (100%) were preloaded into a reaction vial, which was capped, sealed and the headspace flushed with Helium gas. Approximately 1.5 ml of sample water was injected into the vial through the butyl rubber septa using a syringe and left to react for 1 h. The sample tubes were transferred to the Gas Bench and CTC CombiPal Autosampler and the resulting CO_2 in the headspace analysed using a Thermo Delta V Mass Spectrometer. A series of standards and reference samples distributed throughout the run were used to calibrate to the international standard VPDB. Results have a reproducibility of better than 0.1 per mille.

Ventilation ages. Here we report ventilation ages as either deep to shallow sub-surface radiocarbon age offsets, based on benthic-planktonic radiocarbon age differences (B-P), or as deep-atmospheric radiocarbon age offsets (B-Atm), which represent the B-P plus the surface reservoir age (that is, the shallow sub-surface versus atmosphere radiocarbon disequilibrium) that applies to the planktonic radiocarbon age. Although B-P offsets provide useful (and very accurate) information regarding vertical radiocarbon gradients at various locations in the ocean, B-Atm estimates are advantageous by virtue of referencing radiocarbon activities throughout the ocean interior relative to a single reference point: the radiocarbon activity of the atmosphere. They also maintain a constant scaling with respect to radiocarbon disequilibria, unlike relative isotopic offset metrics such as $\Delta^{14}C$ (ref. 37). When mapped throughout the ocean, B-Atm offsets may thus provide a coherent means of inferring the patterns and rates of ocean-atmosphere CO_2 exchange and transport in the ocean interior[38]. However, the uncertainty associated with LGM surface reservoir age estimates makes the derivation of accurate B-Atm offsets challenging.

Surface reservoir ages at the LGM. To accurately reconstruct the distribution of radiocarbon in the Atlantic Ocean at the LGM, benthic radiocarbon activities must be referenced to a single (atmospheric) reference point. Therefore, we need to not only consider the deep- to surface age differences but also the surface to atmosphere age offsets that apply in each instance (alternatively each benthic radiocarbon date can be directly compared with the contemporary atmospheric radiocarbon age, given independent calendar ages). The planktonic foraminifera used in this study calcify in the upper surface ocean (< 100 m deep), but even at these depths the water can be in significant disequilibrium relative to the atmosphere. Today surface reservoir ages range from around 400 to 600 years on the Brazil and Iberian Margins (GLODAP[38]). However, these would have varied in the past due to changes in pCO_2 (affecting air–sea carbon isotope exchange at a given CO_2 solubility), changes in ocean circulation (influencing the mixing/upwelling of 'aged' waters into the surface ocean mixed layer) and, for example, changes in high latitude temperature, salinity and sea-ice cover (affecting CO_2 solubility and gas exchange in sub- mixed-layer source regions). However, determining surface reservoir ages in the past is challenging, and not possible in contexts where independent calendar age or contemporaneous atmospheric radiocarbon age constraints are lacking. The modern surface reservoir age is therefore often used as a best guess, under the tentative assumption that physical conditions affecting radiocarbon exchange and transport have not changed over time[24,39,40].

To accurately determine surface reservoir ages, a calendar age model that is independent of radiocarbon dating and a history of atmospheric radiocarbon variability are required. Although U/Th dating can be used to provide calendar ages for corals[43], chronostratigraphic (including tephrochronological) approaches must be used for sediment cores. In high latitudes, cores can be linked stratigraphically to independently dated ice cores[21,22,41]. However, for low latitude, cores in which chronostratigraphic signals are typically more subdued and synchrony with high-latitude climate changes may be questionable, it is much more difficult to obtain an independent calendar chronostratigraphy. Yet, it is important to note that even at low latitudes, especially in regions with deep mixed layers or upwelling regimes, shallow sub-surface radiocarbon disequilibria are very likely to have changed across the last deglaciation[42,43].

Although it is possible in principle to use model simulations to infer past surface reservoir ages, this approach requires knowledge of past changes in ocean circulation and ocean–atmosphere gas exchange efficiency (for example, because of changes in sea-ice extent or mixed layer depths), which strictly we do not have (indeed, this is typically what we are seeking). Despite this fundamental limitation, efforts have been made to model surface reservoir ages in the past, for example, assuming a constant ocean circulation[44,45]. These studies have shown that at the LGM the pCO_2 difference alone would have caused surface reservoir ages to be around 250 years higher than under modern conditions—a passive response in the ocean that does not reflect the ocean's impact on atmospheric CO_2 but rather the atmosphere's impact on the ocean. Additional changes in the mean exchange rate of CO_2 between the ocean and the atmosphere, because of, for example, increased sea-ice cover and changes in the large-scale overturning circulation, are likely to have further increased these values, especially at high latitudes. Indeed, where shallow sub-surface reservoir age estimates are available for the LGM, they are typically significantly higher than the modern values[21,22,25,42,46–48].

Despite the challenges involved in estimating past shallow sub-surface reservoir age variability, it is clear that ignoring them entirely is not a viable option, especially in a study that seeks to constrain past ocean circulation changes.

Recent work on core MD09-3257, from the Brazil Margin, showed that an independent calendar age model could be established for that core using U-Th dated speleothem records on the adjacent continent[49], based on the fact that increased precipitation is marked by a decrease in $\delta^{18}O$ in South American speleothems[50,51] and increased sedimentary Ti/Ca ratios in marine sediment cores off the northeastern Brazilian coast[52]. However, because of a lack of available calendar age tie points for the LGM, surface reservoir ages based on this calendar age-scale remain sparse. We therefore adopt an estimate for LGM surface reservoir ages on the Brazil Margin based on the modern reservoir age corrected for pCO_2-dependent air–sea gas exchange effects[44,45]. A reservoir age of 750 years (250 years greater than modern), which would have arisen due to globally reduced air–sea gas exchange rates, is used. Because it is unclear how ocean circulation changes may have affected this region we apply large uncertainties to this estimate of ± 250 years. By combining high-resolution radiocarbon dating with calendar age models of centennial precision, it may be possible to reduce this uncertainty in the future.

Because the Iberian Margin cores contain a very clear event stratigraphy[53,54], these cores were stratigraphically aligned to the uranium-series dated speleothem records from Hulu Cave[55] and the layer counted (GICC05) NGRIP dust record[56,57] using the Zr/Sr ratio determined using X-ray fluorescence (XRF; see Supplementary Figs 2 and 3). The cores were aligned to both the Hulu speleothem and NGRIP simultaneously using a series of tie points including two tie-points within the LGM (18–23 1000 years before present (kyrs BP)). Shallow sub-surface reservoir ages were determined directly by subtracting the planktonic age from the contemporaneous atmospheric age, using the calendar ages obtained at tie-points, and the atmospheric radiocarbon calibration curve IntCal13 (ref. 58; Supplementary Table 3). LGM surface reservoir ages on the Iberian Margin are thus estimated to be around 900 years with upper and lower limits of 1,100 years and 700 years, respectively. These ages are consistent with the lowest estimates determined for the Iberian Margin by ref. 22 (that is, excluding their tie-points in HS1 and the LGM where the event stratigraphy is arguably more equivocal).

Age models. The planktonic radiocarbon ages were used to construct age models for each of the cores. In order to do so, an estimated surface reservoir age was subtracted from each planktonic radiocarbon date before conversion to calendar ages using BChron version 3.1.5 (ref. 59) and the IntCal13 calibration curve[58]. These calendar ages were then used to construct sediment depth-age models using the Markov Chain Monte-Carlo method, also using Bchron. As LGM surface reservoir age estimates have not been directly estimated on the Brazil Margin, a 'best guess' value of 750 years, based on isotope exchange constraints backed up by model simulations[44,45], was used. For the Iberian Margin, a surface reservoir age of 900 years was used for the LGM based on the stratigraphic alignments described above.

For compiled data, for cores with age models that do not depend on radiocarbon data, surface and deep reservoir ages were determined by subtracting the contemporary atmospheric radiocarbon age (based on Intcal13) from the planktonic and benthic radiocarbon ages, respectively. For sites with no independent age model, the pCO_2 corrected modern reservoir ages (that is, modern plus 250 years at the LGM) were used and added to the benthic-planktonic age offset to determine the deep ocean-atmospheric age offset (Supplementary Table 2).

Stratigraphic alignments. The cores were stratigraphically aligned to the Hulu speleothem $\delta^{18}O$ record[55] and the NGRIP dust record[57] (on the GICC05 age model[56,60]). The NGRIP dust content changes rapidly and synchronously with changes in ice core $\delta^{18}O$ (ref. 61). NGRIP dust, which is predominantly sourced from East Asian deserts[62–65], increases during cold periods[57]. These increases are caused by atmospheric changes[66,67] and/or intensified sources[68], but the exact mechanism is unconstrained. The Hulu speleothem $\delta^{18}O$ record resembles East Asian monsoon changes and is highly correlated with millennial scale Greenland temperature variations[55,69].

The sediment cores were aligned using elemental ratios determined using high-resolution XRF. Zr/Sr ratios were used to align the Iberian Margin cores. Zr/Sr anti-correlates strongly with Ca/Ti in these cores. The Zr/Sr and Ca/Ti ratios reflect the relative input of biogenic (Ca, Sr) and detrital (Zr, Ti) material, and have been shown to correlate strongly with planktonic $\delta^{18}O$ and alkenone sea-surface temperatures on the Iberian Margin, as well as with millennial scale variations in Greenland ice core $\delta^{18}O$ (ref. 70). Zr/Sr values are generally higher during stadials and lower during interstadials, whereas Ca/Ti ratios show the opposite trend.

A master core, SHAK03-6K was selected for the Iberian Margin through which the other cores were aligned. SHAK03-6K was chosen as the master core based on its high quality (no discontinuities) and age range (SHAK14-4G was also of very high quality but only reached back to $\sim$20 kyrs). SHAK03-6K was aligned to both the Hulu speleothem and the NGRIP dust record using a series of tie points (Supplementary Fig. 2). No tie-points were used in the interval 0–11.4 kyrs BP because of the lack of clear signals in the records. The calendar ages at tie points in SHAK03-6K were then transferred to all the other Iberian Margin cores (Supplementary Fig. 3).

XRF data. The Iberian Margin cores were scanned at the University of Cambridge using an Avaatech XRF core scanner (2nd generation). The surface of the cores was scraped clean then covered with a 4-μm SPEXCertiPrep Ultralene foil to avoid contamination and to prevent the cores drying out and cracking. Each section was measured at three different voltages and currents: 10 kV and 750 μA, 30 kV and 500 μA, and at 50 kV and 1,000 μA. The entire length of each core was analysed at 5-mm resolution with an irradiated surface length and width of 5 mm (downcore) and 12 mm (cross core), respectively. The count time was 60 s for each measurement. Element intensities were obtained by post-processing of the XRF spectra using the Canberra WinAxil software with standard software settings and spectrum-fit models.

References

1. Boyle, E. A. & Keigwin, L. D. Deep Circulation of the North Atlantic over the Last 200,000 Years: geochemical Evidence. *Science* **218,** 784–787 (1982).
2. Duplessy, J. C. *et al.* Deep water source variations during the last climatic cycle and their impact on the global deep water circulation. *Paleoceanography* **3,** 343–360 (1988).
3. Curry, W. & Oppo, D. Glacial water mass geometry and the distribution of $\delta^{13}C$ of Sigma CO_2 in the western Atlantic Ocean. *Paleoceanography* **20,** PA1017 (2005).
4. Marchitto, T. M. & Broecker, W. S. Deep water mass geometry in the glacial Atlantic Ocean: a review of constraints from the paleonutrient proxy Cd/Ca. *Geochem. Geophys. Geosystems* **7,** Q12003 (2006).
5. Lynch-Stieglitz, J. *et al.* Atlantic meridional overturning circulation during the Last Glacial Maximum. *Science* **316,** 66–69 (2007).
6. Lund, D. C., Mix, A. C. & Southon, J. Increased ventilation age of the deep northeast Pacific Ocean during the last deglaciation. *Nat. Geosci.* **4,** 771–774 (2011).
7. Oppo, D. W. & Lehman, S. J. Mid-depth circulation of the subpolar North Atlantic during the Last Glacial Maximum. *Science* **259,** 1148–1152 (1993).
8. Gebbie, G. How much did Glacial North Atlantic Water shoal? *Paleoceanography* **29,** 190–209 (2014).
9. Gebbie, G., Peterson, C. D., Lisiecki, L. E. & Spero, H. J. Global-mean marine ^{13}C and its uncertainty in a glacial state estimate. *Quat. Sci. Rev.* **125,** 144–159 (2015).
10. Lippold, J. *et al.* Strength and geometry of the glacial Atlantic Meridional Overturning Circulation. *Nat. Geosci.* **5,** 813–816 (2012).
11. McManus, J. F., Francois, R., Gherardi, J.-M., Keigwin, L. D. & Brown-Leger, S. Collapse and rapid resumption of Atlantic meridional circulation linked to deglacial climate changes. *Nature* **428,** 834–837 (2004).
12. Keigwin, L. D. & Boyle, E. A. Did North Atlantic overturning halt 17,000 years ago? *Paleoceanography* **23,** PA1101 (2008).
13. Sarnthein, M., Schneider, B. & Grootes, P. M. Peak glacial ^{14}C ventilation ages suggest major draw-down of carbon into the abyssal ocean. *Clim. Past* **9,** 2595–2614 (2013).
14. Zahn, R., Sarnthein, M. & Erlenkeuser, H. Benthic isotope evidence for changes of the Mediterranean outflow during the Late Quaternary. *Paleoceanography* **2,** 543–559 (1987).
15. Charles, C. D. & Fairbanks, R. G. Evidence from Southern Ocean sediments for the effect of North Atlantic deep-water flux on climate. *Nature* **355,** 416–416 (1992).
16. Ninnemann, U. S. & Charles, C. D. Changes in the mode of Southern Ocean circulation over the last glacial cycle revealed by foraminiferal stable isotopic variability. *Earth Planet. Sci. Lett.* **201,** 383–396 (2002).
17. Boyle, E. & Keigwin, L. North Atlantic thermohaline circulation during the past 20,000 years linked to high-latitude surface temperature. *Nature* **330,** 35–40 (1987).
18. Keigwin, L. D. Radiocarbon and stable isotope constraints on Last Glacial Maximum and Younger Dryas ventilation in the western North Atlantic. *Paleoceanography* **19,** PA4012 (2004).
19. Cléroux, C., Demenocal, P. & Guilderson, T. Deglacial radiocarbon history of tropical Atlantic thermocline waters: absence of CO_2 reservoir purging signal. *Quat. Sci. Rev.* **30,** 1875–1882 (2011).
20. Barker, S., Knorr, G., Vautravers, M. J., Diz, P. & Skinner, L. C. Extreme deepening of the Atlantic overturning circulation during deglaciation. *Nat. Geosci.* **3,** 567–571 (2010).
21. Skinner, L. C., Fallon, S., Waelbroeck, C., Michel, E. & Barker, S. Ventilation of the deep southern ocean and deglacial CO_2 rise. *Science* **328,** 1147–1151 (2010).
22. Skinner, L. C., Waelbroeck, C., Scrivner, A. E. & Fallon, S. J. Radiocarbon evidence for alternating northern and southern sources of ventilation of the deep Atlantic carbon pool during the last deglaciation. *Proc. Natl Acad. Sci. USA* **111,** 5480 (2014).
23. Sortor, R. N. & Lund, D. C. No evidence for a deglacial intermediate water [DELTA]^{14}C anomaly in the SW Atlantic. *Earth Planet. Sci. Lett.* **310,** 65 (2011).
24. Keigwin, L. D. & Schlegel, M. A. Ocean ventilation and sedimentation since the glacial maximum at 3 km in the western North Atlantic. *Geochem. Geophys. Geosystems* **3,** 1034 (2002).
25. Burke, A. & Robinson, L. F. The Southern Ocean's role in carbon exchange during the last deglaciation. *Science* **335,** 557–561 (2012).
26. Matsumoto, K. Radiocarbon-based circulation age of the world oceans. *J. Geophys. Res. Oceans* **112,** C09004 (2007).
27. Orsi, A. H., Johnson, G. C. & Bullister, J. L. Circulation, mixing, and production of Antarctic Bottom Water. *Prog. Oceanogr.* **43,** 55–109 (1999).
28. Gutjahr, M., Frank, M., Stirling, C. H., Keigwin, L. D. & Halliday, A. N. Tracing the Nd isotope evolution of North Atlantic Deep and intermediate waters in the western North Atlantic since the Last Glacial Maximum from Blake Ridge sediments. *Earth Planet. Sci. Lett.* **266,** 61–77 (2008).
29. Otto-Bliesner, B. L. *et al.* Last Glacial Maximum ocean thermohaline circulation: PMIP2 model intercomparisons and data constraints. *Geophys. Res. Lett.* **34,** L12706 (2007).
30. Chikamoto, M. O. *et al.* Variability in North Pacific intermediate and deep water ventilation during Heinrich events in two coupled climate models. *Deep Sea Res. Part II* **61-64,** 114–126 (2012).
31. Ferrari, R. *et al.* Antarctic sea ice control on ocean circulation in present and glacial climates. *Proc. Natl. Acad. Sci. USA* **111,** 8753–8758 (2014).
32. Ito, T. & Follows, M. J. Preformed phosphate, soft tissue pump and atmospheric CO_2. *J. Mar. Res.* **63,** 813–839 (2005).
33. Ito, T. & Follows, M. J. Air-sea disequilibrium of carbon dioxide enhances the biological carbon sequestration in the Southern Ocean. *Glob. Biogeochem. Cycles* **27,** 1129–1138 (2013).
34. Freeman, E. *et al.* An Atlantic-Pacific ventilation seesaw across the last deglaciation. *Earth Planet. Sci. Lett.* **424,** 237–244 (2015).
35. Freeman, E., Skinner, L. C., Reimer, R., Scrivner, A. E. & Fallon, S. Graphitization of small carbonate samples for palaeoceanographic research at the Godwin Radiocarbon Laboratory, University of Cambridge. *Radiocarbon* **58,** 89–97 (2015).
36. Stuiver, M. & Polach, H. Reporting of ^{14}C data. *Radiocarbon* **19,** 355–363 (1977).
37. Cook, M. S. & Keigwin, L. D. Radiocarbon profiles of the NW Pacific from the LGM and deglaciation: evaluating ventilation metrics and the effect of uncertain surface reservoir ages. *Paleoceanography* **30,** 174–195 (2015).
38. Key, R. M. *et al.* A global ocean carbon climatology: results from Global Data Analysis Project (GLODAP). *Glob. Biogeochem. Cycles* **18,** GB4031 (2004).

39. De Pol-Holz, R., Keigwin, L., Southon, J., Hebbeln, D. & Mohtadi, M. No signature of abyssal carbon in intermediate waters off Chile during deglaciation. *Nat. Geosci.* **3,** 192–195 (2010).
40. Max, L. *et al.* Pulses of enhanced North Pacific Intermediate Water ventilation from the Okhotsk Sea and Bering Sea during the last deglaciation. *Clim. Past* **10,** 591–605 (2014).
41. Govin, A. *et al.* Evidence for northward expansion of Antarctic Bottom Water mass in the Southern Ocean during the last glacial inception. *Paleoceanography* **24,** PA1202 (2009).
42. de la Fuente, M., Skinner, L., Calvo, E., Pelejero, C. & Cacho, I. Increased reservoir ages and poorly ventilated deep waters inferred in the glacial Eastern Equatorial Pacific. *Nat. Commun* **6,** 7420 (2015).
43. Lindsay, C. M., Lehman, S. J., Marchitto, T. M. & Ortiz, J. D. The surface expression of radiocarbon anomalies near Baja California during deglaciation. *Earth Planet. Sci. Lett.* **422,** 67–74 (2015).
44. Butzin, M., Prange, M. & Lohmann, G. Readjustment of glacial radiocarbon chronologies by self-consistent three-dimensional ocean circulation modeling. *Earth Planet. Sci. Lett.* **317-318,** 177–184 (2012).
45. Galbraith, E. D., Kwon, E. Y., Bianchi, D., Hain, M. P. & Sarmiento, J. L. The impact of atmospheric pCO_2 on carbon isotope ratios of the atmosphere and ocean. *Glob. Biogeochem. Cycles* **29,** 307–324 (2015).
46. Sikes, E. L., Samson, C. R., Guilderson, T. P. & Howard, W. R. Old radiocarbon ages in the southwest Pacific Ocean during the last glacial period and deglaciation. *Nature* **405,** 555–559 (2000).
47. Waelbroeck, C. *et al.* The timing of the last deglaciation in North Atlantic climate records. *Nature* **412,** 724–727 (2001).
48. Skinner, L. *et al.* Reduced ventilation and enhanced magnitude of the deep Pacific carbon pool during the last glacial period. *Earth Planet. Sci. Lett.* **411,** 45–52 (2015).
49. Burckel, P. *et al.* Atlantic Ocean circulation changes preceded millennial tropical South America rainfall events during the last glacial. *Geophys. Res. Lett.* **42,** 411–418 (2015).
50. Cheng, H. *et al.* Climate change patterns in Amazonia and biodiversity. *Nat. Commun.* **4,** 1411–1411 (2013).
51. Cruz, F. W. *et al.* Orbitally driven east-west antiphasing of South American precipitation. *Nat. Geosci.* **2,** 210–214 (2009).
52. Jaeschke, A., Rühlemann, C., Arz, H., Heil, G. & Lohmann, G. Coupling of millennial-scale changes in sea surface temperature and precipitation off northeastern Brazil with high-latitude climate shifts during the last glacial period. *Paleoceanography* **22,** PA4206 (2007).
53. Shackleton, N. J. The 100,000-year ice-age cycle identified and found to lag temperature, carbon dioxide, and orbital eccentricity. *Science* **289,** 1897–1902 (2000).
54. Skinner, L. C., Elderfield, H. & Hall, M. in *Ocean Circulation: Mechanisms and Impacts'Past and Future Changes of Meridional Overturning* 197–208 (American Geophysical Union, 2007).
55. Southon, J., Noronha, A. L., Cheng, H., Edwards, R. L. & Wang, Y. A high-resolution record of atmospheric 14C based on Hulu Cave speleothem H82. *Quat. Sci. Rev.* **33,** 32 (2012).
56. Andersen, K. K. *et al.* The Greenland Ice Core Chronology 2005, 15ka. Part 1: constructing the time scale. *Quat. Sci. Rev.* **25,** 3246–3257 (2006).
57. Ruth, U., Wagenbach, D., Steffensen, J. P. & Bigler, M. Continuous record of microparticle concentration and size distribution in the central Greenland NGRIP ice core during the last glacial period. *J. Geophys. Res. Atmospheres* **108,** 4098 (2003).
58. Reimer, P. *et al.* IntCal13 and Marine13 Radiocarbon Age Calibration Curves 0–50,000 Years cal BP. *Radiocarbon* **55,** 1869–1887 (2013).
59. Parnell, A. C., Haslett, J., Allen, J. R. M., Buck, C. E. & Huntley, B. A flexible approach to assessing synchroneity of past events using Bayesian reconstructions of sedimentation history. *Quat. Sci. Rev.* **27,** 1872–1885 (2008).
60. Rasmussen, S. O. *et al.* A new Greenland ice core chronology for the last glacial termination. *J. Geophys. Res. Atmospheres* **111,** D06102 (2006).
61. Ruth, U. *et al.* Ice core evidence for a very tight link between North Atlantic and east Asian glacial climate. *Geophys. Res. Lett.* **34,** L03706 (2007).
62. Biscaye, P. *et al.* Asian provenance of glacial dust (stage 2) in the Greenland Ice Sheet Project 2 Ice Core, Summit, Greenland. *J. Geophys. Res. Oceans* **102,** 26765–26781 (1997).
63. Kahl, J. D. W. *et al.* Air mass trajectories to Summit, Greenland: a 44-year climatology and some episodic events. *J. Geophys. Res.* **102,** 26861–26875 (1997).
64. Svensson, A., Biscaye, P. & Grousset, F. Characterization of late glacial continental dust in the Greenland Ice Core Project ice core. *J. Geophys. Res.-Atmospheres* **105,** 4637–4656 (2000).
65. Bory, A. J., Biscaye, P. E., Svensson, A. & Grousset, F. E. Seasonal variability in the origin of recent atmospheric mineral dust at NorthGRIP, Greenland. *Earth Planet. Sci. Lett.* **196,** 123–134 (2002).
66. Petit, J. R., M, B. & A, R. Ice age aerosol content from East Antarctic ice core samples and past wind strength. *Nature* **293,** 391–394 (1981).
67. Petit, J. R. *et al.* Palaeoclimatological and chronological implications of the Vostok core dust record. *Nature* **343,** 56–58 (1990).
68. Fuhrer, K., Wolff, E. & Johnsen, S. Timescales for dust variability in the Greenland Ice Core Project (GRIP) ice core in the last 100,000 years. *J. Geophys. Res. Atmospheres* **104,** 31043–31052 (1999).
69. Wang, Y. J. *et al.* a high-resolution absolute-dated late pleistocene monsoon record from Hulu Cave, China. *Science* **294,** 2345–2348 (2001).
70. Hodell, D. *et al.* Response of Iberian Margin sediments to orbital and suborbital forcing over the past 420 ka. *Paleoceanography* **28,** 185–199 (2013).
71. Schlitzer, R. Ocean Data View http://odv.awi.de (2016).

Acknowledgements

We thank James Rolfe and Ian Mather for stable isotope measurements, Simon Crowhurst for XRF core scanning, and Ron Reimer for doing the AMS measurements. This work was made possible by NERC support for the cruise that collected the JC89-SHAK cores (NE/J00653X/1) and by NERC grant NE/L006421/1, as well as support from the Royal Society and the Isaac Newton Trust. MD09 cores were collected on board R/V Marion Dufresne during RETRO cruise III, supported by ESF EUROMARC project RETRO, IPEV and ANR project ANR-09-BLAN-0347. C.W. acknowledges support from the European Research Council grant ACCLIMATE/no 339108. This is LSCE contribution 5579.

Author contributions

E.F. and L.C.S. designed the study. L.C.S., D.H. and C.W. collected the core material. E.F. compiled the published data and obtained all the new data. E.F. and L.C.S. analysed the data. E.F. and L.C.S. wrote the manuscript with contributions from all authors.

Additional information

Competing financial interests: The authors declare no competing financial interests.

10

Evidence for ice-free summers in the late Miocene central Arctic Ocean

Ruediger Stein[1,2], Kirsten Fahl[1], Michael Schreck[3], Gregor Knorr[1], Frank Niessen[1], Matthias Forwick[4], Catalina Gebhardt[1], Laura Jensen[1], Michael Kaminski[5], Achim Kopf[2,6], Jens Matthiessen[1], Wilfried Jokat[1,2] & Gerrit Lohmann[1,6]

Although the permanently to seasonally ice-covered Arctic Ocean is a unique and sensitive component in the Earth's climate system, the knowledge of its long-term climate history remains very limited due to the restricted number of pre-Quaternary sedimentary records. During *Polarstern* Expedition PS87/2014, we discovered multiple submarine landslides along Lomonosov Ridge. Removal of younger sediments from steep headwalls has led to exhumation of Miocene sediments close to the seafloor. Here we document the presence of IP_{25} as a proxy for spring sea-ice cover and alkenone-based summer sea-surface temperatures $>4\,°C$ that support a seasonal sea-ice cover with an ice-free summer season being predominant during the late Miocene in the central Arctic Ocean. A comparison of our proxy data with Miocene climate simulations seems to favour either relatively high late Miocene atmospheric CO_2 concentrations and/or a weak sensitivity of the model to simulate the magnitude of high-latitude warming in a warmer than modern climate.

[1] Alfred Wegener Institute (AWI) Helmholtz Centre for Polar and Marine Research, Am Alten Hafen 26, Bremerhaven 27568, Germany. [2] Department of Geosciences (FB5), University of Bremen, Klagenfurter Strasse 4, Bremen 28359, Germany. [3] Arctic Research Centre, Korea Polar Research Institute, 26 Songdomirae-ro, Yeonsu-gu, Incheon 406-840, Korea. [4] Institute of Geology, University of Tromsø—The Arctic University of Norway, P O Box 6050 Langnes, Tromsø 9037, Norway. [5] Geosciences Department, College of Petroleum Engineering & Geosciences, King Fahd University of Petroleum & Minerals, Dhahran 31261, Saudi Arabia. [6] MARUM-Center for Marine Environmental Sciences, University of Bremen, Leobener Strasse, Bremen 28359, Germany. Correspondence and requests for materials should be addressed to R.S. (email: Ruediger.Stein@awi.de).

There is a general consensus that the polar regions—and in particular the Arctic Ocean and surrounding areas—are at present, and were over historic and geologic time scales, subject to rapid and dramatic environmental changes. Owing to complex feedback processes, collectively known as 'polar amplification', the Arctic is both a contributor to climate change and a region that will be most affected by global warming[1–3]. Despite the importance of the Arctic Ocean in the global climate

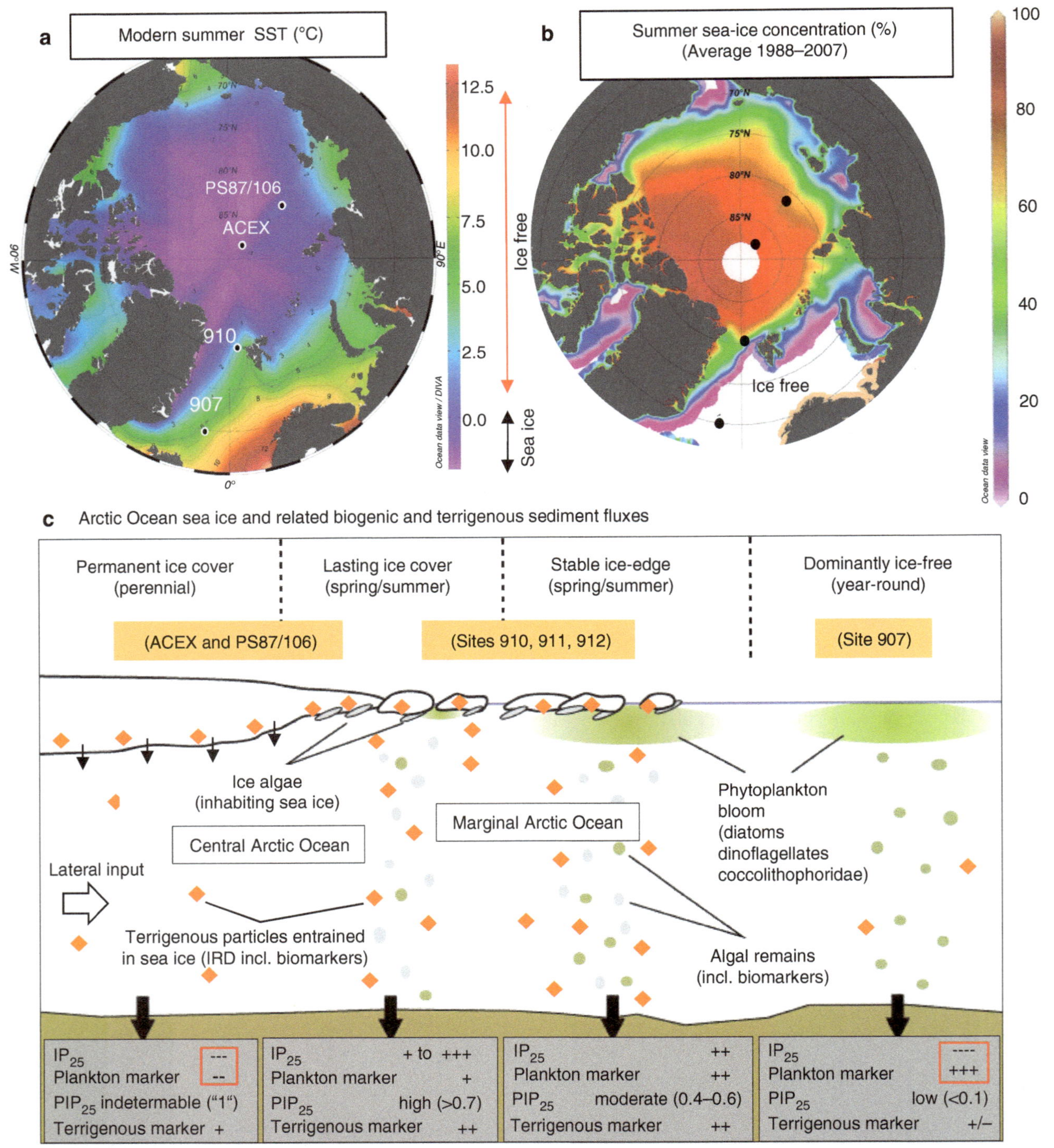

Figure 1 | Modern Arctic Ocean SST and sea-ice conditions. (**a**) Map of modern August SST with locations of ODP Site 907, ODP Site 910 and IODP Expedition 302 (ACEX) Site, as well as Core PS87/106 (August SST data—average of 1955-2012—from World Ocean Atlas; http://odv.awi.de/en/data/ocean/world_ocean_atlas_2013/). (**b**) Map of summer sea-ice concentration—average of 1988-2007—with locations of studied sites (black dots); data source: http://nsidc.org. Maps **a** and **b** produced with Ocean Data View software (source: http://odv.awi.de/). (**c**) Generalized scheme illustrating (1) sea surface conditions and respective (spring/summer) productivities of ice algae and phytoplankton, and (2) sedimentary contents of IP_{25}, terrigenous biomarkers and IRD, phytoplankton-derived biomarkers and PIP_{25} index for different settings in the modern Arctic Ocean (after ref. 14, supplemented). Owing to ice melting and related nutrient and sediment release, a stable ice-edge situation is characterized by high concentrations of IP_{25} and phytoplanktion biomarkers, but also by high concentrations of terrigenous biomarkers and IRD. Modern situations at locations of ACEX and PS87/106, ODP Site 910 (and close-by ODP sites 911 and 912) and ODP Site 907 are indicated.

system, this permanently to seasonally ice-covered region (Fig. 1) is one of the last major physiographic provinces on Earth, whose climate history and its transition from early Cenozoic Greenhouse to late Cenozoic Icehouse conditions remain still poorly known. Only one drill site recovered from the central Arctic Ocean during the Arctic Coring Expedition (ACEX)—the Integrated Ocean Drilling Program (IODP) Expedition 302 in 2004—gives some insight into the early Cenozoic climate[4–6].

Concerning recent climate change, the most prominent example is the dramatic decrease of the extent and thickness of the Arctic sea-ice cover the last decades, a decrease that seems to be by far more rapid than predicted by climate models[1,3]. The scientific community recognized this drastic change with major concern as the Arctic sea ice is a critical component in the global climate system, which contributes to changes in the Earth's albedo, primary productivity and deep-water formation, a driving mechanism for global thermohaline circulation[7]. The causes of these recent changes, that is, natural versus anthropogenic forcings, and their relevance within the global climate system, however, are subject of intense scientific and societal debate. Thus, understanding the processes controlling Arctic sea-ice variability is of overall interest and significance[8,9]. In this context, records of past climate and sea-ice conditions going beyond instrumental records and representing times of different boundary conditions are of major value: such records can be used to assess the sensitivity of the Earth's climate system to changes of different forcing parameters, for example, level of CO_2, and to test the reliability of climate models by evaluating their simulations for conditions very different from the modern climate. This type of records giving detailed information about past Arctic sea-ice conditions are still very rare, especially due to the lack of precise proxies for sea-ice reconstructions (see review in ref. 9).

The ability to (semi-)quantitatively reconstruct paleo-sea-ice distributions has been significantly improved by a biomarker approach based on the determination of a highly branched isoprenoid (HBI) with 25 carbons (C_{25} HBI monoene $= IP_{25}$)[10]. This biomarker is only biosynthesized by specific diatoms living in the Arctic ice[11] and appears to be a specific, sensitive and stable proxy for Arctic sea ice in sedimentary sections representing Pliocene to Pleistocene times[12,13]. When using this proxy, one has to consider that IP_{25} is absent under a permanent sea-ice cover limiting light penetration and, in consequence, sea-ice algal growth (that is, $IP_{25} = 0$). The same consequence applies to totally ice-free conditions (Fig. 1). Müller *et al.*[14,15] overcame this difficulty in interpretating IP_{25} data by the additional use of phytoplankton-derived open-water biomarkers such as brassicasterol or dinosterol (Fig. 1). Furthermore, Müller *et al.*[14] combined the environmental information carried by IP_{25} and phytoplankton biomarkers in a phytoplankton-IP_{25} index, the so-called 'PIP_{25} index' (for calculation of PIP_{25}, see 'Methods'). In general, the PIP_{25} values determined in Arctic Ocean surface sediments correlate reasonably well with the modern sea-ice distribution based on satellite data, although regional differences in the correlation patterns seem to be obvious[14,16]. In the modern central Arctic Ocean characterized by a mostly perennial sea-ice cover, the PIP_{25} values of most of the surface sediments are between 0.65 and 1 (ref. 16). These data suggest that the PIP_{25} index seems to be a promising proxy for more semi-quantitative estimates of the present and past sea-ice extent, although certainly further groundtruth data are needed (see Methods for some more details)[8,16].

Here we apply the new sea-ice biomarker approach together with alkenone-based sea-surface temperatures (SSTs) to sediment cores most recently recovered during *Polarstern* Expedition PS87 (ref. 17), to reconstruct upper Miocene Arctic Ocean sea-ice and SST conditions. The proxy data are combined with climate model simulations using a coupled atmosphere-ocean general circulation model (AOGCM; see Methods) with focus on seasonal changes in the high northern latitudes. Based on our new proxy records, we demonstrate that only a seasonal sea-ice cover has been predominant in the central Arctic Ocean during (most of) the Late Miocene time interval. Furthermore, our combined data/modelling approach seems to indicate either relatively high atmospheric CO_2 concentrations and/or an overly weak sensitivity of the model to simulate the magnitude of warm polar temperatures in the late Miocene. These new findings from the Arctic region provide new benchmarks for groundtruthing global climate reconstructions and modelling.

Results

Large submarine slide scars along Lomonosov Ridge. *Polarstern* Expedition PS87 was scheduled for August–September 2014 to explore the Lomonosov Ridge area (Fig. 2) with the objective of collecting seismic data and sediment cores to reconstruct the short- and long-term climate history and the tectonic evolution of the central Arctic Ocean[17]. More than 3,000 km of high-quality multi-channel seismic (MCS) profiles and ~10,000 km of high-quality multibeam bathymetry and sub-bottom sediment-echosounding (PARASOUND) profiles were acquired along the ship's track and numerous sediment cores were recovered (Fig. 2). One major finding of the expedition was the discovery of numerous submarine slide scars that occurred on both sides of the crest of Lomonosov Ridge over a distance of *ca.* 350 km between 81°07' N and 84°14' N in water depths from ~800 to 1,500 m (Fig. 3a). Single scars are up to several kilometres wide and long, and their head walls are 100–500 m high. Swath-bathymetry data indicate that different processes probably triggered slope failures, that various processes of sediment evacuation took place, and that failures occurred at various times. Slide scars were earlier described from a restricted area on Lomonosov Ridge near 88°N[18]. However, the wide lateral distribution of mass wasting as presented here is a new discovery.

On top of the southern Lomonosov Ridge in areas between the slide scars, we discovered SE–NW oriented, streamlined landforms over distances of >100 km at water depths between 800 and 1,000 m (Fig. 3a). These features are interpreted to be glacial lineations that formed beneath grounded ice[19,20]. Similar unidirectional bed forms have also been identified further east on the East Siberian continental margin where they were related to large and coherent ice masses (East Siberian Chukchi Ice Sheet—ESCIS)[21]. The lineations identified in this study are similar to those on the East Siberian continental margin with respect to their orthogonal orientations to the proposed centre of the former ESCIS. At the present state of knowledge, the most plausible glacial scenario is a larger than originally proposed ESCIS including an ice shelf extending into the Arctic Ocean, which formed an ice rise on the Southern Lomonosov Ridge over areas presently shallower than 1,000 m. The load and erosional behaviour of this ice rise that probably occurred during extended Quaternary glaciations, for example, during Marine Isotope Stage 6 (MIS 6), may have caused physical conditions that triggered the landslides on this part of Lomonosov Ridge. A MIS 6 age of this erosional event is in line with the proposed age of a major glaciation with extended ice sheets/shelves in Eurasia and East Siberia[19–21].

Exhumation of Miocene to lower Quaternary sediments. Sediment removal from the steep slopes of the escarpments exposed older, normally more deeply buried deposits at/near the present seafloor, allowing retrieval of older sediments by gravity

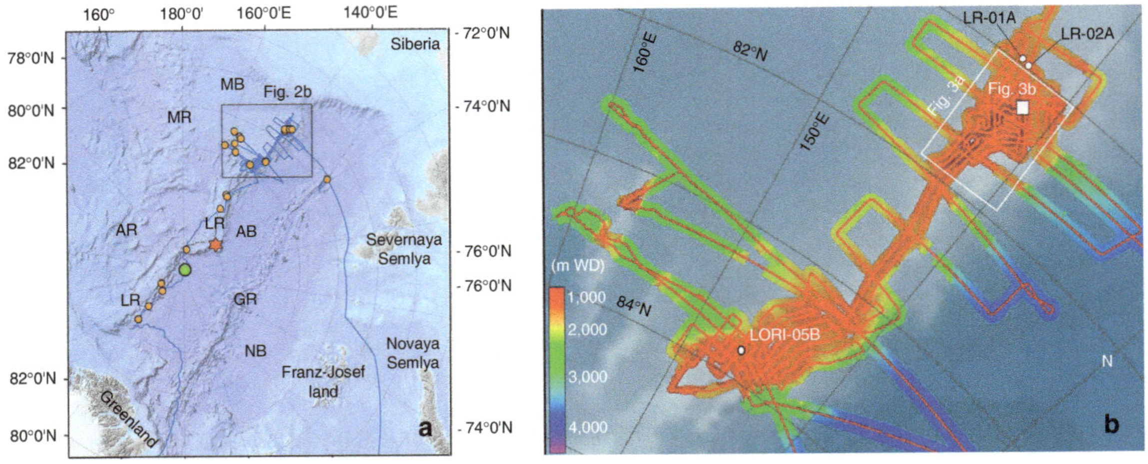

Figure 2 | Cruise track and multibeam bathymetric survey of *Polarstern* Expedition PS87. (**a**) PS87 cruise track (blue line)[17]. AB, Amundsen Basin; AR, Alpha Ridge; GR, Gakkel Ridge; LR, Lomonosov Ridge; MB, Makarov Basin; MR, Mendeleev Ridge; NB, Nansen Basin. Orange circles indicate coring stations, the red asterisk indicates the location of the ACEX Site and the green circle indicates the North Pole (*Polarstern* reached on 26 August 2014 at 10:23 UTC). (**b**) Track lines of multibeam bathymetric survey. Colour bar indicates water depth in metres (m WD). LR-01A, LR-02A and LORI-05B indicate locations of proposed IODP drill sites[51] (*cf.*, IODP Proposal 708; http://www.iodp.org/expeditions). Areas of Fig. 3a,b are indicated.

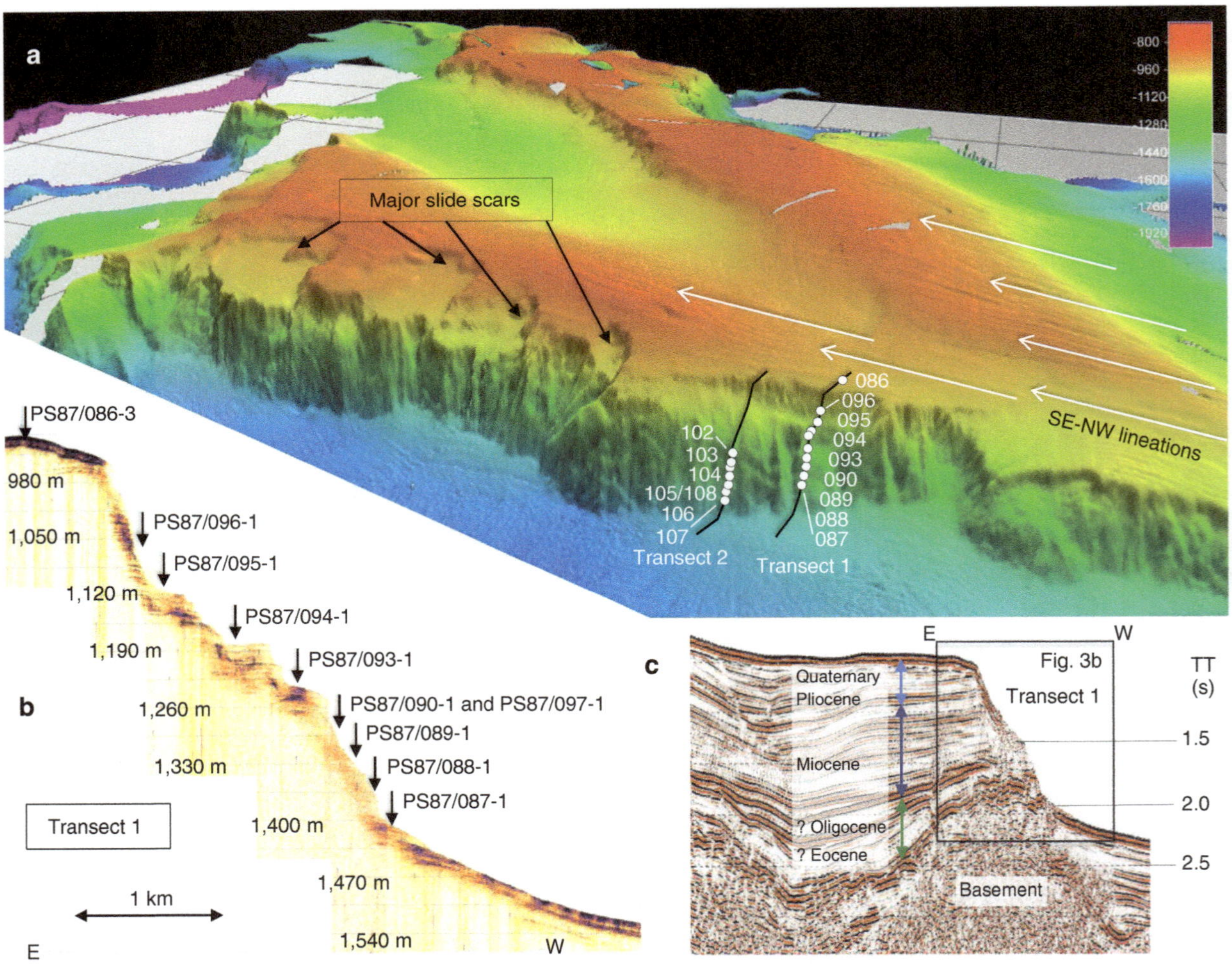

Figure 3 | Bathymetric and acoustic/seismic profiling records from southern Lomonosov Ridge. (**a**) Three-dimensional (3D) image of the swath bathymetry of southern Lomonosov Ridge, showing major slide scars and escarpments, streamlined SE-NW oriented glacial lineations formed beneath grounded ice sheets/streams and transects 1 and 2 with locations of sediment cores (*cf.*, Supplementary Table 1). (**b**) PARASOUND profile across Transect 1 with locations of coring stations. (**c**) Processed multi-channel seismic profile AWI-20140311 across Transect 1, showing prominent reflectors and age assigments based on correlation between regional seismic lines and well data[22,23].

coring from *Polarstern*, rather than expensive drilling. The PARASOUND and MCS profiles suggest that these sedimentary sections are composed of Eocene, Oligocene, Miocene, Pliocene and Quaternary strata (Fig. 3b,c). The age control for the stratigraphic units is based on links of seismic lines to drill site data from the Chukchi Shelf, the ACEX drill site on central Lomonosov Ridge and onshore geology from the New Siberian Islands[22,23]. After evaluation of the multibeam bathymetry and PARASOUND data, we selected two transects across the steep western slope of Lomonosov Ridge for an extensive sediment coring programme (Fig. 3a, Transect 1 and Transect 2). In total, 16 sediment cores were recovered from water depths between 900 m (top of Lomonosov Ridge) and 1,500 m (foot of Lomonosov Ridge)[17] (Fig. 3a,b and Supplementary Table 1).

Whereas most of the sediment cores of Transect 1 are composed of predominantly upper Quaternary (post slide) deposits, some contain prominent unconformities with lower/pre-Quaternary sediments underneath (see Methods and Supplementary Figs 1 and 2). At these unconformities, a 50- to 80-m-thick overburden has been removed, as demonstrated in compaction experiments (see 'Sediment load and compaction experiments'). Unfortunately, the microfossil assemblages (that is, palynomorphs and agglutinated benthic foraminifers) do not allow a precise age determination of the sediments underlying the Quaternary near-surface deposits in most of these cores (Supplementary Table 1). The predominance of Quaternary sediments in the cores of Transect 1, however, seems to be supported by the biomarker composition determined in selected cores. Close to zero concentrations or the absence of specific biomarkers indicative for phytoplankton and sea-ice algae productivity point to surface-water conditions similar to those of the central Arctic Ocean during late Quaternary times[16], characterized by a thick perennial sea-ice cover and SSTs $\ll 0\,°C$ (Supplementary Table 2 and Fig. 1).

The only core providing a clear indication that old sediments are cropping out near the seafloor is Core PS87/106. In this core

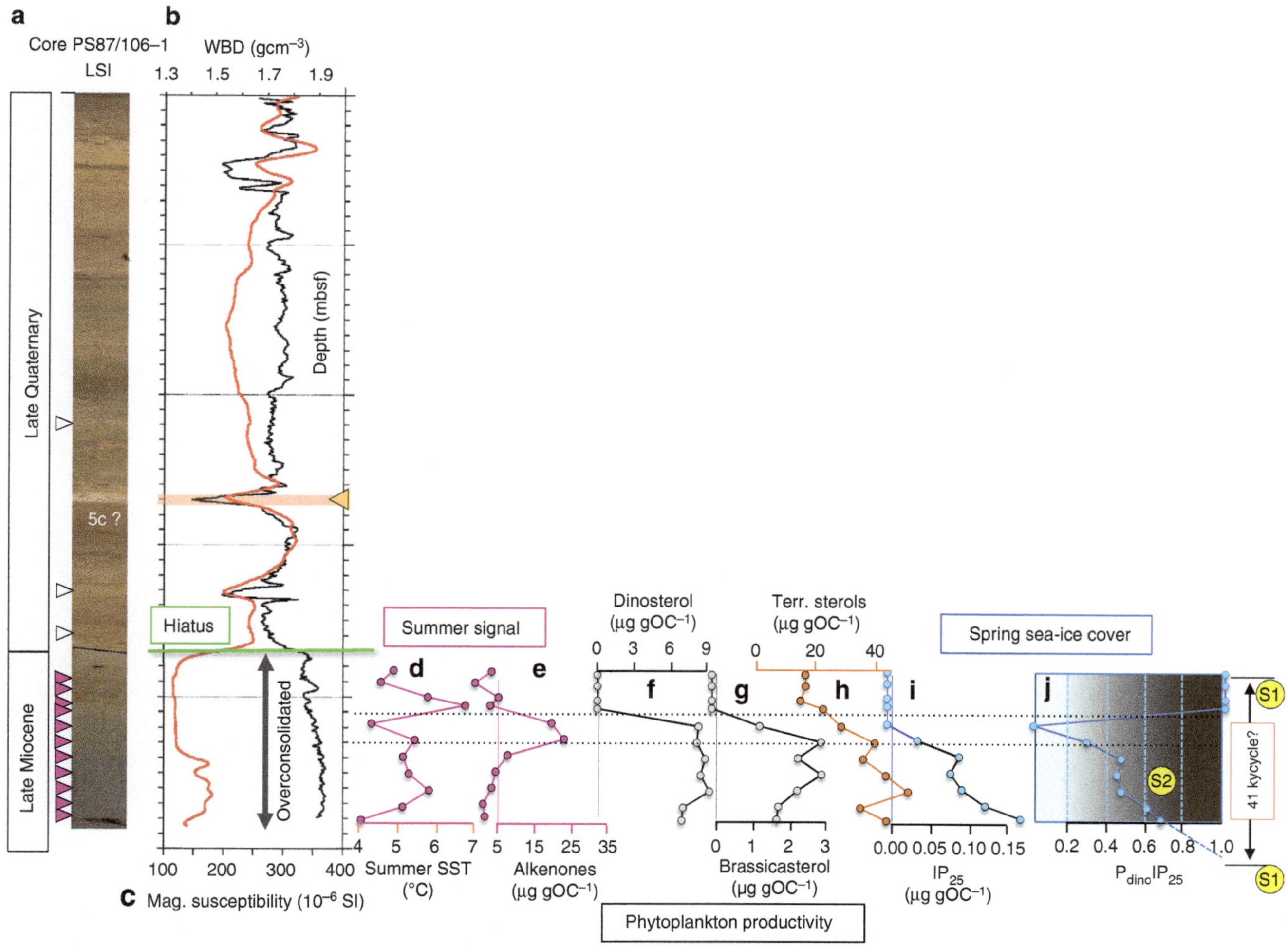

Figure 4 | Proxy evidence from Core PS87/106 for late Miocene Arctic Ocean climate conditions. (**a**) Line-scan image (LSI), (**b**) WBD, (**c**) MS, (**d**) alkenone-based SST, (**e**) concentrations of alkenones, (**f**) dinosterol and (**g**) brassicasterol as proxy for primary productivity, (**h**) concentrations of terrigenous sterols (sum of campesterol and β-sitosterol), (**i**) concentrations of sea-ice proxy IP_{25} and (**j**) sea-ice index $P_{dino}IP_{25}$ (for data see Supplementary Table 2). Purple and open triangles indicate the presence or absence of alkenones, respectively. Green line marks depth of the hiatus. Below the hiatus, sediments are overconsolidated. Zero or near-zero concentrations of IP_{25} and phytoplankton biomarkers are indicative for a closed (spring) sea-ice cover ($P_{dino}IP_{25}$ is indetermable and set to '1'; see Fig. 1 and Methods for further details). PIP_{25} values are used to identify low (0–20%), common (20–50%), ice-edge (50–70%) and extended (>70%) sea-ice concentrations (see Fig. 1). The section below the hiatus represents about one cold ('glacial')-warm ('interglacial') cycle. S1 and S2 indicate two scenarios of different sea-ice concentrations (that is, maximum and transitional/stable ice edge, respectively) within this climate cycle. The interval between the two dashed horizontal lines marks period with minimum sea-ice cover. Using mean sedimentation rates of ~3.2 cm ky^{-1} (as calculated independently from close-by gravity cores[50] and seismic data[51]), the duration of the climate cycle is ~40 ky.

recovered at the lower slope at Transect 2 (Fig. 3a), a sharp unconformity or hiatus occurs at 370 cmbsf. This unconformity is characterized by a marked change in colour, an abrupt increase in wet-bulk density (WBD; related to enhanced sediment consolidation underneath the hiatus), a significant drop in magnetic susceptibility (MS) and a major change in the biomarker composition, and it separates this interval from the overlying young and unconsolidated (upper Quaternary) sediments (Fig. 4 and Supplementary Table 2). The common occurrence of the acritarch *Decahedrella martinheadii* clearly indicates a late Miocene age for the lower part of the sedimentary sequence (Supplementary Fig. 3). The co-occurrence of the dinoflagellate cyst *Nematosphaeropsis labyrinthus* in combination with the absence of the *Batiacasphaera micropapillata* complex, however, may suggest a placement of the PS87/106 sediments within the upper part of its stratigraphic range, that is, upper Tortonian to lower Messinian, when compared with IODP/ACEX Hole M2A and ODP Hole 907A[24–27]. Reworking of these late Miocene palynomorphs is excluded due to excellent preservation of the encountered delicate palynomorph specimens (see Methods for details of the age model of Core PS87/106).

Late Miocene SST and sea-ice records. The biomarker data of Core PS87/106 suggest significantly different late Miocene paleoenvironmental conditions in comparison with those predominant during Quaternary times (Fig. 4 and Supplementary Table 2). In the upper Miocene sediments, elevated concentrations of alkenones and alkenone-derived SST between 4 and 7 °C (or even 6–9 °C in case other calibrations are used; *cf.* Supplementary Table 2) indicate relatively warm, open-water and productive paleoenvironmental conditions in the central Arctic Ocean during the summer season. This is also supported by SST values of »0 °C determined in some samples from the ACEX Site (Supplementary Tables 2 and 3).

Our results reveal for the first time the occurrence of the biomarker sea-ice proxy IP_{25} in sediments as old as late Miocene. This proxy was developed by Belt *et al.*[10] and was before our study only found in Quaternary and Pliocene sediments[12,13]. The presence of IP_{25} in the PS87/106 sediments is indicative for the presence of (spring) sea ice in the late Miocene central Arctic Ocean (Fig. 1). In comparison with IP_{25} values from the Arctic Ocean surface sediments[16] and sediment trap data[28], the absolute IP_{25} concentrations ranging between 0.05 and 0.15 $\mu g\, gOC^{-1}$ (Fig. 4 and Supplementary Table 2) are more than one order of magnitude lower. These differences are caused by an early degradation of biomarkers that already starts in the water column and reaches its maximum in the uppermost centimetres of the sediments[28,29]. On the other hand, both IP_{25} and phytoplankton biomarker concentrations determined in Core PS87/106 are in the same range than those determined in early-mid Holocene Arctic sediments[8,30].

Using the 'PIP_{25} Index' as a more semi-quantitative proxy of paleo-sea-ice cover (see Fig. 1 and Methods), our data from Core PS87/106 point to a variable spring sea-ice coverage of ~20–70% in the lower part and ~100% in the upper part of the sequence

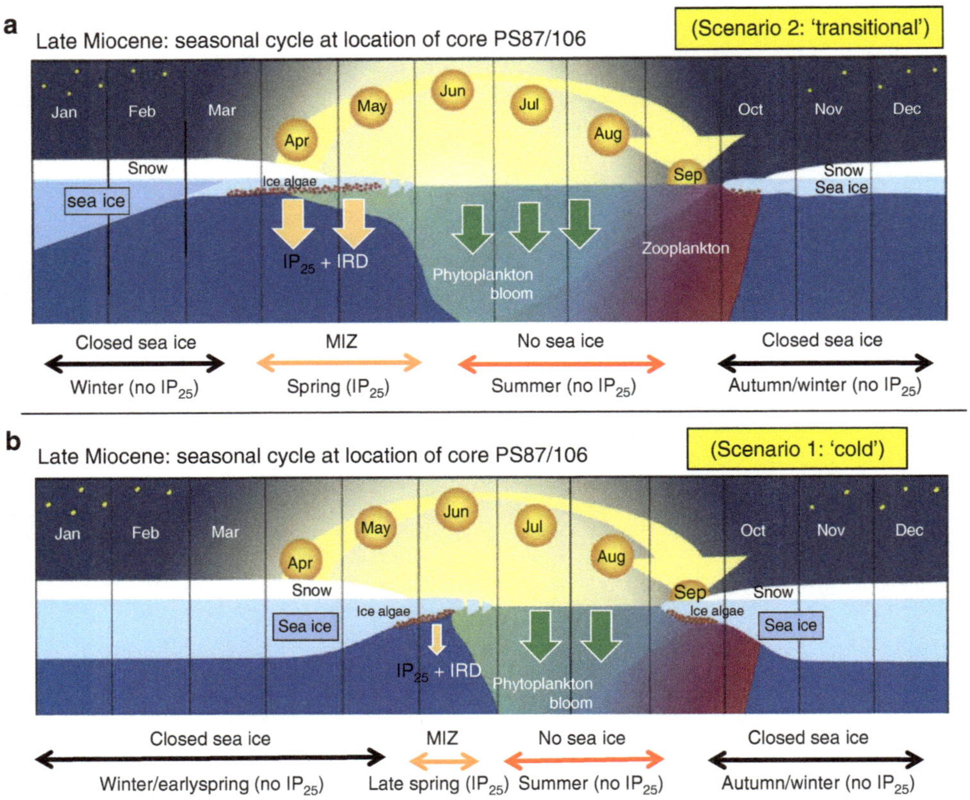

Figure 5 | Schematic illustrations of the seasonal sea-ice cycle during the late Miocene. The seasonal sea-ice cycle and related principal processes controlling productivity and carbon flux at location of central Arctic Ocean Core PS87/106 during the late Miocene are shown for two different scenarios. (**a**) Scenario 2 ('warmer/transitional situation') = extended period of spring sea-ice algae productivity and increased IP_{25} and phytoplankton biomarker fluxes. (**b**) Scenario 1 ('cold situation') = restricted period of late spring sea-ice algae productivity and very reduced fluxes (almost to zero) of IP_{25} and spring phytoplankton biomarkers; *cf.*, Fig. 1). MIZ, Marginal Ice Zone, that is, ice-edge situation. The dark period, height of the sun and changing thickness of snow and ice over the year, as well as phytoplankton, zooplankton and sea-ice productivity are shown. IP_{25} values for the different seasons are indicated (after ref. 72, supplemented).

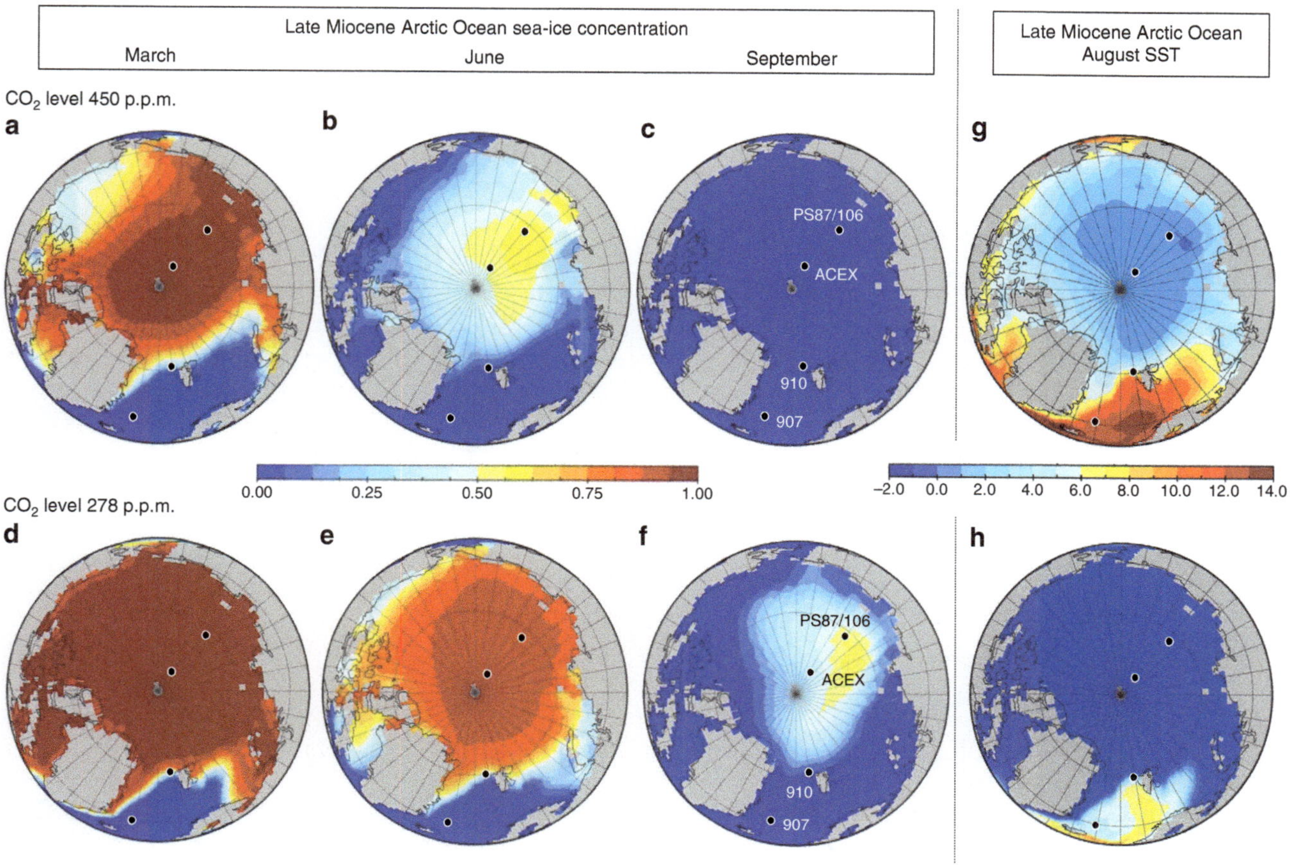

Figure 6 | Late Miocene Arctic Ocean climate simulations for high and low pCO_2 levels using a coupled atmosphere-ocean general circulation model. (a–c,d–f) Sea-ice concentrations for March, June and September at 450 and 278 p.p.m. CO_2 levels, respectively[32]. **(g,h)** Late Miocene August SST at 450 and 278 p.p.m. CO_2 levels, respectively.

(Fig. 4). The combination of IP_{25} and SST data indicates that the central Arctic Ocean must have been relatively warm and ice-free during summer throughout the time interval recovered in the sedimentary section of Core PS87/106 and variable sea ice must have existed during spring when daylight conditions allowed sea-ice algae production (Figs 1 and 5, and Supplementary Table 2). Furthermore, this implies the presence of an extended sea-ice cover during the dark, cold winter season. These new data clearly support that periods with only a seasonal sea-ice coverage must have occurred in the central Arctic Ocean during most of the late Miocene (Fig. 5; see further discussion below).

Simulations of late Miocene Arctic Ocean climate. Model simulations of global climate conditions reconstructing a warm late Miocene climate apply atmospheric CO_2 concentrations in the range of 280–700 p.p.m.[31–34]. Different reconstructions of late Miocene CO_2 levels have narrowed the uncertainties, with atmospheric CO_2 values likely to be below the present day concentrations[35,36]. Nevertheless, the associated uncertainties in the late Miocene CO_2 levels are at least 200 p.p.m.[34,37,38]. Therefore, we have re-analysed Miocene climate simulations[32,39] with CO_2 concentrations of 278 and 450 p.p.m. using a coupled AOGCM. The investigation focus is on seasonal changes in the central Arctic Ocean as simulated with the AOGCM[32,39] (for details and background, see Methods).

Our simulated mean August SST and mean sea-ice concentrations for March, June and September/August for high (450 p.p.m.) and low (278 p.p.m.) CO_2 levels (Fig. 6 and Supplementary Table 4) indicate the following: (1) a winter season with a closed sea-ice cover in the central Arctic Ocean decreasing towards the marginal seas for 450 p.p.m. CO_2 (Fig. 6a) and a closed sea-ice cover in the entire Arctic Ocean including the marginal seas for 278 p.p.m. CO_2 (Fig. 6d); (2) spring season sea-ice concentrations of 20–60% for 450 p.p.m. CO_2 (Fig. 6b) and even 80–90% for 278 p.p.m. CO_2 (Fig. 6e); and (3) a summer season with ice-free conditions and SSTs $>0\,°C$ for 450 p.p.m. CO_2 (Fig. 6c,g) and a reduced but still present sea-ice cover of 10–50% and SSTs $<-1\,°C$ for 278 p.p.m. CO_2 (Fig. 6f,h).

The AOGCM-based monthly mean sea-ice concentrations for 450 and 278 p.p.m. CO_2 levels for late Miocene and preindustrial conditions at the ACEX, PS87/106, ODP 910 and ODP 907 sites are shown in Fig. 7 and—together with SST values—listed in Supplementary Table 4. Based on these results, the central Arctic Ocean was more or less ice free during summer and autumn (that is, between mid-June to mid-November) for 450 p.p.m. CO_2, whereas sea ice still existed during summer times for 278 p.p.m. CO_2 simulations. At the location of ODP Site 907, ice-free conditions occurred throughout the year for both simulations. In addition to the late Miocene and preindustrial simulations, future monthly mean sea-ice concentrations for AD 2100 and AD 2200 (following the Intergovernmental Panel on Climate Change medium stabilization scenario RCP4.5) (ref. 40) are shown (Fig. 7). These simulations suggest that the summer sea-ice concentrations will become significantly reduced to 20–40% during the next two centuries.

In combination with our new SST and sea-ice proxy data, these results suggest that either late Miocene CO_2 levels have been relatively high or, alternatively, the applied model has an overly weak sensitivity especially in the northern high latitudes, as in the

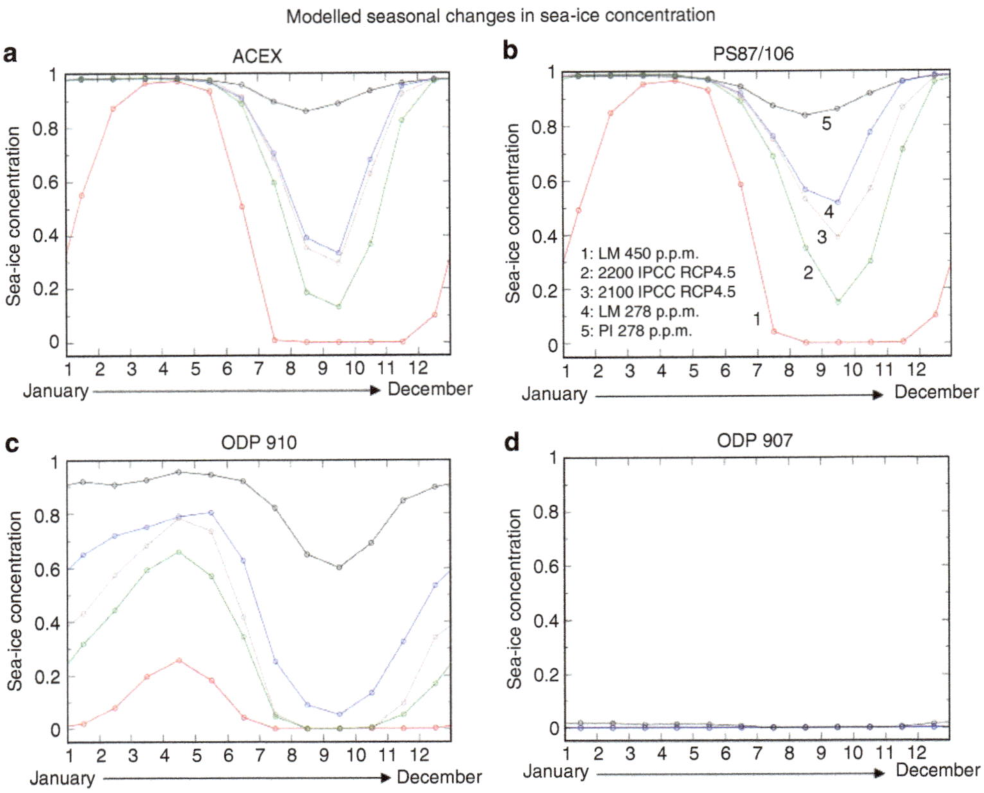

Figure 7 | Modelled monthly sea-ice concentration at the four different core locations. (**a**) ACEX Site, (**b**) Core PS87/106, (**c**) ODP Site 910 and (**d**) ODP Site 907. Climatological monthly mean sea-ice concentrations for late Miocene conditions with atmospheric CO_2 concentrations of 450 p.p.m. (red curve) and 278 p.p.m. (blue curve), modelled sea-ice concentration for preindustrial conditions (black curve) with an atmospheric CO_2 concentration of 278 p.p.m.[32], as well as future conditions (monthly mean 2,085–2,115, brown curve; monthly mean 2,185–2,215, green curve) following the Intergovernmental Panel on Climate Change (IPCC) medium stabilization scenario RCP4.5 (Gierz *et al.*[40]) are shown.

simulation with a relatively low CO_2 level of 278 p.p.m. summer SSTs are too low to explain our sedimentary SST proxy indicative of 4–7 °C. (Fig. 4 and Supplementary Tables 3 and 4). The latter reflects a characteristic challenge of current model approaches to simulate warm climates in the geological past[41,42]. Hence, the new findings from the Arctic region provide an enhanced basis for groundtruthing global climate reconstruction and modelling.

Discussion

Late Miocene climatic conditions significantly warmer than today have been reconstructed from marine and terrestrial proxy records from different localities around the globe[43–47]. However, quantitative SST proxy data from the High Arctic are exclusively restricted to a few terrestrial records[48,49]. With our study, we show for the first time that the late Miocene central Arctic Ocean was relatively warm with SSTs of ~5 °C (Figs 4 and 8) and ice-free during summer, whereas sea ice occurred during spring and autumn/winter. During the late Miocene a general cooling trend is obvious as recorded in the SST record of the sub-Arctic ODP Site 907 (Fig. 8 and Supplementary Table 5). Such a cooling trend seems to be supported by the limited number of SST values available from the ACEX site (Fig. 8). Comparing this general cooling trend with the SST values from Core PS87/106, the sedimentary section of Core PS87/106 below the hiatus probably represents a time span within the upper Tortonian to lower Messinian as supported by palynomorph stratigraphy (Fig. 8). Furthermore, the new Arctic SST data fit in very well with the grander long-term Cenozoic cooling pattern (see Supplementary Fig. 4 and references in the figure legend).

Although on a first view this short sedimentary section of Core PS87/106 only represents a short snapshot of late Miocene Arctic climate, more detailed information about the late Miocene climate on a regional to even global scale can be obtained from our record. Based on the biomarker data, the 1.3-m-thick late Miocene section of Core PS87/106 probably represents almost one cycle with extended and reduced spring sea-ice conditions (Fig. 4). The (almost) absence of phytoplankton biomarkers and IP_{25} (PIP_{25}= '1'; see Fig. 1) and low concentrations of terrigenous biomarkers may be explained by an extended to closed sea-ice cover and a very restricted spring season (Scenario 1 in Figs 4 and 5). Scenario 2, on the other hand, represents a transitional phase with a stable ice edge during an extended and productive spring season, characterized by maximum input of phytoplanktion biomarkers and IP_{25} (resulting in PIP_{25} values of 0.4–0.7), as well as maximum input of terrigenous biomarkers (and ice-rafted debris (IRD)) (Scenario 2 in Figs 4 and 5). The interval between scenarios 1 and 2 is characterized by very low to zero IP_{25} concentrations and increased concentrations of phytoplankton biomarkers, resulting in a distinct PIP_{25} minimum (Fig. 4). This interval is interpreted as a period of minimum spring sea-ice extent. Furthermore, maximum values of alkenones may reflect increased productivity of haptophyte algae during the summer season.

Using mean sedimentation rates of ~3.2 cm ky^{-1} as calculated independently from close-by gravity cores[50] and seismic data[51], the duration of this cycle is about 40 ky, that is, very similar to the 41 ky obliquity cycle (Fig. 4 and also see ref. 52 and references therein). Hence, our record may represent just one obliquity cycle with ice-free conditions during summers in both the cold ('glacial') and the warm ('interglacial') phase of this climate cycle (Figs 4 and 5). As Core PS87/106 probably is of upper Tortonian to lower Messinian age (see above),

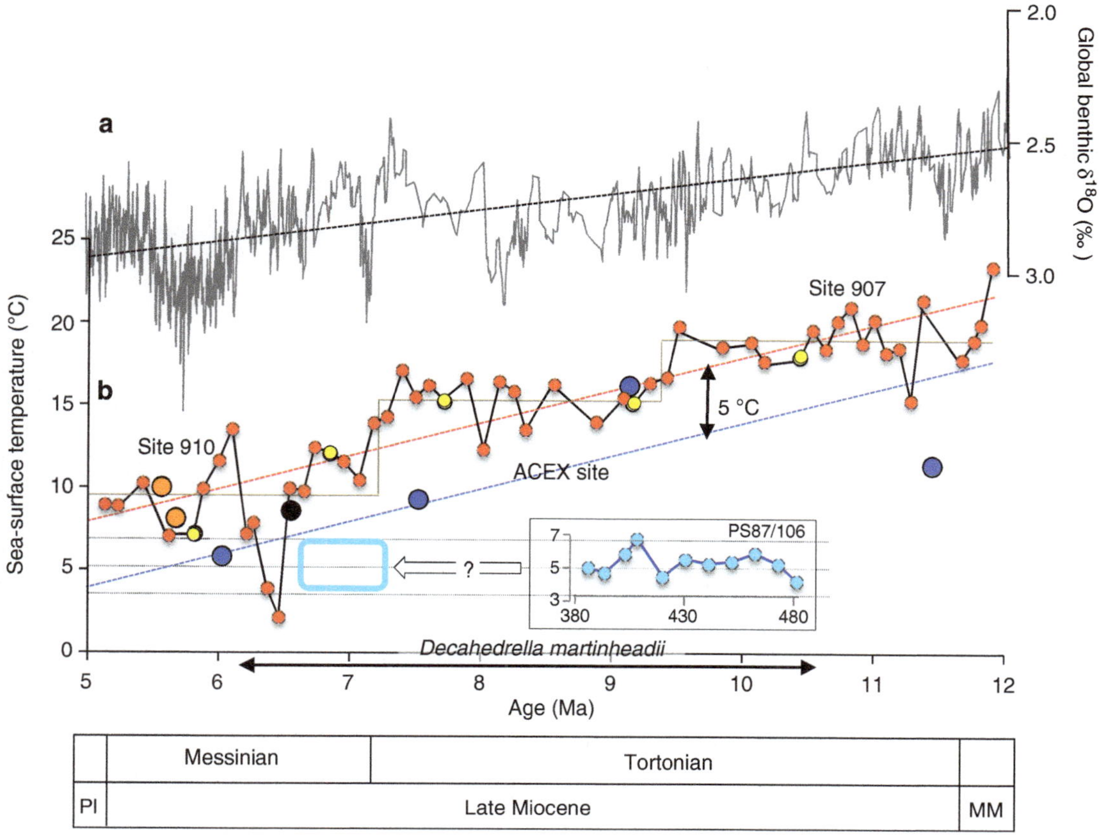

Figure 8 | Global climate (benthic $\delta^{18}O$) stack and (sub-) Arctic SST records for the Late Miocene time interval. (**a**) Global benthic $\delta^{18}O$ stack from ref. 43. (**b**) Late Miocene (Messinian/Tortonian) alkenone-based $U^{k'}_{37}$-SST of Site 907 (yellow circles)[27] and this study (red circles; Supplementary Table 5), TEX_{86}-SST values of Site 910 (orange circles)[12] and alkenone-based $U^{k'}_{37}$-SST of the ACEX Site (blue circles; this study/Supplementary Table 2). The ages of the five ACEX samples are based on ^{10}Be stratigraphy[73]. General cooling trends in the 907 SST record (stippled red line), the ACEX SST record (stippled blue line, assuming a parallel SST trend to Site 907 with an offset of about 5 °C) and the benthic $\delta^{18}O$ record (stippled black line) are indicated. The alkenone-based $U^{k'}_{37}$-SST of Core PS87/106 are plotted versus depth (centimetres below seafloor) (*cf*., Supplementary Table 2). The stratigraphic range of the acritarch *D. martinheadii* in the sub-Arctic/Arctic realm is indicated[24,25]. As the central Arctic Ocean ACEX and the PS87/106 SST values should be similar, the PS87/106 record probably repesents a time span within the upper Tortonian to lower Messinian (light blue box). Pl, Pliocene; MM, Middle Miocene.

ice-free summer conditions should have occurred in the central Arctic Ocean during the warmer Middle Miocene to early Late Miocene time interval *a fortiori* (Fig. 8). In contrast to several previous studies, we therefore propose that a seasonal sea-ice cover was predominant in the central Arctic Ocean during (most of) the late Miocene time interval (see discussion below).

Our new semi-quantitative sea-ice and SST records are an important contribution to the ongoing and controversial debate about the reconstruction of the early (pre-Quaternary) Arctic Ocean sea-ice cover. Within this debate, the distinction between seasonal and perennial sea ice is critical, because year-round sea ice in the central Arctic implies very different climate feedback mechanisms, that is, Earth's albedo and heat exchange conditions, than an environment with ice-free conditions during summer. As outlined above, our new proxy data and modelling data clearly indicate a late Miocene seasonal sea-ice cover. Similar or even warmer climatic conditions also occurred in the Middle Miocene Arctic Ocean when looking at the alkenone-based summer SSTs of 10–13 °C reconstructed from ACEX sediments (Supplementary Fig. 4 and also see ref. 53). Abundant marine palynomorphs and foraminifers found in Miocene and early Pleistocene sediments of the ACEX section also point to at least periods with seasonally ice-free conditions[24,54]. These data are in contrast to an Arctic Ocean perennial sea-ice cover from middle Miocene onwards, as proposed by Darby[55] and Krylov *et al.*[56] based on their provenance studies of IRD in ACEX sediments. Taking the IRD with a North American or East Siberian origin found in the ACEX sediments and using modern sea-ice drift trajectories and velocities, these authors concluded that more than 1 year was needed to transport the sediments entrained in the sea ice to the ACEX location. Hence, the sea ice must have survived the summer melting season to reach the ACEX site.

Using a 'coupled sea-ice slab-ocean model including sediment transport by sea ice', Tremblay *et al.*[57] have most recently demonstrated that sea-ice drift was probably significantly faster under warmer climatic conditions with less or much thinner sea ice than today. They conclude that the presence of ice-rafted sediment of Eurasian and North American origin at the North Pole is not a definite indication of a perennial sea ice cover in the Arctic Ocean. This may demonstrate that the main assumption of Darby[55] and Krylov *et al.*[56] should be regarded critically.

Our proxy and model reconstruction of late Miocene Arctic climate is certainly a substantial step forward to improve the understanding of the pre-Quaternary Arctic Ocean sea-ice and SST history. However, to decipher the pre-Quaternary climate history of this unique and sensitive but still not well-known region on Earth in more detail, long continuous sedimentary records to be obtained only by scientific drilling are needed. These records are planned to be recovered within a new IODP drilling campaign scheduled for 2018 (IODP Proposal 708; http://

www.iodp.org/expeditions). Based on the new PS87 seismic data, ~200 m of Plio–Pleistocene, >600 m of Miocene and >300 m of Oligocene–Eocene may be recovered at the proposed drill sites[51] (see Fig. 2 for locations). The outcome of such a new programme will certainly help to improve our understanding of the complex ocean–atmosphere–ice system in the polar high northern latitudes and its role in the past, modern and future global climate.

Methods

Shipboard surveys and measurements. The methods used on board *Polarstern* during Expedition PS87 are shortly outlined in the following. For a more detailed information about the use and interpretation of the proxy data we refer to the different chapters of the Cruise Report[17].

The bathymetric survey was performed using the hull-mounted ATLAS Hydrographic HYDROSWEEP DS3, a deep-sea multi-beam swath sonar system with a resolution of up to 320 receive beams per ping, a swath width of 4–5 times the water depth and a vertical resolution of ~0.5% of the water depth. It was operated in the chirp mode with a frequency of 14–16 kHz. The mean sound velocity of the water column was calculated from conductivity-temperature-density (CTD), expandable CTD (XCTD) and Valeport Sound Velocity Profiler data.

Sub-bottom profiling data were acquired using the parametric hull-mounted system ATLAS Hydrographic PARASOUND DS III-P70. Primary operating frequencies were 18.75 and 22.95 kHz with a secondary sediment-penetrating frequency of 4.2 kHz, a beam angle of 4° and a pulse length of 2. The vertical resolution is ~0.2 m. As a result of the narrow beam angle, reflections from strata dipping by >4° cannot be received by the vessel. This explains why the thin veneer of post-slide sediments covering the slump scar is not resolved as it is above and below the headwall (Fig. 3b). The headwall has an inclination >4°. In contrast, the older (pre-Quaternary) sediments exhibit nearly horizontal bedding (Fig. 3), which are thus acoustically resolved along the headwall to their near-seafloor location. PARASOUND data visualization and processing was performed using ATLAS PARASTORE-3 software. The vertical scale on profiles has been converted from travel time to metres using a constant sound velocity of 1.5 km s^{-1}, which explains minor differences in water depth between PARASOUND and swath-sonar data.

For the MCS data acquisition, a 3,000-m-long streamer (240 active channels, group interval of 12.5 m) and an air gun array of four G-Guns (total volume of 32 l, fired with 200 bar every 15 s) were used. Processing included sorting, that is, common depth point sorting with 25 m spacing, frequency filtering (20–180 Hz), velocity analysis, multiple suppression and stacking.

Whole-core measurements included non-destructive, continuous determinations of core geometry (diameter), WBD, P-wave velocity (Vp) and loop-sensor MS at 10 mm intervals, using a standard Multi-Sensor Core Logger (GEOTEK Ltd., UK). The principle of logging cores is described in more detail in the GEOTEK manual 'Multi-Sensor Core Logging', which can be downloaded from the web (http://www.geotek.co.uk).

Line-scan images (Supplementary Figs 1 and 2) were acquired with a Jai CV L107 camera with RGB (red-green-blue) channels at 630, 535 and 450 nm, respectively, mounted to an Avaatech XRF core scanner. The camera contains three charge-coupled device sensors and a beam splitter to separate the RGB signal. Images were acquired with a down-core resolution of ~70 µm.

Stratigraphic framework and marine palynology process. The general lithostratigraphic framework and age model of the upper Quaternary sedimentary sections recovered during Expedition PS87 are robust and based on lithostratigraphy, colour imaging, WBD and MS records characterized by very prominent minima and maxima, and correlation with other dated sediment cores from Lomonosov Ridge[17,50]. Based on this concept, MIS 6 to 1 were identified in most of the cores (Supplementary Fig. 1). Some cores from very steep sections of the slope along Transect 1 (Fig. 3b), on the other hand, contain multiple unconformities and a correlation to the reference Core PS87/086 was only partly possible (Supplementary Fig. 2). WBD typically increases sharply below the unconformities due to higher degrees of compaction, supporting a hiatus and 'older' sediments below (Supplementary Fig. 2). For this study dealing with the late Miocene climate history, only the identification and dating of these 'older' (Neogene) sedimentary sections are relevant. For assessing the age of these sediments, assemblages of agglutinated benthic foraminifers and palynomorphs were used and compared with biostratigraphic records obtained from the ACEX and North Atlantic sites[17,24–27] (Supplementary Table 1).

For sample preparation and processing of the agglutinated benthic foraminifers we refer to the PS87 Cruise Report[17] and further references therein. The procedure for studying the palynomorphs are as follows: dinoflagellate cysts and acritarchs were investigated in 25 samples (mainly core catcher) from 15 cores recovered along transects 1 and 2 (Fig. 3a and Supplement Table 1), with special emphasis on Core PS87/106. Sediment was freeze dried, weighed and processed using standard palynological maceration techniques including repeated treatment with cold HCl (10%) and cold HF (38–40%), no oxidation and sieving over a 10-µm-nylon mesh. The residue has been mounted with glycerine jelly on microscope slides, which were then scanned for dinoflagellate cysts and acritarchs using a light microscope at ×400 original magnification.

Age model of the late Miocene section of core PS87/106. Based on microfossil data (that is, palynomorphs and agglutinated benthic foraminifers), the core catcher samples from the sediment cores were barren or give a Pleistocene age (Supplementary Table 1). The only core providing any clear indication that old sediments are cropping out in the shallow sub-seafloor is Core PS87/106. Well-preserved specimens of organic-walled palynomorphs (dinoflagellate cysts and acritarchs) have been recorded in successive samples from the base of the core (including core catcher) up to 420 cmbsf (that is, 50 cm below the hiatus; Supplementary Table 1). Of the encountered species, the acritarch *D. martinheadii* provides evidence that the lower part of the core is composed of sediments of late Miocene age. This species is endemic to the high northern latitudes and its stratigraphic range has been discussed previously based on comprehensive reviews of its occurrence at several DSDP, ODP and IODP sites from the Central Arctic Ocean, Norwegian-Greenland Sea, Labrador Sea, Baffin Bay and Irminger Sea[24–26]. It is restricted to the late Miocene in the Arctic and subarctic realm, and based on the pristine paleomagnetic record of Iceland Sea ODP Site 907, its stratigraphic range is independently calibrated against the astronomically tuned Neogene Timescale, thus providing absolute age control (Supplementary Fig. 3). A near-synchronous highest occurrence at *ca.* 6.3–6.2 Ma has been defined from several northern high latitude sites, suggesting this species to be an excellent marker across the subpolar/polar North Atlantic and Arctic Ocean[25]. In addition, a highest common occurrence is recognized at 6.5 Ma in Iceland Sea ODP Hole 907A and more generally at *ca.* 6.7–6.3 Ma across the Norwegian-Greenland Sea. The lowest occurrence of *D. martinheadii* has been calibrated to 10.5 Ma in ODP Hole 907. According to Schreck *et al.*[25], however, the lowest occurrence is not very well constrained at other sites but certainly younger than 11 Ma across the northernmost North Atlantic and Arctic Ocean.

In Core PS87/106, the consistent occurrence of *D. martinheadii* is accompanied by low numbers of the dinoflagellate cyst *N. labyrinthus*. Such co-occurrence has also been observed within the upper part of its stratigraphic range in IODP Hole M2A[24] and ODP Hole 907 (refs 25,27). Furthermore, the *B. micropapillata* complex has not been recorded in PS87/106 samples. This dinoflagellate cyst dominates assemblages in the late Serravallian of IODP/ACEX Hole M2A, decreases significantly across the Tortonian and disappears close to the Tortonian/ Messinian boundary[26]. In ODP Hole 907A, *B. micropapillata* complex dominates the assemblage until *ca.* 8.2 Ma after which it only occurrs sporadically until its highest common occurrence at *ca.* 4.5 Ma[26,27]. Therefore, the co-occurrence of *D. martinheadii* and *N. labyrinthus* in combination with the absence of *B. micropapillata* complex may allow to place the analysed interval of Core PS87/106 into the upper Tortonian to lower Messinian. However, we note that the highest occurrence derived from ODP Site 907 may represent a minimum age for this bioevent in the Central Arctic Ocean, as successive Neogene cooling may led to an earlier disappearance of species in the higher latitudes.

D. martinheadii and *N. labyrinthus* are both very delicate species that bear processes and trabeculae (ribbon-like bars), which tend to crumple easily. All specimens encountered during palynological analyses, however, are well preserved indicating *in situ* deposition.

Biomarker analyses. Extraction of 5–10 g of freeze-dried sediments was carried out using an accelerated solvent extractor (DIONEX, ASE200; 100 °C, 5 min, 1,000 psi) with dichloromethane:methanol (2:1, v/v) as the solvent. For quantification internal standards, 7-hexylnonadecane (7-HND, 0.076 µg per sample for IP_{25} quantification), squalane (2.4 µg per sample) and cholesterol-d_6 (cholest-5-en-3β-ol-D_6, 10 µg per sample for sterol quantification) were added before analytical treatment. Separation of the hydrocarbon and sterol fractions was carried out via open column chromatography (hydrocarbon fraction with 5 ml *n*-hexane, the sterol fraction with 6 ml *n*-hexane:ethylacetate (5:1, v/v)). The latter fraction was silylated with 500 ml BSTFA (bis-trimethylsilyl-trifluoroacet-amide) (60 °C, 2 h). IP_{25} and sterols were analysed by gas chromatography (GC)/mass spectrometry. Component assignment was based on comparison of GC retention times with those of reference compounds and published mass spectra (Supplementary Figs 5 and 6). The Kovats Index calculated for IP_{25} is 2,086. For the monounsaturated HBI alkene (HBI monoene) most recently found in ancient Arctic sediments and characterized by very similar chromatographic and mass spectral properties[58], the Kovats Index has been calculated as 2,090. As this new HBI monoene is absent in the investigated cores of this study, we have calculated the index from lower Pliocene sediment samples from ODP Site 911 to show that both compounds can clearly be separated by our analytical approach (see Supplementary Fig. 5a,b). The detection limit for quantification of IP_{25} (Agilent 7890B GC, Agilent 5977A Extractor MSD with Performance Turbo Pump) is 0.005 ng μl^{-1} in SIM (selected ion monitoring) mode. To obtain mass spectra in TIC (total ion current) the limit is 0.05 ng μl^{-1}. The retention indices for brassicasterol (as 24-methylcholesta-5, 22E-dien-3β-O-Si(CH3)3), campesterol (as 24-methylcholest-5-en-3β-O-Si(CH3)3) and β-sitosterol (as 24-ethylcholest-5-en-3β-O-Si(CH3)3) were calculated to be 1.018, 1.042 and 1.077 (normalized to cholest-5-en-3β-ol-D_6 set to be 1.000), respectively.

For the quantification of IP_{25}, its molecular ion (*m/z* 350) in relation to the abundant fragment ion *m/z* 266 of the internal standard (7-HND) was used

(SIM mode). The different responses of these ions were balanced by an external calibration (Supplementary Fig. 5c and also see ref. 28). Brassicasterol (24-methylcholesta-5,22E-dien-3β-O-Si(CH_3)$_3$), campesterol (24-methylcholest-5-en-3β-O-Si(CH_3)$_3$) and β-sitosterol (24-ethylcholest-5-en-3β-O-Si(CH_3)$_3$) were quantified as trimethylsilyl ethers using the molecular ions *m/z* 470, *m/z* 472 and *m/z* 486, respectively, in relation to the molecular ion *m/z* 464 of cholesterol-D_6.

More details about the identification and quantification of IP_{25} and the sterols are described elsewhere[8,28,29,58–61].

For more semi-quantitative estimates of the present and past sea-ice coverage, Müller *et al.*[14] combined the sea-ice proxy IP_{25} and phytoplankton biomarkers in a phytoplankton-IP_{25} index, the so-called 'PIP_{25} index' (Fig. 1):

$$PIP_{25} = [IP_{25}]/([IP_{25}] + ([\text{phytoplanktonmarker}] \times c)) \qquad (1)$$

with c = mean IP_{25} concentration/mean phytoplankton biomarker concentration for a specific data set or core. As phytoplankton biomarkers brassicasterol and dinosterol were used, resulting in $P_{bras}IP_{25}$ and $P_{dino}IP_{25}$ values, respectively (see refs 8,16,29,59 for discussion of advantages and limitations of the PIP_{25} approach). Most recently, Smik *et al.*[62] introduced a HBI–III alkene as phytoplankton biomarker replacing the sterols in the PIP_{25} calculation. This modified PIP_{25} approach is far less dependent on the balance factor *c* and based on biomarkers from the same group of compounds (that is, HBIs) with more similar diagenetic sensitivity, certainly an important improvement for paleo-sea-ice reconstructions and comparison of records from different Arctic areas.

Our reconstruction of SST is based on long-chain C_{37} alkenones synthesized by haptophyte algae[63]. The $C_{37:3}$- and $C_{37:2}$-alkenones were present in all samples, whereas the $C_{37:4}$-alkenone was not found. For alkenone ($C_{37:2}$ and $C_{37:3}$) analysis, extraction of additional 6 g of freeze-dried sediment was carried out using the ASE method under same conditions as decribed above but with dichloromethane as the solvent. The separation of compounds was carried out by open column chromatography using 5 ml *n*-hexane, followed by 5 ml *n*-hexane:dichloromethane (1:1, v/v) and 5 ml dichloromethane for eluation of the alkenones. As internal standard *n*-$C_{36:0}$ (10 μg per sample) was added before any analytical treatment. The alkenones were analysed by GC. Individual alkenone ($C_{37:3}$ and $C_{37:2}$) identification is based on retention time and the comparison with an external standard (Supplementary Fig. 7). To exclude a possible coeluation of the alkenones with other compounds, the extracts were measured first as total extract, second after additional column cleaning with dichloromethane and third after saponification[64]. The instrument stability has been continuously controlled by re-runs of an external alkenone standard (extracted from cultures of *Emiliania huxleyi* with known growth temperature) during the analytical sequences. The range of the total analytical error calculated by replicate analyses is <0.4 °C.

For calculation of SST, we used the simplified $U_{37}^{K'}$ Index[63]:

$$Uk_{37}' = C_{37.2}/(C_{37.2} + C_{37.3}) \qquad (2)$$

$U_{37}^{K'}$ was converted to SST according to the World Ocean core top versus annual temperature calibration ($U_{37}^{K'} = 0.033T + 0.044$) (ref. 64), the calibration most often used in the literature. Resulting SSTs vary between 4.2 and 6.7 °C (SST-1; Fig. 4 and Supplementary Table 2). For the central Arctic Ocean, these SSTs certainly have to be interpreted as summer SSTs (instead of annual mean) due to the darkness during late autumn to winter (*cf.*, Fig. 5). In addition, we also have used the Müller *et al.*[64] calibration versus summer SST (SST-2), the Sikes *et al.*[65] calibration versus summer SST obtained from the polar Southern Ocean (SST-3) and the Prahl and Wakeham[63] calibration obtained from cultural experiments (SST-4). Whereas the SST-4 values are more or less the same as the SST-1 values, the SST-2 and SST-3 values (calculated as 'summer SST') are higher and vary between 6 and 9 °C (Supplementary Table 2). Based on these results, we interpret our late Miocene summer SSTs of ~4–7 °C (mean of 5.3 °C) more as minimum values. For the Müller *et al.*[64] calibration, the standard error is reported as ±0.050 $U_{37}^{K'}$ units or ±1.5 °C for the entire temperature range from 0 to 27 °C. In the lower temperature range <10 °C, however, the scatter of the $U_{37}^{K'}$ values is significantly higher than the mean. Thus, one should not overinterpret the SST variability between 4 and 7 °C. In any case and independently of the calibration approach, summer SSTs were significantly higher than zero, preventing sea-ice formation during summer.

The results of the biomarker analyses (that is, alkenones, selected sterols, SSTs, IP_{25} and PIP_{25}) carried out on samples from various PS87 sediment cores, are listed in Supplementary Table 2. All biomarker data (expressed in μg gOC^{-1} and μg gSediment^{-1}) are available online at http://dx.doi.org/10.1594/PANGAEA.855509.

Model simulation. For the investigation of the late Miocene Arctic climatic conditions (that is, climatological sea-ice cover and SST) we have re-analysed Miocene climate simulations[32,39]. The simulations have been performed with a coupled AOGCM. The atmosphere model component ECHAM5 (ref. 66) was used at T31 resolution (~3.75°) with 19 vertical levels. The ocean component MPI-OM[67], including the dynamics of sea ice formulated using viscous-plastic rheology[68], has an average horizontal resolution of 3° × 1.8° with 40 uneven vertical layers. This modelling approach has been used and evaluated for investigations of the Miocene climate[32,39]. For the re-analyses, we have used data from two model runs with the same late Miocene set-up[32], except different atmospheric CO_2 concentrations. One simulation is based on a CO_2 concentration of 278 p.p.m.[32] and one uses a CO_2 concentration of 450 p.p.m.[39]. For further details of the AOGCM model configuration and the boundary conditions, we refer the reader to refs 32,39.

Sediment load and compaction experiments. The preconsolidation stress of a geological sample experienced in the past can be assessed by incrementally loading the specimen in a uniaxial deformation apparatus (so-called 'oedometer'), where deviations in the settling behaviour can be converted to the thickness of the missing overburden. We used this approach to estimate the thickness of sediment removed at prominent unconformities observed in some of our studied sediment cores (Fig. 4 and Supplementary Figs 2 and 8).

We conducted our experiments using a combined GIESA oedometer—direct shear apparatus in which both uniaxial compression tests and shear tests can be conducted in consecutive steps[69] (Supplementary Fig. 8b). The sample cell is a cylindrical volume within a stack of two steel plates. If desired, relative displacement of the plates enforces simple shear deformation in the sample, to measure undrained shear strength. Porous metal frits allow fluid communication with an open pore fluid reservoir (containing distilled water) and dissipation of excess pore pressure. Normal load is applied to the sample with a vertical ram and shear is induced by holding the upper plate fixed, while the lower plate is driven horizontally.

For the experiments in this study, we only used the consolidation function of the system. As is done routinely in such tests[70], we loaded each sample incrementally by starting with applied normal stresses of 10 kPa. This value was assumed to be below the *in situ* stress the samples had experienced before. After 24 h, the normal load is doubled. Tests were run over many days until the desired maximum normal load was reached (10,240 kPa for Core PS87/096 and 5,120 kPa for Core PS87/106). The initial water content of each sample was taken from an aliquot before the specimen was mounted into the GIESA oedometer. We measured the wet weight of the aliquot, subjected it to 48 h of gentle drying in an oven (60 °C) and then measured the dry weight, to calculate the void ratio *e* (where e = volume of voids/volume of solids).

Oedometer results are plotted as effective normal stress versus void ratio, the latter of which is derived from the settling (that is, change in sample thickness as determined using a vertical displacement transducer) after each loading increment. The graph of each experiment show a smooth function when *e* is plotted against normal stress, the latter on a logarithmic axis (Supplementary Fig. 8c). In an undeformed sample, the regular loading results in the so-called 'virgin consolidation curve', whereas samples that were previously subject to loading usually deviate from the smooth curvature and show a distinct change in gradient of the graph. The high-stress end of the graph is generally linear and serves to assess the maximum preconsilidation stress the sample has experienced following the procedure first established by A. Casagrande[69].

The stress measurements are then used to the thickness (h) of the overburden based on the effective load (h = preconsolidation stress/(bulk density of the sample − density of seawater) × g). The degree of overconsolidation of a given sample is then calculated as the ratio between preconsolidation stress and normal stress at that depth below the seafloor.

We have tested one pair of samples in cores PS87/096 and PS87/106, respectively, and loaded these samples to at least 5,120 kPa effective stress. Our results attest that the samples taken above the hiatuses are normally consolidated, while those underneath the discontinuity are overconsolidated. In Core PS87/096, the estimated preconsolidation stress below the hiatus is 590 kPa, which amounts to the removal of a 80-m-thick sediment package (Supplementary Fig. 8c). In Core PS87/106, the preconsolidation stress is between 320 and 480 kPa, which corresponds to 48 and 65 m of overburden, respectively (*cf.*, Fig. 4).

For Core PS87/096, the calculated thickness of removed sediments are more or less identical to those estimated from the ages of the sediments above and below the hiatus (that is, ~100 ka and <2.5 Ma, respectively) and mean sedimentation rates of 3.2 cm ky^{-1} (refs 50,71), resulting in a maximum sediment removal of 80 m (Supplementary Fig. 8a).

References

1. ACIA Members. *Arctic Climate Impact Assessment Report* (Cambridge Univ. Press, 2005).
2. Stocker, T. *et al. Climate Change 2013, The Physical Science Basis* 1535 (Intergovernmental Panel on Climate Change (IPCC), Cambridge Univ. Press, 2013).
3. Stroeve, J., Holland, M. M., Meier, W., Scambos, T. & Serreze, M. Arctic sea ice decline: faster than forecast. *Geophys. Res. Lett.* **34,** L09501 (2007).
4. Backman, J., Moran, K., McInroy, D. B. & Mayer, L. A.the Expedition 302 Scientists. *Proceedings IODP 302* (Integrated Ocean Drilling Program Management International, College Station, 2006).
5. Moran, K. *et al.* The Cenozoic palaeoenvironment of the Arctic Ocean. *Nature* **441,** 601–605 (2006).
6. Stein, R. *et al.* Cenozoic Arctic Ocean climate history: some highlights from the IODP Arctic Coring Expedition (ACEX). *Dev. Marine Geol.* **7,** 259–293 (2014).
7. Thomas, D. N. & Diekmann, G. S. *Sea Ice* (Blackwell Publishing, 2010).
8. Stein, R., Fahl, K. & Müller, J. Proxy reconstruction of Arctic Ocean sea ice history: 'From IRD to IP_{25}'. *Polarforschung* **82,** 37–71 (2012).

9. De Vernal, A., Gersonde, R., Goosse, H., Seidenkrantz, M.-S. & Wolff, E. W. Sea ice in the paleoclimate system: the challenge of reconstructing sea ice from proxies—an introduction. *Quat. Sci. Rev.* **79,** 1–8 (2013).
10. Belt, S. T. *et al.* A novel chemical fossil of palaeo sea ice: IP25. *Org. Geochem.* **38,** 16–27 (2007).
11. Brown, T. A., Belt, S. T., Tatarek, A. & Mundy, C. J. 2014. Source identification of the Arctic sea ice proxy IP_{25}. *Nat. Commun.* **5,** 4197 (2014).
12. Knies, J. *et al.* The emergence of modern sea ice cover in the Arctic Ocean. *Nat. Commun.* **5,** 1–5 (2014).
13. Stein, R. & Fahl, K. Biomarker proxy IP_{25} shows potential for studying entire Quaternary Arctic sea-ice history. *Org. Geochem.* **55,** 98–102 (2013).
14. Müller, J. *et al.* Towards quantitative sea ice reconstructions in the northern North Atlantic: a combined biomarker and numerical modeling approach. *Earth Planet. Sci. Lett.* **306,** 137–148 (2011).
15. Müller, J., Massé, G., Stein, R. & Belt, S. T. Variability of sea-ice conditions in the Fram Strait over the past 30,000 years. *Nat. Geosci.* **2,** 772–776 (2009).
16. Xiao, X., Fahl, K., Müller, J. & Stein, R. Sea-ice distribution in the modern Arctic Ocean: biomarker records from Trans-Arctic Ocean surface sediments. *Geochim. Cosmochim. Acta* **155,** 16–29 (2015).
17. Stein, R. (ed.). *The Expedition PS87 of the Research Vessel Polarstern to the Arctic Ocean in 2014, Reports on Polar and Marine Research* Vol. 688 (Alfred Wegener Institute for Polar and Marine Research. http://epic.awi.de/37728/1/BzPM_0688_2015.pdf, 2015).
18. Kristoffersen, Y., Coakley, B. J., Hall, J. K. & Edwards, M. Mass wasting on the submarine Lomonosov Ridge, central Arctic Ocean. *Mar. Geol.* **243,** 132–142 (2007).
19. Jakobsson, M. *et al.* Arctic Ocean glacial history. *Quat. Sci. Rev.* **92,** 40–67 (2014).
20. Jakobsson, M. *et al.* An Arctic Ocean ice shelf during MIS 6 constrained by new geophysical and geological data. *Quat. Sci. Rev.* **29,** 3505–3517 (2010).
21. Niessen, F. *et al.* Repeated Pleistocene glaciation of the East Siberian Continental Margin. *Nat. Geosci.* **6,** 842–846 (2013).
22. Hegewald, A. & Jokat, W. Tectonic and sedimentary structures in the northern Chukchi region, Arctic Ocean. *J. Geophys. Res.* **118,** 3285–3296 (2013).
23. Weigelt, E., Jokat, W. & Franke, D. Seismostratigraphy of the Siberian Sector of the Arctic Ocean and adjacent Laptev Sea Shelf. *J. Geophys. Res. Solid Earth* **119,** 5275–5289 (2014).
24. Matthiessen, J., Brinkhuis, H., Poulsen, N. & Smelror, M. *Decahedrella martinheadii* Manum 1997—a stratigraphicall and paleoenvironmentally useful Miocene acritarch of the high northern latitudes. *Micropaleontology* **55,** 171–186 (2009).
25. Schreck, M., Matthiessen, J. & Head, M. J. A magnetostratigraphic calibration of Middle Miocene through Pliocene dinoflagellate cyst and acritarch events in the Iceland Sea (Ocean Drilling Program Hole 907A). *Rev. Palaeobot. Palynol.* **187,** 66–94 (2012).
26. Schreck, M. & Matthiessen, J. in *Biological and Geological Perspectives of Dinoflagellates.* (eds Lewis, J., Marret, F. & Bradley, L.) 301–314 (The Micropalaeontological Society, Special Publications: Geological Society, 2013).
27. Schreck, M., Méheust, M., Stein, R. & Matthiessen, J. Response of marine palynomorphs to Neogene climate cooling in the Iceland Sea (ODP Hole 907A). *Mar. Micropaleontol.* **101,** 49–67 (2013).
28. Fahl, K. & Stein, R. Modern seasonal variability and deglacial/Holocene change of central Arctic Ocean sea-ice cover: new insights from biomarker proxy records. *Earth Planet. Sci. Lett.* **351-352C,** 123–133 (2012).
29. Belt, S. T. & Müller, J. The Arctic sea ice biomarker IP_{25}: a review of current understanding, recommendations for future research and applications in palaeo sea ice reconstructions. *Quat. Sci. Rev.* **79,** 9–25 (2013).
30. Müller, J. *et al.* Holocene cooling culminates in sea ice oscillations in Fram Strait. *Quat. Sci. Rev.* **47,** 1–14 (2012).
31. Micheels, A. *et al.* Analysis of heat transport mechanisms from a Late Miocene model experiment with a fully-coupled atmosphere-ocen general circulation model. *Palaeogeogr. Palaeoclimatol. Palaeoecol.* **304,** 337–350 (2011).
32. Knorr, G., Butzin, M., Micheels, A. & Lohmann, G. A warm Miocene climate at low atmospheric CO_2 levels. *Geophys. Res. Lett.* **38,** L20701 (2011).
33. Steppuhn, A. *et al.* The sensitivity of ECHAM4/ML to a double CO_2 scenario for the Late Miocene and the comparison to terrestrial proxy data. *Glob. Planet. Change* **57,** 189–212 (2007).
34. Bradshaw, C. D. *et al.* The relative roles of CO_2 and palaeogeography in determining late Miocene climate: results from a terrestrial model-data comparison. *Clim. Past* **8,** 1257–1285 (2012).
35. Beerling, D. J. & Royer, D. L. Convergent Cenozoic CO_2 history. *Nat. Geosci.* **4,** 418–420 (2011).
36. Van de Wal, R. S. W., de Boer, B., Lourens, L. J., Köhler, P. & Bintanja, R. Reconstruction of a continuous high-resolution CO_2 record over the past 20 million years. *Clim. Past* **7,** 1459–1469 (2011).
37. Breecker, D. O. & Retallack, G. J. Refining the pedogenic carbonate atmospheric CO_2 proxy and application to Miocene CO_2. *Palaeogeogr. Palaeoclimatol. Palaeoecol.* **406,** 1–8 (2014).
38. Zhang, Y. G., Pagani, M., Liu, Z., Bohaty, S. M. & DeConto, R. A 40-million-year history of atmospheric CO_2. *Phil. Trans. R. Soc. A* **371,** 20130096 (2013).
39. Knorr, G. & Lohmann, G. Climate warming during Antarctic ice sheet expension at the Middle Miocene transition. *Nat. Geosci.* **7,** 376–381 (2014).
40. Gierz, P., Lohmann, G. & Wei, W. Response of Atlantic overturning to future warming in a coupled atmosphere-ocean-ice sheet model. *Geophys. Res. Lett.* **42,** 6811–6818 (2015).
41. PALAEOSENS Project-Members. Making sense of palaeoclimate sensitivity. *Nature* **491,** 683–691 (2012).
42. Huber, M. A sensitivity to history. *Nat. Geosci.* **6,** 15–16 (2013).
43. Zachos, J. C., Dickens, G. R. & Zeebe, R. E. An early Cenozoic perspective on greenhouse warming and carbon-cycle dynamics. *Nature* **451,** 281–283 (2008).
44. LaRiviere, J. P. *et al.* Late Miocene decoupling of oceanic warmth and atmospheric carbon dioxide forcing. *Nature* **486,** 97–100 (2012).
45. Zhang, Y. G., Pagani, M. & Liu, Z. A 12-million-year temperature history of the tropical Pacific Ocean. *Science* **344,** 84–87 (2014).
46. Pound, M. J. *et al.* A Tortonian (Late Miocene, 11.61-7.25 Ma) global vegetation reconstruction. *Palaeogeogr. Palaeoclimatol. Palaeoecol.* **300,** 29–45 (2011).
47. Lunt, D. J. *et al.* A methodology for targeting palaeo proxy data aquisition: a case study for the terrestrial late Miocene. *Earth Planet. Sci. Lett.* **271,** 53–62 (2008).
48. Wolfe, J. A. An analysis of Neogene climates in Beringia. *Palaeogeogr. Palaeoclimatol. Palaeoecol.* **108,** 207–216 (1994).
49. Utescher, T., Bondarenko, O. V. & Mosbrugger, V. The Cenozoic Cooling—continental signals from the Atlantic and Pacific side of Eurasia. *Earth Planet. Sci. Lett.* **415,** 121–133 (2015).
50. Stein, R. *et al.* Accumulation of particulate organic carbon at the Eurasian continental margin during late Quaternary times: controlling mechanisms and paleoenvironmental significance. *Glob. Planet. Change* **31,** 87–102 (2001).
51. Stein, R., Jokat, W., Niessen, F. & Weigelt, E. Exploring the long-term Cenozoic Arctic Ocean Climate History—A challenge within the International Ocean Discovery Program (IODP). *Arktos.* doi:10.1007/s41063-015-0012-x (2015).
52. Tian, J., Yang, M., Lyle, M. W., Wilkens, R. & Shackford, J. K. Obliquity and long eccentricity pacing of the Middle Miocene climate transition. *Geochem. Geophys. Geosys* **15,** 1740–1755 (2013).
53. Weller, P. & Stein, R. Paleogene biomarker records from the central Arctic Ocean (IODP Expedition 302): Organic-carbon sources, anoxia, and sea-surface temperature. *Paleoceanography* **23,** PA1S17, doi:10.1029/2007PA001472 (2008).
54. Cronin, T. M., Smith, S. A., Eynaud, F., O'Regan, M. & King, J. Quaternary paleoceanography of the central Arctic based on Integrated Ocean Drilling Program Arctic Coring Expedition 302 foraminiferal assemblages. *Paleoceanography* **23,** PA1S18 (2008).
55. Darby, D. A. Arctic perennial ice cover over the last 14 million years. *Paleoceanography* **23,** PA1S07, doi:10.1029/2007PA001479 (2008).
56. Krylov, A. A. *et al.* A shift in heavy and clay mineral provenance indicates a middle Miocene onset of a perennial sea-ice cover in the Arctic Ocean. *Paleoceanography* **23,** PA1S06, doi:10.1029/2007PA001497 (2008).
57. Tremblay, L.-B., Schmidt, G. A., Pfirman, S., Newton, R. & Derepentigny, P. Is ice-rafted sediment in a North Pole marine record evidence of perennial sea-ice cover? *Phil. Trans. R. Soc. A* **A373,** 20140168 (2015).
58. Cabedo-Sanz, P. & Belt, S. T. Identification and characterisation of a novel mono-saturated highly branched isoprenoid (HBI) alkene in ancient Arctic sediments. *Org. Geochem.* **81,** 34–39 (2015).
59. Belt, S. T. *et al.* Identification of paleo Arctic winter sea ice limits and the marginal ice zone: optimised biomarker-based reconstructions of late Quaternary Arctic sea ice. *Earth Planet. Sci. Lett.* **431,** 127–139 (2015).
60. Belt, S. T. *et al.* A reproducible method for the extraction, identification and quantification of the Arctic sea ice proxy IP_{25} from marine sediments. *Analyt. Methods* **4,** 705–713 (2012).
61. Belt, S. T. *et al.* An inter-laboratory investigation of the Arctic sea ice biomarker proxy IP_{25} in marine sediments: key outcomes and recommendations. *Clim. Past Discuss.* **10,** 155–166 (2014).
62. Smik, L., Cabedo-Sanz & Belt, S. T. Semi-quantitative estimates of paleo Arctic sea ice concentration based on source-specific highly branched isoprenoid alkenes: a further development of the PIP25 index. *Org. Geochem.* **92,** 63–69 (2016).
63. Prahl, F. G. & Wakeham, S. G. Calibration of unsaturation patterns in long-chain ketone compositions for palaeotemperature assessment. *Nature* **330,** 367–369 (1987).
64. Müller, P. J., Kirst, G., Ruhland, G., von Storch, I. & Rosell-Melé, A. Calibration of the alkenone paleotemperature index U^{k}_{37}-based on core-tops from the eastern South Atlantic and the global ocean (60 degrees N -60 degrees S). *Geochim. Cosmochim. Acta* **62,** 1757–1772 (1998).
65. Sikes, E. L., Volkman, J. K., Robertson, L. G. & Pichon, J.-J. Alkenones and alkenes in surface waters and sediments of the Southern Ocean: implications for paleotemperature estimation in polar regions. *Geochim. Cosmochim. Acta* **61,** 1495–1505 (1997).

66. Roeckner, E. *et al.* The atmospheric general circulation model ECHAM5. Part 1. Model description. *Max Planck Inst. Meteorol. Rep.* **349,** ISSN 0937-1060 (2003).
67. Marsland, S. J., Haak, H., Jungclaus, J. H., Latif, M. & Röske, F. The Max-Planck-Institute global ocean/sea ice model with orthogonal curvilinear coordinates. *Ocean Model* **5,** 91–127 (2003).
68. Hibler, W. III. A dynamic thermodynamic sea ice model. *J. Phys. Oceanogr.* **9,** 815–846 (1979).
69. Ikari, M. J. & Kopf, A. J. Cohesive strength of clay-rich sediment. *Geophys. Res. Lett.* **38,** L16309, 5 (2011).
70. Craig, R. F. *Craig's Soil Mechanics* 446 (Spon Press, 1974).
71. Stein, R. *Arctic Ocean Sediments: Processes, Proxies, and Palaeoenvironment* 2, 587 (Elsevier, 2008).
72. Wassmann, P. Arctic marine ecosystems in an era of rapid climate change. Progr. *Oceanography* **90,** 1–17 (2011).
73. Frank, M. *et al.* Beryllium isotopes in central Arctic Ocean sediments over the past 12.3 million years: stratigraphic and paleoclimatic implications. *Paleoceanography* **23,** PA1S02, doi:10.1029/2007PA001478 (2008).

Acknowledgements

We gratefully thank Captain Schwarze and his crew of RV *Polarstern* for the excellent support and cooperation during the entire cruise. We thank the PS87 Geoscience Party for support in getting geophysical and geological shipboard data and sediments during the expedition (see 'Supplementary Note 1'). We thank three anonymous reviewers for numerous constructive suggestions for improvement of the manuscript. Thanks to Simon Belt (Biogeochemistry Research Centre, University of Plymouth/UK) for providing the 7-HND standard for IP_{25} quantification. This publication is a contribution to the Research Programme PACES II, Topic 3 (The earth system from a polar perspective: data, modelling and synthesis) of the Alfred Wegener Institute Helmholtz Centre for Polar und Marine Research (AWI). The study used samples and data provided by AWI (grant number AWI-PS87_01) and the IODP Program. This research is also supported by KOPRI's Basic Research Project (PE15062) and a National Research Foundation of Korea Grant funded by the Korean Government (2015M1A5A1037243).

Author contributions

R.S. and W.J. proposed the main research goals of the PS87 Expedition. R.S. was chief scientist of the PS87 Expedition and leader of the geology group, developed the lithostratigraphic framework of the PS87 sediment cores, and wrote the first draft of the manuscript. K.F. conducted all biomarker analyses, evaluation and quality control. M.S. and M.K. identified the assemblages of palynomorphs and benthic agglutinated foraminifers, respectively, used for biostratigraphy and age control. G.K. and G.L. developed the model and experimental design. G.K. carried out the experiments. F.N. compiled the physical property and PARASOUND data. L.J. aquired and processed multibeam bathymetry data. M.F. was responsible for XRF scanning and line-scan imaging of the sediment cores. J.M. led the offshore geological activities on deck and in the laboratories. W.J. led the offshore geophysical activities and data acquisition. C.G. processed the multi-channel seismic data. A.K. carried out the sediment load and compaction experiments. All authors contributed to the data interpretation and writing of the final version of the manuscript.

Additional information

Competing financial interests: The authors declare no competing financial interests.

11

A better-ventilated ocean triggered by Late Cretaceous changes in continental configuration

Yannick Donnadieu[1], Emmanuelle Pucéat[2], Mathieu Moiroud[2], François Guillocheau[3] & Jean- François Deconinck[2]

Oceanic anoxic events (OAEs) are large-scale events of oxygen depletion in the deep ocean that happened during pre-Cenozoic periods of extreme warmth. Here, to assess the role of major continental configuration changes occurring during the Late Cretaceous on oceanic circulation modes, which in turn influence the oxygenation level of the deep ocean, we use a coupled ocean atmosphere climate model. We simulate ocean dynamics during two different time slices and compare these with existing neodymium isotope data (ε_{Nd}). Although deep-water production in the North Pacific is continuous, the simulations at 94 and 71 Ma show a shift in southern deep-water production sites from South Pacific to South Atlantic and Indian Ocean locations. Our modelling results support the hypothesis that an intensification of southern Atlantic deep-water production and a reversal of deep-water fluxes through the Caribbean Seaway were the main causes of the decrease in ε_{Nd} values recorded in the Atlantic and Indian deep waters during the Late Cretaceous.

[1] Laboratoire des Sciences du Climat et de l'Environnement, LSCE-IPSL, CEA-CNRS-UVSQ, Université Paris-Saclay, 91191 Gif sur Yvette, France. [2] Biogéosciences Dijon, Université Bourgogne-Franche-Comté, UMR CNRS 6282, Dijon 21000, France. [3] Géosciences Rennes, Université de Rennes, UMR CNRS 6118, Rennes 35042, France. Correspondence and requests for materials should be addressed to Y.D. (email: yannick.donnadieu@lsce.ipsl.fr).

In the context of recent warming, modern ocean de-oxygenation that occurs not only on continental margins but also in the tropical oceans worldwide[1] resembles the model initially invoked for the onset of oceanic anoxic events (OAEs)[2]. Recent studies point to a major role of increased nutrient inputs as a trigger for OAEs[3–5]. Nevertheless, ocean circulation, through its impact on oxygen concentration in deep waters, may have affected the thresholds required to trigger an OAE. It has been suggested that Late Cretaceous changes in climate and continental configuration[6–8], namely the widening of the Atlantic Ocean and the deepening of the Central Atlantic (CA) gateway[9], could have induced major changes in oceanic circulation that may have had an impact on the general oxygenation state of the oceanic basins and contributed to the conclusion of these large-scale anoxic events in the deep ocean[10,11].

Nevertheless, no consensus exists on oceanic circulation modes and their possible evolution during the Cretaceous, despite recent improvements of the spatial and temporal coverage of neodymium isotopic data (ε_{Nd}), a proxy for oceanic circulation[11–15]. For instance, a decrease in bottom water ε_{Nd} values during the Late Cretaceous in the Atlantic and Southern Oceans has been interpreted either as reflecting the initiation or intensification of deep-water production in the northern Atlantic[15] or in the South Atlantic (SA) and in the Indian Ocean[10,11,13]. Additional sources for deep waters have been suggested in the North or South Pacific, or even at low latitudes, based on ε_{Nd} or oxygen isotope data[16–18].

General circulation models have also been used to study oceanic circulation during the Cretaceous[7,19,20]. To our knowledge, no modelling studies have reconstructed the evolution of ocean dynamics resulting from the widening of the South and CA Ocean occurring between the Cenomanian and the Maastrichtian. Published simulations either focus on a specific time period[20,21] or are devoted to the impact of the CA gateway opening between the Albian and the Cenomanian[22].

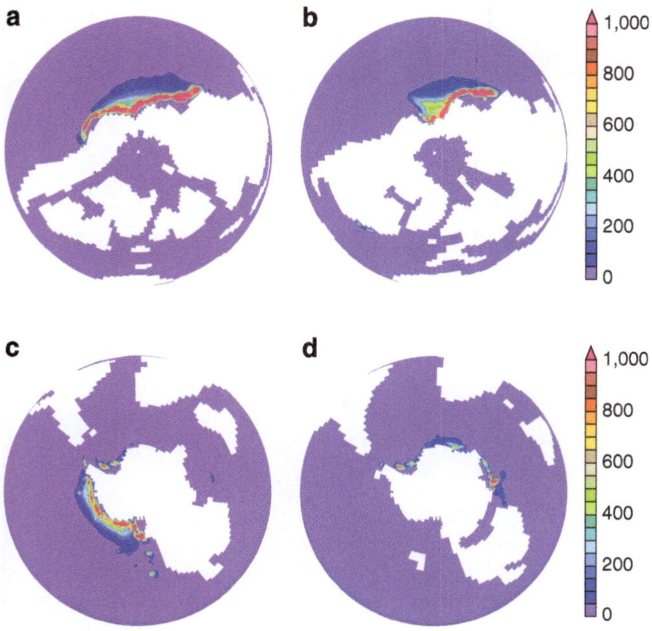

Figure 1 | Annual distribution of convective adjustments across the water column. (**a**) For 94 Ma, 4 × CO_2 from the North Pole. (**b**) For 71 Ma, 4 × CO_2 from the North Pole. (**c**) For 94 Ma, 4 × CO_2 from the South Pole. (**d**) For 71 Ma, 4 × CO_2 from the South Pole. The shading signifies the number of times the water column has undergone convective mixing summed over a year. Regions experiencing a large occurrence of convective adjustments are interpreted to represent site of intermediate and deep-water formation.

Here we use the fast ocean atmosphere model (FOAM), to explore the evolution of oceanic circulation occurring during the Late Cretaceous (see Methods). Our simulations highlight an evolution from a sluggish circulation in the South and CA using a Cenomanian/Turonian land–sea mask towards a much more active circulation in these basins with an early Maastrichtian land–sea mask.

Results

Changes in oceanic circulation at 1,120 p.p.m. The 94 Ma simulation at a CO_2 concentration of 1,120 p.p.m. (four times the pre-industrial atmospheric level) displays a bipolar oceanic circulation characterized by large areas of deep-water formation located both in the north and in the south of the Pacific Ocean (Fig. 1). The production of deep waters along the northwest boundary of the Pacific Ocean is common to many numerical Cretaceous simulations[4,20,22] but not all[23]. The sinking of water masses occurs over the cyclonic subpolar gyres and results from the winter cooling of warm and salty surface waters located within this large region[20]. The use of the early Maastrichtian land–sea mask for the same atmospheric CO_2 level induces substantial changes in the location of deep-water production in the model. The north Pacific area of deep-water formation is reduced to the northwestern Pacific area only and sinking in the south Pacific area completely disappears. Conversely, the Atlantic and the Indian sectors of the Austral Ocean become larger contributors of deep waters and a small area of intermediate-water formation appears along the northeast American margin in the North Atlantic (Fig. 1).

Sensitivity to atmospheric CO_2 levels. Changes in atmospheric CO_2 level from 8 to 2 pre-industrial atmospheric level have a small effect on the distribution of deep-water formation areas in our study (Supplementary Figs 1 and 2). The main features of oceanic circulation modes remain quite similar, with a disappearance of deep-water convection in the southern Pacific and a larger area of deep-water formation appearing in the southern Atlantic when using early Maastrichtian palaeogeography.

Nevertheless, the sensitivity of overturning to CO_2 level remains an open debate, as some models[24,25] found that ocean dynamics vary with CO_2 while others did not[4,26]. We emphasize that all simulations presented here represent an equilibrated climate. The ocean can go through a transient stratified state resulting from a faster warming of the oceanic upper layers when confronted with a CO_2 increase imposed on a short time scale (a few hundred to a few thousand years). Winter convection at high latitudes would then stop, owing to the decrease in vertical density gradient, and the classical thermohaline circulation may stop as well, but only transiently. We thus recognize that there is room for improvement in this area of research and await a modelling inter-comparison project for warm climates such as the one that has been initiated on the warm Eocene climate[27]; however, such a comparison is beyond the scope of this study.

The results of our model suggest at this point that the changes in ocean dynamics simulated between 94 and 71 Ma are driven by changes in palaeogeography rather than by the cooling recorded between the Cenomanian and the Maastrichtian. As only minor changes occur with varying CO_2 levels, only the simulations at 1,120 p.p.m. will be further described and compared for each palaeogeography in the remainder of the study. Our main aim in this study is to describe and understand the changes in ocean dynamics that occur. It is however interesting to note that the atmospheric CO_2 concentration range used in our simulations produces sea-surface and deep-water temperatures that are comparable to oceanic latitudinal thermal gradients reconstructed from palaeoceanographic data for the Cenomanian/Turonian boundary and for the early Maastrichtian (Supplementary Fig. 7).

Pathways of water masses between the oceanic basins. To go a step beyond this first-order description of ocean dynamics, we have computed zonal and/or meridional water transport across several key sections of the ocean basins (Figs 2 and 3). In most previous modelling studies, ocean dynamics have been described through a global overturning function. However, a global overturning function may be dominated by the largest oceanic basin, that is, the Pacific Ocean, and does not allow for the identification of ocean dynamics occurring in the still relatively narrow Atlantic Ocean, in the Tethys or in the Indian Ocean. Our aim here is to provide a better three-dimensional picture of water flow within and between the different oceanic basins.

In the CTRL94Ma run, 7 Sv are transported from the South Pacific to the SA across the D section (Fig. 3c and Table 1). The circulation simulated within the SA Ocean appears quite sluggish at depth, with a very limited northward flow of 4.2 Sv throughout the SA section between 1,200 and 2,750 m (Fig. 3c and Table 1). Conversely, across the East India section, water flow is directed northward down to a depth of 3,000 m and transports 25 Sv to the North (Fig. 3c and Table 1). In the CTRL71Ma run, deep-water production intensifies in the southern Atlantic, extends eastward to the Indian sector and does not occur in the southern Pacific sector any more (Figs 1 and 3). Figure 3d shows a drastic intensification of the northern flow across the SA section for depth below 800 m. These northward-flowing deep waters reach the North Atlantic, as testified by the large flux of water across the CA section (Fig. 3b). Surface-to-intermediate waters across the Central to the Southern Atlantic are also flowing more intensively from the North to the South.

In the CTRL94Ma run and in the North Atlantic, the vertical flux profiles computed for the Caribbean Seaway section, CA section and Mediterranean (Med) section indicate that the surface and intermediate waters down to about 1,000 m flow from the Tethys to the North Atlantic, and then southward across the CA Gateway and westward into the equatorial Pacific via the Caribbean Seaway. Across the Caribbean Seaway below 1,000 m, waters flow from the Pacific to the Atlantic as shown by the eastward water flux from 1,000 to ~2,500 m (Fig. 3). The existence of an estuarine circulation pattern, with the surface layers flowing from the Tethys to the Pacific and the deeper layers flowing in the opposite direction, has been noticed several times in previous modelling studies[4,19,21]. In the CTRL71Ma runs, the model simulates an increase in circum-equatorial transport across

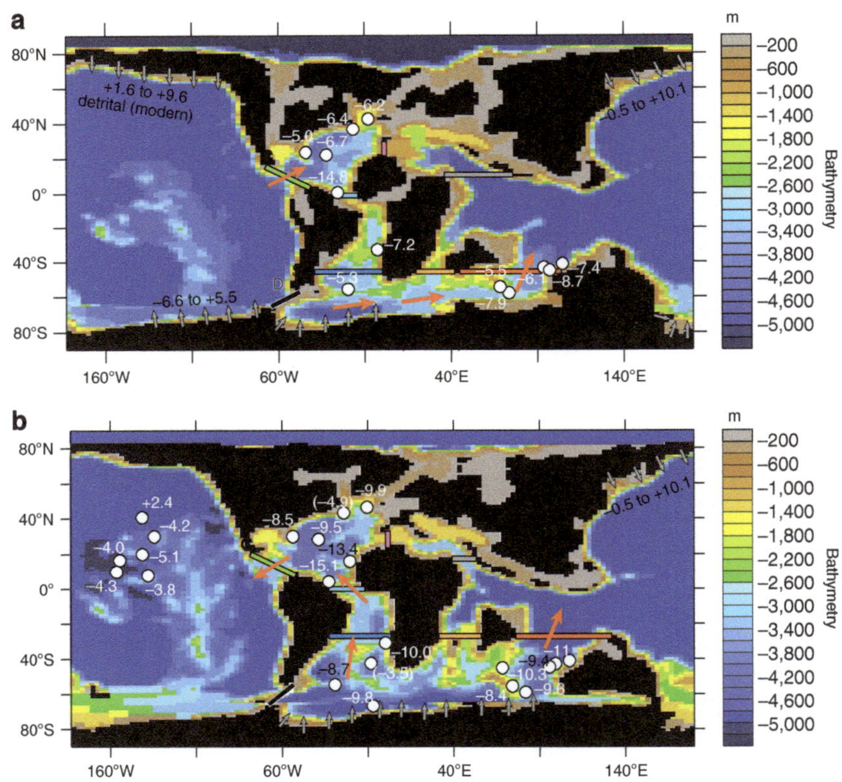

Figure 2 | Land-sea distribution and bathymetry specified in the FOAM simulations. (**a**) Cenomanian/Turonian boundary (94 Ma). (**b**) Early Maastrichtian (71 Ma). Latitudinal and meridional sections across which water transport has been calculated for Fig. 3 are shown. C, Caribbean Seaway; CA, Central Atlantic; D, Drake Passage; EI, East India; Med, Mediterranean; SA, South Atlantic; Tet, Tethys; WI, West India. Supplementary Figs 3 and 4 show the latitude/longitude—depth profile for each section except for the c and d sections for which the bathymetry of the area of interest is shown. The red arrows schematically show the deep-water pathways deduced from our modelling experiments. $\varepsilon_{Nd}(t)$ values of intermediate and deep waters (in white) inferred from fish remains and oxide coatings and averaged for the Turonian (**a**) and Maastrichtian (**b**) are reported at available ODP sites (see Supplementary Fig. 5 for the names of each ODP site). Details on the calculation of average $\varepsilon_{Nd}(t)$ values can be found in Supplementary Data 2. For ODP sites at Demerara Rise (Central Atlantic), $\varepsilon_{Nd}(t)$ values within OAE2 have not been included in the calculation. Values in parentheses indicate a contamination of the samples or a contamination of deep waters by volcanogenic particles[11,12]. Values in orange on the Maastrichtian map (**b**) are from the late Campanian. ε_{Nd} values of the modern detrital fraction (core-top detrital fraction and continental lithogenic sources) in the regions of modelled deep-water production are reported in grey for discussion (details on the specified range of values are available in Supplementary Data 1 and 2, and Supplementary Figs 5 and 6 with associated references).

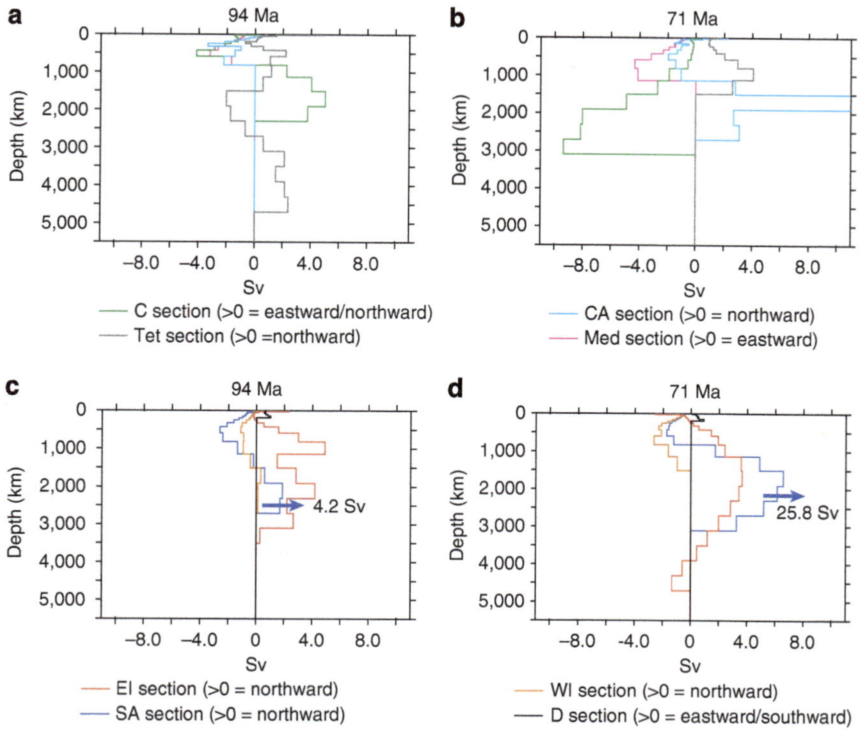

Figure 3 | Water transports. As computed for the segments defined in Fig. 2 for the North Atlantic and the Tethys basins (**a**,**b**) and for the SA and Indian basins (**c**,**d**). The units are in Sv, that is, $10^6\,m^3\,s^{-1}$. Sv is used to quantify the volumetric rate of transport of ocean currents. Values corresponding to the water flow for each vertical level of the ocean model are plotted. The variable corresponds to the integration of the velocity V (U) on the *x* axis (or *y* axis depending on the orientation of the section) and on the *z* axis. Calculated water fluxes account for both the velocity and the area of the section at a given level of the model. For zonal water transports such as the one calculated for the Med segment, positive values indicate eastward transport. For meridional water transports such as those calculated for the Tethyan (Tet), Central Atlantic (CA), South Atlantic (SA), East Indian (EI) and West Indian (WI) segments, positive values indicate northward transport; negative values indicate southward transport. For the specific case of the D section, positive directions are Southward and Eastward, and for the **c** section, positive directions are Northward and Eastward. Blue arrows in **c**,**d** represent water fluxes computed across the SA section for the deeper vertical levels where the transport is northward.

Table 1 | Water fluxes integrated over the water column for each section defined in Figure 1.

	Drake	SA	WI	EI	C	CA	Med	Tet
94 Ma								
DRAKE CLOSED	0	−15 (5.8)	−5	20	−2.3	−15.1	−17.4	17.3
CTRL	**7**	**−11.6 (4.2)**	**−6.4**	**25**	**−5.6**	**−11.7**	**−17.4**	**17.3**
DRAKE408m	16	−9 (3.6)	−5.2	30.2	−7.2	−9.2	−16.5	16.3
SEWALL	17.6	−7 (3.7)	−8.	32.6	−9	−7.	−16.	16
CAS560m	17.2	−4.6 (3.9)	−8.4	30.2	−16.2	−4.7	−21.1	21
CAS-Islands	16	−8.8 (3.8)	−4.5	29.3	−8.1	−8.9	−17	16.9
71 Ma								
DRAKE CLOSED	0	12.3 (35.4)	−9.5	21.8	−34.9	12.3	−22.5	21.7
CTRL	**6.25**	**14.9 (25.8)**	**−21.9**	**13.3**	**−39.5**	**14.9**	**−24.8**	**23.8**
DRAKE408m	12.4	18.7 (23.7)	−19.8	13.5	−41.1	18.7	−22.4	22.1
SEWALL	14.4	15.6 (16.2)	−15.3	14.1	−33.6	15.6	−18.1	18.6
CAS560m	10.1	−14 (1.1)	−9.8	33.8	2	−14	−12.1	11.6
CAS-Islands	14.6	17.7 (22.1)	−18.5	15.4	−41.1	17.7	−23.4	23.5

C, Caribbean Seaway; CA, Central Atlantic; CAS560m, Caribbean gateway at 560 m; D, Drake Passage; DRAKE408m, Drake Passage at 408 m; EI, East India; Med, Mediterranean; SA, South Atlantic; Tet, Tethys; WI, West India.

Additional numbers in SA column correspond to the deep flow going northward in the South Atlantic across the SA section. Numbers in bold put forward the main simulations discussed in the paper, that is, with a shallow Drake Passage at 145 m.

the Caribbean Seaway and the Med gateway. Deep waters are now flowing from the Atlantic to the Pacific at all depths across the Caribbean Seaway, which marks the end of estuarine circulation in the Atlantic.

Comparison with available ε_{Nd}. Nd in seawater ultimately derives from continents and the rocks eroded around an area of deep-water production imprint the surface waters with a distinct isotopic composition that depends on the age and lithology of the rock[28]. This signature is then exported to depth as the water sinks. The residence time of Nd in the ocean is shorter than the oceanic mixing time, which prevents complete homogenization but is long enough for Nd to be transported by deep-water masses along their pathways[28]. The Nd isotope composition of deep waters has thus been used to track oceanic circulation patterns in both modern and ancient oceans[10,28–30].

Available data for the Late Cretaceous highlight a major decrease in deep-water ε_{Nd} recorded both in the North and South Atlantic and in the Indian Ocean[10,14,31], with more unradiogenic (lower) ε_{Nd} values recorded during the Maastrichtian (on average in the -8.5 to -11 ε-units range) compared with the Turonian (on average in the -5 to -8.5 ε-units; Fig. 2). This decrease has been interpreted to reflect initiation of deep-water production in an area receiving unradiogenic Nd inputs from the nearby continents, with the northern Atlantic[15], or the Atlantic or Indian sector of the Southern Ocean proposed as possible sources[10,11]. Indeed at present, ε_{Nd} values as low as -14 to -26 ε-units recorded in Baffin Bay and Labrador Sea waters reflect the unradiogenic Nd continental supply from nearby Archean terranes[32–34] that were already present during the Cretaceous. In the southern Ocean, unradiogenic detrital inputs occur at present (Fig. 2, Supplementary Fig. 5 and Supplementary Data 1). The presence of similar unradiogenic inputs during the Late Cretaceous is supported by Campanian–Maastrichtian ε_{Nd} values

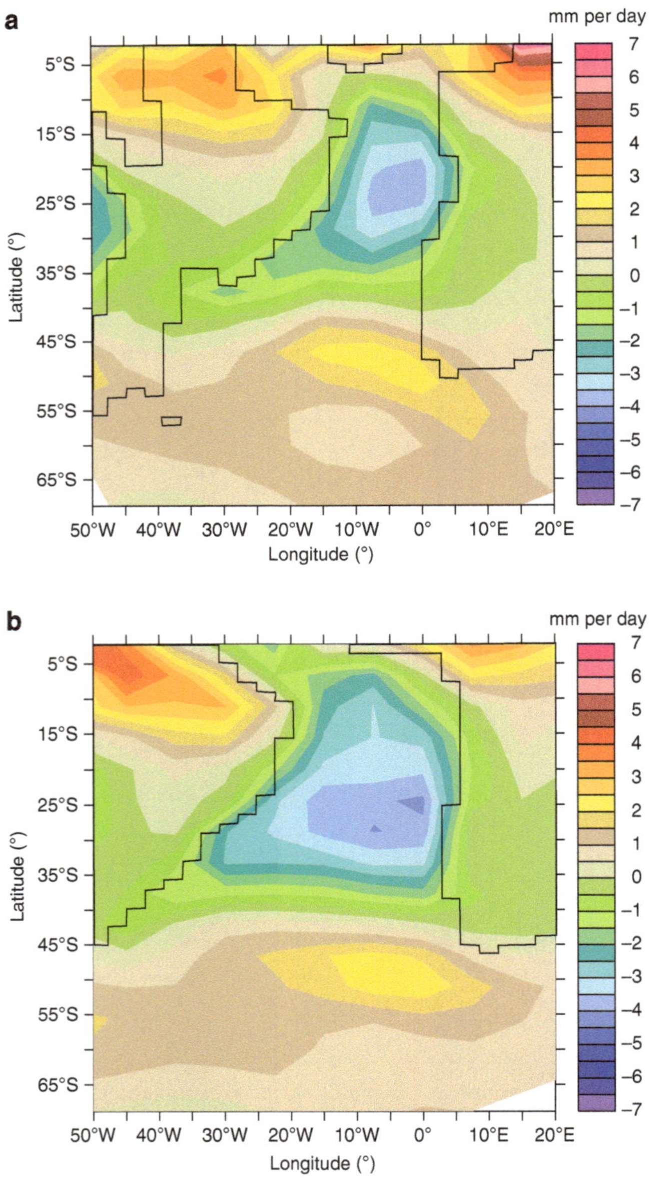

Figure 4 | Hydrological budget over the Central and South Atlantic. Precipitation minus evaporation for 94 (**a**) and 71 Ma (**b**) runs.

of detrital material of around -10 ε-units at ocean drilling program (ODP) site 690 (ref. 12). Because of these unradiogenic detrital inputs from the Antarctic continent, the more radiogenic ε_{Nd} values (higher) recorded in the SA and Indian deep waters during the mid-Cretaceous have been interpreted[10] to reflect at that time additional inputs of radiogenic Nd from abundant volcanic dust in the context of a sluggish circulation within a restricted basin.

It has also been suggested that the subsidence of Rio Grande Rise and other large igneous provinces (for example, Kerguelen or the Madagascar Plateau) during the Late Cretaceous could have diminished such inputs of radiogenic Nd to the southern Atlantic and Indian surface waters, contributing to decrease the ε_{Nd} values of the deep waters sourced in this region[31].

The oceanic circulation modelled here is in agreement with existing ε_{Nd} and part of the proposed scenarios. In the CTRL94Ma runs, the model simulates deep-water formation in the northern and southern Pacific. In the modern ocean, the southern Pacific receives inputs of quite radiogenic Nd eroded from the nearby Antarctic continent (Fig. 2, Supplementary Fig. 5 and Supplementary Data 1). $^{40}Ar/^{39}Ar$ ages of detrital hornblende grains from West Antarctica[35] support the presence of intrusive and volcanic rocks during the Cretaceous (and as early as the Jurassic) that probably provided radiogenic material to the nearby surface waters[36], imprinting the deep water produced there with a radiogenic signature. Deep water produced in the North Pacific region should also have a radiogenic composition, because weathering of radiogenic young circum-Pacific volcanic arcs linked to the subduction of the Kula and Farallon plates was already active during the mid-Cretaceous[37]. This is supported by the high ε_{Nd} values of neritic seawater inferred from Late Cretaceous fish remains recovered from the northwestern Pacific, (typically in the -3 to $+1$ ε-units range)[38]. In the North Atlantic, deep waters below 1,000 m are conveyed in our simulation from the equatorial eastern Pacific through the Caribbean Seaway and result from the mixing of radiogenic deep waters formed both in the North and in the South Pacific. Therefore, the ocean dynamics simulated by the model are expected to result in quite radiogenic ε_{Nd} values of deep waters in the North Atlantic, in agreement with existing ε_{Nd} (Fig. 2 and Supplementary Data 2).

In the CTRL94Ma run, the deep waters bathing the southern Atlantic and Indian Oceans originate in the area located near the modern Weddell Sea, a region that most probably received unradiogenic continental supply from Antarctica as discussed above (Fig. 2). It has been suggested[10,31] that the southern Atlantic may have received during the mid-Cretaceous additional inputs of radiogenic Nd from subaerially exposed continental volcanic provinces, exposure of submarine Large Igneous Provinces and hot spots, and from associated volcanic activity. The sluggish circulation simulated in this work in the SA in the CTRL94Ma run (Fig. 3 and Table 1) would have favoured seawater-particle exchange processes with such radiogenic volcanic particles[10], resulting in a quite radiogenic composition of SA and Indian deep waters as depicted in the available Nd data set (Fig. 2).

In the CTRL71Ma run, convection in the southern Pacific Ocean completely disappears and is reduced in the North Pacific to the northwest area, whereas convection in the SA intensifies and extends eastward to Australia (Figs 2 and 3). The development of sites of deep-water formation near areas of probable unradiogenic continental supply in the southern Atlantic is expected to result in a relatively unradiogenic composition of SA and Indian Ocean waters. This is in agreement with available ε_{Nd} that display quite negative ε_{Nd} values of deep waters during the latest Cretaceous in the southern Atlantic and Indian Oceans, in the range of about -9.5 to -10.5 ε-units at ODP site 690 for the Late Campanian–Maastrichtian interval[12]

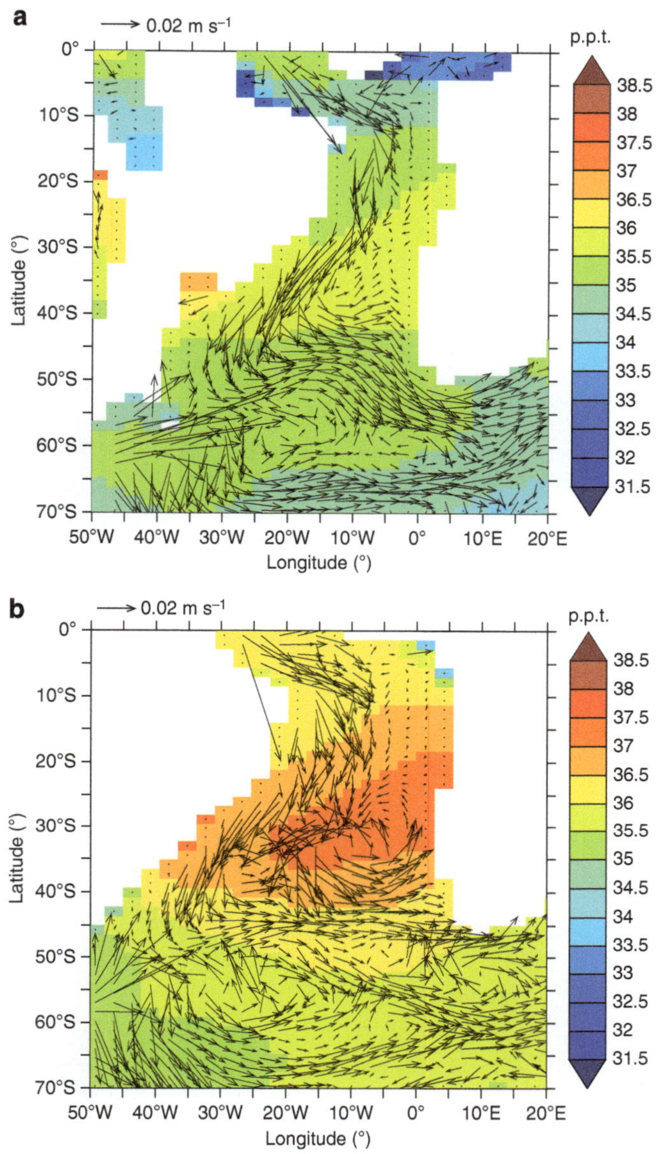

Figure 5 | Absolute salinity fields. Salinity and current velocity (m s^{-1}) averaged over the first 300 m of the ocean for 94 (**a**) and 71 Ma (**b**) runs.

and of −8.4 to −11 ε-units on average for the Maastrichtian at several Indian Ocean ODP sites[10,12,13] (Fig. 2).

Following the simulated circulation patterns, this unradiogenic deep-water composition would then be exported to the North Atlantic through the deeper CA gateway. In addition, with the inversion of the deep-water flux through the Caribbean Seaway in the model, the radiogenic Pacific deep waters would no longer enter the North Atlantic. A westward flux of deep waters throughout the Caribbean Seaway during the Campanian and Maastrichtian is supported by ε_{Nd} values of the detrital fraction higher than that of the bottom waters at site 152 (Nicaragua Rise)[39]. The oceanic circulation simulated for the Maastrichtian should thus generate a much less radiogenic Nd isotope signature of North and SA deep waters, as well as in the Indian Ocean down to 3,000 m of depth, in agreement with ε_{Nd} data (Fig. 2).

Our work thus supports an intensification of deep-water production in the southern Atlantic and Indian Oceans along with a reversal of the deep-water flux through the Caribbean Seaway as a driver for the decrease in deep-water ε_{Nd} depicted during the Late Cretaceous. The subsidence of large igneous provinces such as the Kerguelen plateau could also have contributed to reducing the inputs of radiogenic Nd to the surface waters of the southern Atlantic and Indian Oceans[31]. As Nd is not incorporated as a tracer into our model, the impact of reduced inputs of radiogenic Nd linked to this mechanism on the decrease recorded by deep-water ε_{Nd} values cannot be further discussed here.

The simulated circulation during the Maastrichtian remains difficult to reconcile with the highly non-radiogenic values recorded at Cape Verde and Demerara Rise[14,16] in the low-latitude Atlantic Ocean. It has been suggested that intermediate to deep waters could have been generated in the area of Demerara Rise ('Demerara Rise Bottom Waters')[16]. Nevertheless, the Late Cretaceous ε_{Nd} from Demerara Rise stands in marked contrast to the data from other North Atlantic sites[11]. Demerara Rise bottom waters may have been restricted to intermediate depths, similar to Med outflow water, and thus would not have greatly influenced deep-water masses in the North Atlantic[11]. Our results show no evidence for water sinking in this area (Fig. 1). One way to reconcile the ε_{Nd} of Demerara Rise and Cape Verde with the modelled circulation could be to invoke the impact of boundary exchange processes that are known to occur along continental margins and can modify the Nd isotope composition of the local bottom waters[32,40]. Considering the proximity of the two sites to very old, unradiogenic terranes (Guyana and African Shields), such a process may have locally lowered the Nd isotopic composition of the water masses. The increase in Nd isotope values recorded at this site at the end of the Maastrichtian when they reach values similar to the other Atlantic sites may then reflect the progressive opening of the Demerara Rise region to the remaining of the Atlantic.

Mechanisms driving the changes in oceanic circulation. In the model, the shift in the location of deep-water formation from the Pacific to the Atlantic/Indian sector is clearly due to modifications of the land–sea mask. In detail, it is the hydrological cycle over the SA basin that seems to be responsible for the shift in the deep-water formation area between the Cenomanian and the early Maastrichtian. During the early Maastrichtian, the fall in sea level and the retreat of the sea from the tropical continents induce more tropical monsoons and less vertical advection of moist air, which result in a more negative P − E (precipitation minus evaporation) budget in the CA Ocean (Fig. 4). When integrated over the latitudes 40°S–10°N, corresponding to the Central and SA, the P − E budget is more negative in the 71 Ma run with a value around—1.5 mm per day (−0.14 mm per day for the 94 Ma run). These changes will favour convection in the SA during the early Maastrichtian, because the saltier the surface waters are, the denser and heavier they will be when they reach the southern high latitudes (Fig. 5).

Sensitivity to the Drake Passage and to the Caribbean Seaway. The results presented here are based on the two palaeogeographies published in Sewall *et al.*[9], in which the depth of the Drake Passage has been reduced to 145 m. Major uncertainties exist on the configuration of the Drake Passage during the Late Cretaceous, with reported depths ranging from over 1,000 m[9] to a shallow[41] or completely closed passage[42]. Similarly, the depth and configuration of the Caribbean gateway is not well constrained. Island arc volcanism was already present across the Caribbean Seaway during the Late Cretaceous, possibly partly impeding deep or intermediate water communications across the seaway[43,44]. Nevertheless, the configuration of these two seaways have been shown to have important implications for oceanic circulation in more recent periods[45,46].

To test the impact of the configuration of the Drake Passage and of the Caribbean Seaway on oceanic circulation,

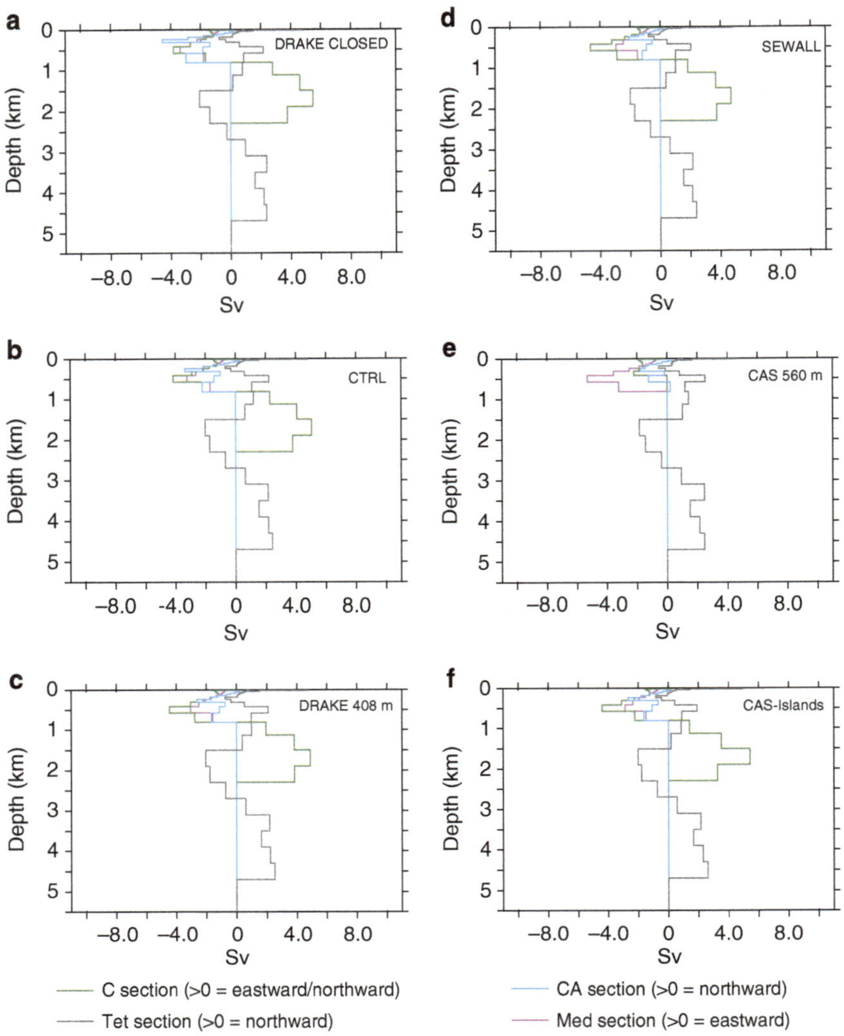

Figure 6 | Water transports for the North Atlantic and the Tethys basins at 94 Ma. (**a**) Drake Passage closed (DRAKE CLOSED), (**b**) Drake Passage at 145 m (CTRL), (**c**) Drake Passage at 408 m (DRAKE408m), (**d**) Drake Passage as in Sewall *et al.*[9] (SEWALL), (**e**) Caribbean gateway at 560 m (CAS560m) and (**f**) continental plate across the Caribbean gateway (CAS-Islands).

we performed a series of sensitivity experiments. In a first set of simulations, we modified the Cenomanian/Turonian and the early Maastrichtian palaeogeographies of Sewall *et al.*[9] using the same horizontal configuration but varying the depth of the Drake Passage, using the depth originally specified by Sewall *et al.*[9], which is around 1,000 m (SEWALL simulations), a depth of 408 m (DRAKE408m simulations), a depth of 145 m (CTRL simulations) and a closed Drake Passage (DRAKE CLOSED simulation). In a second set of simulations, two runs for each palaeogeography were performed with a very shallow Caribbean Gateway, fixed at 560 m depth (CAS560m simulations) and a small continental plate across the Caribbean Seaway (Supplementary Fig. 8) to simulate the presence of islands (CAS-Islands simulation). The results of these experiments are presented in Figs 6–9 and in Table 1.

Most features described above for the two land–sea masks in the CTRL runs remain the same independently of the depth of the Drake Passage (Figs 6–9). These features are the inversion of the direction of deep-water flow across the Caribbean Seaway for depths > 1,000 m and the increase in northward water fluxes at depth across the SA basin. Summed over the whole water column, water flows globally from the North Atlantic to the SA during the Cenomanian–Turonian and in the opposite direction during the early Maastrichtian. However, the depth of the Drake Passage substantially affects the intensity of these flows. The main changes occur when going from a closed passage to a depth of 408 m. Most notably, the deeper branch of the thermohaline circulation coming from the Weddell Sea decreases its volume transport from 5.8 to 3.7 Sv for the 94 Ma experiments and from 35.4 to 16.2 Sv for the 71 Ma experiments (Table 1). Nevertheless, the trend from a sluggish circulation to a more active circulation at depth in the South and CA with the early Maastrichtian palaeogeography holds independently of the configuration of the Drake Passage.

Sensitivity experiments assessing the effect of Caribbean Seaway depth on circulation yield contrasting results, especially for the early Maastrichtian simulations. A shallow gateway across the Caribbean Seaway (CAS560m simulations) completely modifies deep circulation and results in a shut down of thermohaline circulation in the SA (Figs 8 and 9). By contrast, the same configuration but for the 94 Ma experiments has a minor impact on the general pattern of oceanic circulation (Figs 6 and 7). Finally, experiments with a small continental plate across the Caribbean Seaway result in only minor changes in the modelled oceanic circulation for both palaeogeographies.

The circulation modelled with a shallow Caribbean Seaway (CAS560 m) for the Cenomanian/Turonian is less consistent with

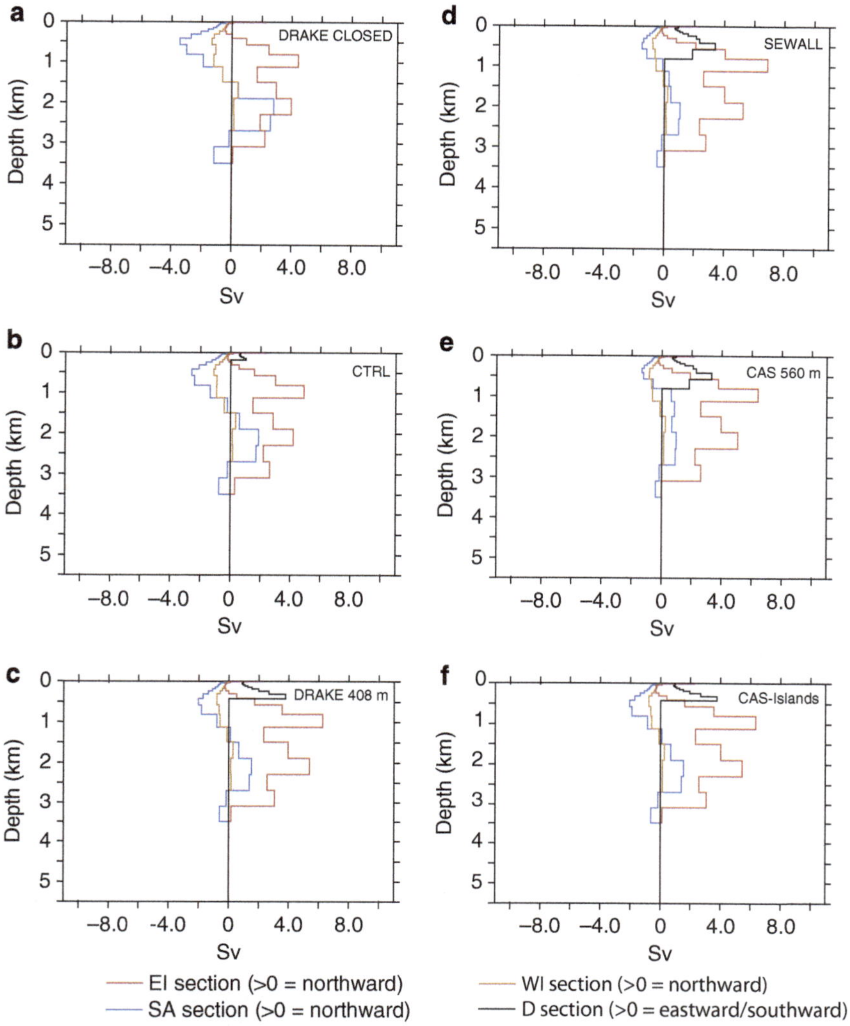

Figure 7 | Water transport for the SA and Indian basins at 94 Ma. (**a**) Drake Passage closed (DRAKE CLOSED), (**b**) Drake Passage at 145 m (CTRL), (**c**) Drake Passage at 408 m (DRAKE408m), (**d**) Drake Passage as in Sewall *et al.*[9] (SEWALL), (**e**) Caribbean gateway at 560 m (CAS560m) and (**f**) Continental plate across the Caribbean gateway (CAS-Islands).

the ε_{Nd} , because a deep-water influx of radiogenic Pacific waters through the Caribbean Seaway as observed in the other simulations is in better agreement with the radiogenic composition of North Atlantic deep waters. In the early Maastrichtian, the weaker fluxes of surface and intermediate waters across the Tethys, Med and Caribbean Seaway modelled in the CAS560m experiment are also less consistent with an intense equatorial circumglobal current that has been suggested based on the large-scale deposition of phosphorites along the southern Tethyan margin[47].

These additional sensitivity experiments reveal a major role of the circulation pattern across the Caribbean Seaway on the overall modelled thermohaline circulation and shed new light on the mechanisms driving the change in oceanic circulation observed between the simulations using the Cenomanian/Turonian versus early Maastrichtian palaeogeographies. The hydrological cycle changes described above contribute to the intensification of deep-water production in the southern Atlantic and Indian Oceans. Changes in land–sea configuration and the associated atmospheric feedbacks induce a saltier subtropical SA Ocean during the early Maastrichtian, which in turn results in a larger area of deep-water formation around the eastern Antarctic (Fig. 1). Indeed, the saltier the surface waters are, the denser and heavier they will be when they reach the southern high latitudes. An additional important factor appears to be the strong flow of Atlantic water into the Pacific at all depths during the Maastrichtian, observed in all experiments, except the CAS 560 m simulation. This flow creates a large divergence area in the CA, drawing in waters from the Med gateway and from the SA (Table 1). The transition from an estuarine circulation during the Cenonamian–Turonian to the one characterizing the early Maastrichtian as simulated here is probably the main mechanism explaining the shift from a sluggish to an active deep circulation. This transition occurs in response to continental drift and the subsequent modification of the geometry of the Caribbean Seaways. The sensitivity of oceanic circulation across the Caribbean Seaways has already been investigated many times and results from a complex interplay between local processes, such as wind-driven circulation and salinity contrast[45], and remote processes such as the geometry of other gateways[46,48].

Discussion

Our simulations highlight an evolution from a sluggish circulation in the SA and CA using the Cenomanian/Turonian land–sea mask towards a much more active circulation in these basins with the early Maastrichtian land–sea mask. Alongside changes in the location of deep-water production areas at high latitudes, these

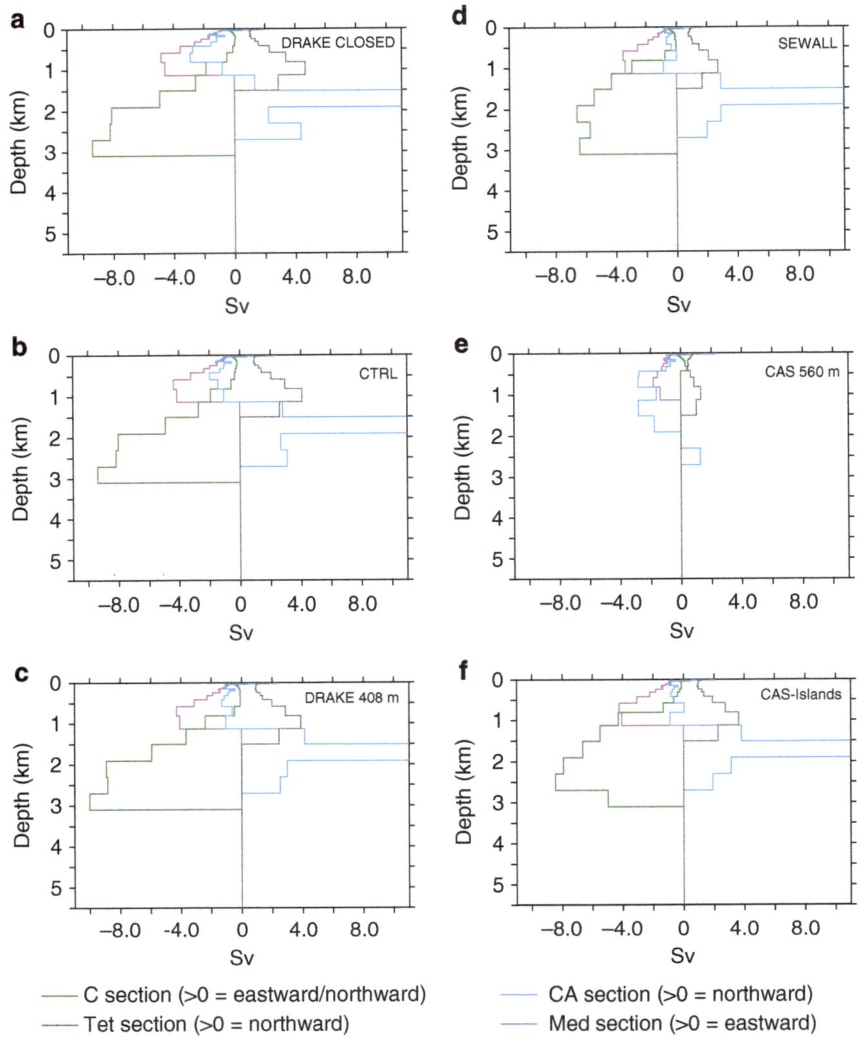

Figure 8 | Water transport for the North Atlantic and the Tethys basins at 71 Ma. (**a**) Drake Passage closed (DRAKE CLOSED), (**b**) Drake Passage at 145 m (CTRL), (**c**) Drake Passage at 408 m (DRAKE408m), (**d**) Drake Passage as in Sewall *et al.*[9] (SEWALL), (**e**) Caribbean gateway at 560 m (CAS560m) and (**f**) continental plate across the Caribbean gateway (CAS-Islands).

features may have paved the way for a well-oxygenated deep ocean. The absence of large-scale anoxia in deep waters during abrupt warming events after the Late Cretaceous, such as the Paleocene–Eocene transition[49] ($\sim$55 Ma), supports this hypothesis. The experiments conducted here emphasize the potentially important role of land–sea configuration as a pre-conditioning factor that affects the ease with which OAEs can develop. In that sense, palaeogeography represents one of the required conditions for the occurrence of OAEs but would not be sufficient on its own to trigger these events, implying additional causal factors[50,51]. As a result of the palaeogeographic changes occurring during the Late Cretaceous, our results suggest that worldwide anoxic conditions in deep waters were more likely to develop as a result of the same triggering factor (for example, increased nutrient input into the oceans[2–4]) during the Cenomanian/Turonian compared with the early Maastrichtian. Two recent geochemical modelling studies have emphasized the need for a sluggish circulation in the Atlantic basin, to explain the large spatial extent of OAE-2 (ref. 52) and a substantial sensitivity of the onset of anoxia/euxinia to oceanic overturning[53]. Based on the latter study, we predict that in a sluggish state such as the one simulated at the Cenomanian/Turonian boundary with <5 Sv of deep water flowing northward in the SA, a 20–40% increase in nutrient input to the ocean could generate deep-water euxinia. In contrast, a 225% increase in nutrient input is required to provoke euxinia with the early Maastrichtian palaeogeography in which 25.8 Sv feed the deep-water flowing northward (Fig. 3c,d). Our results imply that thresholds for the ocean-climate system to shift towards a state of global anoxia in deep waters are likely to be much higher at present than during the Cenomanian/Turonian because of the strength of the modern thermohaline circulation.

In summary, this study provides a clear description of oceanic circulation changes occurring during the Late Cretaceous and is able to broadly explain available published ε_{Nd} data. Our work points to changes in continental configuration as the major driver for the depicted ocean circulation change, through a modification of the hydrological cycle over the SA basin and the development of a strong westward flow of water at all depths throughout the Caribbean Seaway, that both favour an intensification of deep-water production in the southern Atlantic and Indian region. We note that our knowledge of the geometry of Cretaceous oceanic gateways is contentious and a definitive history of ocean circulation during this time must await further constraints on the state of the various oceanic connections. We also speculate on the consequences of long-term oceanic circulation changes on the likelihood that the Earth System will be affected by an OAE. More explicit modelling including

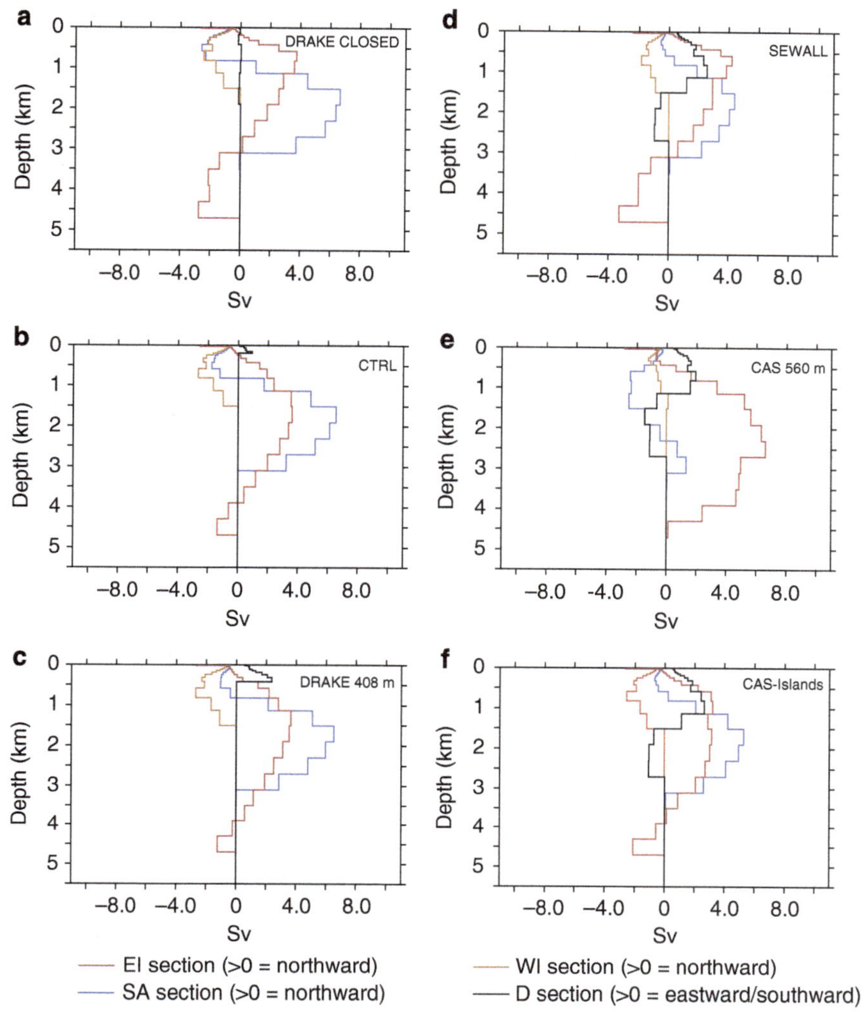

Figure 9 | Water transport for the SA and Indian basins at 71 Ma. (**a**) Drake Passage closed (DRAKE CLOSED), (**b**) Drake Passage at 145 m (CTRL), (**c**) Drake Passage at 408 m (DRAKE408m), (**d**) Drake Passage as in Sewall *et al.*[9] (SEWALL), (**e**) Caribbean gateway at 560 m (CAS560m) and (**f**) continental plate across the Caribbean gateway (CAS-Islands).

complex climate models and biogeochemistry will help reveal the extent to which the tectonic and climatic background state can influence the development of oceanic anoxic conditions.

Methods

Code availability. The code of the model FOAM is available on request by e-mail to the first author.

Description of the model. The model experiments were completed using the FOAM, developed by Jacob[54]. This model combines a low spectral resolution R15 (48 × 40 grid) atmosphere model counting 18 altimetric levels with a highly efficient medium resolution (128 × 128 grid) ocean module composed by 24 bathymetric levels. FOAM successfully simulates many aspects of the present-day climate and compares well with other contemporary medium-resolution climate models. It was previously used to investigate numerous past climate changes, ranging from the Neoproterozoic glaciations to the onset of the ACC at the Eocene–Oligocene boundary[55,56].

Description of the experiments. The two sets of simulations (called hereafter CTRL94Ma and CTRL71Ma) developed for this study are based on existing palaeogeographic reconstructions for the Cenomanian/Turonian boundary and for the early Maastrichtian[9] (Fig. 2). For the sake of simplicity, we choose to use 94 Ma though the accurate age of the boundary is 93.9 Ma[57]. Based on existing palaeogeographies[41,58,59], the depth of the Drake Passage in the CTRL simulations has been restricted to a shallower depth (that is, 145 m) than in the original bathymetry in which the depth reached more than a thousand metres. Indeed, first evidences for the deepening of the Drake Passage correspond to the creation of oceanic crust and are generally dated between the Eocene and the Oligocene[60,61]. During the Late Cretaceous, pull-apart basins were formed between the Antarctica Peninsula and southern South America[62], which represent the onset of the opening of the Scotia Sea. Therefore, the Drake Passage most probably formed a narrow and shallow marine domain at that time[63]. To account for existing uncertainties on the configuration of the Drake Passage and of the Caribbean Seaway, sensitivity experiments assessing the effect of the depth specified for these two marine gateways are also presented in the discussion. Data and model-based atmospheric CO_2 level estimates for the Late Cretaceous are between two and eight times the pre-industrial atmospheric CO_2 concentration (280 p.p.m.)[64–69]. Accordingly, three atmospheric CO_2 concentrations were tested for the two palaeogeographies: 560, 1,120 and 2,240 p.p.m. The solar luminosity was specified as 99% of modern. Orbital parameters and other greenhouse gases were set to present-day values. River routing is specified using the Sewall *et al.* reconstruction.

The experiments were integrated for 2,000 years without flux corrections or deep ocean acceleration. During the last 100 years of model integration, there is no apparent drift in the upper ocean (between the surface and 300 m depth) and <0.1 °C per century change in globally averaged ocean temperature. All model results have been averaged over the last 50 model years.

References

1. Keeling, R. E., Körtzinger, A. & Gruber, N. Ocean deoxygenation in a warming world. *Annu. Rev. Marine Sci.* **2,** 199–229 (2010).
2. Jenkyns, H. C. Geochemistry of oceanic anoxic events. *Geochem. Geophys. Geosyst.* **11** doi:10.1029/2009GC002788 (2010).
3. Meyer, K. M. & Kump, L. R. Oceanic euxinia in earth history: causes and consequences. *Annu. Rev. Earth Planet Sci.* **36,** 251–288 (2008).
4. Monteiro, F. M., Pancost, R. D., Ridgwell, A. & Donnadieu, Y. Nutrients as the dominant control on the spread of anoxia and euxinia across the

Cenomanian-Turonian oceanic anoxic event (OAE2): model-data comparison. *Paleoceanography* **27** doi:10.1029/2012PA002351 (2012).
5. Föllmi, K. B. Early Cretaceous life, climate and anoxia. *Cretaceous Res.* **35,** 230–257 (2012).
6. Friedrich, O., Norris, R. D. & Erbacher, J. Evolution of middle to Late Cretaceous oceans--A 55 m.y. record of Earth's temperature and carbon cycle. *Geology* **40,** 107–110 (2011).
7. Poulsen, C. J., Gendaszek, A. S. & Jacob, R. L. Did the rifting of the Atlantic Ocean cause the Cretaceous thermal maximum? *Geology* **31,** 115–118 (2003).
8. Pucéat, E. *et al.* Thermal evolution of Cretaceous Tethyan marine waters inferred from oxygen isotope composition of fish tooth enamels. *Paleoceanography* **18** doi:10.1029/2002PA000823 (2003).
9. Sewall, J. O. *et al.* Climate model boundary conditions for four Cretaceous time slices. *Climate of the Past* **3,** 647–657 (2007).
10. Robinson, S. A., Murphy, D. P., Vance, D. & Thomas, D. J. Formation of 'Southern Component Water' in the Late Cretaceous: evidence from Nd-isotopes. *Geology* **38,** 871–874 (2010).
11. Robinson, S. A. & Vance, D. Widespread and synchronous change in deep-ocean circulation in the North and South Atlantic during the Late Cretaceous. *Paleoceanography* **27** doi:10.1029/2011PA002240 (2012).
12. Voigt, S. *et al.* Tectonically restricted deep-ocean circulation at the end of the Cretaceous greenhouse. *Earth Planet Sci. Lett.* **369-370,** 169–177 (2013).
13. Murphy, D. P. & Thomas, D. J. Cretaceous deep-water formation in the Indian sector of the Southern Ocean. *Paleoceanography* **27,** PA1211 (2012).
14. Martin, E. E., MacLeod, K. G., Berrocoso, A. J. & Bourbon, E. Water mass circulation on Demerara Rise during the Late Cretaceous based on Nd isotopes. *Earth Planet Sci. Lett.* **327,** 111–120 (2012).
15. MacLeod, K. G., Londono, C. I., Martin, E. E., Berrocoso, A. J. & Basak, C. Changes in North Atlantic circulation at the end of the Cretaceous greenhouse interval. *Nat. Geosci.* **4,** 779–782 (2011).
16. MacLeod, K. G., Martin, E. E. & Blair, S. W. Nd isotopic excursion across Cretaceous ocean anoxic event 2 (Cenomanian-Turonian) in the tropical North Atlantic. *Geology* **36,** 811–814 (2008).
17. Friedrich, O., Erbacher, J., Moriya, K., Wilson, P. A. & Kuhnert, H. Warm saline intermediate waters in the Cretaceous tropical Atlantic Ocean. *Nat. Geosci.* **1,** 453–457 (2008).
18. Hague, A. M. *et al.* Convection of North Pacific deep water during the early Cenozoic. *Geology* **40,** 527–530 (2012).
19. Poulsen, C. J., Seidov, D., Barron, E. J. & Peterson, W. H. The impact of paleoceanographic evolution on the surface oceanic circulation and the marine environment within the mid-Cretaceous Tethys. *Paleoceanography* **13,** 546–559 (1998).
20. Otto-Bliesner, B. L., Brady, E. C. & Shields, C. Late Cretaceous ocean: coupled simulations with the national center for atmospheric research climate system model. *J. Geophys. Res.* **107,** Art. No. 4019 (2002).
21. Trabucho Alexandre, J. *et al.* The mide-Cretaceous North Atlantic nutrient trap: Black shales and OAEs. *Paleoceanography* **25** doi:10.1029/2010PA001925 (2010).
22. Poulsen, C. J., Barron, E. J., Arthur, M. A. & Peterson, W. H. Response of the mid-Cretaceous global oceanic circulation to tectonic and CO_2 forcings. *Paleoceanography* **16,** 576–592 (2001).
23. Brady, E. C., DeConto, R. & Thompson, S. L. Deep water formation and poleward ocean heat transport in the warm climate extreme of the Cretaceous (80 Ma). *Geophys. Res. Lett.* **25,** 4205–4208 (1998).
24. Lunt, D. J. *et al.* CO2-driven ocean circulation changes as an amplifier of Paleocene-Eocene thermal maximum hydrate destabilization. *Geology* **38,** 875–878 (2010).
25. Poulsen, C. J. & Zhou, J. Sensitivity of Arctic climate variability to mean state: insights from the Cretaceous. *J. Clim.* **26,** 7003–7022 (2013).
26. Flögel, S. *et al.* Simulating the biogeochemical effects of volcanic CO2 degassing on the oxygen-state of the deep ocean during the Cenomanian/Turonian Anoxic Event (OAE2). *Earth Planet Sci. Lett.* **305,** 371–384 (2011).
27. Lunt, D. J. *et al.* A model-data comparison for a multi-model ensemble of early Eocene atmosphere-ocean simulations: EoMIP. *Clim. Past Discuss.* **8,** 1229–1273 (2012).
28. Frank, M. Radiogenic isotopes: tracers of past ocean circulation and erosional input. *Rev. Geophys.* **40** doi:10.1029/2000RG000094 (2002).
29. Goldstein, S. L. & Jacobsen, S. B. The Nd and Sr isotopic systematics of river-water dissolved material: Implications for the sources of Nd and Sr in seawater. *Chem. Geol.* **66,** 245–272 (1987).
30. Thomas, D. J. Evidence for deep-water production in the North Pacific Ocean during the early Cenozoic warm interval. *Nature* **430,** 65–68 (2004).
31. Murphy, D. P. & Thomas, D. J. The evolution of Late Cretaceous deep-ocean circulation in the Atlantic basins: Neodymium isotope evidence from South Atlantic drill sites for tectonic controls. *Geochem. Geophys. Geosyst.* **14** doi:1002/2013GC004889 (2013).
32. Jeandel, C., Arsouze, T., Lacan, F., Téchiné, P. & Dutay, J. -C. Isotopic Nd compositions and concentrations of the lithogenic inputs into the ocean: A compilation, with an emphasis on the margins. *Chem. Geol.* **239,** 156–164 (2007).
33. Lacan, F. & Jeandel, C. Acquisition of the neodymium isotopic composition of the North Atlantic Deep Water. *Geochem. Geophys. Geosyst.* **6** doi:10.1029/2005GC000956 (2005).
34. Vance, D. & Burton, K. Neodymium isotopes in planktonic foraminifera: a record of the response of continental weathering and ocean circulation rates to climate change. *Earth Planet Sci. Lett.* **173,** 365–379 (1999).
35. Roy, M., van de Flierdt, T., Hemming, S. R. & Goldstein, S. L. 40Ar/39Ar ages of hornblende grains and bulk Sm/Nd isotopes of circum-Antarctic glacio-marine sediments: implications for sediment provenance in the southern ocean. *Chem. Geol.* **244,** 507–519 (2007).
36. Dalziel, I. W. D. Antarctica; a tale of two supercontinents? *Annu. Rev. Earth Planet Sci.* **20,** 501–526 (1992).
37. Larson, R. L. & Pitman, W. C. World-wide correlation of Mesozoic magnetic anomalies, and its implications. *Geol. Soc. Am. Bull.* **83,** 3645–3662 (1972).
38. Moiroud, M. *et al.* Evolution of neodymium isotopic signature of neritic seawater on a northwestern Pacific margin: new constraints on possible end-members for the composition of deep-water masses in the Late Cretaceous ocean. *Chem. Geol.* **356,** 160–170 (2013).
39. Moiroud, M. *et al.* Evolution of neodymium isotopic signature of seawater during the Late Cretaceous: implications for intermediate and deep circulation. *Gond. Res.* http://dx.doi.org/10.1016/j.gr.2015.08.005 (2015).
40. Carter, P., Vance, D., Hillenbrand, C. D., Smith, J. A. & Shoosmith, D. R. The neodymium isotopic composition of water masses in the eastern Pacific sector of the Southern Ocean. *Geochem. Cosmochim. Acta* **79,** 41–59 (2012).
41. Vrielynck, B. & Bouysse, P. *The Changing Face of the Earth: The Breakup of Pangea and Continental Drift Over the Past 250 Million Years in Ten Steps* (Commission de la Carte Géologique du Monde and Unesco, 2003).
42. Hay, W. W. *et al.*in *Evolution of the Cretaceous Ocean-Climate System* Vol. 332 (eds Barrera, E. & Johnson, C. C.) 1–47 (Geological Society of America Special Paper, 1999).
43. Giunta, G., Marroni, E., Padoa, E. & Pandolfi, L. in *The Circum-Gulf of Mexico and the Carribean: Hydrocarbon habitats, basin formation, and plate tectonics* Vol. 79 (eds Bartolini, C., Buffer, R. & Blickwede, J.) 104–125 (2003).
44. Itturalde-Vinent, M. in *From Greenhouse to Icehouse: The Marine Eocene-Oligocene Transition.* (eds Prothero, D., Ivany, L. C. & Nesbitt, E.) **Ch. 22,** 386–396 (Colombia Univ. Press, 2003).
45. Sepulchre, P. *et al.* Consequences of shoaling of the Central American Seaway determined from modeling Nd isotopes. *Paleoceanography* **29** doi:10.1002/2013PA002501 (2014).
46. von der Heydt, A. S. & Dijkstra, H. A. Effect of ocean gateways on the global ocean circulation in the late Oligocene and early Miocene. *Paleoceanography* **21** doi:10.1029/2005PA001149 (2006).
47. Soudry, D., Glenn, C. R., Nathan, Y., Segal, I. & VonderHaar, D. L. Evolution of Tethyan phosphogenesis along the northern edges of the Arabian-African shield during the Cretaceous-Eocene as deduced from temporal variations of Ca and Nd isotopes and rates of P accumulation. *Earth Sci. Rev.* **78,** 27–57 (2006).
48. Omta, A. W. & Dijkstra, H. A. A physical mechanism for the Atlantic-Pacific flow reversal in the early Miocene. *Glob. Planet Change* **36,** 265–276 (2003).
49. Zachos, J. C., Dickens, G. R. & Zeebe, R. E. An early Cenozoic perspective on greenhouse warming and carbon-cycle dynamics. *Nature* **451,** 279–283 (2008).
50. Tejada, M. L. G. *et al.* Ontong Java Plateau eruption as a trigger for the early Aptian oceanic anoxic event. *Geology* **37,** 855–858 (2009).
51. Turgeon, S. C. & Creaser, R. A. Cretaceous oceanic anoxic event 2 triggered by a massive magmatic episode. *Nature* **454,** 323–327 (2008).
52. Ruvalcaba Baroni, I., Topper, R. P. M., van Helmond, N. A. G. M., Brinkhuis, H. & Slomp, C. P. Biogeochemistry of the North Atlantic during oceanic anoxic event 2: role of changes in ocean circulation and phosphorus input. *Biogeosciences* **11,** 977–993 (2014).
53. Ozaki, K., Tajima, S. & Tajika, E. Conditions required for oceanic anoxia/euxinia: constraints from a one-dimensional ocean biogeochemical cycle model. *Earth Planet Sci. Lett.* **304,** 270–279 (2011).
54. Jacob, R. L. *Low Frequency Variability in a Simulated Atmosphere Ocean System* (Wisconsin, 1997).
55. Poulsen, C. J. & Jacob, R. L. Factors that inhibit snowball Earth simulation. *Paleoceanography* **19** doi:10.1029/2004PA001056 (2004).
56. Lefebvre, V., Donnadieu, Y., Sepulchre, P., Swingedouw, D. & Zhang, Z. S. Deciphering the role of southern gateways and carbon dioxide on the onset of the Antarctic Circumpolar Current. *Paleoceanography* **27** doi:10.1029/2012PA002345 (2012).
57. Gradstein, F. M., Ogg, J. G., Schmitz, M. D. & Ogg, G. M. *The Geologic Time Scale 2012* Vol. 2 (Elsevier, 2012).

58. Hay, W. W. Cretaceous paleoceanography. *Geol. Carp.* **46,** 257–266 (1995).
59. Vérard, C., Flores, K. & Stampfli, G. Geodynamic reconstructions of the South America-Antarctica plate system. *J. Geodyn.* **53,** 43–60 (2012).
60. Lagabrielle, Y., Godderis, Y., Donnadieu, Y., Malavieille, J. & Suarez, M. The tectonic history of Drake Passage and its possible impacts on global climate. *Earth Planet Sci. Lett.* **279,** 197–211 (2009).
61. Maldonado, A. *et al.* A model of oceanic development by ridge jumping: opening of the Scotia Sea. *Glob. Planet Change* **123,** 152–173 (2014).
62. Martos, Y. M., Catalan, M., Galindo-Zaldivar, J., Maldonado, A. & Bohoyo, F. Insights about the structure and evolution of the Scotia Arc from a new magnetic data compilation. *Glob. Planet Change* **123,** 239–248 (2014).
63. Eagles, G. The age and origin of the central Scotia Sea. *Geophys. J. Int.* **183,** 587–600 (2010).
64. Royer, D. L., Pagani, M. & Beerling, D. J. Geobiological constraints on Earth system sensitivity to CO2 during the Cretaceous and Cenozoic. *Geobiology* **10,** 298–310 (2012).
65. Donnadieu, Y., Godderis, Y. & Bouttes, N. Exploring the climatic impact of the continental vegetation on the Mezosoic atmospheric CO2 and climate history. *Clim. Past* **5,** 85–96 (2009).
66. Fletcher, B. J., Beerling, D. J., Brentnall, S. J. & Royer, D. L. Fossil bryophytes as recorders of ancient CO2 levels: experimental evidence and a Cretaceous case study. *Global Biogeochem. Cycles* **19** doi:10.1029/2005GB002495 (2005).
67. Franks, P. J., Royer, D. L., Johnson, K. R., Miller, I. & Enquist, B. J. New constraints on atmopsheric CO2 concentration for the Phanerozoic. *Geophys. Res. Lett.* **41,** 4685–4694 (2014).
68. Haworth, M., Hesselbo, S. P., McElwain, J. C., Robinson, S. A. & Brunt, J. W. Mid-Cretaceous pCO(2) based on stomata of the extinct conifer *Pseudofrenelopsis* (Cheirolepidiaceae). *Geology* **33,** 749–752 (2005).
69. Passalia, M. G. Cretaceous pCO2 estimation from stomatal frequency analysis of gymnosperm leaves of Patagonia, Argentina. *Palaeogeogr. Palaeoclimatol. Palaeoecol.* **273,** 17–24 (2009).

Acknowledgements

This work was supported by a funding from the ANR project Anox-Sea. We thank three anonymous reviewers for their constructive comments that greatly contributed to improve this manuscript. We thank the CEA/CCRT for providing access to the HPC resources of TGCC under the allocation 2014-012212 made by GENCI.

Author contributions

Y.D. and E.P. conceived the project. E.P., M.M. and J.-F.D. built the neodymium database and performed the analysis. Y.D. conducted all numerical climate modelling. Y.D. and E.P. wrote the manuscript with contributions from all authors.

Additional information

Competing financial interests: The authors declare no competing financial interests.

12

Correlation and anti-correlation of the East Asian summer and winter monsoons during the last 21,000 years

Xinyu Wen[1], Zhengyu Liu[1,2], Shaowu Wang[1], Jun Cheng[3] & Jiang Zhu[2]

Understanding the past significant changes of the East Asia Summer Monsoon (EASM) and Winter Monsoon (EAWM) is critical for improving the projections of future climate over East Asia. One key issue that has remained outstanding from the paleo-climatic records is whether the evolution of the EASM and EAWM are correlated. Here, using a set of long-term transient simulations of the climate evolution of the last 21,000 years, we show that the EASM and EAWM are positively correlated on the orbital timescale in response to the precessional forcing, but are anti-correlated on millennial timescales in response to North Atlantic melt water forcing. The relation between EASM and EAWM can differ dramatically for different timescales because of the different response mechanisms, highlighting the complex dynamics of the East Asian monsoon system and the challenges for future projection.

[1]Department of Atmospheric and Oceanic Sciences & Laboratory for Climate and Ocean-Atmosphere Studies, School of Physics, Peking University, Beijing 100871, China. [2]Department of Atmospheric and Oceanic Sciences & Center for Climatic Research, Nelson Institute for Environmental Studies, University of Wisconsin-Madison, WI 53706, USA. [3]Polar Climate System and Global Change Laboratory, Nanjing University of Information Science and Technology, Nanjing 210044, China. Correspondence and requests for materials should be addressed to X.W. (email: xwen@pku.edu.cn).

The East Asia monsoon system consists of a southwesterly summer monsoon and a northerly winter monsoon. The co-variability of EASM and EAWM on multiple timescales significantly affect East Asia's climate in the past and future[1,2]. The last decade has seen an explosive growth of high-resolution proxy records of EASM[3–12] during the glacial-interglacial cycles, notably from cave $\delta^{18}O$ records in southeastern China and nearby regions[3–9,13]. These records consistently suggest an EASM evolution that follows the Northern Hemisphere summer insolation while punctuated by abrupt millennial events. In comparison, high-resolution proxy records are less abundant for EAWM[14–17] and the records available seem to show inconsistent results[14,16–19], leaving a debate on the co-variability between the EASM and EAWM ambiguous. The first high-resolution EAWM index constructed with Titanium content in the sediment of Lake Huguang Maar in South China suggested an anti-correlation between the EASM and EAWM over the past 16,000 years, especially over the major millennial periods of Bølling-Allerød (BA), Younger Drays (YD) and the mid-to-late Holocene[14]. However, the lake sediment record has been argued to be a proxy for local hydrology rather than the large-scale EAWM[20]. In the meanwhile, the anti-correlation between the EASM and the EAWM is inconsistent with Chinese historical records during some key centennial periods in the last millennium[21]. Furthermore, this anti-correlation is now challenged by the reconstructed EAWM sea surface temperature (SST) indices in the South China Sea[15,16] and lake-level changes in central Asia[18], which suggest a positive correlation between the EAWM and the EASM in the last deglaciation on the orbital timescale. Another independent and robust evidence indicate that EAWM, reconstructed from loess grain sizes in Northwest China, was closely linked with Greenland's isotope records on millennial and glacial-to-interglacial timescales[18], leaving the precessional-scale winter monsoon changes not well resolved because of the intrinsic feature of loess data. All the debate and inconsistence among proxy records raise a fundamental question: How is the EASM and EAWM correlated in response to past climate changes?

In this study, we revisit the co-variability of EASM and EAWM during the last 21,000 years by combining long-term transient model simulations and reconstructed indices. We conclude that a changing monsoon correlation, from a positive correlation at orbital timescale to a negative correlation at millennial timescale, was the key responses to solar insolation at northern hemisphere (NH) mid-latitudes and the fresh water fluxes in the North Atlantic. Greenhouse gases and continental ice sheets also have minor effects in modulating East Asia monsoons.

Results

Simulated and observed monsoons. Here we study the EASM–EAWM co-variability in a transient simulation of the climate evolution of the last 21,000 years (TRACE21) using the National Center for Atmospheric Research (NCAR) Community Climate System Model version 3 (CCSM3)[21]. TRACE21 is forced by the realistic climate forcing that consists of the orbital insolation, atmospheric concentration of greenhouse gases (GHGs), continental ice sheets and meltwater fluxes (Methods) and has been shown to simulate deglacial climate changes consistent with many proxy records at global and regional scales[21–24]. In particular, CCSM3 simulates both the EASM and EAWM reasonably well in the atmospheric wind field, with the EASM characterized by a southeasterly penetrating deep into northern China and the EAWM featured by a northeasterly sweeping across southern China and South China Sea (SCS; Fig. 1f,g and Supplementary Fig. 1).

TRACE21 simulates an EASM evolution largely consistent with the observation. Consistent with the classical view that a stronger EASM is accompanied by a deeper moisture penetration into northern China[25], we will use the summer southerly averaged in East China as an index for the EASM in the model[26] (Fig. 1f). Overall, the simulated EASM intensifies from the LGM (21 ka) towards the early Holocene (10 ka), and then weakens towards the late Holocene; this orbital scale response is punctuated by millennial events with an abrupt intensification at the Bølling onset and a subsequent weakening towards YD (Fig. 1c, red). This evolution pattern is largely consistent with the proxy records available[27,28]. The model-data consistency is reinforced by a direct comparison of the $\delta^{18}O$ in East China between the cave records[3–5] (Fig. 1c, black) and the isotope simulations[26] (Fig. 1c, blue circles), the latter being derived from the atmospheric precipitation $\delta^{18}O$ simulated in a series of 21 snapshot experiments in an isotope-enabled atmosphere model forced by the SST of TRACE21 (Methods).

TRACE21 also simulates an EAWM evolution in agreement with the observation in terms of a SST gradient index (Fig. 1d, red and black; Methods). Both the model and proxy EAWM SST indices show a distinct orbital scale response in phase with EASM, such that both EAWM and EASM intensify from LGM towards the early Holocene and then weaken towards the late Holocene. This positive EAWM–EASM correlation, however, seems to become negative during the millennial events of H1-BA-YD. This SST gradient index is derived with the southeastern SST subtracting the northwestern SST in the SCS. Physically, the EAWM northerly advects cold and dry continental air to the southwestern SCS along the Vietnam coast (Fig. 1g), forming a cold tongue and in turn a positive west-to-east SST gradient. A stronger SST gradient is therefore forced by a stronger EAWM northerly[15,16]. Our model also simulates a consistent SST gradient index and EAWM (blue and red in Fig. 1d), supporting the SST gradient as an index for EAWM. In contrast to the SST index that is determined by large-scale dynamics closely related with the EAWM, Lake Huguang Maar is located in a region of complex terrain and the lake deposition can be affected by the river's catchment and various local hydrological processes[20]. Our model also simulates diverse evolution patterns of the local wind surrounding the lake site, none of which is consistent with the Lake record throughout the last 21,000 years (Supplementary Fig. 2). The Lake Huguang Ti content shows a relatively consistent variability as compared with the SST index (Fig. 1d, blue) during the deglaciation, strengthening during H1, weakening in BA and strengthening again towards YD (Fig. 1e, black). However, after the YD towards the early-to-mid Holocene (12–8 ka), opposite to the maximum response in the proxy SST index and the model EAWM, the lake record declines to a minimum (Fig. 1e), destroying the positive correlation with EASM at orbital scale. We speculate that the linkage of lake record with the large-scale EAWM circulation might be interrupted by local hydrological processes after YD event.

Responses to orbital and North Atlantic meltwater forcing. Our model-data comparison suggests a changing monsoon correlation, from a positive correlation at orbital timescale to a negative correlation at millennial timescale. The opposite monsoon correlations at orbital and millennial timescales, we hypothesize, is caused by different forcing mechanisms, with the former dominated by the precessional forcing, whereas the latter by the meltwater forcing. This hypothesis is confirmed in two sensitivity experiments that are integrated through the last 19,000 years the same as TRACE21, but forced by the orbital forcing (ORB) (Fig. 2b) and meltwater forcing (MWF) (Fig. 2c) individually (Methods). It is

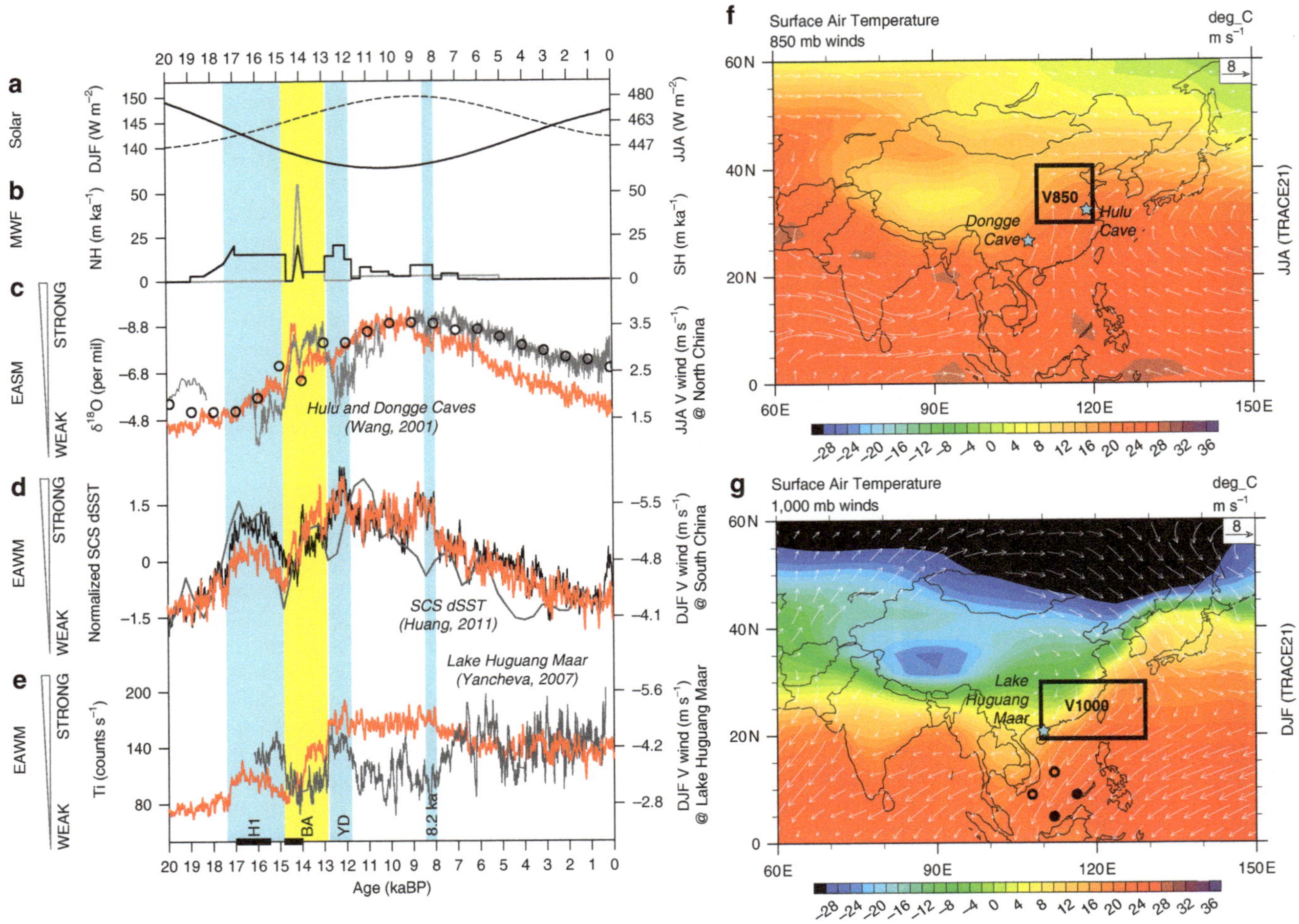

Figure 1 | TRACE simulation and observations for EASM and EAWM in the last 20,000 years. (**a**) December-January-February (DJF) (solid line) and June-July-August (JJA) (dashed line) insolation at 45° N. (**b**) Melting water flux (in equivalent global sea level per 1,000 years) into the North Atlantic (black) and Southern Ocean (grey) in TRACE21. (**c**) δ^{18}O (grey) from Dongge and Hulu caves (Wang *et al.*[3]), precipitation δ^{18}O simulated in isotope-enable snapshot experiments (blue circles, Liu *et al.*[26]) and EASM wind index in TRACE21 (red). (**d**) SCS SST gradient index (dSST) for EAWM in the reconstruction (grey, Huang *et al.*[16]) and in TRACE21 (blue), EAWM wind index in TRACE21 (red). (**e**) Sediment Ti content from Lake Huguang Maar (grey) that was considered as an EAWM indicator by Yancheva *et al.*[14], and the simulated meridional wind speed (red) near Lake Huguang Maar in TRACE21. (**f**) Modern June-August climatology of surface temperature and 850 mb winds in TRACE21, the locations for Hulu and Dongge caves (cyan stars), the model domain (110-120 E, 30-40 N) for calculating the EASM meridional wind index at 850 mb. (**g**) Modern December-February climatology of surface temperature and 1,000 mb winds in TRACE21, the location of Lake Huguang Maar (cyan star) and the sites for Huang's (2011) SST index (two solid and two open circles), and the model domain (110-130 E, 20-30 N) for calculating the EAWM meridional wind index at 1,000 mb. Three cold events, Heinrich 1 (H1), Younger Dryas (YD) and 8.2 ka, are marked as the blue vertical panels. The warm event Bølling-Allerød (BA) is marked as the yellow vertical panel.

seen that the EASM and EAWM are largely correlated in experiment ORB (Fig. 2b), but anti-correlated in experiment MWF, especially for the major millennial events during H1-BA-YD (Fig. 2c).

The monsoon correlation at orbital scale is dominated by the classical monsoon response to precessional forcing[29,30](Fig. 1a). Figure 3 shows the difference between the strong (10–8 ka) and weak (2–0 ka) EASM (EAWM) conditions in terms of the summer (winter) surface wind as well as the surface temperature, pressure and precipitation in TRACE21. The differences reflect mainly the orbital scale evolution (Figs 1 and 2) and can be discussed in the context of the slow evolution from the LGM towards the early Holocene[31]. As perihelion shifts from boreal winter to summer, the insolation is enhanced (reduced) in the NH in summer (winter). This leads to a warming (cooling) and lower (higher) pressure over the East Asia relative to the ocean and, in turn, a stronger southerly EASM (northerly EAWM) wind and increased precipitation downstream of the monsoon wind moisture transport (Fig. 3).

In contrast, the anti-correlation monsoon response for millennial events is caused by the meltwater flux into the North Atlantic and the resulted change in the Atlantic Meridional Overturning Circulation (AMOC; Supplementary Fig. 3, ALL and MWF). A meltwater flux into the North Atlantic weakens the AMOC and the associated poleward oceanic heat transport, leading to a cooling that extends from the North Atlantic across the Eurasian continent throughout the year[21,32]. This response can be illustrated in the difference between H1 and BA. The colder northern China generates a higher pressure over land and in turn an anomalous northerly wind that weakens EASM (Fig. 4a,b), but enhances EAWM (Fig. 4c,d). The enhanced northerly in winter is somewhat weak in TRACE21, because of the effect of other forcing. Indeed, for experiments that are forced by the meltwater flux alone, the anomalous northerly wind becomes more significant in winter as well as summer (Fig. 2c and Supplementary Fig. 4).

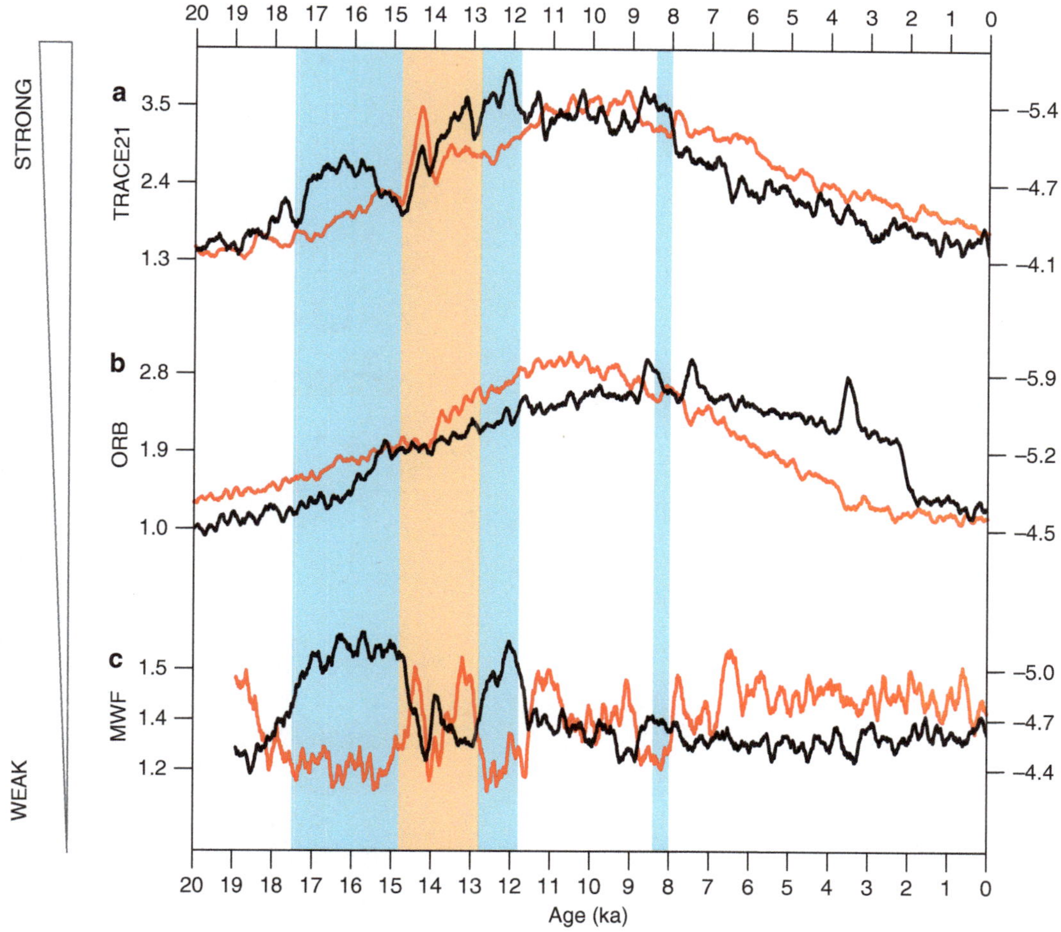

Figure 2 | Monsoon index in TRACE and single forcing experiments. EASM (red) and EAWM (blue) wind indices in (**a**) TRACE21, and single forcing experiments (**b**) ORB and (**c**) MWF. The EAWM and EAWM indices are defined in Fig.1f,g after a 200-year running-average. Three cold events, Heinrich 1 (H1), Younger Dryas (YD) and 8.2 ka, are marked as the blue vertical panels. The warm event Bølling-Allerød (BA) is marked as the yellow vertical panel.

Responses to other minor forcing. The deglacial global climate evolution is also forced by rising GHGs (Supplementary Fig. 5a) and retreating ice sheets (Supplementary Fig. 5b). For the East Asia monsoon, however, the impact of GHG and ice sheet forcing seem to compensate each other. This is shown in two additional sensitivity experiments forced by the GHGs (CO2) and ice sheets (ICE) individually (Methods). The rising GHGs during the early deglacial period strengthen the EASM but weakens the EAWM, leading to an anti-correlation in experiment CO2 (Supplementary Fig. 5a,c). In contrast, the retreating ice sheets in experiment ICE lead to a weakening EASM and strengthening EAWM, also an anti-correlated monsoon response (Supplementary Fig. 5b,d), but with the opposite signs to that in CO2. As such, the effects of GHG and ice sheet tend to cancel out, leaving little residual after the deglaciation.

The anti-correlation monsoon response in experiment CO2 seem to differ from the monsoon response to future global warming, likely contributed by the indirect impact of the AMOC change. In response to future global warming, current climate models show a robust intensification of the EASM wind[1,33–36] (Supplementary Fig. 6a), consistent with experiment CO2. This EASM intensification is caused by the direct radiative warming effect, which forces a stronger warming and lower pressure over land than over ocean and, in turn, an intensified southerly over East Asia (Supplementary Fig. 7). In comparison, there is no clear response of EAWM across these models[1,35] (Supplementary Fig. 6b). The robust EAWM response and, in turn, the anti-correlation monsoon response in CO2 can be contributed by the strengthening AMOC[37] in CCSM3 (Supplementary Fig. 3; Methods), which, opposite to the case of meltwater forcing (Fig. 2d and Supplementary Fig. 4), warms the Eurasian Continent, enhances the EASM, but weakens EAWM, contributing to the anti-correlation monsoon response. Finally, the anti-correlation monsoon response to ice sheet retreat seems to be caused by the shift of the westerly jet[38] in response to the lowering ice sheets and their global climate impact (Methods).

Discussion

Our study suggests that the evolution of EASM and EAWM is correlated at the orbital scale in response to the precessional forcing, but anti-correlated at millennial timescale in response to the meltwater forcing and the resulting AMOC change. The monsoon correlation is also affected by the rising GHGs and retreating ice sheets, but the two responses might be opposite such that there is little residual signal left after the deglaciation. Another, the current simulation might be too short to address the slow-varying impacts from ice sheets and GHGs. Their coupled effects over monsoon variability need more investigations on the glacial-interglacial timescale.

The new understanding here, from a modelling framework, has reconciled the conflicting interpretations on the co-variability between EASM and EAWM in proxy records, and demonstrated that their relationship is influenced by multiple physical processes

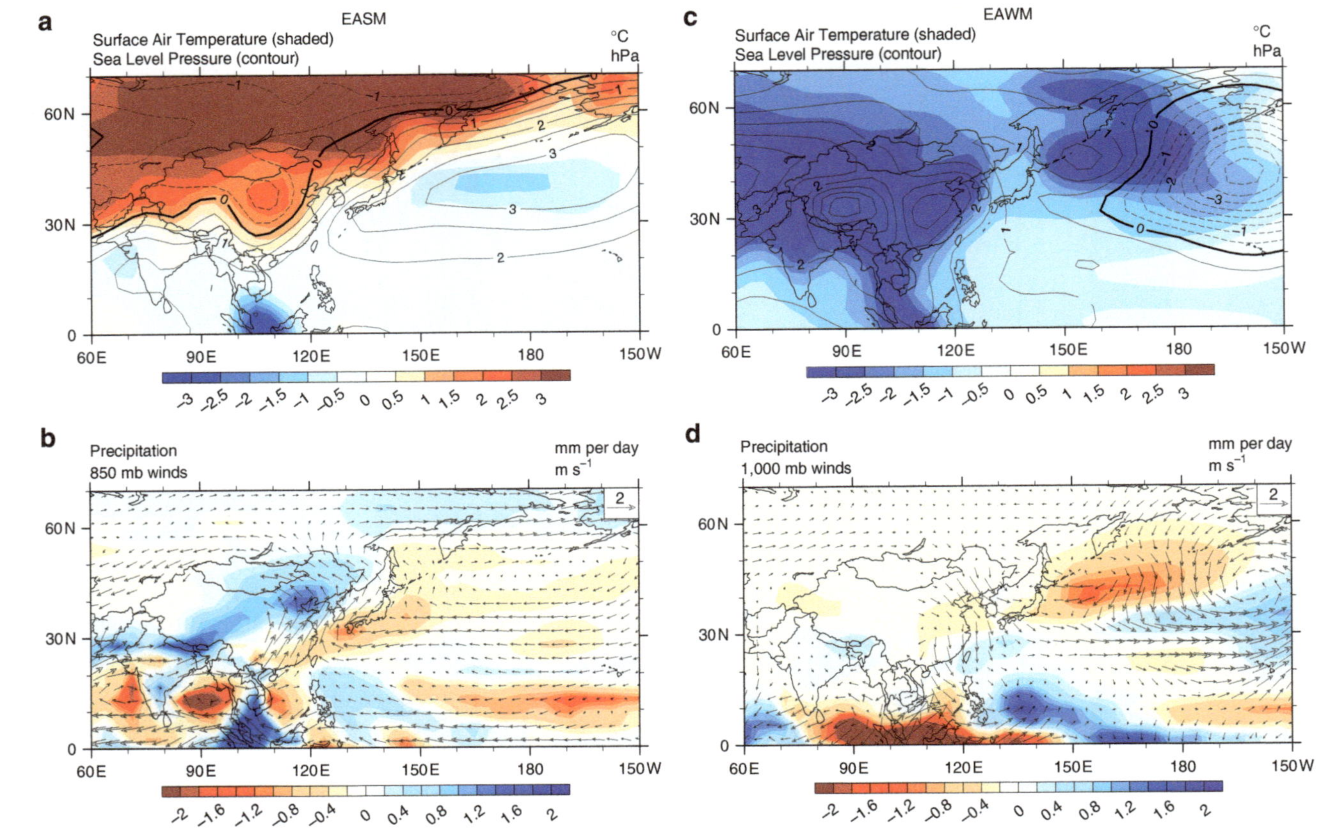

Figure 3 | Monsoon response to insolation forcing on the orbital timescale. The differences of four key variables between the early Holocene (10–8 ka) and the late Holocene (2–0 ka) for EASM (**a**,**b**) and EAWM (**c**,**d**). Surface air temperature (colour) and sea level pressure (contour) are shown in **a**,**c**. Winds and total precipitation (colour) are shown in **b**,**d**.

on diverse timescales. Our findings indicate that any single correlation/anti-correlation, as concluded in previous studies, is not sufficient to describe the complicated nature of East Asia monsoon system. This work presents great challenge and potential to understand the response of the East Asian monsoon system to global climate changes in the past and the future.

Methods

TRACE21 and single forcing experiments. TRACE21 is a transient simulation of the global climate of the last 21,000 years in CCSM3 of the NCAR[39] forced by the realistic climatic forcing, including orbital insolation[40], atmospheric GHGs[41], melting water discharge[42] (Fig. 1b), the continental ice sheets derived from ICE-5G[43] and the modification of coastlines and bathymetry at 13.1, 12.9, 7.6 and 6.2 ka for the Barents Sea, the Bering Strait, Hudson Bay and the Indonesian through-flow, respectively. More details can be found in ref. 44.

Four single forcing sensitivity experiments are also integrated the same as TRACE21, except that each sensitivity experiment is forced by only one single transient forcing with the other forcing fixed at their values at 19 ka. The ORB, CO2, MWF and ICE experiments are forced by the changing orbital insolation, GHG concentration, meltwater flux and the ice sheet, respectively[45,46].

Snapshot simulations in a water isotope enabled CAM3. For a direct model-data comparison of water isotopes, a set of time slice simulations are performed in NCAR Community Atmosphere Model (CAM) version 3 with the incorporation of a water isotope module (isoCAM3)[47]. We conducted 23 snapshot sensitivity experiments in the last 21,000 years: 21 experiments are 1,000 years apart, at 20, 19, ..., 0 ka, and 2 experiments are at the time of special events, 14.5 ka (BA) and 12.1 ka (YD). Each slice is integrated for 50 years with the forcing of land-use, land-sea mask, sea surface temperature (SST) and sea-ice fraction that taken from TRACE21. The first 20-year results are removed as the spin-up and the remaining 30-year results are averaged for the analysis. A more complete description of these experiments can be found in Liu *et al.*[26].

Selection of EAWM index. The paleoclimatic proxies suggest that the EAWM index can be effectively derived from two major places. One is Northwest China[17] and another is South China[14,16]. The winds in Northwest China mostly reflect the intensity of westerlies, which is more sensitive to AMOC rather than East Asian Meridional circulation. The winds in South China represents EAWM's intensity over North-to-South China and surrounding coastal region, as well as the land–sea sea level pressure gradient. Thus, we select the meridional wind at 1,000 hPa (V1000) over South China to serve as the EAWM index (Fig. 1g).

Monsoon responses in experiments CO2 and ICE. Experiments CO2 and ICE show that the monsoon responses to the increasing GHGs and retreating ice sheets are the opposite and therefore tend to compensate each other. In CO2, EASM strengthens and EAWM weakens, generating an anti-correlation monsoon response. In ICE, EASM weakens and EAWM strengthens, also an anti-correlation monsoon response, but of the opposite sign to that in CO2 (Supplementary Fig. 5).

In response to increased GHGs in CO2, the strengthened EASM is consistent with the future global warming experiments in current climate models[1] (Supplementary Fig. 6a). This robust EASM intensification is caused by the direct radiative warming effect, which forces a stronger warming and lower pressure over land than over ocean and, in turn, an intensified southerly over East Asia(Supplementary Fig. 7). In contrast, the weakening of EAWM wind in CO2 (Supplementary Fig. 6b, black dots) is no robust across various model in global warming experiments[1,36] (Supplementary Fig. 6b). The difference in the EAWM responses between CO2 and global warming experiments, we speculate, is caused by an indirect impact of the AMOC change. In CO2, the AMOC is intensified in response to the slow rise of GHGs (Supplementary Fig. 3, CO2), which reduces sea ice cover and therefore increases the heat loss out of the glacial North Atlantic, and therefore intensifies the deep convection and in turn AMOC[37]. As discussed earlier on the monsoon response to meltwater forcing in Fig. 2d, the intensified AMOC warms the North Hemisphere continent, which weakens the EAWM, and reinforces the EASM, contributing to the anti-correlation monsoon response in CO2.

The causes for the responses of both EASM and EAWM to ice sheet retreat seem complex. In response to the lowering of ice sheets, especially the large reduction of Laurentide Ice sheet at ~14 ka, the westerly jet shifts northward over the North Atlantic[48], which leads to a cooling over the North Pacific because of the sea ice expansion[38], an intensified Aleutian Low and a weakened western Pacific Subtropical High year round. (Supplementary Fig. 8a,c). This leads to a weakening of the EASM southerly wind and intensified EAWM northerly wind

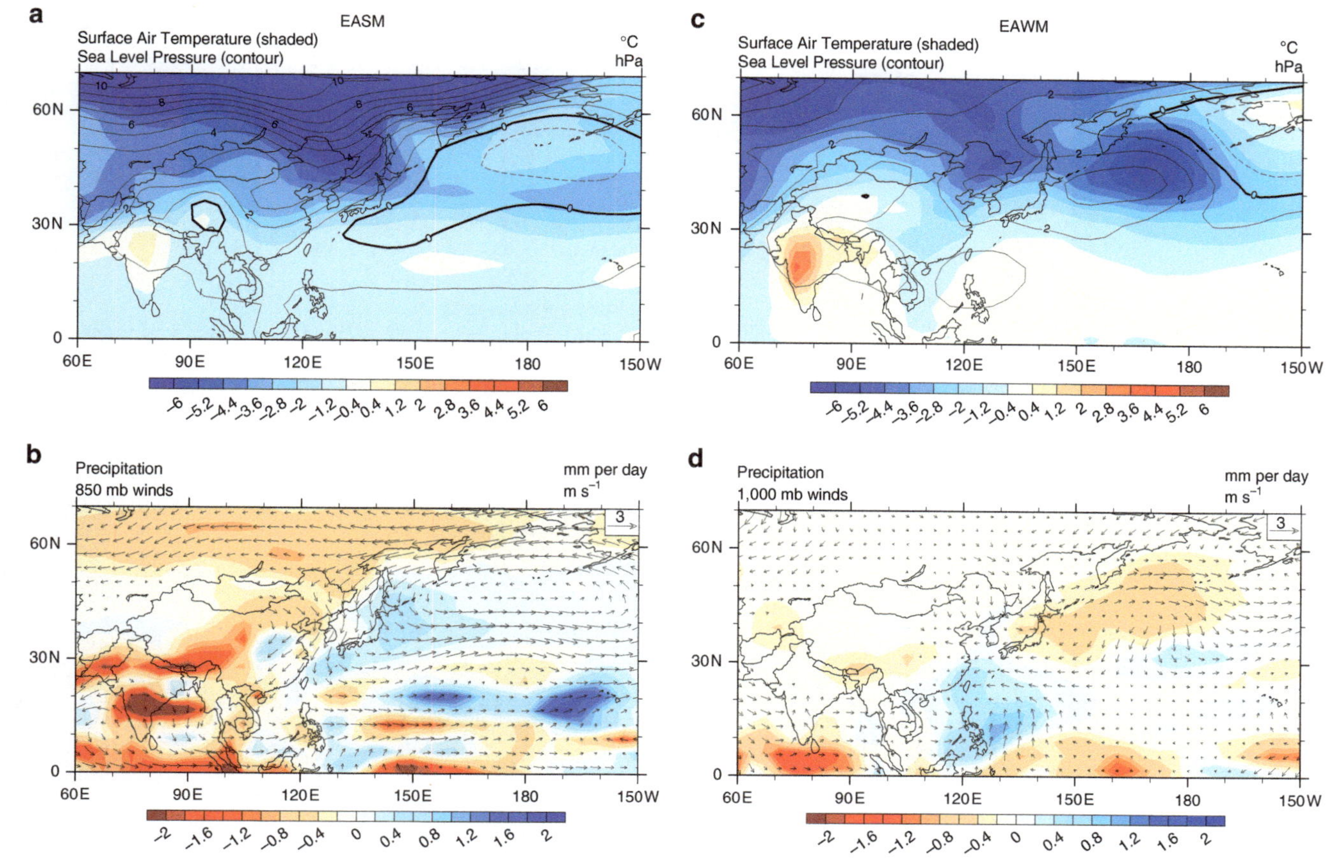

Figure 4 | Monsoon response to melting water on the millennial timescale. Differences of key meteorological variables between H1 (17.0–15.5 ka) and Bølling-Allerød (14.8–14.0 ka) for EASM (**a**,**b**) and EAWM (**c**,**d**). Surface air temperature (colour) and sea level pressure (contour) are shown in **a**,**c**. Winds and total precipitation are shown in **b**,**d**.

(Supplementary Fig. 8b,d), and in turn an anti-correlation monsoon response. In the meantime, AMOC is also weakened because of the sea ice expansion over the North Atlantic associated with the jet shift[39] (Supplementary Fig. 3, ICE). The weakening AMOC should also contribute to a cooling over the Northern Hemisphere and in turn a weakening EASM and strengthening EAWM, contributing to the anti-correlation monsoon response in ICE.

References

1. Christensen, J. H. *et al.* in *Climate Change 2013: The Physical Science Basis. Contribution of Working Group I to the Fifth Assessment Report of the Intergovernmental Panel on Climate Change* (eds Stocker, T. F. *et al.*) 1217–1308 (Cambridge Univ., 2013).
2. Zhou, T. *et al.* Detecting and understanding the multi-decadal variability of the East Asian Summer Monsoon-Recent progress and state of affairs. *Meteorol. Zeitschrift* **18,** 455–467 (2009).
3. Wang, Y. J. *et al.* A high-resolution absolute-dated late Pleistocene monsoon record from Hulu Cave. *Science* **294,** 2345–2348 (2001).
4. Wang, Y. J. *et al.* The Holocene Asian monsoon: links to solar changes and North Atlantic climate. *Science* **308,** 854–857 (2005).
5. Wang, Y. J. *et al.* Millennial- and orbital-scale changes in the East Asian monsoon over the past 224,000 years. *Nature* **451,** 1090–1093 (2008).
6. Yuan, D. X. *et al.* Timing, duration, and transitions of the last interglacial Asian monsoon. *Science* **304,** 575–578 (2004).
7. Yang, Y. *et al.* Precise dating of abrupt shifts in the Asian monsoon during the last deglaciation based on stalagmite data from Yamen Cave, Guizhou Province, China. *Sci. China Ser. D Earth Sci.* **55,** 633–641 (2010).
8. Chen, S. T. *et al.* A possible younger Dryas-type event during Asian monsoonal termination 3. *Sci. China Ser. D Earth Sci.* **49,** 982–990 (2006).
9. Shakun, J. D. *et al.* A high-resolution absolute dated deglacial speleothem record of Indian Ocean climate from Socotra Island, Yemen. *Earth Planet Sci. Lett.* **259,** 442–456 (2007).
10. Lu, H. *et al.* Variation of East Asian monsoon precipitation during the past 21ky and potential CO2 forcing. *Geology* **41,** 1023–1026 (2013).
11. Wang, P. X. *et al.* The global monsoon across timescales: coherent variability of regional monsoons. *Clim. Past* **10,** 2007–2052 (2014).
12. An, Z. *et al.* Global monsoon dynamics and climate change. *Annu. Rev. Earth Planet. Sci.* **43,** 29–77 (2015).
13. Cheng, H. *et al.* The Global Paleomonsoon as seen through speleothem records from Asia and the Americas. *Clim. Dyn.* **39,** 1045–1062 (2012).
14. Yancheva, G. *et al.* Influence of the intertropical convergence zone on the East Asian monsoon. *Nature* **445,** 74–77 (2007).
15. Tian, J., Huang, E. & Pak, D. K. East Asian winter moon variability over the last glacial cycle: Insights from a latitudinal sea-surface temperature gradient across the South China Sea. *Palaeogeogr. Palaeoclimatol. Palaeoecol.* **292,** 319–324 (2010).
16. Huang, E., Tian, J. & Steinke, S. Millennial-scale dynamics of the winter cold tongue in the southern South China Sea over the past 26 ka and the East Asian winter monsoon. *Quat. Res.* **75,** 196–204 (2011).
17. Sun, Y. *et al.* Influence of Atlantic meridional overturning circulation on the East Asian winter monsoon. *Nat. Geosci.* **5,** 46–49 (2012).
18. Li, Y. & Morrill, C. Multiple factors causing Holocene lake-level change in monsoonal and arid central Asia as identified by model experiments. *Clim. Dyn.* **35,** 1119–1132 (2010).
19. Zhang, D. & Lu, L. Anti-correlation of summer/winter monsoons? *Nature* **250,** E7–E9 (2007).
20. Zhou, H. Y., Guan, H. Z. & Chi, B. Q. Record of winter monsoon strength. *Nature* **450,** E10–E11 (2007).
21. Liu, Z. *et al.* Transient simulation of Last Deglaciation with a new mechanism for Bolling-Allerod warming. *Science* **325,** 310–314 (2009).
22. Liu, Z. *et al.* Younger dryas cooling and the Greenland climate response to CO2. *Proc. Natl Acad. Sci. USA* **109,** 11101–11104 (2012).
23. Shakun, J. *et al.* Global warming preceded by increasing carbon dioxide concentrations during the last deglaciation. *Nature* **484,** 49–54 (2012).

24. Otto-Bliesner, B. *et al.* Coherent changes of southeastern equatorial and northern African rainfall during the last deglaciation. *Science* **346**, 1223–1227 (2014).
25. Ding, Y. & Chan, J. The East Asian summer monsoon: an overview. *Meteo. Atmos. Phys.* **89**, 117–142 (2005).
26. Liu, Z. *et al.* Chinese cave records and the East Asia Summer Monsoon. *Quat. Sci. Rev.* **83**, 115–128 (2014).
27. Zhou, W. *et al.* Terrestrial evidence for a spatial structure of tropical-polar interconnections during the Younger Dryas Episode. *Earth Planet. Sci. Lett.* **191**, 231–239 (2001).
28. Yu, G., Xue, B. & Li, Y. in *The Encyclopedia of Quaternary Science* 2nd edn (ed Elias, S. A.) 506–523 (Elsevier, 2013).
29. Kutzbach, J. E. Monsoon climate of the early Holocene: climate experiment with the earth's orbital parameters for 9000 years ago. *Science* **214**, 59–61 (1981).
30. Kutzbach, J. E., Liu, X. & Liu, Z. Simulation of the evolutionary response of global summer monsoons to orbital forcing over the past 280,000 years. *Clim. Dyn.* **30**, 567–579 (2008).
31. Jiang, D. *et al.* Considerable model–data mismatch in temperature over China during the mid-Holocene: results of PMIP simulations. *J. Clim.* **25**, 4135–4153 (2012).
32. Crowley, T. J. North Atlantic deep water cools the Southern Hemisphere. *Paleoceanography* **7**, 489–497 (1992).
33. Kitoh, A. *et al.* Monsoons in a changing world: A regional perspective in a global context. *J. Geophy. Res.* **118**, 3053–3065 (2013).
34. Wang, B. & Lin, H. Rainy season of the Asian-Pacific summer monsoon. *J. Clim.* **15**, 386–398 (2002).
35. Lee, J.-Y. & Wang, B. Future change of global monsoon in the CMIP5. *Clim. Dyn.* **42**, 101–119 (2014).
36. Jiang, D. & Tian, Z. East Asian monsoon change for the 21st century: results of CMIP3 and CMIP5 models. *Chinese Sci. Bull* **58**, 1427–1435 (2013).
37. Zhu, J. *et al.* AMOC response to global warming: dependence on the background climate and response timescale. *Clim. Dyn.* **44**, 3449–3468 (2014).
38. Zhu, J. *et al.* Linear weakening of the AMOC in response to receding glacial ice sheets in CCSM3. *Geophys. Res. Lett.* **41**, 6252–6258 (2014).
39. Collins, W. *et al.* The community climate system model version 3 (CCSM3). *J. Clim.* **19**, 2122–2143 (2006).
40. Berger, A. Long-term variations of daily insolation and quaternary climatic changes. *J. Atmos. Sci.* **35**, 2362–2367 (1978).
41. McManus, J. *et al.* Collapse and rapid resumption of Atlantic meridional circulation linked to deglacial climate changes. *Nature* **428**, 834–837 (2004).
42. Joos, F. & Spahni, R. Rates of change in natural and anthropogenic radiative forcing over the past 20,000 years. *Proc. Natl Acta. Sci. USA* **105**, 1425–1430 (2008).
43. Peltier, W. Global glacial isostasy and the surface of the ice-age earth: the ICE-5G (VM2) model and GRACE. *Annu. Rev. Earth Planet. Sci.* **32**, 111–149 (2004).
44. He, F. *Simulating Transient Climate Evolution of the Last Deglaciation with CCSM3* (PhD thesis, Univ. Wisconsin-Madison, 2011).
45. He, F. *et al.* Northern Hemisphere forcing of Southern Hemisphere Climate during the last deglaciation. *Nature* **494**, 81–85 (2013).
46. Liu, Z. *et al.* Evolution and forcing mechanisms of El Nino over the past 21,000 years. *Nature* **515**, 550–553 (2014).
47. Noone, D. & Sturm, C. in *Isoscapes: Understanding Movement, Patterns, and Process on Earth through Isotope Mapping* (eds West, J. *et al.*) 487 (Springer, 2010).
48. Lu, Z., Liu, Z. & Zhu, J. Abrupt intensification of ENSO forced by deglacial ice sheet retreat in CCSM3. *Clim. Dyn.* **46**, 1877–1891 (2016).

Acknowledgements

We are grateful to Dr Esther Brady for helpful discussions on numerical experiments design. This work is supported by the National Science Foundation of China No. 41130105, 41130962, 41005035 and 41206024, and the US NSF P2C2 Program and Office of Science (BER), DOE. The computation is carried out at Oak Ridge National Lab/DOE and the NCAR supercomputing facility.

Author contributions

X.W., Z.L. and S.W. conceived the study, X.W. and Z.L. wrote the paper, X.W. performed the analysis, J.C. and J.Z. contributed to the numerical experiments. All authors discussed the results and provided inputs to the paper.

Additional information

Competing financial interests: The authors declare no competing financial interests.

13

Local biomass burning is a dominant cause of the observed precipitation reduction in southern Africa

Øivind Hodnebrog[1], Gunnar Myhre[1], Piers M. Forster[2], Jana Sillmann[1] & Bjørn H. Samset[1]

Observations indicate a precipitation decline over large parts of southern Africa since the 1950s. Concurrently, atmospheric concentrations of greenhouse gases and aerosols have increased due to anthropogenic activities. Here we show that local black carbon and organic carbon aerosol emissions from biomass burning activities are a main cause of the observed decline in southern African dry season precipitation over the last century. Near the main biomass burning regions, global and regional modelling indicates precipitation decreases of 20–30%, with large spatial variability. Increasing global CO_2 concentrations further contribute to precipitation reductions, somewhat less in magnitude but covering a larger area. Whereas precipitation changes from increased CO_2 are driven by large-scale circulation changes, the increase in biomass burning aerosols causes local drying of the atmosphere. This study illustrates that reducing local biomass burning aerosol emissions may be a useful way to mitigate reduced rainfall in the region.

[1]Center for International Climate and Environmental Research-Oslo (CICERO), P.O. Box 1129 Blindern, N-0318 Oslo, Norway. [2]School of Earth and Environment, University of Leeds, Leeds LS2 9JT, UK. Correspondence and requests for materials should be addressed to Ø.H. (email: oivind.hodnebrog@cicero.oslo.no).

Black carbon (BC) aerosols, in the atmosphere, absorb solar radiation and may affect the hydrological cycle through changes in the atmospheric and surface energy budgets[1–5]. In contrast to most greenhouse gases, BC is removed from the atmosphere only after a few days[6,7]. Absorbing aerosols such as biomass burning BC modify cloud distribution, mainly by suppressing convective activity, albeit with some debate existing in the literature[8–12]. On a global scale, the altitude of added absorbing aerosol concentration is shown to influence the magnitude and sign of global average precipitation changes, with a net decrease in precipitation expected unless BC is added close to the surface[3,13–17]. Organic carbon (OC) aerosols are emitted together with BC and have approximately the same lifetime, but OC scatters sunlight to a much larger extent than BC, and therefore cools the atmosphere-surface system[6]. Biomass burning is a major source of BC and OC aerosols and normally takes place during the dry season, which is therefore the focus of this study.

Observational data reveal changes in precipitation patterns since the middle of the past century, with some regions in the world getting increased precipitation while other regions are experiencing less precipitation[18,19]. In general, increases in anthropogenic emissions have led to more precipitation on a global scale due to surface temperature warming[20]. However, the detailed effects of changes in greenhouse gases and aerosols on the hydrological cycle are complex[2,21]. Precipitation responses depend on the climate forcing mechanism, and also differ with timescale[15,16,22]. Responses in precipitation often show strong regional differences. This is mainly due to large-scale changes in circulation patterns and land-ocean contrasts, but could also arise from strictly local interactions, which are often caused by constituents that have short residence times in the atmosphere, such as BC.

Southern Africa is a region where observations indicate reduced precipitation and increased dryness since the 1950s (refs 18,19). The region is highly vulnerable to climate change (see ref. 23 and references therein), and has strong connections between climate and water, energy and food security[24]. Global climate models predict a further decrease in precipitation and a strong increase in the number of consecutive dry days near the end of the century for a business-as-usual scenario, due to widening of the Hadley Circulation[25]. While previous work on observed precipitation has focused mostly on annual trends, the focus in this study is on the dry season during which emissions from biomass burning are particularly strong. Although long-term observational data are sparse in the region, previous studies have shown evidence of declining trends in annual precipitation for several stations. In the latter half of the 20th century, significant declining precipitation trends have been shown in Zambia and southwest Tanzania, in addition to a more widespread increase in the annual maximum of consecutive dry days[26]. Annual precipitation has also decreased in tropical rainforest regions, such as in Congo, the Democratic Republic of the Congo and Gabon[27,28].

The present study focuses on explaining causes of observed precipitation decline during the dry season in southern Africa. We use a global precipitation observation data set (Global Precipitation Climatology Centre (GPCC)[29,30]) and a global (CESM1.0.4/CAM4 (refs 31,32)) and regional climate model (Weather Research and Forecasting (WRF)[33]). Our results show that the combination of BC/OC from local anthropogenic biomass burning activities, and increases in the global CO_2 concentration, can explain the precipitation decline in the region. These findings are further supported by analysis of the regional energy budget, which shows a local drying of the atmosphere from added biomass burning aerosols.

Results

Spatial distributions of precipitation changes. Some of the largest BC and OC emissions in the world take place in southern Africa due to extensive biomass burning during the dry season[34,35], approximately from June to September each year. Due to human activities, BC biomass burning emissions in the region almost doubled during the 20th century, with the majority of the increase in the latter half[35] (Supplementary Fig. 1), leading to a 60% increase in the BC burden[36] (Table 1). The increase in OC burden over the period is about 70% (Table 1).

In the present study, we show the modelled precipitation impact of increases in local BC and OC emissions due to biomass burning over the past one-and-a-half centuries, and compare them with the possible influence of increases in atmospheric CO_2 concentrations over the same period. We quantify and interpret the results from global and regional model results in terms of the regional energy budget[37] to uncover the mechanisms behind modelled changes in precipitation.

Throughout the manuscript, we show differences between model experiments (see Table 1) and denote 'BB' as the impact of changes in biomass burning emissions of BC and OC from 1850 to 2000 (experiments BASE − BB1850), 'CO2' as the impact of changes in CO_2 volume mixing ratios over the same time period (BASE − CO21850) and 'CO2BB' as the combined impact of BC/OC biomass burning and CO_2 concentration changes (BASE − CO2BB1850). Due to large variability in the precipitation changes, five-member ensemble runs were performed for each scenario.

Observations over southern Africa in the past century indicate a precipitation decline during the dry season from June to September (Fig. 1a,b), with a stronger decline in the latter part of the period, coinciding with the period of strongly increasing biomass burning emissions (Supplementary Fig. 1b).

A strong impact of anthropogenic biomass burning particles on precipitation during the dry season can be seen in the regional and the global climate model (Fig. 1c,d; Supplementary Fig. 2c,d), with precipitation decreases in the order of 10–40% in large parts of the region. In both models the strongest decline in absolute precipitation occurs in the northwest, which also is the region with the largest increase in atmospheric biomass burning aerosol

Table 1 | Overview of model experiments.

Experiment*	Short description	Global CO_2 vmr (p.p.m.)	BC burden† ($mg\,m^{-2}$)	OC burden† ($mg\,m^{-2}$)
BASE	Year 2000 conditions	367	1.85	11.8
BB1850	Biomass burning emissions of BC/OC for 1850	367	1.15	6.79
CO21850	CO_2 concentrations for 1850	285	1.85	11.8
CO2BB1850	Both CO_2 and BB changed to 1850	285	1.15	6.79

*BB = BASE − BB1850; CO_2 = BASE − CO21850; CO2BB = BASE − CO2BB1850.
†Burdens are averaged over southern Africa (subregion shown by dashed rectangle in Fig. 1) and over the June to September season.

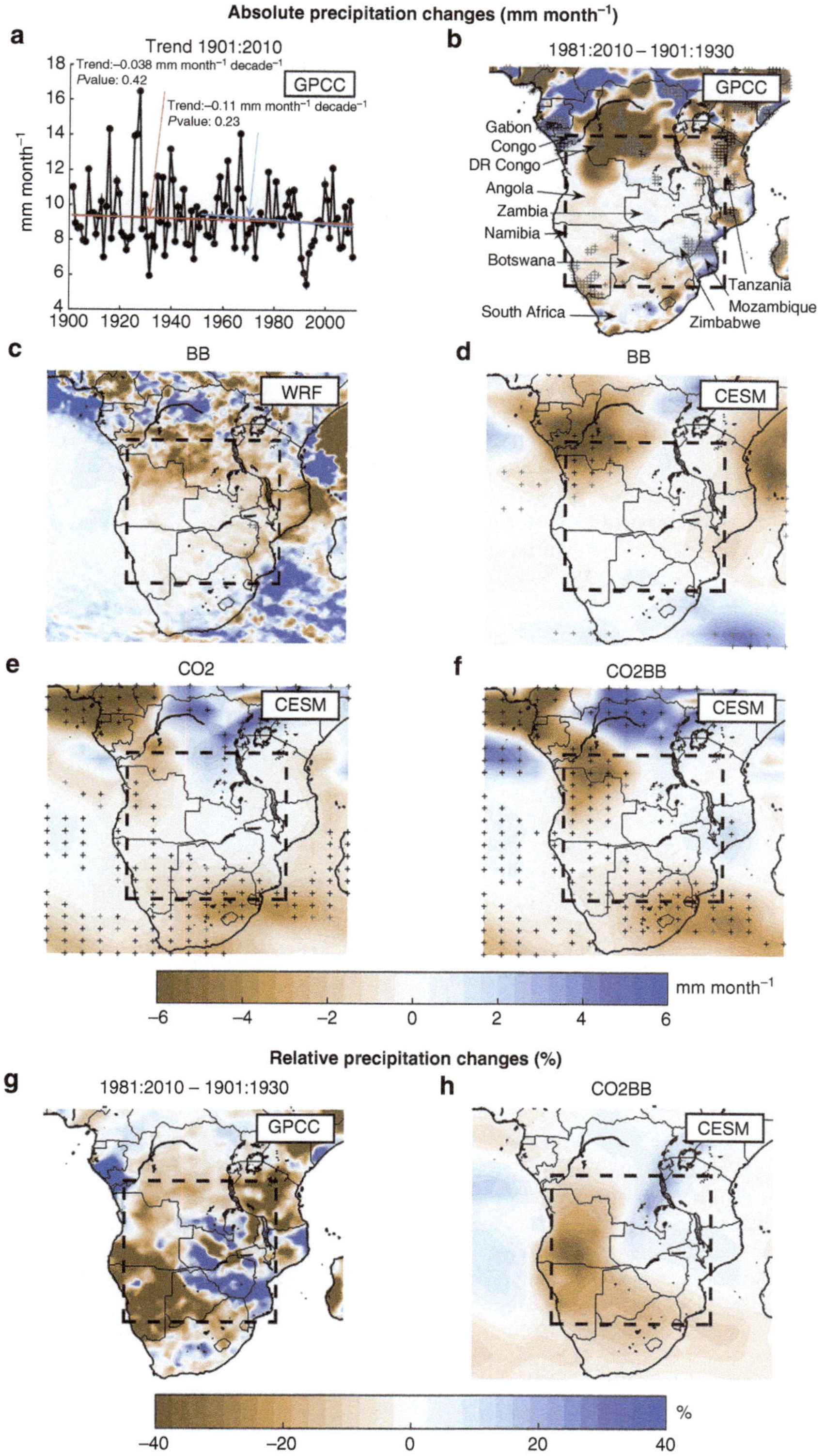

Figure 1 | Distributions of observed and modelled precipitation changes. Trend in observed precipitation (mm month^{-1}) over the June to September season (**a**), absolute precipitation changes (mm month^{-1}) for the same season from observations and various model experiments (**b–f**), and relative precipitation changes (%) from observations and the CO2BB model experiment (**g–h**). Values in (**a**) are averaged over the subregion shown by the dashed rectangle in (**b**). Panels **a**,**b** and **g** are from observations (GPCC) whereas the five other panels are from models results, one with the regional WRF model (**c**) and four from the global CESM model (**d–f**,**h**). Model results are shown as differences from the BASE simulation and show differences between 2000 and 1850 conditions in each case; see text and Table 1 for notation of the experiments. Observed precipitation trends in (**a**) are shown for the whole GPCC time period from 1901 to 2010 (red) and from 1951 to 2010 (blue). Observations in (**b**) are shown as the difference 1981:2010 – 1901:1930 because 30-year averages are needed to reduce noise (that is, the influence of natural variability). Note that in most parts of the region, the number of stations per grid cell is low in the GPCC data. In plots (**b–f**), symbols denote grid boxes where changes are significant (P value <0.05) according to a two-tailed Student's t-test (grey '+' symbols), and when also accounting for multiple statistical testing according to ref. 54 (black '+' symbols).

burden (Supplementary Fig. 1c). In this region, the global (Community Earth System Model (CESM)) model BB and CO2BB simulations show statistically significant reductions in precipitation. The BB simulation in the regional (WRF) model supports this finding, showing precipitation reduction in the same area, although not statistically significant due to a smaller sample of years. An overall decrease in precipitation can be seen in southern Africa, but with some smaller regions experiencing increased precipitation due to biomass burning. Such regions of increased precipitation downwind of regions where BB leads to precipitation decreases can be expected and have been seen before[11]. The regional model, with higher spatial resolution than the global model, shows a more inhomogeneous distribution with more local scale features.

The modelled impact of industrial-era changes in CO_2 concentrations on precipitation is an increase in the northeast of southern Africa and a decrease of up to 30% in most of the remaining part (Fig. 1e; Supplementary Fig. 2e). Thus, our results indicate that the known changes in both BB and CO_2 concentrations may have led to a precipitation decrease since 1850, but with slightly different patterns. This decrease is enhanced in the simulation with combined changes in BB and CO_2 concentrations (Fig. 1f,h). The precipitation pattern in the subregion in CO2BB resembles that of the sum of BB and CO2 (not shown), indicating that the two effects are to a large extent additive. Distributions of changes in the number of dry days (that is, days with precipitation <1 mm) and maximum number of consecutive dry days show similar tendencies of a drying due to BB and increased CO_2 concentrations (Supplementary Figs 3 and 4).

Modelled precipitation changes compare relatively well with what has been observed over the previous century (Fig. 1; Supplementary Fig. 2). Notably, observed precipitation decreases close to regions where biomass burning emissions have increased most are well reproduced by both models' BB experiment, particularly in Angola and south of the Democratic Republic of the Congo. Correlation coefficients for the subregion confirm that observed spatial patterns of absolute precipitation change correlate well with modelled precipitation changes induced by increased biomass burning (correlation coefficients >0.5 at 2.5° resolution, and significant at $P<0.05$ level), while there is no correlation with modelled precipitation change due to increased CO_2 (correlation coefficient ~ 0.0) (Supplementary Table 1). Uncertainties usually increase when going from larger to smaller scales in studies of climate change impacts, and a small displacement in simulated patterns may therefore lead to poor correlation. We note here that correlation coefficients calculated after averaging to coarser resolution are much higher (for example, 0.7–0.8 for BB at 5° resolution). It is also encouraging that the models are able to reproduce the spatial patterns of present-day precipitation climatology, with significant (P value<0.05) correlation coefficients of 0.97 and 0.78 for WRF and CESM, respectively (Supplementary Fig. 5; Supplementary Table 1).

The domination of BB aerosols in driving precipitation changes is unique for the dry season in the southern Africa region. Anthropogenic biomass burning aerosols are not found to dominate precipitation changes in other regions or seasons. In central Africa, biomass burning activities mainly take place from December to February, but the increase in biomass burning aerosol emissions since pre-industrial times has not led to a clear impact on precipitation (Supplementary Fig. 6). The third region of the world, in addition to southern and central Africa, with strong increases in biomass burning activities is the central part of South America. Here observations indicate a precipitation decline during its dry season from June to September, and global model results show that this is caused by changes in CO_2 rather than changes in BB (Supplementary Fig. 7).

Changes in mean precipitation and number of dry days. In the following we focus on the southern African subregion indicated by the dashed rectangle in Fig. 1. This region has had a strong increase in biomass burning emissions, and is a region with minimal precipitation during the June to September period, implying that a climatic change in precipitation could have important societal effects.

The reduction in precipitation in the subregion due to BB and anthropogenic CO_2 emissions is evident from Fig. 2, with a decrease in monthly mean precipitation and an increase in the number of dry days. This precipitation reduction is also observed, and although the absolute precipitation change of around 1 mm month^{-1} seems small, the relative change is as much as 5–10% (Supplementary Fig. 8). It is also an average over a relatively large area, locally observed changes are much larger (Fig. 1b,g). With the exception of one ensemble member for the BB case and one for the CO2BB case, the decrease in mean precipitation and increase in the number of dry days are consistent for all cases, between all ensemble members, and across the two models (Fig. 2; Supplementary Fig. 8). In the BB and CO2BB cases, the results are statistically significant (P value<0.05) for all ensemble members aggregated (Fig. 2). The WRF model result is based on fewer years and therefore not statistically significant at the same level, but is still in general agreement with the mean of the CESM ensembles.

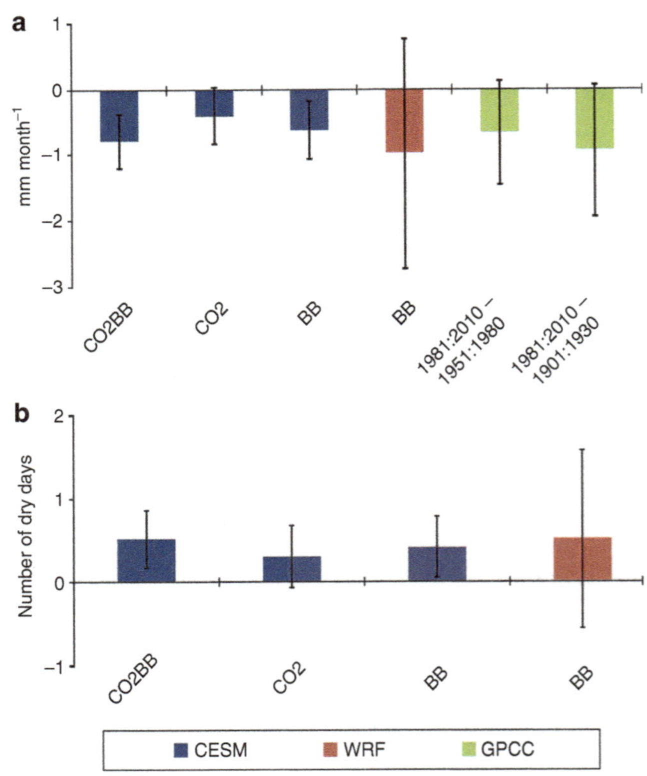

Figure 2 | Southern Africa average change in precipitation. Change in mean precipitation (mm month^{-1}) (**a**) and number of dry days (**b**) over the June to September season within the subregion (region indicated by dashed rectangle in Fig. 1) for various model experiments. Model results are shown as differences from the BASE simulation, where CESM denotes the global climate model and WRF the regional model. GPCC observations are both shown as the difference 1981:2010 – 1901:1930 and 1981:2010 – 1951:1980. Error bars show the 95% confidence interval derived from a two-tailed Student's t-test.

Precipitation and the regional atmospheric energy budget. Analysis of the energy budget, including changes in horizontal energy transport, have proved useful in understanding global and regional precipitation changes[21,37]. The local atmospheric energy budget representing the difference between two climate situations can be approximated by

$$L\delta P = \delta Q - \delta H_{\text{dry}} = \delta LH + \delta H_{\text{moist}}, \qquad (1)$$

where L is the latent heat of condensation, δP is the change in surface precipitation flux, δQ is the change in diabatic cooling of the atmosphere, δH_{dry} and δH_{moist} denote the change in net inward horizontal dry and moist static energy transport, respectively, and δLH is the net upward latent heat flux from the surface (see Methods section and refs 21,37 for details). The diabatic cooling term (δQ) can be written $\delta Q = -(\delta SW + \delta LW + \delta SH)$, where δSW and δLW is the net atmospheric shortwave and longwave radiative heating, respectively, and δSH is the upward sensible heat flux from the surface. Equation 1 shows that precipitation changes are constrained by the atmospheric energy balance[21].

Regional energy budgets for each model experiment are shown in Fig. 3 and Supplementary Fig. 9. Net atmospheric absorption of solar radiation from added BC in southern Africa is strong and exceeds $2\,\text{W}\,\text{m}^{-2}$ (Fig. 3a,b). The level of agreement between the two models and between the various ensemble members is high relative to the other terms that experience much larger natural year-to-year variability.

The atmospheric radiative heating due to solar absorption by BC, or in other words, the reduced diabatic cooling term (δQ) caused by BC, leads to a decrease in surface precipitation flux (assuming L is constant). However, this is partly balanced by reduced sensible heat flux from the surface to the atmosphere caused by reduced solar radiation reaching the surface. Scattering by OC particles also contributes to the reduced sunlight at the surface in the BB case. Absorbing aerosols have previously been shown to cause circulation changes, due to the induced temperature gradients, and this depends principally on the vertical location and magnitude of the aerosols[38]. Here, in the global model, increased net horizontal transport of dry static energy outward from the atmospheric column, presumably resulting from circulation changes, draws in the same direction as the sensible heat term (Fig. 3b). In the regional model, the horizontal transport terms are negligible (Fig. 3a), and this difference seems to be the main reason for the larger precipitation reduction in the regional model, compared with the global model, seen in the BB case in Fig. 2a.

The local energy budget for increased CO_2 concentrations is fundamentally different from that for the BB case (Fig. 3c). Increased CO_2 leads to increasing global sea surface temperatures, thus feedback effects dominate the atmospheric energy budget. The main change in the local atmospheric energy budget is from the LW radiative cooling term, caused by increased surface temperature, partly offset by atmospheric absorption[39], and contributes to increased precipitation. However, this precipitation increase is counterbalanced by increases in atmospheric SW heating caused mainly by increased solar absorption by water vapour[37], sensible heat flux from the surface, and net inward horizontal energy flux due to large scale circulation changes. Increase in CO_2 leads to a precipitation decline; despite a small increase in latent heat flux from the surface to the atmosphere, a net loss of moist static energy in the horizontal direction leads to an overall decrease. Although the effect of increased CO_2 concentrations is a precipitation increase on a global scale, this local decrease in precipitation is not unexpected as the spatial pattern of precipitation changes due to anthropogenic climate change is to a large extent affected by circulation changes[37]. In fact, analysis of modelled vertical velocity shows a widening of the Hadley cell in the increased CO_2 case, consistent with what has been observed[40]. This widening leads to more subsidence, and presumably less precipitation, south of $\sim 18°$S in the subregion, and slightly less subsidence north of this latitude (not shown), largely consistent with modelled precipitation changes (Fig. 1e; Supplementary Fig. 2e).

When both adding biomass burning aerosols and increasing CO_2, the decrease in precipitation is driven by shortwave atmospheric absorption and horizontal transport of both dry and moist static energy (Fig. 3d). Overall the regional energy budget explains the precipitation changes illustrated in Figs 1 and 2, with dissimilar impact on the energy budget from CO_2 and BB. Comparison of an additional transient simulation that includes changes in BC/OC from biomass burning and increased CO_2 concentrations, and a transient simulation including all anthropogenic climate forcers, indicate that increases in CO_2 and biomass burning BC/OC are the main drivers of changes to the atmospheric energy budget since pre-industrial time (Supplementary Fig. 10).

Discussion

It should be noted that long-term measurements of precipitation in most of southern Africa are sparse[23], and different observed precipitation products show relatively large discrepancies[41]. There may also be other drivers additional to BB and CO_2, such as non-CO_2 greenhouse gases and other aerosols, which may have contributed to the observed precipitation changes. However, reflective aerosols such as sulphate are not likely to contribute substantially because emissions of its main precursor, SO_2, are relatively small in and around the surrounding region[35]. Furthermore, remote influences from the mid-latitudes cause only small effects in southern Africa[42].

Previous studies have also found a clear impact of biomass burning aerosols on the hydrological cycle over southern and central Africa[5,11,43,44]. Although different experimental setups and focuses make it difficult to compare, there are some contradictory results in previous studies; some studies indicate more precipitation[5,43] from biomass burning aerosols and others indicate less precipitation[11,44], as is found here. Impacts of BC on clouds and precipitation are known to depend strongly on the altitude of added BC[3,14], and this could be one reason for the differing results. The BC and OC aerosol concentrations in this study are from a model that includes a plume rise assumption for biomass burning emissions (OsloCTM2), and this leads to a portion of the aerosols being emitted above the boundary layer. Consequently, the main atmospheric heating from biomass burning BC aerosols occurs at an altitude high enough to inhibit vertical instability under certain conditions, and biomass burning BC may therefore suppress convective activity in our simulations. While we acknowledge that uncertainties in BC vertical distribution are large, it is well established that biomass burning emissions are rapidly lifted to high altitudes[45]. OsloCTM2 vertical profiles of extinction coefficients compare very well with the CALIOP satellite instrument over the southern Africa region (see Fig. 4 in ref. 46).

Although the findings in this study are supported by a regional and a global model, observations, and analysis of the regional energy budget, uncertainties exist when studying changes in precipitation on a regional scale. A multi-model study would be required to provide further robustness to the findings.

The global climate model calculations in the present study were performed using a slab ocean setup, to reduce the required computational resources. We do not expect that our results would differ largely if a full ocean rather than a slab ocean model had

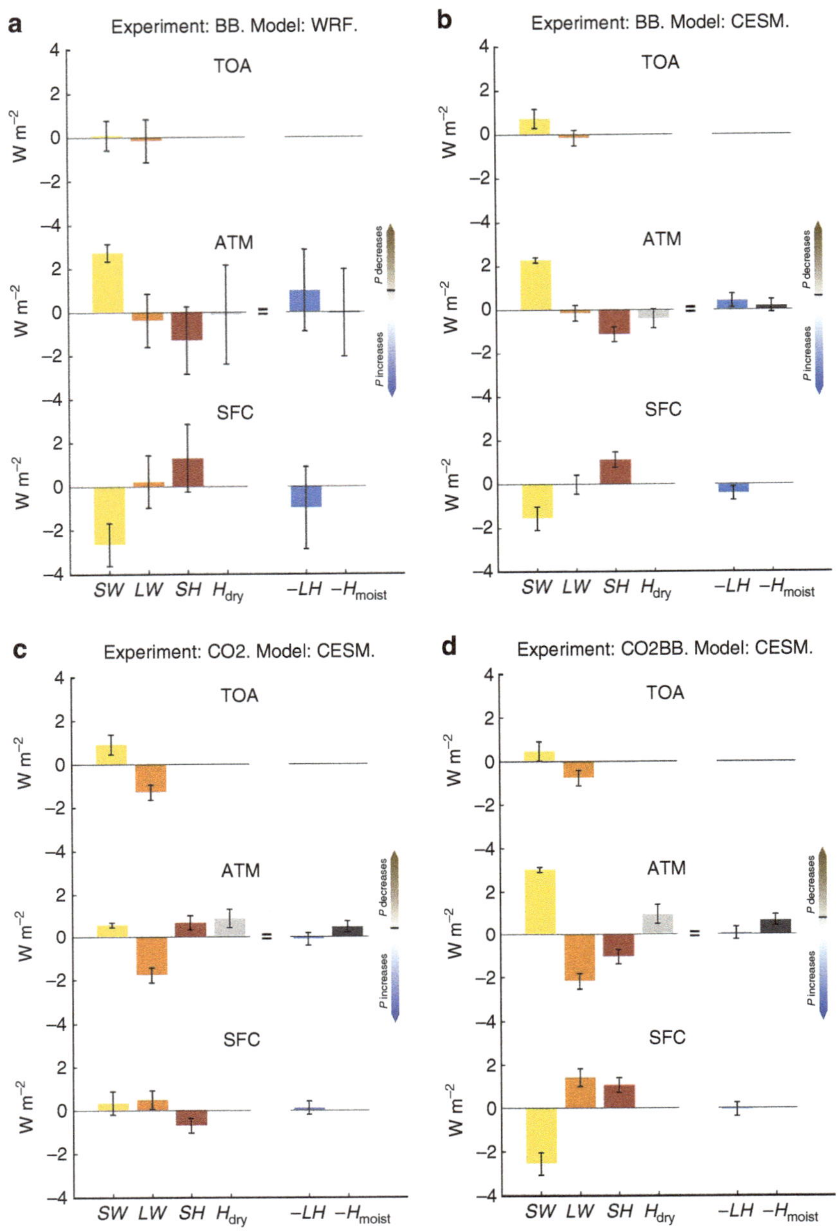

Figure 3 | Regional atmospheric energy budget for southern Africa. Atmospheric energy budget over the June to September season and averaged over the subregion (shown by the dashed rectangle in Fig. 1) for different models and experiments. Results are shown as differences from the BASE simulation, where CESM denotes the global climate model and WRF the regional model. At top-of-atmosphere (TOA) and surface (SFC), net downward energy fluxes have positive values. The energy budget of the atmosphere (ATM) is defined as the difference between the energy balance at TOA and SFC, except for the horizontal transport terms (H_{dry} and H_{moist}), which are positive when net energy transport is into the atmospheric column from the outside. Direction of brown and blue arrows on the right-hand-side of each plot denotes contribution towards decreased or increased surface precipitation according to equation 1 ($-L\delta P = \delta SW + \delta LW + \delta SH + \delta H_{dry} = -\delta LH - \delta H_{moist}$), and the total change in precipitation in each experiment is indicated by a black line embedded on these arrows. Error bars show the 95% confidence interval derived from a two-tailed Student's *t*-test.

been used. In a study using an earlier version of the CESM model[47], a fully coupled ocean model simulation led to a slightly higher climate sensitivity. Hence, a slightly stronger response in the hydrological cycle could possibly have been expected in the increased CO_2 case if a full ocean model would have been used. The impact of BC on the hydrological cycle is however dominated by rapid adjustments[15] and is therefore not likely to be strongly affected by the choice of slab ocean versus full ocean.

Our results show that local BC and OC aerosol emissions from anthropogenic biomass burning activities and increasing global CO_2 concentrations are the main causes of the observed decline in southern African dry season precipitation over the last century. This implies that, in addition to improving air quality, reducing biomass burning emissions of BC and OC may be a useful way to mitigate reduced rainfall in the southern Africa region. This conclusion is strengthened by the suggestion that these aerosols may be part of a positive feedback loop where increased drought could increase the risk of igniting more fires[11]. Precipitation in the region is expected to decrease further in the future if greenhouse gas emissions continue unabated[25]. Hence, initiation of stringent air quality legislation, involving BC mitigation, may help to counteract this precipitation decline in the short term.

Methods

Observation data. We have used long-term observation data from the GPCC[29,30], which provides monthly values of precipitation from 1901 to 2010 from 67,200 stations world-wide, gridded to a horizontal resolution of 0.5° × 0.5°. This observation dataset has been chosen because it gives a better coverage than other long-term gridded precipitation observation datasets, such as CRU and GHCN - see TFE.1, Fig. 2 in ref. 48. In regions where GPCC coverage overlaps with CRU and GHCN, they give similar estimates of annual trends. The trend analysis in Fig. 1a has been calculated using the Mann–Kendall method[49], and P values in the observed trends are calculated using a two-tailed z-test.

Global climate model. Global climate model simulations have been carried out using the National Center for Atmospheric Research (NCAR) CESM1.0.4 (ref. 31). The atmospheric component, Community Atmosphere Model (CAM4) (ref. 32), was in our simulations run with 1.9° × 2.5° horizontal resolution and 26 vertical layers, and coupled with the Community Land Model (CLM4) (ref. 50). Five different ensemble members have been run for 60 years each, whereof the first 10 years were considered as spin-up and are not part of the analysis. The ensemble members were started from different initial conditions but the same climatological state, and we have used a configuration where the atmospheric component is coupled with a slab ocean model[51]. Results are shown as averages of all ensemble members. Atmospheric distributions of aerosol concentrations have been prescribed for black and organic carbon, and sulphate, using monthly mean 3-D concentrations from the chemistry-transport model OsloCTM2 from a previous study[36]. This model has previously shown good agreement with observed pattern and magnitude of aerosol optical depth in southern Africa during the biomass burning season[34], and with vertical aerosol extinction coefficient profiles over the same region[46]. In the CESM model experiments, both the CO_2 and BC/OC concentrations have been changed globally. To test the effect of global versus local perturbation of biomass burning BC/OC, the first two ensemble members of the BB1850 simulation were re-run, but with biomass burning BC/OC concentrations changed to 1850 levels only during June to September and only in and around southern Africa. The results in terms of southern Africa dry season precipitation change were within 25% of the results obtained when biomass burning BC/OC were changed globally and annually.

Regional climate model. Additional climate model simulations have been conducted using the regional WRF model[33] version 3.5. The model has been run in 50 × 50 km horizontal resolution over Africa, and with 50 vertical layers from the surface and up to 10 hPa. The extent of the horizontal domain is the same as defined in CORDEX-Africa[52]. WRF has been run for the 30 year period, 1984–2013, in 1 year time slices, each re-initialized in December of the previous year to allow one month of spin-up. Meteorological initial and boundary conditions, as well as sea-surface temperatures, have been taken from the 6-hourly 1° × 1° resolution final reanalysis data of the Global Forecast System (GFS) model of the National Centers for Environmental Prediction (NCEP) from year 2000 onward, and NCEP/NCAR 2.5° × 2.5° resolution reanalysis for the preceding years[53]. The same radiation scheme (CAM) as in the CESM has been used in WRF. In the standard version of WRF, however, the CAM scheme only allows one fixed column value of each aerosol species to represent the entire domain for the whole year. Here, the WRF code was modified to enable prescribed monthly mean 3-D concentration fields of BC, organic carbon and sulphate from OsloCTM2, for consistency with what is used in the CESM.

Energy budget. In the calculation of δH_{dry}, we have used the same approach as in ref. 37 and calculated it as the residual of equation 1 ($\delta H_{dry} = L\delta P - \delta Q$), assuming $L = 29\,W\,m^{-2}\,mm^{-1}$ day, rather than a direct calculation of the total dry static energy flux. This approximation has been shown to introduce only small errors[37]. In Fig. 3, we have represented differences in horizontal transport of energy in and out of the atmospheric region both as dry static energy flux (δH_{dry}) and as moist static energy flux (δH_{moist}). The latter term is again calculated as a residual, $\delta H_{moist} = \delta LH\text{-}L\delta P$, and both LH and P are available from the model output.

References

1. Ramanathan, V., Crutzen, P. J., Kiehl, J. T. & Rosenfeld, D. Atmosphere - Aerosols, climate, and the hydrological cycle. *Science* **294,** 2119–2124 (2001).
2. Allen, M. R. & Ingram, W. J. Constraints on future changes in climate and the hydrologic cycle. *Nature* **419,** 224–232 (2002).
3. Ban-Weiss, G. A., Cao, L., Bala, G. & Caldeira, K. Dependence of climate forcing and response on the altitude of black carbon aerosols. *Clim. Dyn.* **38,** 897–911 (2012).
4. Kawase, H. *et al.* Physical mechanism of long-term drying trend over tropical North Africa. *Geophys. Res. Lett.* **37,** L09706 (2010).
5. Randles, C. A. & Ramaswamy, V. Direct and semi-direct impacts of absorbing biomass burning aerosol on the climate of southern Africa: a Geophysical Fluid Dynamics Laboratory GCM sensitivity study. *Atmos. Chem. Phys.* **10,** 9819–9831 (2010).
6. Boucher, O. *et al.* in *Climate Change 2013: The Physical Science Basis. Contribution of Working Group I to the Fifth Assessment Report of the Intergovernmental Panel on Climate Change* (eds Stocker, T. F. *et al.*) Ch. 7 571–658 (Cambridge University Press, 2013).
7. Samset, B. H. *et al.* Modelled black carbon radiative forcing and atmospheric lifetime in AeroCom Phase II constrained by aircraft observations. *Atmos. Chem. Phys.* **14,** 12465–12477 (2014).
8. Koren, I., Kaufman, Y. J., Remer, L. A. & Martins, J. V. Measurement of the effect of Amazon smoke on inhibition of cloud formation. *Science* **303,** 1342–1345 (2004).
9. Feingold, G., Jiang, H. L. & Harrington, J. Y. On smoke suppression of clouds in Amazonia. *Geophys. Res. Lett.* **32,** L02804 (2005).
10. Rosenfeld, D. *et al.* Flood or drought: How do aerosols affect precipitation? *Science* **321,** 1309–1313 (2008).
11. Tosca, M. G., Diner, D. J., Garay, M. J. & Kalashnikova, O. V. Human-caused fires limit convection in tropical Africa: first temporal observations and attribution. *Geophys. Res. Lett.* **42,** 6492–6501 (2015).
12. Koch, D. & Del Genio, A. D. Black carbon semi-direct effects on cloud cover: review and synthesis. *Atmos. Chem. Phys.* **10,** 7685–7696 (2010).
13. Ming, Y., Ramaswamy, V. & Persad, G. Two opposing effects of absorbing aerosols on global-mean precipitation. *Geophys. Res. Lett.* **37,** L13701 (2010).
14. Samset, B. H. & Myhre, G. Climate response to externally mixed black carbon as a function of altitude. *J. Geophys. Res.-Atmos.* **120,** 2913–2927 (2015).
15. Kvalevåg, M. M., Samset, B. H. & Myhre, G. Hydrological sensitivity to greenhouse gases and aerosols in a global climate model. *Geophys. Res. Lett.* **40,** 1432–1438 (2013).
16. Andrews, T., Forster, P. M., Boucher, O., Bellouin, N. & Jones, A. Precipitation, radiative forcing and global temperature change. *Geophys. Res. Lett.* **37,** L14701 (2010).
17. Hodnebrog, Ø., Myhre, G. & Samset, B. H. How shorter black carbon lifetime alters its climate effect. *Nat. Commun.* **5,** 5065 (2014).
18. Hartmann, D. L. *et al.* in *Climate Change 2013: The Physical Science Basis. Contribution of Working Group I to the Fifth Assessment Report of the Intergovernmental Panel on Climate Change* (eds Stocker, T. F. *et al.*) Ch. 2 159–254 (Cambridge University Press, 2013).
19. Greve, P. *et al.* Global assessment of trends in wetting and drying over land. *Nat. Geosci.* **7,** 716–721 (2014).
20. Knutti, R. & Sedlacek, J. Robustness and uncertainties in the new CMIP5 climate model projections. *Nat. Clim. Change* **3,** 369–373 (2013).
21. O'Gorman, P. A., Allan, R. P., Byrne, M. P. & Previdi, M. Energetic constraints on precipitation under climate change. *Surv. Geophys.* **33,** 585–608 (2012).
22. Bala, G., Caldeira, K. & Nemani, R. Fast versus slow response in climate change: implications for the global hydrological cycle. *Clim. Dyn.* **35,** 423–434 (2010).
23. Niang, I. *et al.* in *Climate Change 2014: Impacts, Adaptation, and Vulnerability. Contribution of Working Group II to the Fifth Assessment Report of the Intergovernmental Panel on Climate Change* (eds Barros, V. R. *et al.*) Ch. 22 1199–1265 (Cambridge University Press, 2014).
24. Conway, D. *et al.* Climate and southern Africa's water-energy-food nexus. *Nat. Clim. Change* **5,** 837–846 (2015).
25. Collins, M. *et al.* in *Climate Change 2013: The Physical Science Basis. Contribution of Working Group I to the Fifth Assessment Report of the Intergovernmental Panel on Climate Change* (eds Stocker, T. F. *et al.*) Ch. 12 1029–1136 (Cambridge University Press, 2013).
26. New, M. *et al.* Evidence of trends in daily climate extremes over southern and west Africa. *J. Geophys. Res.-Atmos.* **111,** D14102 (2006).
27. Aguilar, E. *et al.* Changes in temperature and precipitation extremes in western central Africa, Guinea Conakry and Zimbabwe, 1955-2006. *J. Geophys. Res.-Atmos.* **114,** D02115 (2009).
28. Malhi, Y. & Wright, J. Spatial patterns and recent trends in the climate of tropical rainforest regions. *Philos. Trans. R. Soc. Lond. Ser. B-Biol. Sci* **359,** 311–329 (2004).
29. Schneider, U. *et al.*GPCC Full Data Reanalysis Version 6.0 at 0.5°: Monthly Land-Surface Precipitation from Rain-Gauges built on GTS-based and Historic Data (2011).
30. Becker, A. *et al.* A description of the global land-surface precipitation data products of the global precipitation climatology centre with sample applications including centennial (trend) analysis from 1901–present. *Earth Syst. Sci. Data* **5,** 71–99 (2013).
31. Gent, P. R. *et al.* The community climate system model version 4. *J. Clim.* **24,** 4973–4991 (2011).
32. Neale, R. B. *et al.*Description of the NCAR Community Atmosphere Model (CAM 4.0). (NCAR Technical Report, NCAR/TN-485 + STR (2010).
33. Skamarock, W. C. & Klemp, J. B. A time-split nonhydrostatic atmospheric model for weather research and forecasting applications. *J. Comput. Phys.* **227,** 3465–3485 (2008).

34. Myhre, G. *et al.* Modeling the solar radiative impact of aerosols from biomass burning during the Southern African Regional Science Initiative (SAFARI-2000) experiment. *J. Geophys. Res.-Atmos.* **108,** 8501 (2003).
35. Lamarque, J. F. *et al.* Historical (1850–2000) gridded anthropogenic and biomass burning emissions of reactive gases and aerosols: methodology and application. *Atmos. Chem. Phys.* **10,** 7017–7039 (2010).
36. Skeie, R. B. *et al.* Anthropogenic radiative forcing time series from pre-industrial times until 2010. *Atmos. Chem. Phys.* **11,** 11827–11857 (2011).
37. Muller, C. J. & O'Gorman, P. A. An energetic perspective on the regional response of precipitation to climate change. *Nat. Clim. Change* **1,** 266–271 (2011).
38. Lee, S. S. *et al.* Effect of gradients in biomass burning aerosol on shallow cumulus convective circulations. *J. Geophys. Res.-Atmos.* **119,** 9948–9964 (2014).
39. Pendergrass, A. G. & Hartmann, D. L. The atmospheric energy constraint on global-mean precipitation change. *J. Clim.* **27,** 757–768 (2014).
40. Seidel, D. J., Fu, Q., Randel, W. J. & Reichler, T. J. Widening of the tropical belt in a changing climate. *Nat. Geosci.* **1,** 21–24 (2008).
41. Nikulin, G. *et al.* Precipitation climatology in an ensemble of CORDEX-Africa regional climate simulations. *J. Clim.* **25,** 6057–6078 (2012).
42. Shindell, D. T., Voulgarakis, A., Faluvegi, G. & Milly, G. Precipitation response to regional radiative forcing. *Atmos. Chem. Phys.* **12,** 6969–6982 (2012).
43. Tummon, F., Solmon, F., Liousse, C. & Tadross, M. Simulation of the direct and semidirect aerosol effects on the southern Africa regional climate during the biomass burning season. *J. Geophys. Res.-Atmos.* **115,** D19206 (2010).
44. Tosca, M. G., Randerson, J. T. & Zender, C. S. Global impact of smoke aerosols from landscape fires on climate and the Hadley circulation. *Atmos. Chem. Phys.* **13,** 5227–5241 (2013).
45. Freitas, S. R. *et al.* Including the sub-grid scale plume rise of vegetation fires in low resolution atmospheric transport models. *Atmos. Chem. Phys.* **7,** 3385–3398 (2007).
46. Koffi, B. *et al.* Application of the CALIOP layer product to evaluate the vertical distribution of aerosols estimated by global models: AeroCom phase I results. *J. Geophys. Res.-Atmos.* **117,** D10201 (2012).
47. Danabasoglu, G. & Gent, P. R. Equilibrium climate sensitivity: is it accurate to use a slab ocean model? *J. Clim.* **22,** 2494–2499 (2009).
48. Stocker, T. F. *et al.* in *Climate Change 2013: The Physical Science Basis. Contribution of Working Group I to the Fifth Assessment Report of the Intergovernmental Panel on Climate Change* (eds Stocker, T. F. *et al.*) Ch. TS 33–115 (Cambridge University Press, 2013).
49. Sen, P. K. Estimates of the regression coefficient based on Kendall's Tau. *J. Am. Stat. Assoc.* **63,** 1379–1389 (1968).
50. Lawrence, D. M. *et al.* Parameterization improvements and functional and structural advances in version 4 of the Community land model. *J. Adv. Model. Earth Sy.* **3,** M03001 (2011).
51. Kiehl, J. T., Shields, C. A., Hack, J. J. & Collins, W. D. The climate sensitivity of the community climate system model version 3 (CCSM3). *J. Clim.* **19,** 2584–2596 (2006).
52. Kim, J. *et al.* Evaluation of the CORDEX-Africa multi-RCM hindcast: systematic model errors. *Clim. Dyn.* **42,** 1189–1202 (2014).
53. Kalnay, E. *et al.* The NCEP/NCAR 40-year reanalysis project. *Bull. Amer. Meteorol. Soc.* **77,** 437–471 (1996)077<0437:tnyrp>2.0.co;2 (1996).
54. Ventura, V., Paciorek, C. J. & Risbey, J. S. Controlling the proportion of falsely rejected hypotheses when conducting multiple tests with climatological data. *J. Clim.* **17,** 4343–4356 (2004).

Acknowledgements

GPCC precipitation data are provided by the NOAA/OAR/ESRL PSD, Boulder, Colorado, USA, from their web site at http://www.esrl.noaa.gov/psd/. Funding from the Research Council of Norway through the NAPEX (no. 229778) and HYPRE (no. 243942) projects is acknowledged. P.M.F. acknowledges IAGP grant EP/I014721/ and a Royal Society Wolfson Merit Award.

Author contributions

Ø.H. wrote the paper, but with contributions from all the authors. G.M. developed the idea. G.M. and Ø.H. designed the GCM and RCM model experiments and Ø.H. carried out the experiments. All authors participated in analysis of the model results.

Additional information

Competing financial interests: The authors declare no competing financial interests.

14

Large deglacial shifts of the Pacific Intertropical Convergence Zone

A.W. Jacobel[1,2], J.F. McManus[1,2], R.F. Anderson[1,2] & G. Winckler[1,2]

The position of the Intertropical Convergence Zone (ITCZ) is sensitive to changes in the balance of heat between the hemispheres which has fundamental implications for tropical hydrology and atmospheric circulation. Although the ITCZ is thought to experience the largest shifts in position during deglacial stadial events, the magnitude of shifts has proven difficult to reconstruct, in part because of a paucity of high-resolution records, particularly those including spatial components. Here we track the position of the ITCZ from 150 to 110 ka at three sites in the central equatorial Pacific at sub-millennial time resolution. Our results provide evidence of large, abrupt changes in tropical climate during the penultimate deglaciation, coincident with North Atlantic Heinrich Stadial 11 (~136–129 ka). We identify this event both as a Northern Hemisphere increase in aeolian dust and as a shift in the mean position of the ITCZ a minimum of 4° southwards at 160° W.

[1] Department of Earth and Environmental Sciences, Columbia University, New York, New York 10027, USA. [2] Lamont-Doherty Earth Observatory, Palisades, New York 10964, USA. Correspondence and requests for materials should be addressed to A.W.J. (email: jacobel@ldeo.columbia.edu).

The Intertropical Convergence Zone (ITCZ) is a band of vigorous atmospheric convection and precipitation that varies position seasonally in connection with the latitude of maximum local insolation and on longer timescales in response to changes in the thermal equator[1,2]. This sensitivity has been reproduced in model experiments[3,4] and is indicated by a variety of palaeoclimate reconstructions covering the Last Glacial Maximum, last deglaciation and Holocene[5–7]. Because the ITCZ monitors the hemispheric thermal balance, its past positions record the (a)symmetry of climate changes and can provide insight into the hemispheric origin of millennial and orbital climate events. As an active component of the global hydrological system, the ITCZ also propagates changes at high latitudes into the tropics by altering albedo, regional precipitation and wind fields[8]. Understanding the forcings and dynamics behind these atmospheric shifts is important for predicting future changes in the ITCZ/monsoon complex with significant implications for communities living in hydrologically sensitive environments[9]. On longer timescales, studies have suggested that the ITCZ may exert an influence on the positions of the Southern Hemisphere subtropical jet and westerly winds[10,11], which it is hypothesized are an essential element of deglacial Southern Ocean upwelling[12]. Accounting for past changes in ITCZ position is thus critical to our understanding of global ocean-atmosphere dynamics, tropical hydrology and the ocean's role in regulating atmospheric pCO_2.

As expected, given the role of ITCZ precipitation in scavenging particles from the atmosphere, previous work has found that Pacific Ocean core sites underlying the ITCZ display a latitudinal gradient in dust flux[13] and provenance[14,15]. When the amount of suspended dust is the same, greater precipitation yields more dust to the underlying sediments and therefore relative dust flux can be used as a proxy for ITCZ position. Specifically, the edges of the ITCZ will have higher dust fluxes than sites within the ITCZ due to lower residual dust availability towards the interior of the ITCZ[13]. However, it is essential to recognize that a site exhibiting a temporal maximum in dust flux does not necessarily mean that the ITCZ has moved over that site, since the amount of suspended dust is not always the same above all core sites. Because the ITCZ is a barrier to cross-hemispheric dust transport[14,16], variations in ITCZ latitude can force substantial changes in the provenance and amount of dust above a site. This nuance requires that we use independent dust flux data, in this case from Antarctica[17], to inform our interpretations. Combined with our latitudinal transect and attention to relative dust flux variations between core sites, this approach allows us to differentiate between changes in atmospheric dust abundance and shifts in ITCZ position.

The central equatorial Pacific's Line Islands are well positioned to be sensitive to changes in the ITCZ because at this longitude the precipitation structure is well defined, with little deflection due to continental effects. Our records are derived from a latitudinal transect of three sediment cores: ML1208-37BB (37BB) at 7.04° N, 161.63° W and 2,798 m water depth; ML1208-31BB (31BB) at 4.68° N, 160.05° W and 2,857 m depth; and ML1208-17PC (17PC) at 0.48° N, 156.45° W and 2,926 m depth (Fig. 1). Core 37BB is located near the local modern boreal summer maximum of the ITCZ (~7° N), and 17PC is south of the local boreal winter minimum of ~5° N (ref. 18).

We present high-resolution records of dust flux from 150 to 110 ka at three sites in the Line Islands to constrain the ITCZ position during the penultimate deglaciation (Termination II or TII). These records place latitudinal constraints on the millennial-scale movement of the Pacific ITCZ over this time period. Our data reconstruct a large, abrupt change in tropical Pacific climate over TII, characterized by both an increase in atmospheric dust abundance and by a shift of the ITCZ of at least 4° southwards. This tropical event is coincident with Heinrich Stadial 11 (HS11). After TII, dust fluxes declined to interglacial levels and the ITCZ shifted northwards of its previous glacial position.

Results

Sedimentation rates. Sedimentation rates vary between cores and within cores with average rates for the interval studied (110–150 ka) of 1.1, 4.8 and 3.0 cm ka^{-1} for cores 37BB, 31BB and 17PC respectively. A higher level of confidence is placed on conclusions when the sedimentation rate is higher, notably for the deglacial interval key to our primary interpretations in this paper. Details on the development of the age models from which these sedimentation rates are derived may be found in Methods.

Dust flux data. Dust fluxes were reconstructed using $^{230}Th_{xs,0}$-normalized ^{232}Th fluxes (Supplementary Data 1) as described in Methods. Data for sites 37BB, 31BB and 17PC (Fig. 2b) show similar trends in time from Marine Isotope Stage (MIS) 6 through MIS5. During MIS6 (oldest part of record to ~140 ka) all cores show high dust deposition (>0.25 g m^{-2} yr^{-1}) relative to the interglacial. At the peak glacial (as marked by the maxima in oxygen isotope values; Fig. 2a) the northernmost core is the first to show a small increase in dust flux, followed by the two more southerly cores between ~136 and 135 ka. Deglacial dust flux maxima occur in cores 31BB and 17PC between ~133.4 and 134.5 ka, and after that values in all cores decline precipitously reaching a minimum (0.1 to 0.15 g m^{-2} yr^{-1}) during the MIS5 interglacial. This interglacial dust flux value is ~2.5 times lower than glacial values. For the remainder of the records all three cores show gradually increasing dust fluxes.

Oxygen isotope data. Oxygen isotope data generated on the planktonic foraminifera *Globigerinoides ruber* in ML1208 cores 37BB, 31BB and 37BB (Fig. 2a) have been published elsewhere[19], but here we provide additional interpretation of these data in the context of our inferences about MIS6-5. Planktonic oxygen isotope ($\delta^{18}O$) data record a combination of surface water conditions including ambient temperature, salinity and the global $\delta^{18}O$ of sea water. These variables in turn reflect changes in sea-surface temperature, local precipitation, upwelling and global ice volume. Variations in $\delta^{18}O$ may reflect either local processes or the signature of waters advected to our study sites, or a combination of both. Although we cannot parse the contributions of each of these variables to observed changes in the records, it is instructive to consider the relationship between $\delta^{18}O$ data from the three cores and how it changes over time.

From 150 and 140 ka, oxygen isotope values for cores 37BB and 31BB are between −1.28 and −0.8‰, representing substantially fresher or warmer conditions than at site 17PC (−0.5 to −0.02‰). As the deglaciation begins (~140 ka) $\delta^{18}O$ values freshen and warm at all three sites in response to rising global temperatures and reduced ice volume. However, during HS11 (136–129 ka), site 17PC shows the largest change (1.07‰), indicating that freshening and/or warming was most strongly felt at the southernmost site. At the peak of MIS5e (~120 ka), the $\delta^{18}O$ recorded by foraminifera at sites 37BB, 31BB and 17PC is similar.

Discussion

Few dust flux records exist from the central equatorial Pacific, but a low-resolution record from TT013-72PC (72PC) at 0.1° N, 139.4° W (ref. 20) provides an important comparison for our sediment cores. Data from 72PC show a similar decrease in dust flux from 145 to 118 ka, followed by an increase from 118 to 110 ka, in agreement with our results. Given the paucity of high-resolution marine dust flux records covering TII, the

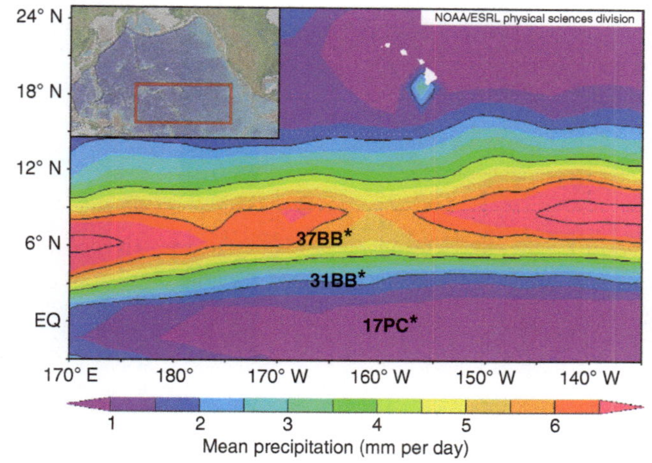

Figure 1 | Map of study area. ML1208 core sites 37BB, 31BB and 17PC (black stars) are shown relative to Hawaii (white islands), and the modern mean annual position of the ITCZ, illustrated by the maximum mean annual precipitation in mm per day (coloured contours). Inset shows the location of this map (red rectangle) in the broader context of the Pacific Ocean. Basemap generated using the global precipitation climatology product (GPCP) version 2.2 from National Oceanic and Atmospheric Administration's (NOAA's) Physical Sciences Division, Earth System Research Laboratory. Inset generated in GeoMapApp.

Antarctic EPICA Dome C (EDC) dust record is a valuable companion data set (Fig. 2c). Antarctic and equatorial dust fluxes co-vary at coarse resolution between hemispheres during the late Pleistocene ice-age cycles[20], and in this context it is the differences between our high-resolution records and the EDC record that are the most significant. Specifically, the decline in EDC dust flux at the MIS5/6 boundary (TII at ~134 ka; from ~0.036 to 0.007 $g\,m^{-2}\,yr^{-1}$) trends oppositely to the increase in dust flux observed at sites 37BB, 31BB and 17PC during HS11 (grey shading in Fig. 2). Because of the relatively high resolution of the EDC ice core and a comparable decline in dust flux in at least one other high-resolution, Southern Hemisphere record[21], we hypothesize that the absence of a HS11 dust peak in the ice core record indicates that the observed dust peak in the central equatorial Pacific does not reflect an increase in global or Southern Hemisphere dust abundance, but rather an increase in northern hemisphere sources alone. Independent behaviour of Northern and Southern Hemisphere dust sources has been repeatedly observed for Heinrich stadials over the last glacial cycle[17,22] (Supplementary Fig. 1c,d), and we propose that a similar hemispheric dust flux asymmetry characterized TII during HS11. Although no high-resolution dust records exist from the Northern Hemisphere covering HS11, we suggest that the increase in Northern Hemisphere dust that marked previous Heinrich stadials also characterized HS11 (ref. 23), especially since the relatively long duration of HS11 (ref. 24) would likely have allowed for the full expression of stadial conditions. Other indicators of stadial conditions[25] have also identified HS11, including high ice-rafted debris (IRD) abundance in North Atlantic sediments (Fig. 2d) and cold sea-surface temperatures (Fig. 2e). While not causally related to dust fluxes, these stadial indicators are repeatedly seen to change in concert with increases in Northern Hemisphere dust abundance. Recurrent intervals of high dust flux during Heinrich stadials, in combination with the clear evidence for decreased dust abundance in the Southern Hemisphere, strongly support the idea that the dust flux increases in our central equatorial Pacific cores are due to a Northern Hemisphere dust increase.

Although atmospheric dust abundance clearly varies on deglacial timescales, it cannot explain the time-variant differences in dust flux among our three sites. Instead, we attribute inter-core differences to changes in the position of the ITCZ (Fig. 3a). Our interpretations of ITCZ movement take into account our new dust flux records, planktonic oxygen isotope data from the same three sites, variations in Southern Hemisphere dust abundance as recorded in Antarctica[17] and temperature proxy records that indicate the thermal gradient between poles (Fig. 2e, f and Supplementary Note 2).

During MIS6 (~145 ka) dust fluxes at the three sites do not show a sufficiently clear relationship to infer confidently, on dust flux data alone, the position of the ITCZ. Instead we suggest that the similar oxygen isotope values seen at sites 37BB and 31BB indicate an ITCZ positioned approximately equidistant between the two cores. Fresher conditions at these two sites would be consistent with ITCZ-derived precipitation located directly above the sites. Colder and/or more saline conditions at 17PC are also consistent with this interpretation, as cross equatorial winds could have driven enhanced upwelling at this southernmost site as they do seasonally at present when the ITCZ is at its northernmost position[26].

Between ~136 and 135 ka, coincident with the onset of HS11 in the North Atlantic (grey bar Fig. 2), dust fluxes increase in 31BB and 17PC, which we interpret as indicative of a southward shift in the northern edge of the ITCZ. Because dust fluxes decrease in the Southern Hemisphere from 138.5 ka into MIS5, and because an increase in dustiness is repeatedly associated with Heinrich stadials, increasing dust fluxes in cores 31BB and 17PC indicate that they are likely receiving dust from the Northern Hemisphere. Since the ITCZ acts as a barrier to dust, the observed dust increases at 31BB and 17PC imply that the ITCZ must have been south of both of these sites. Additional evidence for an ITCZ south of 0.48° N during HS11 comes from contrasting the 24% increase in dust flux at the southernmost site (17PC), with the record from the intermediate core (31BB); this shows only a 9% increase, despite a higher sedimentation rate in core 31BB during this interval of ~3.9 $cm\,ka^{-1}$ relative to a rate of ~2.9 $cm\,ka^{-1}$ in core 17PC over the same time interval (see Methods). These increases in dust flux represent mean values for a range of data points before HS11 and during the dust flux maximum, with a maximum change in core 17PC of 32% and a minimum of 15%, and a maximum for core 31BB of 21% and a minimum of 1%. Here more efficient dust scavenging at 17PC suggests that the northern edge of the ITCZ was closer to that site, since at this time the source of dust to all three cores was most likely in the Northern Hemisphere. The hemispheric thermal gradient (Supplementary Note 2) supports this interpretation, as it shows rapid, deglacial Antarctic warming at a time when North Atlantic foraminifera assemblages and alkenone unsaturation indices record cold sea-surface temperatures (cf. Fig. 2e, f). This millennial-scale response of the ITCZ to changes in the hemispheric thermal gradient is consistent with models that predict ITCZ shifts in response to North Atlantic Heinrich stadials[4].

Planktonic oxygen isotope data from the three sites are consistent with our inference of a southward-shifted ITCZ during the deglaciation. All three cores show a shift towards values indicating warmer, fresher conditions, but the change is largest at 17PC where a southward-shifted ITCZ may have warmed that site and/or more ITCZ precipitation may have freshened surface waters. In contrast, sites 31BB and 37BB experienced more muted freshening and warming over this time period as might be expected if the $\delta^{18}O$ signal from decreasing precipitation opposed the effect of warming deglacial temperatures and decreasing ice volume.

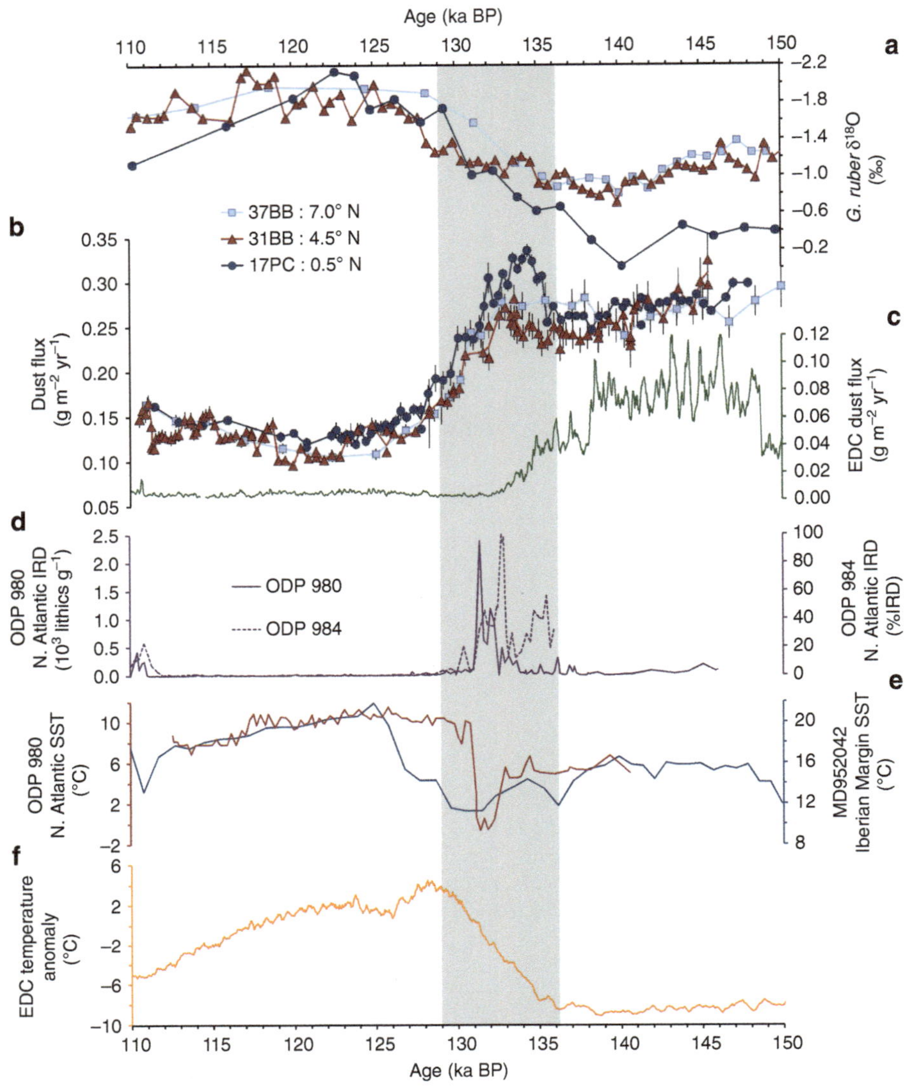

Figure 2 | Climate of the penultimate deglaciation. Planktonic foraminifera $\delta^{18}O$ stratigraphy for cores 37BB (light blue squares), 31BB (red triangles) and 17PC (navy circles)[19] (**a**). Dust flux records from 37BB, 31BB and 17PC. Symbols as in **a**, error bars are 2 s.d. Lines connect data points and the mean value of replicates (**b**). Dust flux from Antarctic ice core EDC[17] (75°06′ S, 123° 21′ E) on AICC2012 chronology[41] (5 pt smoothed) (**c**). North Atlantic (N. Atlantic) ice-rafted debris (IRD) abundance from sites ODP 980 (55° 29′ N, 14° 42′ W) and ODP 984 (61° 25′ N, 24° 4′ W)[42] (**d**). North Atlantic sea surface temperature (SST) proxy records from ODP 980 (ref. 29) on the timescale of Mokeddem *et al.*[42] (red line), and MD952042 (ref. 43; navy blue line) (**e**). Antarctic EDC temperature anomaly record[44] on AICC2012 chronology[41] (5 pt smoothed), where the anomaly is relative to the last 1,000 years (**f**). Grey shading represents Weak Monsoon Interval II (WMI-II)[27] correlated with HS11 in the North Atlantic. Principles underlying the interpretation of proxies in **e** and **f** are discussed in Supplementary Note 2.

The small lag in the onset of dust flux increases in cores 31BB and 17PC relative to the onset of HS11 is likely an artefact of uncertainty in our age models (see Methods) and we therefore do not interpret this offset as a climate system lag. Interestingly, core 37BB does not appear to show an increase in dust flux coincident with the records from our other two cores. We suggest two possible reasons for this observation. First, core 37BB may be seeing relatively little dust deposition during this otherwise high-dust-abundance time period. Because it would lie significantly ($\sim 7°$) north of the ITCZ at this time, it is possible that dust is inefficiently scavenged from the air column above the site. Alternatively, it could be that core 37BB experienced a significant increase in dust flux during HS11, an increase that has been subsequently obscured due to relatively lower accumulation rates and the effects of bioturbation (Supplementary Note 1).

Dust levels remain above the interglacial baseline in the southernmost core until $\sim$126 ka, which could indicate that the ITCZ was in its southernmost position until that time. However, we suggest that it is more likely that the ITCZ actually began to shift northwards around 129 ka when dust fluxes dropped below pre-stadial values, coincident with interglacial warming with stronger warming in the Northern Hemisphere. Our data display a slightly longer duration of HS11, probably due to the smoothing effects of bioturbation. This would cause HS11 to appear longer in our dust records than reconstructed from other proxies such as speleothems[27], which have better age control and higher resolution but do not directly record dust fluxes or shifts in the ITCZ.

During MIS5e, all three cores record a decrease in dust flux to minimum values between 123 and 119 ka. On the basis of near-identical dust flux data for the two more northerly cores

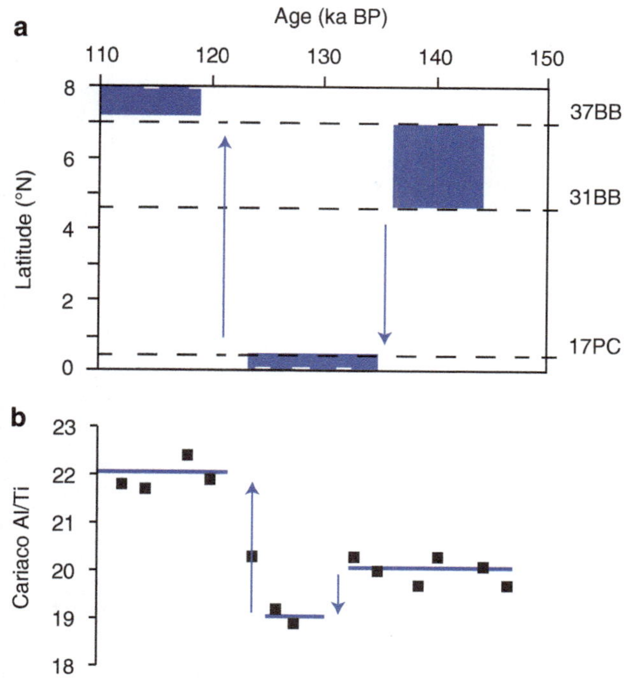

Figure 3 | Termination II ITCZ shifts. Blue boxes indicate range of possible latitudes of the mean ITCZ position, black dashed lines denote core latitudes (**a**). Cariaco Basin Al/Ti record (black squares)[28] with blue bars indicating estimates of the relative ITCZ position through time (**b**).

37BB and 31BB, we suggest that at MIS5e the ITCZ was northwards of 7° N and that it was not influencing any of our sites during the peak interglacial. Planktonic oxygen isotope values which are indistinguishable at the three sites during MIS5e, are consistent with this argument, as an ITCZ located north of all three cores would be unlikely to differentially influence surface hydrography or temperature at any of the three sites. It is notable that after the ITCZ retreats northward during MIS5e that the $\delta^{18}O$ at site 17PC does not show a 0.8‰ enrichment relative to cores 37BB and 31BB, as it did during the peak of MIS6. As suggested previously by the work on a larger set of ML1208 $\delta^{18}O$ records[19], this could be due to differences in the salinity of water carried by the North Equatorial Counter Current during glacial versus interglacial times. If the North Equatorial Counter Current carried fresher water from the western Pacific warm pool during MIS6 than during MIS5, a larger $\delta^{18}O$ gradient would be predicted between equatorial and more northerly cores.

Our interpretation of an ITCZ located north of 7° N during MIS5e is consistent with a strong hemispheric thermal gradient, with the North Atlantic experiencing peak warmth at this time. At the end of MIS5e (~119 ka) interglacial temperatures subsided in both hemispheres and dust fluxes began to rise, but the differential rate of cooling appears to have been small enough that the ITCZ was not displaced southward of 37BB, since 37BB and 31BB show comparable dust fluxes to the end of our record at 110 ka.

Only one other record of ITCZ movement exists for the penultimate deglaciation, a record of variations in the Al/Ti ratio from the Cariaco Basin[28]. The Cariaco Basin records ITCZ-driven changes in terrestrial runoff with higher (lower) ratios indicating more (less) runoff and therefore a relatively more northerly (southerly) ITCZ (Fig. 3b). Given that the Cariaco record is interpreted as recording relative movement of the terrestrial ITCZ, the agreement between our data sets is noteworthy and is consistent with the idea that ITCZ movement is likely coherent across large distances[2] even on short timescales.

Model reconstructions of past ITCZ migrations have been challenging in part because of the sensitivity of the ITCZ to small changes in cloud feedbacks and energy balances[2]. Here we have shown that palaeo reconstructions can constrain the magnitude of even abrupt ITCZ movement, demonstrating the potential to relate ITCZ changes to the magnitude of thermal forcing and to investigate thermal and hydrological components of other climate change events. Our results provide evidence that an abrupt, millennial climate oscillation was a prominent feature of the penultimate deglaciation, both at high and low latitudes. During HS11, in concert with North Atlantic cooling[29], ice rafting[25], increases in northern hemisphere dust abundance, and a shift in the interhemispheric thermal gradient[29,30], the ITCZ migrated southwards by at least 4° relative to MIS6, and 6.5° south of its present latitude. This shift likely had significant consequences for global atmospheric circulation[10] and could thus have contributed to tipping the climate system over the threshold for deglaciation.

Methods

Age model development and uncertainty. Age–depth relationships for cores ML1208-37BB, 31BB and 17PC were developed on the basis of planktonic oxygen isotope stratigraphies[19] tied to the LR04 benthic stack[31], with the assistance of a modified Monte-Carlo-enabled cross-correlation maximization scheme and random walk algorithm (MonteXCM)[32]. A record of coeval planktonic and benthic oxygen isotopes values was previously reconstructed in the Eastern Equatorial Pacific[33], which showed zero offset between the timing of deglaciation in the two stratigraphies. This observation gives us confidence that our use of planktonic oxygen isotope stratigraphies is not introducing a significant systematic bias into the timing of our events relative to data from the ice cores.

The use of the random walk algorithm in MonteXCM allows us to quantify age uncertainty at each data point. On average, over the interval studied, 2 s.d. uncertainties are 1.5, 1.5 and 1.7 ka for cores 37BB, 31BB and 17PC, respectively. Specifically, for the HS11 interval (129–136 ka), uncertainties are ~1.3, 1.4 and 1.4 ka for the same three cores. The magnitude of these uncertainties means that it is probable that the small offsets in event timing observed between our three sediment cores are artefacts of the age model and do not represent true differences in the timing of the reconstructed dust flux changes or ITCZ shifts. However, our age constraints are sufficient to indicate that it is unlikely that these events are temporally disconnected, and a lag between the onset of HS11 and our inferred initiation of tropical changes is not inconsistent with the interpretations we present.

Dust flux reconstruction. We reconstruct dust fluxes through the quantification of $^{230}Th_{xs,0}$-normalized ^{232}Th fluxes in marine sediments, which act as a proxy for dust flux[13,20]. Terrestrial sediments contain ^{232}Th at a mean concentration of 10.7 p.p.m., which varies 1 p.p.m. or less over a range of dust provenance[13]. A more recent summary of ^{232}Th concentrations in dust source areas found evidence that the fine fraction (<5 μm) of dust samples from a range of provenance had ^{232}Th concentrations closer to 14 ± 1 p.p.m. (ref. 34). We note that changing the ^{232}Th concentration used to reconstruct dust fluxes would decrease the absolute magnitude of the dust fluxes reconstructed at our study sites but does not alter our interpretations. As our core locations are a great distance from continental sources of lithogenic material, we expect ^{232}Th at our sites to be delivered predominately by aeolian processes. Indeed, grain size analyses show that the average size of lithogenic sediments at ~1° N 131° W is 3.1 μm (ref. 35), unlikely to have been derived from hemipelagic sedimentation. Here, as in other studies[13,36], dust fluxes based on ^{232}Th are evaluated by normalization to $^{230}Th_{xs,0}$ (ref. 37).

Uranium and thorium geochemistry. Sediment samples from ML1208-37BB, 31BB and 17PC were analysed for uranium and thorium isotopes by isotope dilution and inductively coupled plasma mass spectrometry (Element XR, ICP-MS) at the Lamont–Doherty Earth Observatory of Columbia University. Between 100 and 200 mg of sediment per sample was spiked with ^{236}U and ^{229}Th before sediment dissolution and digestion with HNO_3, $HClO_4$ and HF (ref. 38). Anion-exchange column chemistry was also used to isolate the U/Th fraction[39].

Samples were run with an external in-house standard (mirroring the physical and geochemical properties of our samples) to determine measurement reproducibility, yielding relative s.d.s of 0.74% for ^{238}U, 1.94% for ^{230}Th and 0.87% for ^{232}Th. Background contamination was evaluated in each run via blanks, which were spiked and digested, but contained no sediment. Results show ^{238}U blanks (0.4 ng), ^{230}Th blanks (0.009 pg) and ^{232}Th blanks (1.1 ng) all of which are <1% of even the lowest sample values. Errors reported in Fig. 2 represent 2 s.d. uncertainty estimates unique to each sample, including the propagated error due to uncertainty in counting statistics, mass bias corrections (from measurements of a natural

$^{238}U/^{235}U$ standard), counting gain corrections, spike measurements, the lithogenic fraction $(0.7 \pm 0.1)^{40}$ and a 1% age uncertainty. The average s.d. of replicate dust flux values is ~6.4% of sample values. Samples for which dust flux values differed from a same-depth replicate, or an average of four nearest-neighbour data points by more than 4 s.d., have been excluded from the data set. The number of data points excluded using this method is <5% of the total number of data points reported.

Data archiving. Line Islands dust flux data are archived at the National Oceanic and Atmospheric Administration National Centers for Environmental Information (NCEI) database and are also available as a supplement to this manuscript (Supplementary Data 1).

References

1. McGee, D., Donohoe, A., Marshall, J. H. & Ferreira, D. Changes in ITCZ location and cross-equatorial heat transport at the Last Glacial Maximum, Heinrich Stadial 1, and the mid-Holocene. *Earth Planet. Sci. Lett.* **390,** 69–79 (2014).
2. Schneider, T., Bischoff, T. & Haug, G. H. Migrations and dynamics of the intertropical convergence zone. *Nature* **513,** 45–53 (2014).
3. Donohoe, A., Marshall, J. H., Ferreira, D. & McGee, D. The relationship between ITCZ location and cross-equatorial atmospheric heat transport: from the seasonal cycle to the last glacial maximum. *J. Clim.* **26,** 3597–3618 (2013).
4. Chiang, J. C. H. & Bitz, C. M. Influence of high latitude ice cover on the marine Intertropical Convergence Zone. *Clim. Dynam.* **25,** 477–496 (2005).
5. Arbuszewski, J. A., deMenocal, P. B., Cléroux, C., Bradtmiller, L. & Mix, A. Meridional shifts of the Atlantic intertropical convergence zone since the Last Glacial Maximum. *Nat. Geosci.* **6,** 1–4 (2013).
6. Haug, G. H., Hughen, K. A., Sigman, D. M. & Peterson, L. C. Southward migration of the Intertropical Convergence Zone through the Holocene. *Science* **293,** 1304–1308 (2001).
7. Wang, X. *et al.* Millennial-scale precipitation changes in southern Brazil over the past 90,000 years. *Geophys. Res. Lett.* **34,** L23701 (2007).
8. Seager, R., Kushnir, Y., Herweijer, C., Naik, N. & Velez, J. Modeling of tropical forcing of persistent droughts and pluvials over Western North America: 1856-2000*. *J. Clim.* **18,** 4065–4088 (2005).
9. Stager, J. C., Ryves, D. B., Chase, B. M. & Pausata, F. S. R. Catastrophic drought in the Afro-Asian monsoon region during Heinrich Event 1. *Science* **331,** 1299–1302 (2011).
10. Ceppi, P., Hwang, Y. T., Liu, X., Frierson, D. M. & Hartmann, D. L. The relationship between the ITCZ and the Southern Hemispheric eddy driven jet. *J. Geophys. Res. Atmos.* **118,** 5136–5146 (2013).
11. Lee, S.-Y., Chiang, J. C. H., Matsumoto, K. & Tokos, K. S. Southern Ocean wind response to North Atlantic cooling and the rise in atmospheric CO_2: modeling perspective and paleoceanographic implications. *Paleoceanography* **26,** PA1214 (2011).
12. Anderson, R. F. *et al.* Wind-driven upwelling in the Southern Ocean and the deglacial rise in atmospheric CO2. *Science* **323,** 1443–1448 (2009).
13. McGee, D., Marcantonio, F. & Lynch-Stieglitz, J. Deglacial changes in dust flux in the eastern equatorial Pacific. *Earth Planet. Sci. Lett.* **257,** 215–230 (2007).
14. Xie, R. C. & Marcantonio, F. Deglacial dust provenance changes in the Eastern Equatorial Pacific and implications for ITCZ movement. *Earth Planet. Sci. Lett.* **317-318,** 386–395 (2012).
15. Pichat, S., Abouchami, W. & Galer, S. J. G. Lead isotopes in the Eastern Equatorial Pacific record Quaternary migration of the South Westerlies. *Earth Planet. Sci. Lett.* **388,** 293–305 (2014).
16. Husar, R. B., Prospero, J. M. & Stowe, L. L. Characterization of tropospheric aerosols over the oceans with the NOAA advanced very high resolution radiometer optical thickness operational product. *J. Geophys. Res.* **102,** 16889–16909 (1997).
17. Lambert, F. *et al.* Dust-climate couplings over the past 800,000 years from the EPICA Dome C ice core. *Nature* **452,** 616–619 (2008).
18. Adler, R. F. *et al.* The Version-2 Global Precipitation Climatology Project (GPCP) monthly precipitation analysis (1979–present). *J. Hydrometeorol.* **4,** 1147–1167 (2015).
19. Lynch-Stieglitz, J. *et al.* Glacial-interglacial changes in central tropical Pacific surface seawater property gradients. *Paleoceanography* **30,** 423–438 (2015).
20. Winckler, G., Anderson, R. F., Fleisher, M. Q., McGee, D. & Mahowald, N. Covariant glacial-interglacial dust fluxes in the Equatorial Pacific and Antarctica. *Science* **320,** 93–96 (2008).
21. Martinez-Garcia, A. *et al.* Iron fertilization of the Subantarctic Ocean during the last Ice Age. *Science* **343,** 1347–1350 (2014).
22. Ruth, U. *et al.* Ice core evidence for a very tight link between North Atlantic and east Asian glacial climate. *Geophys. Res. Lett.* **34,** L03706 (2007).
23. McGee, D., Broecker, W. & Winckler, G. Gustiness: the driver of glacial dustiness? *Quatern. Sci. Rev.* **29,** 2340–2350 (2010).
24. Marino, G. *et al.* Bipolar seesaw control on last interglacial sea level. *Nat. Geosci.* **522,** 197–201 (2015).
25. McManus, J. F., Oppo, D. W. & Cullen, J. L. A 0.5-million-year record of millennial-scale climate variability in the North Atlantic. *Science* **283,** 971–975 (1999).
26. Koutavas, A. & Lynch-Stieglitz, J. Variability of the Marine ITCZ over the Eastern Pacific during the past 30,000 years: regional perspective and global context. *Adv. Glob. Change Res.* **21,** 347–369 (2004).
27. Cheng, H. *et al.* Ice Age terminations. *Science* **326,** 248–252 (2009).
28. Yarincik, K. M., Murray, R. W. & Peterson, L. C. Climatically sensitive eolian and hemipelagic deposition in the Cariaco Basin, Venezuela, over the past 578,000 years: Results from Al/Ti and K/Al. *Paleoceanography* **15,** 210–228 (2000).
29. Oppo, D. W., McManus, J. F. & Cullen, J. L. Evolution and demise of the Last Interglacial warmth in the subpolar North Atlantic. *Quatern. Sci. Rev.* **25,** 3268–3277 (2006).
30. Parrenin, F. The EDC3 chronology for the EPICA Dome C ice core. *Clim. Past* **3,** 485–497 (2007).
31. Lisiecki, L. E. & Raymo, M. E. A Pliocene-Pleistocene stack of 57 globally distributed benthic $\delta^{18}O$ records. *Paleoceanography* **20,** PA1003 (2005).
32. Hoffman, J. S., Clark, P. U., Pisias, N. G., Marcott, S. A. & Shakun, J. D. Estimating age model uncertainties for the last interglaciation. Fall Meeting, American Geophysical Union, abstract #PP41D-1432 (2015).
33. Lea, D. W. *et al.* Paleoclimate history of Galápagos surface waters over the last 135,000 yr. *Quatern. Sci. Rev.* **25,** 1152–1167 (2006).
34. McGee, D. *et al.* Tracking eolian dust with helium and thorium: impacts of grain size and provenance. *Geochim. Cosmochim. Acta* **175,** 47–67 (2016).
35. Rea, D. K. & Hovan, S. A. Grain size distribution and depositional processes of the mineral component of abyssal sediments: Lessons from the North Pacific. *Paleoceanography* **10,** 251–258 (2010).
36. Anderson, R. F., Fleisher, M. Q. & Lao, Y. Glacial–interglacial variability in the delivery of dust to the central equatorial Pacific Ocean. *Earth Planet. Sci. Lett.* **242,** 406–414 (2006).
37. Francois, R., Frank, M., Rutgers van der Loeff, M. M. & Bacon, M. P. ^{230}Th normalization: an essential tool for interpreting sedimentary fluxes during the late Quaternary. *Paleoceanography* **19,** PA1018 (2004).
38. Fleisher, M. Q. & Anderson, R. F. Assessing the collection efficiency of Ross Sea sediment traps using ^{230}Th and ^{231}Pa. *Deep Sea Res. II* **50,** 693–712 (2003).
39. Kraus, K. A., Moore, G. E. & Nelson, F. Anion-exchange Studies. XXI. Th (IV) and U (IV) in Hydrochloric Acid. Separation of Thorium, Protactinium and Uranium1, 2. *J. Am. Chem. Soc.* **78,** 2692–2695 (1956).
40. Bradtmiller, L. I., Anderson, R. F., Sachs, J. P. & Fleisher, M. Q. A deeper respired carbon pool in the glacial equatorial Pacific Ocean. *Earth Planet. Sci. Lett.* **299,** 417–425 (2010).
41. Veres, D., Bazin, L. & Landais, A. The Antarctic ice core chronology (AICC2012): an optimized multi-parameter and multi-site dating approach for the last 120 thousand years. *Clim. Past* **9,** 1733–1748 (2013).
42. Mokeddem, Z., McManus, J. F. & Oppo, D. W. Oceanographic dynamics and the end of the last interglacial in the subpolar North Atlantic. *Proc. Natl Acad. Sci. USA* **111,** 11263–11268 (2014).
43. Pailler, D. & Bard, E. High frequency palaeoceanographic changes during the past 140 000 yr recorded by the organic matter in sediments of the Iberian Margin. *Palaeogeogr. Palaeoclimatol. Palaeoecol.* **181,** 431–452 (2002).
44. Jouzel, J. *et al.* Orbital and millennial Antarctic climate variability over the past 800,000 years. *Science* **317,** 793–796 (2007).

Acknowledgements

We thank the core repository at Lamont–Doherty Earth Observatory for providing samples, Marty Fleisher for assistance in the lab and R. Jacobel for helpful discussions. Special thanks to Jeremy Hoffman whose generous assistance with the MonteXCM algorithm improved ML1208 age models and permitted robust quantification of chronostratigraphic uncertainty. We also wish to acknowledge the contributions of Jean Lynch-Stieglitz, Pratigya Polissar, and the ML1208 science party, including the planning for and collection of the sediment cores used in this study. The manuscript was improved through the thoughtful comments of Sylvain Pichat and two anonymous reviewers. This research was funded in part by an award from the US National Science Foundation AGS-1502889 (J.F.M. and G.W.), and benefited from previous support from NSF OCE-1003374, OCE-1159053, and OCE-1158886, and the Comer Science and Education Foundation.

Author contributions

A.W.J. collected all data, interpreted results and prepared the manuscript and figures; J.F.M. designed and supervised the project, and aided in interpretation and editing of the manuscript; R.F.A. and G.W. discussed interpretation of results, and contributed to the editing of the manuscript.

Additional information

Competing financial interests: The authors declare no competing financial interests.

15

North Atlantic Deep Water Production during the Last Glacial Maximum

Jacob N.W. Howe[1], Alexander M. Piotrowski[1], Taryn L. Noble[1,2], Stefan Mulitza[3], Cristiano M. Chiessi[4] & Germain Bayon[5]

Changes in deep ocean ventilation are commonly invoked as the primary cause of lower glacial atmospheric CO_2. The water mass structure of the glacial deep Atlantic Ocean and the mechanism by which it may have sequestered carbon remain elusive. Here we present neodymium isotope measurements from cores throughout the Atlantic that reveal glacial-interglacial changes in water mass distributions. These results demonstrate the sustained production of North Atlantic Deep Water under glacial conditions, indicating that southern-sourced waters were not as spatially extensive during the Last Glacial Maximum as previously believed. We demonstrate that the depleted glacial $\delta^{13}C$ values in the deep Atlantic Ocean cannot be explained solely by water mass source changes. A greater amount of respired carbon, therefore, must have been stored in the abyssal Atlantic during the Last Glacial Maximum. We infer that this was achieved by a sluggish deep overturning cell, comprised of well-mixed northern- and southern-sourced waters.

[1] Department of Earth Sciences, University of Cambridge, Cambridge CB2 3EQ, UK. [2] Institute for Marine and Antarctic Studies (IMAS), University of Tasmania, Hobart, Tasmania 7001, Australia. [3] MARUM-Center for Marine Environmental Sciences, University of Bremen, Leobener Strasse, D-28359 Bremen, Germany. [4] School of Arts, Sciences and Humanities, University of São Paulo, Av. Arlindo Bettio 1000, CEP03828-000 São Paulo SP, Brazil. [5] Institut Français de Recherche pour l'Exploitation de la Mer (IFREMER), Unité de Recherche Géosciences Marines, F-29280 Plouzané, France. Correspondence and requests for materials should be addressed to J.N.W.H. (email: jacob.howe@cantab.net).

Changes in Atlantic Meridional Overturning Circulation are important in controlling glacial–interglacial climatic shifts due to their role in regulating heat transport in the surface ocean and carbon storage in the deep ocean[1,2]. Nutrient-based proxy reconstructions suggest that overturning circulation in the Atlantic during the Last Glacial Maximum (LGM) was different to that of the modern ocean[3–5]. These nutrient proxy reconstructions are often interpreted as indicating that during the LGM North Atlantic Deep Water (NADW) shoaled to form Glacial North Atlantic Intermediate Water (GNAIW) and was completely replaced by southern-sourced water in the deep Atlantic (>2.5 km)[6,7]. However, elucidation of the water mass distribution in such reconstructions is complicated by the fact that the nutrient content of deep waters may vary independently of their source and dynamics[8]; meanwhile, modelling studies of glacial Atlantic overturning have produced conflicting results[9].

In the modern ocean Antarctic Bottom Water (AABW), which is formed in the high latitude Southern Ocean, represents a major inefficiency in the biological pump—the name given to the biologically mediated processes which sequester carbon in the deep ocean. In this region, deep waters are upwelled to the surface, allowing dissolved carbon to be outgassed to the atmosphere, but these waters sink again before their nutrient load can be fully consumed by marine organisms, thus they are said to have a high-preformed nutrient concentration[10]. The formation of NADW and AABW, and their resultant proportions in the deep ocean thus controls the preformed nutrient concentration of the deep ocean, which, in part, controls atmospheric pCO_2 (ref. 11). Ice core records show that atmospheric CO_2 concentrations rose by ~90 p.p.m. between the LGM and the Holocene[12]. However, considered in isolation, the replacement of high-preformed nutrient AABW with low-preformed nutrient NADW, which has been suggested to have occurred across the last deglaciation, would have increased the efficiency of the biological pump and thereby decreased atmospheric CO_2 (refs 11,13). If our current understanding of glacial–interglacial circulation changes is correct then other parts of the carbon cycle must have counteracted the effects of switching from the GNAIW to NADW mode of deep water formation across the deglaciation. Alternatively, nutrient proxy reconstructions of the Atlantic may be recording carbon cycle changes rather than differences in water mass sourcing.

Neodymium isotopes are a water mass tracer independent of biological processes, allowing deconvolution of changes in water mass sourcing and nutrient regeneration[14]. Modern NADW has a characteristic ε_{Nd} value (normalized $^{143}Nd/^{144}Nd$ ratio in parts per ten thousand) of −13.5 (ref. 15) while seawater in the deep Southern Ocean shows ε_{Nd} values around −8.5 (ref. 16), because it has a greater proportion of Pacific Deep Water (PDW) which has an ε_{Nd} of −3.5 (ref. 17). Away from continental margins water masses throughout the deep ocean reflect the quasi-conservative mixing of these deep water masses[14]. Planktic foraminifera from sea-floor sediment cores have been shown to successfully preserve bottom water ε_{Nd}, thereby offering an archive of past seawater ε_{Nd} (ref. 18).

Here we present Holocene and LGM seawater ε_{Nd} reconstructions, based on foraminiferal ε_{Nd} measurements from 24 cores (Supplementary Table 1, Supplementary Fig. 1) spanning from 46° S to 40° N in the Atlantic Ocean. These reconstructions are used to decipher the nature of changes in Atlantic water mass distributions between glacial and interglacial conditions. Although the LGM reconstruction shows more radiogenic neodymium isotope values than the Holocene profile, the observation of less radiogenic values in the deep North Atlantic than the deep South Atlantic indicates that NADW was produced under glacial conditions. When compared with benthic foraminiferal carbon isotope values, the neodymium isotope measurements reveal that more respired organic carbon was stored in the deep Atlantic Ocean during the LGM than in the Holocene.

Results

Core top foraminiferal ε_{Nd} versus seawater ε_{Nd}. Figure 1 cross plots the Holocene foraminiferal ε_{Nd} values (Supplementary Tables 2 and 3) against the nearest available published seawater ε_{Nd} measurements. Data points were only included in the cross plot if the seawater measurements were within both 10° of latitude and longitude and within 500 m depth of the core site. Most of the data points on the cross plot are within error of the 1:1 line, indicating that foraminifera are faithfully preserving seawater ε_{Nd}. Despite the distance criteria outlined above being applied, the three data points which are outside of error of the 1:1 line are likely due to the seawater data not being of sufficiently close proximity to represent the water mass bathing the core sites. This effect is likely to be particularly prominent in site locations that are near water mass boundaries.

Holocene Atlantic ε_{Nd} reconstruction. We gridded our Holocene foraminiferal ε_{Nd} measurements (Supplementary Table 2) with suitable published data (Fig. 2c). The published data includes results from measurements made on foraminifera, fish debris, high resolution crusts and leachates (Supplementary Tables 3 and 4). As some measurements made on leachates have been shown to be susceptible to contamination by the detrital fraction[19,20], results from sites where the core top leachate ε_{Nd} values deviated significantly from nearby seawater values were excluded from the reconstruction. The data points included from the crusts BM1969.05 and TR079 D-14 (Supplementary Table 4) were dated outside of the definition of the LGM used for selecting sediment core data (23–18 ka); however, the records from these crusts were used to argue for the stability of seawater ε_{Nd} across glacial–interglacial cycles in their respective locations[21]. The slightly older glacial data points from these crusts were therefore deemed appropriate for inclusion in the LGM time slice.

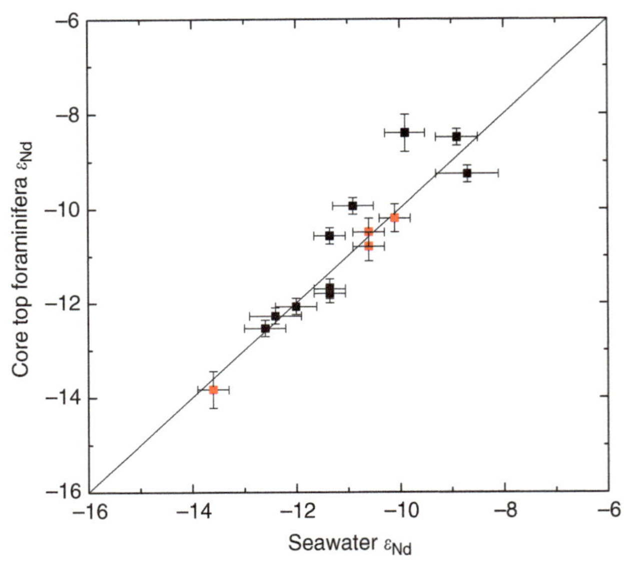

Figure 1 | Core top foraminiferal ε_{Nd} versus seawater ε_{Nd}. Modern open Atlantic seawater[15,16,55,56] ε_{Nd} cross plotted against Holocene Atlantic ε_{Nd} measurements made on uncleaned planktic foraminifera in this work (black squares; Supplementary Table 2) together with previously reported foraminiferal values used in this work (red squares; Supplementary Table 3)[18,56]. Error bars are 2σ external errors. A 1:1 line is also plotted.

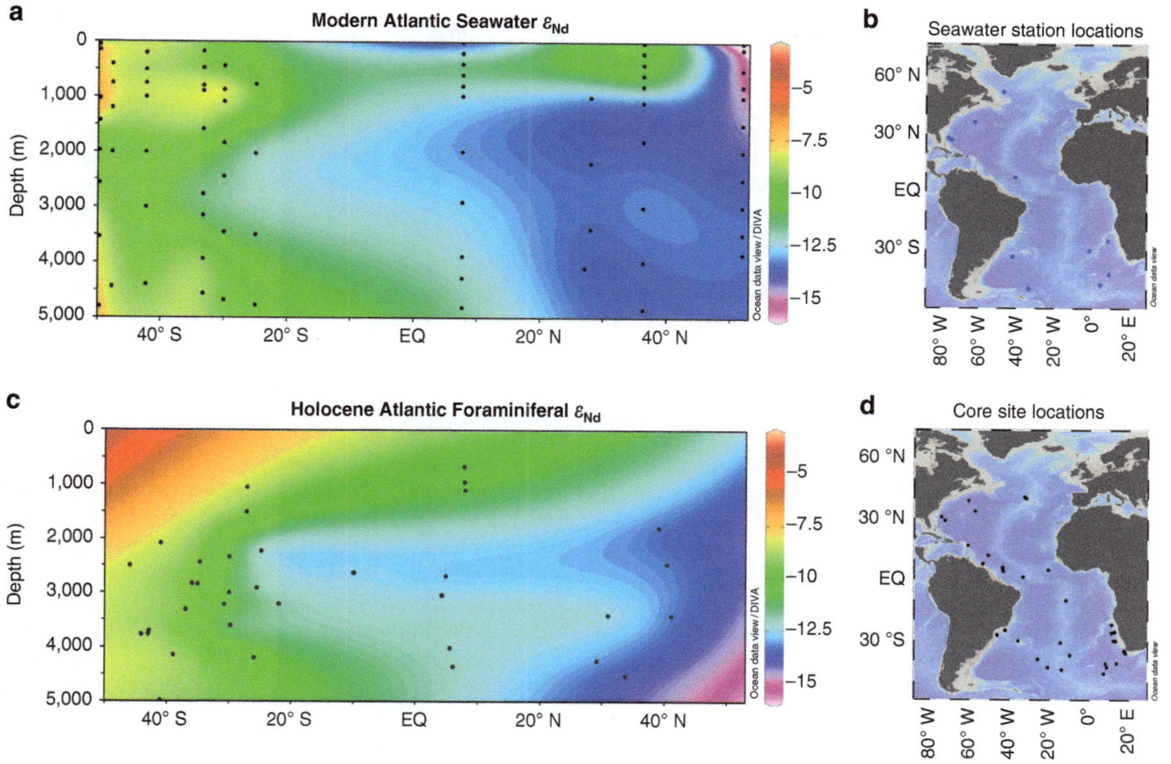

Figure 2 | Atlantic seawater ε_{Nd} and Holocene ε_{Nd} reconstruction. (**a**) Modern seawater ε_{Nd} profile from 50° S to 53° N for the western Atlantic and the eastern Atlantic Ocean south of the Walvis Ridge with data points indicated by large black dots[15,16,55]. (**b**) Map showing location of seawater ε_{Nd} profiles. (**c**) Reconstruction of Holocene Atlantic ε_{Nd} with data points indicated by large black dots (data is listed in Supplementary Tables 2,3 and 4). (**d**) Map showing location of sites used in the Holocene and LGM reconstructions. Figure created using Ocean Data View software[22].

DIVA gridding was performed using Ocean Data View[22] with the 'signal-to-noise ratio' variable set to a value of 25. This variable determines how much a single data point is able to influence the overall plot and this level was chosen to reduce the influence of single data points, as was done by Curry and Oppo[4].

The Holocene reconstruction shows the most unradiogenic ε_{Nd} values, around -13, below 1,500 m at the northernmost extent of the plot (Fig. 2c). These values become more radiogenic to the south, with the greatest propagation of this unradiogenic signal centred at 2,000–4,000 m water depth. This is surrounded by areas with more radiogenic ε_{Nd} values, from -8 to -10, at all depths south of 25° S; below 4,000 m from 25° S to the equator; and above 1,500 m at all latitudes. Although the data set is not as extensive as that used in $\delta^{13}C$ reconstructions[4], this reconstruction replicates the salient features of the seawater ε_{Nd} (Fig. 2a). The only significant mismatches are above 1,000 m at all latitudes and below 4,000 m north of 40° N; in these regions the gridding procedure extrapolates from the nearest data point to the edge of the profile due to a lack of data based constraints. These areas should therefore be interpreted with caution and for this reason the depth range of 0–1,000 m is excluded from later reconstructions.

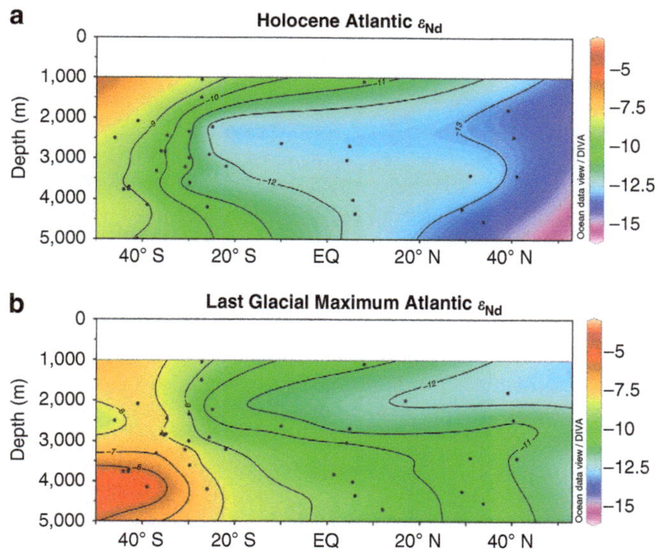

Figure 3 | Holocene and LGM Atlantic ε_{Nd} reconstructions. (**a**) Holocene and (**b**) LGM (23–18 ka) reconstructions of ε_{Nd} of the Atlantic Ocean from 50° S to 53° N measured on uncleaned foraminifera in this study and published authigenic data (Supplementary Tables 2,3 and 4)[18,21,25,26,32,56-59]. Core sites are given by the black dots; locations are plotted in Fig. 2d and Supplementary Fig. 1. Depths from 0 to 1,000 m have been left blank due to a lack of data. Figure created using Ocean Data View software[22].

Last Glacial Maximum Atlantic ε_{Nd} reconstruction. The good correlation between seawater ε_{Nd} and the salinity profile of the modern Atlantic[23] demonstrates that the modern seawater ε_{Nd} profile is the result of mixing between northern- and southern-sourced water masses. Collectively, these observations provide confidence that this selection of cores can be used to constrain past Atlantic water mass distributions below 1,000 m. The LGM ε_{Nd} reconstruction (Fig. 3b) shows the most unradiogenic values around -12.5 between 1,500 and 2,000 m extending from the North Atlantic to ~10° N which then transition to values around -8 at 40° S. In contrast, the deep North Atlantic is occupied by a homogeneous water mass with an ε_{Nd} around -10.5; ε_{Nd} values become more radiogenic to the south, reaching -5.5 at 45° S.

The largest differences between the LGM and modern Atlantic profiles occur below 2,500 m (Fig. 3); the deep North Atlantic shifts from ε_{Nd} values around -10.5 in the glacial to -13.5 in the modern ocean, whilst the deep South Atlantic shifts from -5.5 to -8.5. This coherent spatial pattern of changes in ε_{Nd} values in the Atlantic between the LGM and the Holocene (Fig. 3b) is consistent with a change in water mass advection and indicates there was a greater influence of southern-sourced waters in the deep Atlantic under glacial conditions.

Discussion

The change observed in the deep South Atlantic between the LGM and Holocene (Fig. 3) is consistent with a lower flux of NADW into the Southern Ocean under cooler, glacial conditions[24,25]. This lower flux of NADW resulted in the deep Southern Ocean being filled with a greater proportion of water of Pacific and Indian Ocean origin that was more radiogenic in composition than Atlantic-sourced deep water[26–28]. The deep South Atlantic, however, remained less radiogenic than the deep Pacific[26,28], requiring a source of unradiogenic neodymium to the deep South Atlantic under glacial conditions. Furthermore, the deep North Atlantic was less radiogenic than the deep South Atlantic during the LGM (Fig. 3b); this implies that northern-sourced water with its unradiogenic ε_{Nd} must have been exported to abyssal depths under glacial climate conditions. The mixing of GNAIW with denser southern-sourced waters is deemed an unlikely explanation given the large amount of energy required to mix these intermediate waters down to abyssal depths in the North Atlantic. Rather, these observations suggest that two glacial North Atlantic-sourced water masses existed during the LGM: GNAIW at depths above 2,500 m and denser Glacial NADW (GNADW)[29] below 2,500 m.

The inference of a lower flux of GNADW under glacial conditions[24] than NADW in the Holocene resulting in a more radiogenic composition of Glacial AABW (GAABW) than AABW would cause the deep Atlantic to exhibit more radiogenic ε_{Nd} values without a change in the water mass mixing proportions in the deep Atlantic. By taking this end-member change of deep southern-sourced water ε_{Nd} into account, however, our results can be used to calculate the proportion of the Atlantic ventilated by NADW (%NADW) both in the Holocene and during the LGM (Supplementary Table 5), and thus elucidate changes in water mass mixing proportions from the effect of the change in the AABW end-member composition. In each case three end-members were required; namely NADW; AABW and Antarctic Intermediate Water, or their glacial counterparts (Supplementary Fig. 2). A binary mixing calculation was performed between a northern- and a southern-sourced water mass at each core site using equation (1). The choice of southern-sourced water varied with depth and latitude according to the observed boundary between Antarctic Intermediate Water and Lower Circumpolar Deep Water as defined by a salinity of 34.7 psu (ref. 30) and the gradient in neodymium concentration (but not ε_{Nd}) between them in the Atlantic sector of the modern Southern Ocean[16].

$$\%\mathrm{NADW} = \frac{[\mathrm{Nd}]_{\mathrm{SSW}}\left(\varepsilon_{\mathrm{Nd(NADW)}} - \varepsilon_{\mathrm{Nd(SSW)}}\right)}{[\mathrm{Nd}]_{\mathrm{NADW}}\varepsilon_{\mathrm{Nd(NADW)}} - [\mathrm{Nd}]_{\mathrm{SSW}}\varepsilon_{\mathrm{Nd(SSW)}} - \varepsilon_{\mathrm{Nd}}\varepsilon_{\mathrm{Nd(NADW)}} + \varepsilon_{\mathrm{Nd(SSW)}}\varepsilon_{\mathrm{Nd}}} \times 100 \quad (1)$$

The modern seawater ε_{Nd} and [Nd] end-member values were taken from published seawater data (Supplementary Fig. 2). For the glacial ocean, the ε_{Nd} and [Nd] of NADW and PDW were kept constant; support for the stability of the ε_{Nd} of both comes from crust data[31] whilst there is, at present, no proxy for past neodymium concentration available. The more radiogenic ε_{Nd} of GAABW was taken from a South Atlantic record of benthic foraminifera[32] and the concentration calculated by conservative mixing of NADW and PDW (ref. 16). Unlike records from the deep South Atlantic[26,32], the new intermediate depth glacial foraminiferal ε_{Nd} data from the South Atlantic measured in this work exhibit no changes between the LGM and the Holocene (GeoB2107-3 and GeoB2104-3; Supplementary Table 2), so the neodymium composition of Glacial Antarctic Intermediate Water was assumed to be the same as in the modern ocean. This observation clearly demonstrates that the intermediate and deep South Atlantic were isotopically distinct in terms of neodymium during the LGM, reinforcing the need for two distinct southern-sourced end-member water masses in the mixing calculations.

Figure 4 shows the %NADW values calculated for the Holocene and LGM Atlantic gridded into meridional sections using the same technique employed for the ε_{Nd} data profiles plotted in Fig. 2. In the LGM ε_{Nd} profile (Fig. 3b), the most radiogenic values, around -5.5, are observed at 4,000 m depth in the South Atlantic, whereas less radiogenic values, around -6.5, are seen at ~5,000 m at the same latitude. The more radiogenic values come from the Mid-Atlantic Ridge; the less radiogenic from the Cape Basin, which lies to the east of the Mid-Atlantic Ridge. The ε_{Nd} difference between these sites likely reflects a longitudinal mixing gradient between radiogenic, Pacific-derived, circumpolar waters[28] and less radiogenic North Atlantic derived waters[25]. A similar phenomenon is seen in modern seawater data from the South Atlantic although the offset is less pronounced[16]. The results from the deep Cape Basin[26] were, therefore, excluded from the mixing proportion plots in Fig. 4 as the longitudinal gradient in ε_{Nd} observed in the deep South Atlantic obscures the latitudinal mixing gradient which is the primary interest here.

The gridded profile of the Holocene %NADW results (Fig. 4a) shows the Atlantic north of 20° N and below 1,500 m depth is mostly (>90%) NADW. %NADW decreases at all depths to the

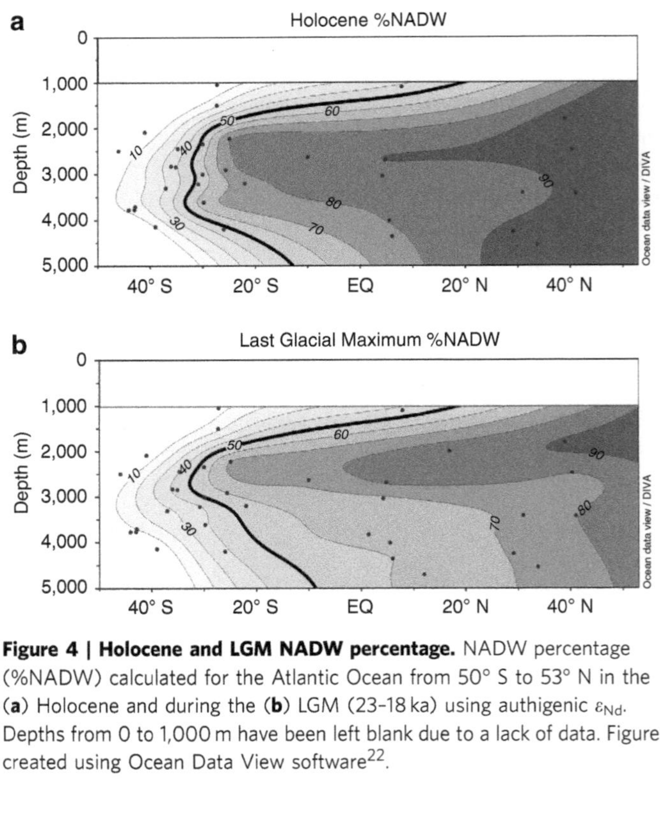

Figure 4 | Holocene and LGM NADW percentage. NADW percentage (%NADW) calculated for the Atlantic Ocean from 50° S to 53° N in the (**a**) Holocene and during the (**b**) LGM (23-18 ka) using authigenic ε_{Nd}. Depths from 0 to 1,000 m have been left blank due to a lack of data. Figure created using Ocean Data View software[22].

south, with the 50% mixing line occurring near 10° S at 5,000 m, but not until 30° S from 2,000 to 4,000 m. The LGM %NADW contours, including the 50% contour, (Fig. 4b) are similar to the Holocene profile above 2,500 m (Fig. 4a). The greatest difference between the two plots occurs below 2,500 m in the North Atlantic, which has significantly less NADW in the LGM profile (55–80% NADW) than in the Holocene profile (>90% NADW). From this observation it is clear that there was a greater proportion of southern-sourced waters in the deep Atlantic during the LGM than in the Holocene; however, it is also clear that the deep Atlantic was not occupied solely by southern-sourced waters during the LGM.

Although changes in the proportion of NADW in the deep Atlantic between glacial and interglacial conditions are the simplest explanation for the results in Figs 3 and 4, other alternatives must be considered. The sensitivity of the calculated %NADW for the glacial deep Bermuda Rise ε_{Nd} of -10.4 (OCE326-GGC6; 33.7° N, 57.6° W, 4,540 m)[18] to the neodymium composition of end-member water masses is shown in Fig. 5. The ε_{Nd} of AABW, considered to be known with reasonable confidence[32], was held constant. The ε_{Nd} of GNADW was varied from a near maximum possible value of -10.5 to a minimum of -16.5, similar to the least radiogenic value of -16 that was used in similar calculations performed for a nearby site[33]. As the %NADW result is dependent on the relative neodymium concentrations of the northern and southern end-members, the ratio of $[Nd]_{GNADW}$ to $[Nd]_{GAABW}$ was varied from 0.2 to 1. The latter value was used in similar calculations performed for a South Atlantic site[32] and is used as an upper limit in this work as ratios >1 would require more neodymium dissolved in the deep Atlantic than the deep Pacific, a situation that is not consistent with the observation that neodymium concentrations increase with water mass age in the modern ocean[34].

Using the glacial end-member values assigned in Fig. 4, the deep Bermuda Rise site was calculated to be bathed by 72% NADW during the LGM (black box, Fig. 5). In contrast an $\varepsilon_{Nd(GNADW)}$ of -10.5 and concentration ratio of 0.2 yields nearly 100% NADW, whilst an $\varepsilon_{Nd(GNADW)}$ of -16.5 and a concentration ratio of 1.0 means just 44% NADW is required to explain the ε_{Nd} observed at the Bermuda Rise during the LGM. Although the resultant uncertainty in the %NADW at the Bermuda Rise of ±28% is significant, all end-member configurations yield %NADW in the glacial deep North Atlantic values >44%. This is a robust finding that directly contradicts the notion that the deep Atlantic was dominated by southern-sourced waters during the LGM. Furthermore, the lower estimate of 44% is deemed unlikely to be realistic as less radiogenic $\varepsilon_{Nd(NADW)}$ values appear restricted to interstadials, and have been interpreted as pulses of Labrador Sea Water formation during these warm intervals[33] or unradiogenic weathering pulses during ice sheet retreat[35]. Rather, most evidence suggests that $\varepsilon_{Nd(GNADW)}$ values were similar to or slightly more radiogenic than NADW in the modern Atlantic[21] yielding higher %NADW values (Fig. 5). The influence of analytical error on the calculated %NADW values is small; using the external error bounds of 0.5 epsilon units[18] gives a range of 67–77% NADW at the deep Bermuda Rise during the LGM.

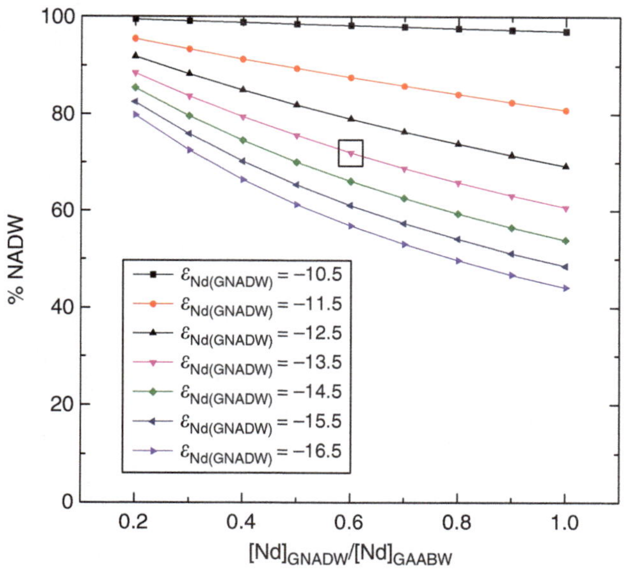

Figure 5 | Sensitivity test of NADW percentage calculation. Results of the sensitivity tests performed on the %NADW calculated for the Bermuda Rise ε_{Nd} of -10.4 during the LGM (OCE326-GGC6; 33.7° N, 57.6° W, 4,540 m)[18]. $\varepsilon_{Nd(GAABW)}$ was held constant at -5.5 (ref. 32), whilst $\varepsilon_{Nd(GNADW)}$ and the ratio of $[Nd]_{GNADW}$ to $[Nd]_{GAABW}$ were varied as shown. The result from end-members used in Fig. 4 ($\varepsilon_{Nd(GNADW)} = -13.5$ and $[Nd]_{GNADW}/[Nd]_{GAABW} = 0.60$) is highlighted by a black box.

Although these uncertainties do limit the certainty of the exact proportion of the deep Atlantic that was ventilated by northern-sourced waters during the LGM, the need for Glacial NADW appears inescapable. If localized processes, such as boundary exchange, were controlling the glacial ε_{Nd} profile, one would expect to see a heterogeneous profile controlled by regional detrital inputs. Furthermore, modelling studies have shown that boundary exchange processes are not able to explain the magnitude of the observed glacial–interglacial shifts in deep ocean ε_{Nd} and thus changes in water mass advection must be invoked to explain them[36]. Reconstructions of glacial deep water mass ventilation from radiocarbon[37], and carbonate ion concentration from B/Ca[38,39] also show a younger, better ventilated, water mass in the deep North Atlantic than the deep South Atlantic. These proxies, therefore, also provide evidence for a significant proportion of northern-sourced deep waters in the deep North Atlantic during the LGM, supporting the results shown in Fig. 4.

Reconciling the glacial ε_{Nd} values in the Atlantic with nutrient proxy based reconstructions requires a greater amount of respired organic carbon in the deep glacial Atlantic relative to the modern situation[4,8]. Circulation versus biological sources of carbon can be differentiated by cross plots of ε_{Nd} against benthic foraminiferal $\delta^{13}C$ for the Holocene and LGM Atlantic (Fig. 6). Deep water mass mixing end-members were ascribed to each plot as detailed in the Supplementary Information. The South Atlantic foraminiferal values (grey regions Fig. 6) were corrected for the Mackensen Effect, the phenomenon where benthic foraminifera display $\delta^{13}C$ values lower than that of the overlying deep water due to a phytodetrital layer of light organic carbon on the sea floor[40]. The composition of NADW is similar in the Holocene and LGM cross plots with $\delta^{13}C$ and ε_{Nd} values ~1.4‰ and -13.5, respectively, whereas glacial AABW $\delta^{13}C$ is more depleted (-0.5‰) and ε_{Nd} more radiogenic (-5.5) than its modern counterpart (0.4‰, -8.5).

The blue curves show the values expected for conservative mixing between these end-members for each time slice; those points falling along this mixing line have seen little input of biological carbon. During the LGM, data points lying close to the mixing line come from cores shallower than 3,000 m. Most of the glacial data points from cores located below 3,000 m sit well below the mixing curve which we interpret as being caused by the addition of respired organic matter with a low $\delta^{13}C$ (usually ~ -20‰)[41]. It is important to note that these cores represent locations within both northern- and southern- sourced glacial deep water, so this signal was not advected from either source but was acquired by biological processes along the advection pathway; this highlights the shortcomings of using $\delta^{13}C$ in isolation as a proxy for water mass mixing. The lower than expected $\delta^{13}C$ values could also in part be due to the use of mixed *Cibicidoides* species for some of the benthic foraminiferal $\delta^{13}C$ measurements. The variable depth habitats of different *Cibicidoides* species could lead to offsets of the measured foraminiferal calcite $\delta^{13}C$ from bottom water $\delta^{13}C$ (ref. 42). The inference of organic matter remineralization being the dominant cause of the offset, however, is supported by a recent study that modelled the glacial Atlantic and predicted a biological regenerative imprint in the glacial deep Atlantic $\delta^{13}C$ (ref. 8). The same study calculated that, after allowing for this remineralisation of organic matter, the glacial deep North Atlantic was ventilated by 50–80% NADW, which compares well with the ε_{Nd} derived values of 55–80% NADW produced in this work (Fig. 4b).

The accumulation of respired organic carbon in the deep Atlantic during the LGM could be due to higher surface productivity and thus export of organic carbon to the deep ocean under glacial climate conditions. Although glacial productivity reconstructions in the northwestern Atlantic are sparse, nearby regions show elevated export production during the LGM relative to the Holocene[43]. Higher glacial surface productivity, however, cannot explain the depth dependence of the $\delta^{13}C$ offset in the LGM (Fig. 6b), nor can it explain the chemocline seen at ~2,500 m in nutrient proxy reconstructions[3,4]. Thus it seems likely that the greater amount of respired organic matter in the deep Atlantic during the LGM must have also been at least partially due to a longer residence time of seawater in the glacial deep Atlantic than in the modern Atlantic. This conclusion is supported by a modelling study of a $^{231}Pa/^{230}Th$ data compilation which concluded that the glacial Atlantic had rapid overturning in the shallow cell but slower overturning at depth[44].

Our findings are summarized in hypothetical overturning schematics for the Atlantic Ocean during glacials and interglacials in Fig. 7. In the glacial scenario (Fig. 7b) there are two distinct northern-sourced water masses, separated by density differences. The glacial northern-sourced intermediate depth water mass was

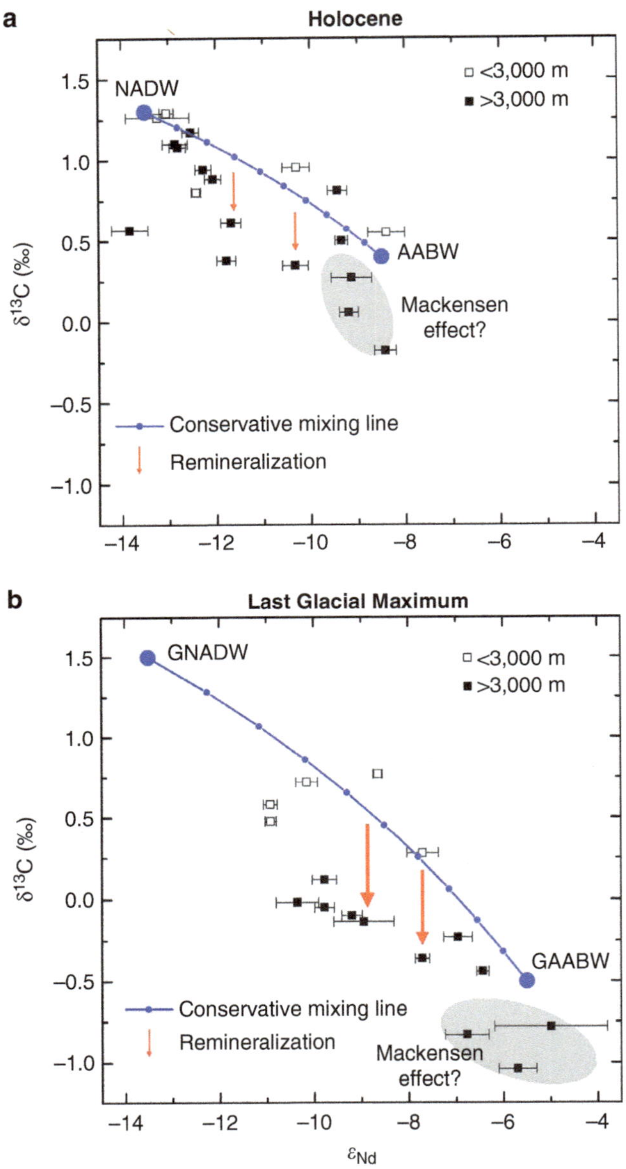

Figure 6 | Cross plots of Atlantic benthic foraminiferal $\delta^{13}C$ against foraminiferal ε_{Nd}. Cross plots of benthic foraminiferal $\delta^{13}C$ against foraminiferal ε_{Nd} for the deep Atlantic Ocean during the (**a**) Holocene and the (**b**) LGM. Water mass end-members labelled are NADW, AABW, GNADW and GAABW. The blue curve shows the values expected for conservative mixing between these water masses in the corresponding cross plot. Offsets from this line are attributed to either the remineralization of organic matter (red arrows) or a possible Mackensen Effect (shaded grey regions)[40]. Details of the data used are given in Supplementary Table 7, and how end-member values were assigned in Supplementary Note 1 and Supplementary Table 8. Error bars show the 2σ external error of the ε_{Nd} measurements.

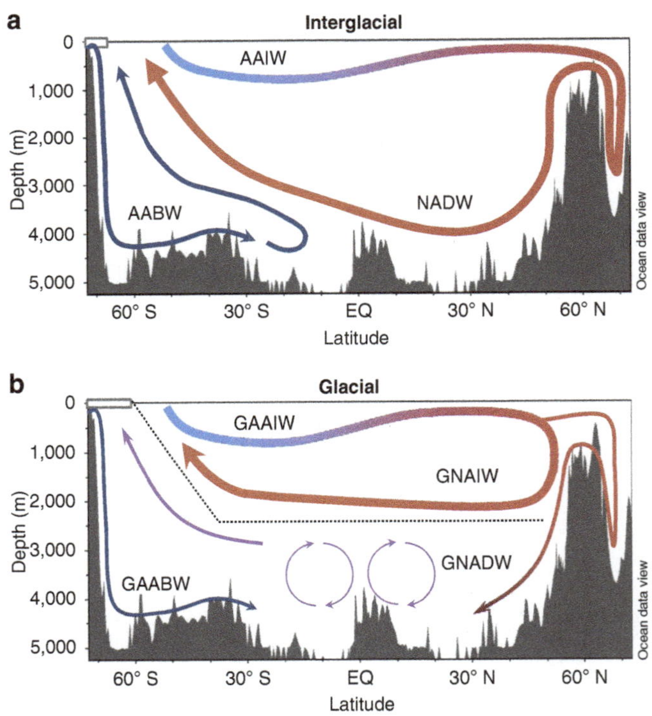

Figure 7 | Interglacial and Glacial Atlantic overturning schematics. Schematics of the structure of Atlantic Meridional Overturning Circulation during (**a**) interglacial and (**b**) glacial periods assuming the Holocene and LGM are typical of each. Water masses depicted are Antarctic Intermediate Water (AAIW), NADW and AABW and their glacial counterparts as well as GNAIW. Thin arrows associated with GNADW and GAABW in the glacial schematic represent the inferred lower flux of these water masses during glacial periods relative to interglacials. The dotted black line in the glacial schematic shows the location of the chemocline in nutrient proxy reconstructions at the interface between the two overturning cells. Figure partly created using Ocean Data View software[22].

likely formed by convection south of Iceland[45] and is inferred to have overturned rapidly[44]. The deep Atlantic cell is surmised to have overturned more slowly and may have been supplied by NADW formed in the Nordic Seas during seasonal ice-free periods coming over the Greenland-Scotland Ridge[46]. Brine rejection from sea ice may also have contributed to this process[47]. A lower flux of GNADW during the LGM than NADW in the Holocene in to the deep North Atlantic would then explain how AABW was able to penetrate much further north during the LGM (Fig. 4b) than in the modern Atlantic Ocean where it has little influence north of the equator (Fig. 4a). In contrast, the deep South Atlantic shows less of a change in water mass mixing proportions between the LGM and the Holocene (Fig. 4) because it is dominated by southern-sourced waters in both climate states (Fig. 7).

Many modelling studies and hypothesized glacial overturning schemes invoke GNAIW being incorporated into deep waters elsewhere in the glacial ocean to ventilate the deep ocean with low-preformed nutrient concentration water[10,13,48]. Here, we have presented evidence that instead the deep Atlantic was ventilated directly from the North Atlantic resolving the difficulty in reconciling glacial CO_2 drawdown with the observed proxy data. Another important prediction of our work is that switching from the glacial to interglacial mode of North Atlantic circulation (Fig. 7) would flush respired organic carbon from the deep Atlantic and would therefore be expected to raise atmospheric CO_2 without invoking changes in nutrient utilization in the high latitude Southern Ocean[13]. Although a full carbon cycle model would be required to quantify this effect, the resumption of strong NADW production at the start of the Bølling-Allerød inferred from ε_{Nd} and $^{231}Pa/^{230}Th$ records, coincides with a ~12 p.p.m. increase in atmospheric CO_2 (Supplementary Fig. 3)[12,18,49]. This increase likely reflects the flushing of respired carbon from the deep Atlantic by strong NADW production during the deglaciation[50].

Methods

Site selection. ε_{Nd} measurements were made on uncleaned planktic foraminifera from Holocene and LGM samples of cores from throughout the western Atlantic basin or selected regions within the eastern Atlantic (Fig. 2d); all core names and locations are listed in Supplementary Table 1 and plotted in Supplementary Fig. 1. Eastern Atlantic cores were limited to sites south of the Walvis Ridge or sites on bathymetric rises above the sill depth of the Mid-Atlantic Ridge (3,750 m)[51]. This selection was based on the criteria outlined by Curry and Oppo[4], and was intended to avoid the regions of the eastern Atlantic which are ventilated through fracture zones in the Mid-Atlantic Ridge and thus do not display the same latitudinal water mass mixing gradient as the western Atlantic[52]. Cores north of 45° N were excluded as they have been shown to be susceptible to the influence of volcanic ash and IRD in the North Atlantic[19,53].

Age controls. Age controls for cores used in this work range from planktic or benthic foraminiferal $\delta^{18}O$ records to radiocarbon dates (Supplementary Table 1) and came from published age models, with the exception of the Ceara Rise cores. The radiocarbon-based age models developed in this work for the latter cores, ODP 925E, ODP 928B and ODP 929B, are presented in Supplementary Table 6. For all cores, depths which were assigned calendar ages between 23 and 18 ka were included in the LGM reconstruction.

Sample preparation. Samples were prepared for analysis following the methods of Roberts *et al.*[18] and references therein. In short, where possible, ~80 mg of mixed planktic foraminifera were picked from the coarse fraction (>63 μm) for neodymium isotope measurements. After picking, foraminifera tests were broken open between two glass plates, rinsed, sonicated and any clays removed. The samples were then dissolved in 1 mol l^{-1} reagent grade acetic acid. The REEs were extracted from the dissolved sample using Eichrom TRUspec resin in 100 ml Teflon columns. Neodymium was then separated from the other rare earth elements using Eichrom LNspec resin on volumetrically calibrated Teflon columns.

Neodymium isotopic measurements. Neodymium isotopes were analysed using the Nu Plasma HR or Neptune Plus multi-collector inductively coupled plasma mass spectrometers at the University of Cambridge. $^{146}Nd/^{144}Nd$ was normalized to 0.7219 and samples were bracketed with a concentration-matched solution of reference standard JNdi-1, the measured composition of which varied between runs but was corrected to the accepted value of $^{143}Nd/^{144}Nd = 0.512115$ (ref. 54). The ε_{Nd} of each sample is reported with the external error (2σ) of the bracketing standards from the corresponding measurement session, unless the internal error was larger than the external error, in which case the combined internal and external error (2σ) is reported.

Eight complete procedural blanks for the process from foraminiferal dissolution through to column chemistry were run on a TIMS Sector 54 at the University of Cambridge using a ^{150}Nd spike. For the typical sample, of at least 15 ng, the average blank of 66 pg of neodymium is <0.5% of the total neodymium.

Benthic foraminiferal stable isotopes. Benthic foraminifera (mixed *Cibicidoides* species; between 3 and 7 tests) were picked for stable isotope analysis from the coarse fraction (>125 μm) of samples without published $\delta^{13}C$ data. Samples were analysed by the Godwin Laboratory at the University of Cambridge using either a Micromass Multicarb Sample Preparation System attached to a VG SIRA or a Thermo Kiel device attached to a Thermo MAT253 Mass Spectrometer in dual inlet mode. Isotopic ratios are presented relative to standard Vienna PeeDee Belemnite; external precision was ±0.06‰ for $\delta^{13}C$.

References

1. Broecker, W. S. Glacial to interglacial changes in ocean chemistry. *Prog. Oceanogr.* **11**, 151–197 (1982).
2. Denton, G. H. *et al.* The last glacial termination. *Science* **328**, 1652–1656 (2010).
3. Lynch-Stieglitz, J. *et al.* Atlantic meridional overturning circulation during the Last Glacial Maximum. *Science* **316**, 66–69 (2007).
4. Curry, W. B. & Oppo, D. W. Glacial water mass geometry and the distribution of $\delta^{13}C$ of ΣCO_2 in the western Atlantic Ocean. *Paleoceanography* **20**, PA1017 (2005).
5. Marchal, O. & Curry, W. B. On the abyssal circulation in the glacial Atlantic. *J. Phys. Oceanogr.* **38**, 2014–2037 (2008).
6. Ferrari, R. *et al.* Antarctic sea ice control on ocean circulation in present and glacial climates. *Proc. Natl Acad. Sci. USA* **111**, 8753–8758 (2014).
7. Negre, C. *et al.* Reversed flow of Atlantic deep water during the Last Glacial Maximum. *Nature* **468**, 84–88 (2010).
8. Gebbie, G. How much did Glacial North Atlantic Water shoal? *Paleoceanography* **29**, 190–209 (2014).
9. Weber, S. L. *et al.* The modern and glacial overturning circulation in the Atlantic ocean in PMIP coupled model simulations. *Clim. Past* **3**, 51–64 (2007).
10. Sigman, D. M., Hain, M. P. & Haug, G. H. The polar ocean and glacial cycles in atmospheric CO_2 concentration. *Nature* **466**, 47–55 (2010).
11. Marinov, I. *et al.* Impact of oceanic circulation on biological carbon storage in the ocean and atmospheric pCO2. *Global Biogeochem. Cycles* **22**, GB3007 (2008).
12. Marcott, S. A. *et al.* Centennial-scale changes in the global carbon cycle during the last deglaciation. *Nature* **514**, 616–619 (2014).
13. Hain, M. P., Sigman, D. M. & Haug, G. H. Carbon dioxide effects of Antarctic stratification, North Atlantic Intermediate Water formation, and subantarctic nutrient drawdown during the last ice age: Diagnosis and synthesis in a geochemical box model. *Global Biogeochem. Cycles* **24**, GB4023 (2010).
14. Frank, M. Radiogenic isotopes: tracers of past ocean circulation and erosional input. *Rev. Geophys.* **40**, doi:10.1029/2000RG000094 (2002).
15. Piepgras, D. J. & Wasserburg, G. J. Rare earth element transport in the western North Atlantic inferred from Nd isotopic observations. *Geochim. Cosmochim. Acta* **51**, 1257–1271 (1987).
16. Stichel, T., Frank, M., Rickli, J. & Haley, B. A. The hafnium and neodymium isotope composition of seawater in the Atlantic sector of the Southern Ocean. *Earth Planet. Sci. Lett.* **317–318**, 282–294 (2012).
17. Amakawa, H., Sasaki, K. & Ebihara, M. Nd isotopic composition in the central North Pacific. *Geochim. Cosmochim. Acta* **73**, 4705–4719 (2009).
18. Roberts, N. L., Piotrowski, A. M., McManus, J. F. & Keigwin, L. D. Synchronous deglacial overturning and water mass source changes. *Science* **327**, 75–78 (2010).
19. Elmore, A. C., Piotrowski, A. M., Wright, J. D. & Scrivner, A. E. Testing the extraction of past seawater Nd isotopic composition from North Atlantic deep sea sediments and foraminifera. *Geochem. Geophys. Geosyst.* **12**, Q09008 (2011).

20. Wilson, D. J., Piotrowski, A. M., Galy, A. & Clegg, J. A. Reactivity of neodymium carriers in deep sea sediments: implications for boundary exchange and paleoceanography. *Geochim. Cosmochim. Acta* **109**, 197–221 (2013).
21. Foster, G. L., Vance, D. & Prytulak, J. No change in the neodymium isotope composition of deep water exported from the North Atlantic on glacial-interglacial time scales. *Geology* **35**, 37–40 (2007).
22. Schlitzer, R. Ocean Data View. http://odv.awi.de (2016).
23. Blanckenburg, F. V. Tracing past ocean circulation? *Science* **286**, 1862b–1863b (1999).
24. Rutberg, R. L., Hemming, S. R. & Goldstein, S. L. Reduced North Atlantic Deep Water flux to the glacial Southern Ocean inferred from neodymium isotope ratios. *Nature* **405**, 935–938 (2000).
25. Wei, R., Abouchami, W., Zahn, R. & Masque, P. Deep circulation changes in the South Atlantic since the Last Glacial Maximum from Nd isotope and multi-proxy records. *Earth Planet. Sci. Lett.* **434**, 18–29 (2016).
26. Piotrowski, A. M. *et al.* Reconstructing deglacial North and South Atlantic deep water sourcing using foraminiferal Nd isotopes. *Earth Planet. Sci. Lett.* **357–358**, 289–297 (2012).
27. Wilson, D. J., Piotrowski, A. M., Galy, A. & Banakar, V. K. Interhemispheric controls on deep ocean circulation and carbon chemistry during the last two glacial cycles. *Paleoceanography* **30**, PA2707 (2015).
28. Noble, T. L., Piotrowski, A. M. & McCave, I. N. Neodymium isotopic composition of intermediate and deep waters in the glacial southwest Pacific. *Earth Planet. Sci. Lett.* **384**, 27–36 (2013).
29. Völker, C. & Köhler, P. Responses of ocean circulation and carbon cycle to changes in the position of the Southern Hemisphere westerlies at Last Glacial Maximum. *Paleoceanography* **28**, 726–739 (2013).
30. Orsi, A. H., Whitworth, T. & Nowlin, W. D. On the meridional extent and fronts of the Antarctic Circumpolar Current. *Deep Sea Res. Part I Oceanogr. Res. Pap.* **42**, 641–673 (1995).
31. Pena, L. D. & Goldstein, S. L. Thermohaline circulation crisis and impacts during the mid-Pleistocene transition. *Science* **345**, 318–322 (2014).
32. Skinner, L. C. *et al.* North Atlantic versus Southern Ocean contributions to a deglacial surge in deep ocean ventilation. *Geology* **41**, 667–670 (2013).
33. Böhm, E. *et al.* Strong and deep Atlantic meridional overturning circulation during the last glacial cycle. *Nature* **517**, 73–76 (2015).
34. Elderfield, H., Whitfield, M., Burton, J. D., Bacon, M. P. & Liss, P. S. The oceanic chemistry of the rare-earth elements [and Discussion]. *Philos. Trans. R. Soc. A Math. Phys. Eng. Sci.* **325**, 105–126 (1988).
35. Crocket, K. C., Vance, D., Gutjahr, M., Foster, G. L. & Richards, D. A. Persistent Nordic deep-water overflow to the glacial North Atlantic. *Geology* **39**, 515–518 (2011).
36. Rempfer, J., Stocker, T. F., Joos, F. & Dutay, J.-C. Sensitivity of Nd isotopic composition in seawater to changes in Nd sources and paleoceanographic implications. *J. Geophys. Res.* **117**, C12010 (2012).
37. Skinner, L. C., Waelbroeck, C., Scrivner, A. E. & Fallon, S. J. Radiocarbon evidence for alternating northern and southern sources of ventilation of the deep Atlantic carbon pool during the last deglaciation. *Proc. Natl. Acad. Sci. USA* **111**, 5480–5484 (2014).
38. Yu, J., Elderfield, H. & Piotrowski, A. M. Seawater carbonate ion-$\delta^{13}C$ systematics and application to glacial–interglacial North Atlantic ocean circulation. *Earth Planet. Sci. Lett.* **271**, 209–220 (2008).
39. Yu, J. *et al.* Deep South Atlantic carbonate chemistry and increased interocean deep water exchange during last deglaciation. *Quat. Sci. Rev.* **90**, 80–89 (2014).
40. Mackensen, A., Hubberten, H., Bickert, T. & Ftitterer, D. K. The $\delta^{13}C$ in benthic foraminifera tests of *Fontbotia wuellerstorfi* (Schwager) relative to the $\delta^{13}C$ of dissolved inorganic carbon in Southern Ocean deep water: implications for glacial ocean circulation models. *Paleoceanography* **8**, 587–610 (1993).
41. Kroopnick, P. The distribution of ^{13}C of ΣCO_2 in the world oceans. *Deep. Sea Res. Part A Oceanogr. Res. Pap.* **32**, 57–84 (1985).
42. Hodell, D. A., Venz, K. A., Charles, C. D. & Ninnemann, U. S. Pleistocene vertical carbon isotope and carbonate gradients in the South Atlantic sector of the Southern Ocean. Geochemistry. *Geophys. Geosyst.* **4**, 1–19 (2003).
43. Kohfeld, K. E., Le Quéré, C., Le, Harrison, S. P. & Anderson, R. F. Role of Marine Biology in Glacial-Interglacial CO_2 Cycles. *Science* **308**, 74–78 (2005).
44. Lippold, J. *et al.* Strength and geometry of the glacial Atlantic Meridional Overturning Circulation. *Nat. Geosci.* **5**, 813–816 (2012).
45. Labeyrie, L. D. *et al.* Changes in the vertical structure of the North Atlantic Ocean between glacial and modern times. *Quat. Sci. Rev.* **11**, 401–413 (1992).
46. Millo, C., Sarnthein, M., Voelker, A. & Erlenkeuser, H. Variability of the Denmark Strait Overflow during the Last Glacial Maximum. *Boreas* **35**, 50–60 (2006).
47. Dokken, T. & Jansen, E. Rapid changes in the mechanism of ocean convection during the last glacial period. *Nature* **401**, 458–461 (1999).
48. Kwon, E. Y. *et al.* North Atlantic ventilation of 'southern-sourced' deep water in the glacial ocean. *Paleoceanography* **27**, 1–12 (2012).
49. McManus, J. F., Francois, R., Gherardi, J.-M., Keigwin, L. D. & Brown-Leger, S. Collapse and rapid resumption of Atlantic meridional circulation linked to deglacial climate changes. *Nature* **428**, 834–837 (2004).
50. Chen, A. T. *et al.* Synchronous centennial abrupt events in the ocean and atmosphere during the last deglaciation. *Science* **349**, 1537–1542 (2015).
51. Metcalf, W. G., Heezen, B. C. & Stalcup, M. C. The sill depth of the Mid-Atlantic Ridge in the equatorial region. *Deep Sea Res. Oceanogr. Abstr.* **11**, 1–10 (1964).
52. McCartney, M. S., Bennett, S. L. & Woodgate-Jones, M. E. Eastward flow through the Mid-Atlantic Ridge at 11° N and its influence on the abyss of the eastern basin. *J. Phys. Oceanogr.* **21**, 1089–1121 (1991).
53. Roberts, N. L. & Piotrowski, A. M. Radiogenic Nd isotope labeling of the northern NE Atlantic during MIS 2. *Earth Planet. Sci. Lett.* **423**, 125–133 (2015).
54. Tanaka, T. *et al.* JNdi-1: a neodymium isotopic reference in consistency with LaJolla neodymium. *Chem. Geol.* **168**, 279–281 (2000).
55. Jeandel, C. Concentration and isotopic composition of Nd in the South Atlantic Ocean. *Earth Planet. Sci. Lett.* **117**, 581–591 (1993).
56. Huang, K.-F., Oppo, D. W. & Curry, W. B. Decreased influence of Antarctic intermediate water in the tropical Atlantic during North Atlantic cold events. *Earth Planet. Sci. Lett.* **389**, 200–208 (2014).
57. Jonkers, L. *et al.* Deep circulation changes in the central South Atlantic during the past 145 kyrs reflected in a combined $^{231}Pa/^{230}Th$, Neodymium isotope and benthic $\delta^{13}C$ records. *Earth Planet. Sci. Lett.* **419**, 14–21 (2015).
58. Gutjahr, M., Frank, M., Stirling, C. H., Keigwin, L. D. & Halliday, A. N. Tracing the Nd isotope evolution of North Atlantic Deep and Intermediate Waters in the western North Atlantic since the Last Glacial Maximum from Blake Ridge sediments. *Earth Planet. Sci. Lett.* **266**, 61–77 (2008).
59. Lang, D. C. *et al.* Incursions of southern-sourced water into the deep North Atlantic during late Pliocene glacial intensification. *Nat. Geosci.* **9**, 375–379 (2016).

Acknowledgements

Sample material was provided by the Godwin Laboratory for Paleoclimate Research at the University of Cambridge, the International Ocean Discovery Program the GeoB Core Repository at the MARUM—Center for Marine Environmental Sciences, University of Bremen and Petrobras. We thank Jo Kerr and Aurora Elmore for providing additional samples. We thank Thiago Pereira dos Santos for providing the unpublished age model data for GL1090; we thank Jo Clegg and Vicky Rennie for technical support and Natalie Roberts for helpful discussions. Radiocarbon analyses were supported by NERC radiocarbon grant 1752.1013 and Nd isotope analyses by NERC grants NERC NE/K005235/1 and NERC NE/F006047/1 to A.M.P. J.N.W.H. was supported by a Rutherford Memorial Scholarship. S.M. was supported through the DFG Research Center/Cluster of Excellence 'The Ocean in the Earth System'. C.M.C. acknowledges financial support from FAPESP (grant 2012/17517-3).

Author contributions

J.N.W.H. and A.M.P. designed the study, J.N.W.H. performed the work, and J.N.W.H. and A.M.P. wrote the manuscript with contributions from all other authors. T.L.N. provided unpublished data from ODP 1088; S.M., C.M.C. and G.B. provided sample material and unpublished data.

Additional information

Competing financial interests: The authors declare no competing financial interests.

16

Decadal predictions of the North Atlantic CO_2 uptake

Hongmei Li[1], Tatiana Ilyina[1], Wolfgang A. Müller[1] & Frank Sienz[1]

As a major CO_2 sink, the North Atlantic, especially its subpolar gyre region, is essential for the global carbon cycle. Decadal fluctuations of CO_2 uptake in the North Atlantic subpolar gyre region are associated with the evolution of the North Atlantic Oscillation, the Atlantic meridional overturning circulation, ocean mixing and sea surface temperature anomalies. While variations in the physical state of the ocean can be predicted several years in advance by initialization of Earth system models, predictability of CO_2 uptake has remained unexplored. Here we investigate the predictability of CO_2 uptake variations by initialization of the MPI-ESM decadal prediction system. We find large multi-year variability in oceanic CO_2 uptake and demonstrate that its potential predictive skill in the western subpolar gyre region is up to 4-7 years. The predictive skill is mainly maintained in winter and is attributed to the improved physical state of the ocean.

[1]Max Planck Institute for Meteorology, Bundesstraße 53, 20146 Hamburg, Germany. Correspondence and requests for materials should be addressed to H.L. (email: hongmei.li@mpimet.mpg.de) or to T.I. (email: tatiana.ilyina@mpimet.mpg.de).

The world's oceans currently take over about 25–30% of anthropogenic CO_2 from the atmosphere, and the North Atlantic is a key component of this oceanic carbon sink[1,2]. Steered by deep convection, most of the North Atlantic CO_2 uptake takes place in the subpolar gyre (SPG) region, a region also characterized by pronounced variability in CO_2 fluxes[3–5]. The variability of carbon uptake in the subpolar North Atlantic ocean was found to be largely related to variations of the North Atlantic Oscillation (NAO), the Atlantic meridional overturning circulation (AMOC), sea surface temperature (SST) and ocean-mixing strength, and all these processes are intrinsically related[3,5–9]. The SPG region experienced several abrupt warming and cooling events in the twentieth century, with SST increasing or decreasing by more than 1 °C in only a few years[10–13]. These changes are related to variations of the NAO and the associated AMOC, heat transports[11,14], SPG circulation and Labrador Sea convection[15,16]. The key elements such as AMOC and SST have been shown to be predictable several years ahead by initialization of Earth system models (ESMs)[10–12,17]. However, no such investigation of predictability has been performed for the CO_2 uptake so far. Are variations in the North Atlantic CO_2 uptake represented in decadal prediction simulations with an ESM? And if so, to what extent is the CO_2 uptake predictable?

We address these questions by using the Max Planck Institute-ESI (MPI-ESM) decadal prediction system[18], which includes an assimilation simulation by nudging observations into the atmospheric and oceanic components of the ESM (see the Methods section) and a set of initialized retrospective prediction simulations (that is, starting from the assimilation run). For comparison, a set of uninitialized simulations is performed without any observations nudged. No carbon cycle observations are used in the assimilation run, so that the ocean biogeochemical processes perform as a passive component. Such an assimilation run, including both anthropogenic trends and natural fluctuations, allows us to investigate the spatial and temporal variability of the oceanic carbon uptake and its connection to local and large-scale physical processes. The initialized retrospective prediction simulations enable us to assess the predictability of carbon uptake. Here we mainly focus on interannual and longer timescale and if not stated otherwise use multi-year and annual mean data set. In this study, we focus on the SPG region because first it plays a prominent role in the oceanic carbon budget. Second, both the physical and the biogeochemical states of the ocean in the SPG region exhibit pronounced natural variability, which is governed by processes acting on decadal scales. Third, this region demonstrates robust predictive skill for the physical fields offering promise for predictions of the oceanic carbon uptake. Our analysis indicates that the North Atlantic CO_2 uptake shows large spatial and temporal variations on decadal scale. Furthermore, these variations are better represented in the initialized simulations than in the uninitialized simulations. Direct comparison to observations indicates improved representation of the ocean surface pCO_2 by initialization. We find the potential prediction skill of the western SPG CO_2 uptake until lead time of 4–7 years. This predictability of CO_2 uptake is related to the initialization of SST and AMOC by observations.

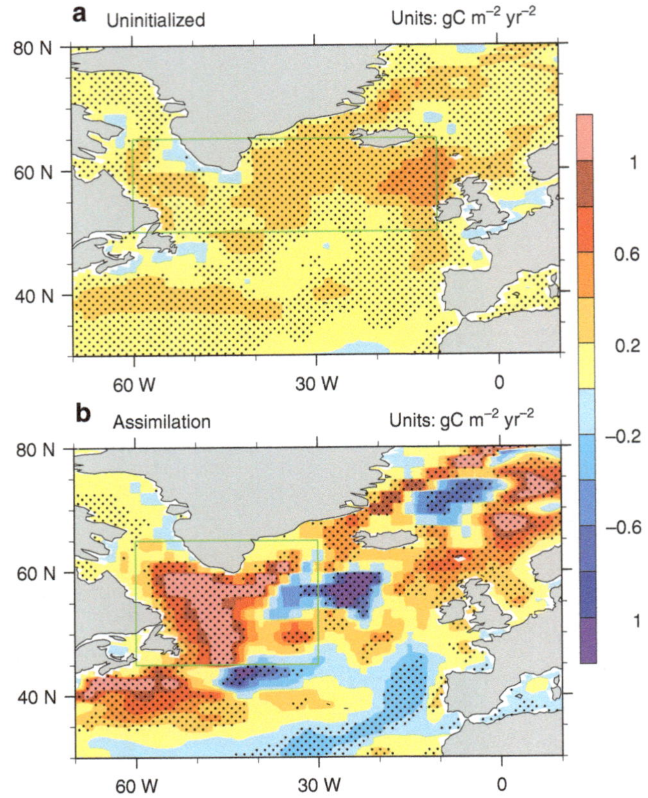

Figure 1 | Trends of annual mean CO_2 flux into the ocean from 1970 to 1995. Ensemble mean of uninitialized simulations (**a**) and assimilation simulation (**b**) (units: gC m^{-2} per year^2). The trends are calculated based on linear regression. The dots show grid points where the trends are significant at 95% level, based on a two-sided *t*-test. The green box in **a** denotes the SPG region where the mid-1990s abrupt warming occurred[11], and the green box in **b** denotes the western SPG region where the CO_2 flux increases the most.

Results

Spatial variations of CO_2 flux in the North Atlantic. Spatial distributions of trends in CO_2 flux during the 25 years before the mid-1990s SPG abrupt warming (that is, from 1970 to 1995) show a remarkably different pattern between the uninitialized and assimilation runs (Fig. 1). The oceanic CO_2 uptake in the uninitialized simulations increases at about 0.2 gC m^{-2} per year^2, following the increase in atmospheric CO_2 due to rising fossil fuel carbon emissions, with a somewhat larger increase in the eastern part of the North Atlantic. The CO_2 flux in the assimilation run does not increase uniformly in the North Atlantic; the largest increase is found in the western SPG region (Fig. 1b), and decreasing trends are found in the eastern SPG region. A zonal dipole distribution in delta pCO_2 anomalies was also found in a previous study based on the simulations with an ocean only model[6]. This dipole trend pattern of CO_2 uptake can be attributed to the NAO-related western–eastern heat loss gradient[19] and the related ocean-mixing strength changes (Supplementary Fig. 1 and Supplementary Note 1). Regarding the mid-1990s warming, further decomposition of the SPG region by a previous study[20] suggests a longer predictive skill in the western SPG region, which involves AMOC and meridional heat transport changes in response to persistent positive NAO phase from 1988 to 1995, and a shorter predictive skill in the eastern SPG region, which involves gyre circulation adjustment in response to NAO phase switch from positive to negative in 1995. Here we focus on the western SPG region with the largest trend of CO_2 uptake and with longer prediction skill of the ocean physical state.

Temporal evolution of CO_2 uptake and related processes. To further explore the temporal variability of the oceanic carbon uptake and how it is related to prominent local hydrodynamic processes in the SPG region, we compare the CO_2 flux calculated in the assimilation run with the observed NAO index[21], observed

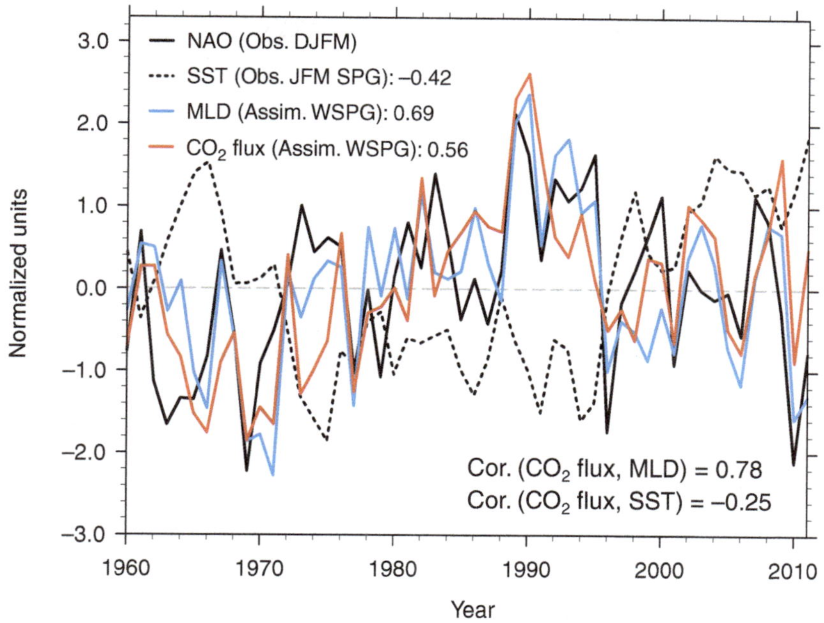

Figure 2 | Normalized time series of CO_2 flux and related physical variables. The observed (Obs.) DJFM NAO[21], observed JFM SST[22] in SPG region, western SPG (WSPG) MLD and CO_2 flux calculated with the assimilation run are shown with black solid line, black dashed line, blue solid line and red solid line, respectively. The location of SPG and WSPG refers to green box in Fig. 1a,b, respectively. The correlation (Cor.) coefficients between time series and NAO (CO_2 flux) are shown on the top (bottom) of the figure. The normalized time series are calculated by dividing individual variable anomalies with their respective standard deviation.

SST[22] and mixed layer depth (MLD) in the assimilation run (Fig. 2). The western SPG MLD and CO_2 flux from our model results are highly correlated (with a correlation coefficient of 0.78), and both are highly correlated (with correlation coefficients of 0.69 and 0.56, respectively) with the observed December–January–February–March (DJFM) NAO index. The correlations are significantly different from zero with P values below 0.01. Consistent with previous model studies[6], the close connection between the NAO and the North Atlantic CO_2 uptake at interannual and decadal timescales is also reproduced in the assimilation run. On the one hand, the heat loss related to a positive NAO leads to an enhancement of the ocean mixing and deep-water formation in the North Atlantic[19]; this is indicated by the relatively high correlation (0.69) between NAO and MLD. Moreover, the SST anomalies in the SPG region with deep convection recur from winter to winter through a re-emergence process[23,24], which works as follows. The winter thermal anomalies related to positive NAO remain at depth below the shallow summer mixed layer, these anomalies persist through summer and are partially re-entrained into the following winter mixed layer. On the other hand, the North Atlantic atmospheric circulation changes can also affect CO_2 uptake through the associated changes in AMOC and the corresponding meridional heat transport changes. Enhanced northward heat transport leads to increased SST thereby affecting the oceanic carbon uptake in the SPG. These processes lag the NAO changes by several years[14]. Accordingly, given all the processes, the correlation between CO_2 flux and SST is not significant in our simulations.

Potential predictive skill of CO_2 flux and SST. The state-of-the-art decadal prediction systems consistently identify the North Atlantic as a key region with pronounced forecast skill for different parameters of the climate system[25–31]. Several abrupt climate events in the twentieth century have been shown to be predictable several years ahead[10–13]. The SPG abrupt warming and cooling events are also well captured in our initialized simulations, whereas the uninitialized simulations only capture the long-term warming trend and display large spread among ensemble members (Supplementary Fig. 2 and Supplementary Note 2). Given the robust predictive skill of the physical state of the ocean, can variations in the North Atlantic CO_2 uptake be predicted as well? A previous study explored multi-year predictability of tropical marine productivity and found a predictive skill of 3 years, whereas the predictive skill of SST there is only 1 year[32]. In our analysis, as indicated by the initialized time series at different lead times (Supplementary Fig. 3 and Supplementary Note 3), the variability of CO_2 flux, which is absent in the uninitialized simulations, can be reproduced by the model with initialization several years in advance and implies predictability of CO_2 uptake.

Owing to lack of observational data, we first use model fields calculated in the assimilation run as a proxy to quantify the potential predictive skill of SST and of the CO_2 flux into the ocean (Fig. 3, see the Methods section). The correlation coefficients of the initialized 4-year mean SST exceed the uninitialized correlations for lead times of 1–4 and 2–5 years. The correlation goes down in the intermediate lead time, however, it recovers from a lead time of 5–8 years. The results suggest improved potential predictive skills of the background ocean physical fields. The correlation coefficient of initialized CO_2 flux is significantly higher than the uninitialized correlation until a lead time of 4–7 years. The P values are lower than 0.05 until the lead time of 4–7 years suggesting significant improvement of potential predictive skill due to initialization of the physical states constrained by observations.

Further investigation of the potential predictive skill with seasonal and monthly time series (Fig. 3c,d) reveals that the high predictive skill of CO_2 flux in the initialized simulations is mainly maintained during winter months. This is related to the seasonal shift of processes in regulating the CO_2 uptake and the fact of assimilation for the decadal prediction system. In winter, the CO_2 uptake in western SPG region is regulated primarily by physical process such as the ocean-mixing strength and ocean

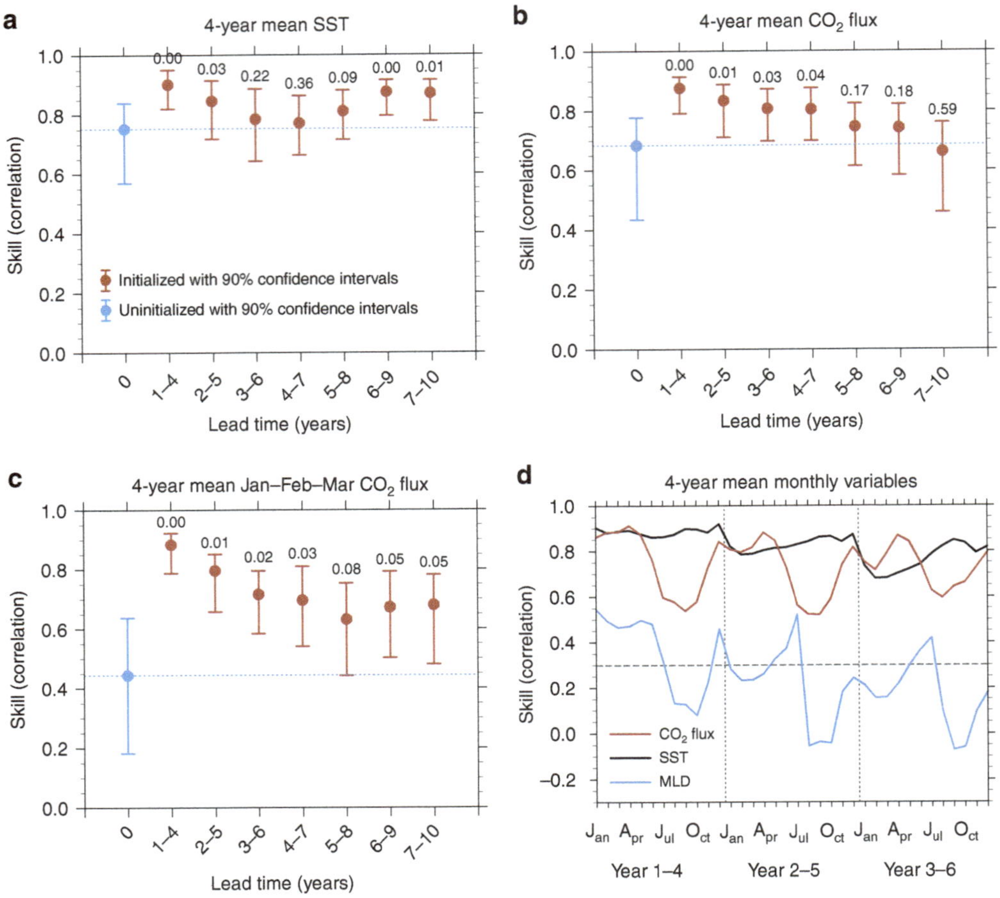

Figure 3 | Potential predictive skill of CO_2 flux and related physical variables. Correlation skill of ensemble mean of 4-year mean SST (**a**), 4-year mean CO_2 flux into the ocean (**b**), seasonally stratified 4-year mean CO_2 flux into the ocean (late winter, that is, January-February-March) (**c**) and monthly stratified 4-year mean CO_2 flux, SST and MLD (**d**) in the western SPG region. Shown are uninitialized (blue dot in **a–c**) and initialized (red dots in **a–c**) simulations at different lead time verified against assimilation. The correlations are calculated from 4-year mean predictions (that is, years 1-4, 2-5 and so on) of initialized simulations, and 4-year running mean of the corresponding assimilation and uninitialized time series. To ensure that the number of validation years is the same for both initialized and uninitialized simulations, we use the common time period from 1967-1970 mean to 2008-2011 mean. The blue dashed line in **a–c** extends the uninitialized correlation for easy comparison. The vertical lines in **a–c** provide 90% confidence intervals based on a bootstrap approach[46]. The numbers on the top of the bars in **a–c** show the *P* values based on the hypothesis that the difference of correlations between the initialized and uninitialized simulations is smaller or equal to zero based on 1,000 bootstrapped resamples. The potential predictive skills of monthly stratified 4-year mean CO_2 flux, SST and MLD in **d** are shown with red, black and blue curves, respectively. Three-month running mean is applied before monthly correlation estimation in **d**. The horizontal dashed grey line in **d** shows the 95% level of significance based on a two-sided *t*-test. The vertical dashed grey lines in **d** are added to better distinguish different lead year results.

circulation[33–35]. The variations of ocean physical fields are well represented in the initialized simulations, as the corresponding initial states are constrained by observations through assimilation. However, in spring and early summer, when the ocean warms, the biological primary production draws down seawater pCO_2 and regulates the oceanic CO_2 uptake in the western SPG region[33–35]; the correlation of the initialized simulations goes down from May, as no ocean biological observations are assimilated into the system.

We further explore possible causes of the potential predictive skill of CO_2 uptake in comparison with that of the SST and MLD (Fig. 3d). The SST correlations peak in December and go down from late winter, which is coherent with that of the MLD and the CO_2 uptake. It suggests that both the evolution of the ocean thermal state and the local mixing strength contribute to the predictive skill of CO_2 uptake in winter. Although the correlation skill of MLD is generally lower than that of CO_2 uptake, both share similar seasonal cycle of predictive skill, indicating effects of MLD changes in maintaining the predictive skill of CO_2 uptake. Moreover, the potential predictive skill of CO_2 uptake is related to the AMOC variability. As revealed by previous studies, the predictive skill of SPG SST is assured by initialization of the AMOC variability[27], and the AMOC shows predictive skill up to 4 years[17]. The potential predictive skill of AMOC in our system is up to 2–5 years (figure omitted). The AMOC at lead time of 1 year is highly correlated with CO_2 flux at lead time of 1 year and onwards (Supplementary Fig. 4 and Supplementary Note 4). An observational study also suggested a close connection between the AMOC and the CO_2 uptake in SPG region[3].

From these our results suggest that the potential predictive skill of CO_2 uptake in the western SPG is up to 4–7 years. The predictive skill is mainly attributed to the combination of improved ocean physical states and circulation variability, primarily in winter.

Evaluation of model predictions against observations. The high potential predictability of CO_2 uptake provides a basis for assessing our predictions against observations using the surface ocean CO_2 atlas (SOCAT) measurement[36]. Although observational data from SOCAT are sparse in the SPG region,

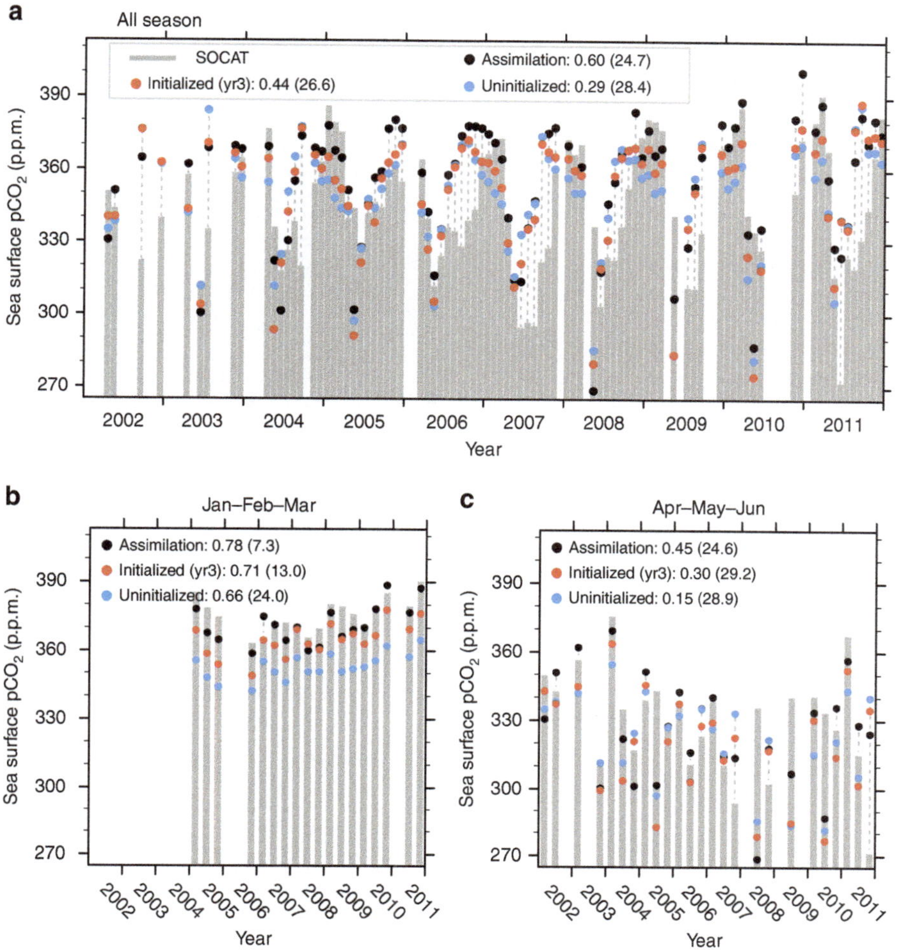

Figure 4 | Observed and model simulated monthly mean ocean surface pCO_2 in western SPG region. (**a**) All season time series, (**b**) only January-February-March are shown and (**c**) only April-May-June are shown. The SOCAT observations[36] are shown with grey bars, and model output of assimilation, uninitialized simulations and initialized simulations at a lead time of 3 years (yr3) are shown in black, blue and red dots, respectively. The grey dashed lines connect outlier dots with the grey bars to make the comparison clearer. The numbers in legend are correlation coefficients and root mean squared error (in brackets) between model simulations and SOCAT observations. For the all season time series in **a** the correlations of the assimilation and initialized simulations at lead time of 3 years are significantly larger than that of the uninitialized simulations at 95% significant level based on a bootstrap test. Note the SOCAT observational data are not continuously distributed in space and time; the model simulations are averaged over the grid points where SOCAT data are available.

these are the best ocean surface observations we can get for this region. The ocean surface pCO_2 in the SPG region peaks in winter as a result of the enhanced vertical supply of carbon from the intermediate waters by deep convection; surface pCO_2 values reach a minimum in summer due to biological draw down[33–35] (Fig. 4). The initialized predictions produce ocean surface pCO_2 closer to SOCAT observation than the uninitialized simulations as indicated by the correlations and root mean squared errors. The correlations of assimilation (0.60) and initialized simulation at a lead time of 3 years (0.44) are significantly larger than the correlation of initialized simulation (0.29) at 95% significance level. The root mean squared error is lower in the assimilation and initialized simulations than in the uninitialized simulation. As we use monthly data due to lack of continuous observations, the better performances of initialized simulations are partially due to better representation of the seasonal cycle in the initialized simulations. We further separate the time series seasonally, and find that in addition to the seasonal cycle there is an improvement of the initialized run against the uninitialized run particularly in the winter months when the pCO_2 is high. The root mean squared error of ocean surface pCO_2 is much smaller in the assimilation (7.3 p.p.m.) and initialized simulations at a lead time of 3 years (13.0 p.p.m.) than in the uninitialized simulations (24.0 p.p.m.; Fig. 4b). The coherences between model simulations and observations are generally lower in spring months (Fig. 4c). The improvement of prediction in individual seasons further demonstrates that the interannual variations of oceanic carbon cycle are improved in the initialized simulations. The higher correlations between SOCAT and initialized simulations and the lower root mean squared error of the initialized simulations against SOCAT confirm that the oceanic carbon cycle can be predicted several years ahead by initialization of the ESM (Supplementary Fig. 5 and Supplementary Note 5). However, owing to temporal and spatial gaps in observations, the precise prediction skill with respect to observations cannot be estimated. Comparison of simulations against SOCAT observations only confirm that prediction of the state of the oceanic carbon cycle is improved by initialization of the physical ocean state with observations.

Discussion

We have focused here on the impacts of the ocean physical state on the variability and predictability of the North Atlantic CO_2

uptake. The drawdown of pCO_2 by biological processes such as carbon fixation during seasonal phytoplankton growth, with consequent export of carbon into the deep ocean is also critical for the air–sea carbon exchange in the nutrient-rich SPG region[37]. These marine biogeochemical processes themselves are influenced by variability in climate and circulation. For instance, an increase in ocean temperature leads to a decrease of CO_2 uptake through decrease in solubility. Biological production starts in spring when the ocean warms up. This draws down the ocean surface pCO_2 and enhances CO_2 uptake. However, the growth of phytoplankton in the North Atlantic is more limited by nutrient supply than by temperature. The nutrient supply, in turn, is constrained by the enhanced thermal stratification associated with the ocean warming, thus resulting in a decrease of primary production[38]. Therefore, while the biological CO_2 drawdown regulates the seasonal cycle of CO_2 flux, the physical regulation is more important for interannual and decadal variations of CO_2 flux in the SPG region[39]. Our findings show that there is an improved representation of the oceanic carbon flux because the background oceanic circulation and thermal state are well reproduced. Physical processes such as the ocean thermal state, local mixing and large-scale circulation contribute to the predictability of CO_2 uptake. It requires further large ensembles of multi-model decadal prediction simulations and sensitivity experiments to disentangle the relative role of an individual physical mechanism on the predictability of CO_2 uptake.

We find that beside the trend due to CO_2 emissions, predictions of the oceanic uptake and storage of carbon show considerable decadal variations that are not fully captured in the uninitialized simulations with modern ESMs. We demonstrate that variations in the oceanic carbon cycle can be predicted several years ahead. Hence, predictions of the North Atlantic carbon sink considering both anthropogenic change and natural fluctuations are important to understand the evolution of climate and ocean acidification and to reduce uncertainties in CO_2 uptake estimates. Predictions of the oceanic carbon sink are also necessary to provide information to monitoring programs aimed at the present and future oceanic carbon sink.

Methods

Model description. We use the MPI-ESM[40] to conduct historical (uninitialized) experiments and retrospective decadal prediction (initialized) experiments for the period 1961–2013. A low-resolution configuration (MPI-ESM-LR) is used with the resolution of the ocean model being 1.5° on average with 40 vertical levels. The ocean component of MPI-ESM-LR use bipolar configuration, with one pole over Greenland and another over Antarctica, so the resolution in the North Atlantic SPG region is from ~20 km in the north (65° N) to ~70 km in the south (45° N). MPI-ESM-LR is capable of producing North Atlantic deep convection confined to the Labrador Sea, which is close to observation[41]. The ocean biogeochemistry component of MPI-ESM is represented by the Hamburg Ocean Carbon Cycle Model (HAMOCC)[42].

Numerical simulations. An ensemble of 10 uninitialized historical simulations and RCP4.5 scenario simulations is performed for the periods 1850–2005 and 2006–2013, respectively. The uninitialized experiments are started from a preindustrial control simulation and forced with historical greenhouse gas and aerosol concentrations together with solar variability and volcanic eruptions. The initial conditions of the uninitialized run ensemble members differ based on the preindustrial control simulation starting from every 50 years. On the basis of an assimilation run, an ensemble of 10 initialized simulations is started on 1 January in every year over the period 1960–2012. The ensemble members of the initialized simulations are generated with lagged 1-day initialization, that is, the runs start from 10 consecutive days centred on 1 January. Each initialized simulation has a length of 10 years[18]. The assimilation experiment is also performed with MPI-ESM-LR. For the oceanic component, monthly temperature and salinity anomalies from the ECMWF ocean reanalysis system 4 (ORAS4)[43] are added to the model climatology and nudged into the model state at every time step. For the atmospheric component, globally full-field temperature, vorticity, divergence and surface pressure data from the ECMWF ERA40[44] and ERA-Interim[45] reanalyses are nudged.

Predictive skill assessment. We mainly show ensemble mean results in this study. For the uninitialized experiments, the anomalies are calculated by removing the climatology of the ensemble mean. For the initialized experiments, the anomalies are calculated in terms of the climatology of the ensemble member and the lead time. The observed temporal evolution of the DJFM NAO index (station based)[21] and the SST[22] in SPG is investigated together with the mixing and CO_2 flux around the SPG region in the assimilation run. We use the fields in the assimilation run as a proxy to quantify the historical variability and to estimate the potential predictive skill of CO_2 uptake. The potential predictions are verified with skill scores based on anomaly correlation coefficients against the assimilation run. As short-term/small-scale variability may add noise and hence reduce predictive skill, temporal and spatial averaging was recommended[46] and has been used in most decadal prediction studies[25,31,47]. A set of 4-year mean (that is, 1–4, 2–5, ..., 7–10 years) predictions is verified in this study against corresponding running mean of the assimilation and uninitialized simulations.

Bootstrap approach for significance test. Statistical tests and confidence intervals of the correlations in Figs 3 and 4 are calculated through a bootstrapping approach[46]. The resampling scheme considers uncertainty arising from both ensemble size and hindcast period. A block size of 2 is used to account for autocorrelation in the SOCAT time series (Fig. 4). Note that the significance of correlation differences in Fig. 3 remains unaltered for the same block size.

Conversion of ocean surface CO_2 observation data. The gridded observations of the SOCAT[36] surface ocean fugacity of CO_2 (fCO_2) from 2002 to 2011 are used to verify model simulations. The fugacity is converted to partial pressure[48]. As observational data are not continuously distributed in space and time, for the model evaluation presented in Fig. 4 and Supplementary Fig. 5 we only consider those model grid points where SOCAT data are available.

Code availability. The MPI-ESM model is freely available to the scientific community. The code can be accessed with a license agreement on the Max Planck Institute for Meteorology model distribution website: http://www.mpimet.mpg.de/en/science/models/license/

References

1. Sabine, C. L. *et al.* The oceanic sink for anthropogenic CO_2. *Science* **305,** 367–371 (2004).
2. Le Quéré, C. *et al.* Global carbon budget 2015. *Earth Syst. Sci. Data* **7,** 349–396 (2015).
3. Pérez, F. F. *et al.* Atlantic Ocean CO_2 uptake reduced by weakening of the meridional overturning circulation. *Nat. Geosci.* **6,** 146–152 (2013).
4. McKinley, G. A., Fay, A. R., Takahashi, T. & Metzl, N. Convergence of atmospheric and North Atlantic carbon dioxide trends on multidecadal timescales. *Nat. Geosci.* **4,** 606–610 (2011).
5. Corbière, A., Metzl, N., Reverdin, G., Brunet, C. & Takahashi, A. Interannual and decadal variability of the oceanic carbon sink in the North Atlantic subpolar gyre. *Tellus Ser. B* **59,** 168–178 (2007).
6. Thomas, H. *et al.* Changes in the North Atlantic Oscillation influence CO_2 uptake in the North Atlantic over the past 2 decades. *Global Biogeochem. Cycles* **22,** GB4027 (2008).
7. Metzl, N. *et al.* Recent acceleration of the sea surface fCO_2 growth rate in the North Atlantic subpolar gyre (1993-2008) revealed by winter observations. *Global Biogeochem. Cycles* **24,** GB4004 (2010).
8. Tjiputra, J. F., Olsen, A., Assmann, K., Pfeil, B. & Heinze, C. A model study of the seasonal and long-term North Atlantic surface pCO_2 variability. *Biogeosciences* **9,** 907–923 (2012).
9. Halloran, P. R. *et al.* The mechanisms of North Atlantic CO_2 uptake in a large Earth System Model ensemble. *Biogeosciences* **12,** 4497–4508 (2015).
10. Robson, J., Sutton, R. & Smith, D. Decadal predictions of the cooling and freshening of the North Atlantic in the 1960s and the role of ocean circulation. *Clim. Dyn.* **42,** 2353–2365 (2014).
11. Robson, J. I., Sutton, R. T. & Smith, D. M. Initialized decadal predictions of the rapid warming of the North Atlantic Ocean in the mid 1990s. *Geophys. Res. Lett.* **39,** L19713 (2012).
12. Müller, W. A., Pohlmann, H., Sienz, F. & Smith, D. Decadal climate prediction for the period 1901-2010 with a coupled climate model. *Geophys. Res. Lett.* **41,** 2100–2107 (2014).
13. Yeager, S., Karspeck, A., Danabasoglu, G., Tribbia, J. & Teng, H. A decadal prediction case study: Late twentieth-century North Atlantic Ocean heat content. *J. Clim.* **25,** 5173–5189 (2012).
14. Robson, J., Sutton, R., Lohmann, K., Smith, D. & Palmer, M. D. Causes of the rapid warming of the North Atlantic Ocean in the mid-1990s. *J. Clim.* **25,** 4116–4134 (2012).
15. Hakkinen, S. & Rhines, P. B. Decline of subpolar North Atlantic circulation during the 1990s. *Science* **304,** 555–559 (2004).

16. Hátún, H., Sando, A. B., Drange, H., Hansen, B. & Valdimarsson, H. Influence of the Atlantic subpolar gyre on the thermohaline circulation. *Science* **309,** 1841–1844 (2005).
17. Matei, D. *et al.* Multiyear prediction of monthly mean atlantic meridional overturning circulation at 26.5°N. *Science* **335,** 76–79 (2012).
18. Pohlmann, H. *et al.* Improved forecast skill in the tropics in the new MiKlip decadal climate predictions. *Geophys. Res. Lett.* **40,** 5798–5802 (2013).
19. Eden, C. & Willebrand, J. Mechanism of interannual to decadal variability of the North Atlantic circulation. *J. Clim.* **14,** 2266–2280 (2001).
20. Barrier, N., Deshayes, J., Treguier, A.-M. & Cassou, C. Heat budget in the North Atlantic subpolar gyre: impacts of atmospheric weather regimes on the 1995 warming event. *Prog. Oceanogr.* **130,** 75–90 (2015).
21. Hurrell, J. W. Decadal trends in the North Atlantic oscillation: regional temperatures and precipitation. *Science* **269,** 676–679 (1995).
22. Rayner, N. A. *et al.* Global analyses of sea surface temperature, sea ice, and night marine air temperature since the late nineteenth century. *J. Geophy. Res. Atmos.* **108,** 4407 (2003).
23. Polkova, I., Koehl, A. & Stammer, D. Impact of initialization procedures on the predictive skill of a coupled ocean-atmosphere model. *Clim. Dyn.* **42,** 3151–3169 (2014).
24. Deser, C., Alexander, M. A. & Timlin, M. S. Understanding the persistence of sea surface temperature anomalies in midlatitudes. *J. Clim.* **16,** 57–72 (2003).
25. Meehl, G. A. *et al.* Decadal climate prediction: an update from the trenches. *Bull. Am. Meteorol. Soc.* **95,** 243–267 (2014).
26. Smith, D. M. *et al.* Skilful multi-year predictions of Atlantic hurricane frequency. *Nat. Geosci.* **3,** 846–849 (2010).
27. Matei, D. *et al.* Two tales of initializing decadal climate prediction experiments with the ECHAM5/MPI-OM model. *J. Clim.* **25,** 8502–8523 (2012).
28. Müller, W. A. *et al.* Forecast skill of multi-year seasonal means in the decadal prediction system of the Max Planck Institute for Meteorology. *Geophys. Res. Lett.* **39,** L22707 (2012).
29. Pohlmann, H., Jungclaus, J. H., Koehl, A., Stammer, D. & Marotzke, J. Initializing decadal climate predictions with the GECCO Oceanic synthesis: effects on the North Atlantic. *J. Clim.* **22,** 3926–3938 (2009).
30. Doblas-Reyes, F. J. *et al.* Initialized near-term regional climate change prediction. *Nat. Commun.* **4,** 1715 (2013).
31. Keenlyside, N. S., Latif, M., Jungclaus, J., Kornblueh, L. & Roeckner, E. Advancing decadal-scale climate prediction in the North Atlantic sector. *Nature* **453,** 84–88 (2008).
32. Séférian, R. *et al.* Multiyear predictability of tropical marine productivity. *Proc. Natl Acad. Sci. USA* **111,** 11646–11651 (2014).
33. Körtzinger, A., Send, U., Wallace, D. W. R., Karstensen, J. & DeGrandpre, M. Seasonal cycle of O_2 and pCO_2 in the central Labrador Sea: atmospheric, biological, and physical implications. *Global Biogeochem. Cycles* **22,** GB1014 (2008).
34. Takahashi, T. *et al.* Global sea-air CO_2 flux based on climatological surface ocean pCO_2, and seasonal biological and temperature effects. *Deep Sea Res. Part II* **49,** 1601–1622 (2002).
35. Ullman, D. J., McKinley, G. A., Bennington, V. & Dutkiewicz, S. Trends in the North Atlantic carbon sink: 1992-2006. *Global Biogeochem. Cycles* **23,** GB4011 (2009).
36. Bakker, D. C. E. *et al.* An update to the Surface Ocean CO_2 Atlas (SOCAT version 2). *Earth Syst. Sci. Data* **6,** 69–90 (2014).
37. Takahashi, T. *et al.* Climatological mean and decadal change in surface ocean pCO_2, and net sea-air CO_2 flux over the global oceans. *Deep Sea Res. Part I* **56,** 554–577 (2009).
38. Behrenfeld, M. J. *et al.* Climate-driven trends in contemporary ocean productivity. *Nature* **444,** 752–755 (2006).
39. Bennington, V., McKinley, G. A., Dutkiewicz, S. & Ulman, D. What does chlorophyll variability tell us about export and air-sea CO_2 flux variability in the North Atlantic? *Global Biogeochem. Cycles* **23,** GB3002 (2009).
40. Giorgetta, M. A. *et al.* Climate and carbon cycle changes from 1850 to 2100 in MPI-ESM simulations for the Coupled Model Intercomparison Project phase 5. *J. Adv. Model. Earth Syst.* **5,** 572–597 (2013).
41. Jungclaus, J. H. *et al.* Characteristics of the ocean simulations in the Max Planck Institute Ocean Model (MPIOM) the ocean component of the MPI-Earth system model. *J. Adv. Model. Earth Syst.* **5,** 422–446 (2013).
42. Ilyina, T. *et al.* The global ocean biogeochemistry model HAMOCC: model architecture and performance as component of the MPI-Earth System Model in different CMIP5 experimental realizations. *J. Adv. Model. Earth Syst.* **5,** 287–315 (2013).
43. Balmaseda, M. A., Mogensen, K. & Weaver, A. T. Evaluation of the ECMWF ocean reanalysis system ORAS4. *Q. J. R. Meteorol. Soc.* **139,** 1132–1161 (2012).
44. Uppala, S. M. *et al.* The ERA-40 re-analysis. *Q. J. R. Meteorol. Soc.* **131,** 2961–3012 (2005).
45. Dee, D. P. *et al.* The ERA-Interim reanalysis: Configuration and performanceof the data assimilation system. *Q. J. R. Meteorol. Soc.* **137,** 553–597 (2011).
46. Goddard, L. *et al.* A verification framework for interannual-to-decadal predictions experiments. *Clim. Dyn.* **40,** 245–272 (2013).
47. Smith, D. M. *et al.* Improved surface temperature prediction for the coming decade from a global climate model. *Science* **317,** 796–799 (2007).
48. Zeebe, R. E. & Wolf-Gladrow, D. *CO2 in Seawater: Equilibrium, Kinetics, Isotopes* Vol. 65 (Elsevier, 2001).

Acknowledgements

We thank J. Marotzke, H. Pohlmann and C. Li for their helpful comments and suggestions on the manuscript. We acknowledge funding from the Federal Ministry of Education and Research in Germany (BMBF) through the research programme 'MiKlip' (FKZ: 01LP1144A; 01LP1145A; 01LP1519A; 01LP1157B; and 01LP1517B). This study has also received funding from EU FP7 project CARBOCHANGE (grant 264879) and the European Union's Horizon 2020 research and innovation programme (grant 641816). The simulations were performed at the German Climate Computing Center (DKRZ).

Author contributions

H.L., T.I. and W.M. jointly designed the study and wrote the manuscript; H.L. analysed the data; F.S. provided suggestions and comprehensive support on statistical analyses; all authors discussed the results and commented on the manuscript.

Additional information

Competing financial interests: The authors declare no competing financial interests.

17

Top-down assessment of the Asian carbon budget since the mid 1990s

R.L. Thompson[1], P.K. Patra[2], F. Chevallier[3], S. Maksyutov[4], R.M. Law[5], T. Ziehn[5], I.T. van der Laan-Luijkx[6], W. Peters[6,7], A. Ganshin[8,9], R. Zhuravlev[8,9,10], T. Maki[11], T. Nakamura[12], T. Shirai[4], M. Ishizawa[4], T. Saeki[2], T. Machida[4], B. Poulter[13], J.G. Canadell[14] & P. Ciais[3]

Increasing atmospheric carbon dioxide (CO_2) is the principal driver of anthropogenic climate change. Asia is an important region for the global carbon budget, with 4 of the world's 10 largest national emitters of CO_2. Using an ensemble of seven atmospheric inverse systems, we estimated land biosphere fluxes (natural, land-use change and fires) based on atmospheric observations of CO_2 concentration. The Asian land biosphere was a net sink of -0.46 (-0.70–0.24) PgC per year (median and range) for 1996–2012 and was mostly located in East Asia, while in South and Southeast Asia the land biosphere was close to carbon neutral. In East Asia, the annual CO_2 sink increased between 1996–2001 and 2008–2012 by 0.56 (0.30–0.81) PgC, accounting for $\sim$35% of the increase in the global land biosphere sink. Uncertainty in the fossil fuel emissions contributes significantly (32%) to the uncertainty in land biosphere sink change.

[1] Norsk Institutt for Luftforskning (NILU), Kjeller, Norway. [2] Department of Environmental Geochemical Cycle Research, Japan Agency for Marine Earth Science and Technology (JAMSTEC), Yokohama 236-0001, Japan. [3] Laboratoire des Sciences du Climat et l'Environnement (LSCE, CEA-CNRS-UVSQ), Gif-sur-Yvette, France. [4] National Institute for Environmental Studies (NIES), Tsukuba 305-8506, Japan. [5] Oceans and Atmosphere, Commonwealth Scientific and Industrial Research Organisation (CSIRO), 3195 Aspendale, Australia. [6] Department of Meteorology and Air Quality, Environmental Sciences Group, Wageningen University (WU), 6708 PB Wageningen, The Netherlands. [7] University of Groningen, Centre for Isotope Research, 9747 AG Groningen, The Netherlands. [8] Department of Upper Atmospheric Layers Physics, Central Aerological Observatory (CAO), Moscow 141700, Russia. [9] National Research Tomsk State University (TSU), 634050 Tomsk, Russia. [10] Department of Atmospheric Physics and Microwave Diagnostics, Institute of Applied Physics of the Russian Academy of Sciences, Nizhny Novgorod 603950, Russia. [11] Atmospheric Environment and Applied Meteorology Research Department, Meteorological Research Institute (MRI), Tsukuba 305-0052, Japan. [12] Japan Meteorological Agency (JMA), Global Environment and Marine Department, Tokyo 100-8122, Japan. [13] Institute on Ecosystems and Department of Ecology, Montana State University (MSU), 59717 Bozeman, Montana, USA. [14] Global Carbon Project, Commonwealth Scientific and Industrial Research Organisation (CSIRO), 2601 Canberra, Australia. Correspondence and requests for materials should be addressed to R.L.T. (email: rona.thompson@nilu.no).

Increasing atmospheric carbon dioxide (CO_2) is the single most important driver of climate change, and CO_2 accounts for ~64% of the radiative forcing from well-mixed greenhouse gases[1]. Hence reducing CO_2 emissions is one of the principal concerns of efforts to mitigate anthropogenic climate change. The global mean atmospheric concentration of CO_2 has increased from 277 parts-per-million (p.p.m.) in 1750 (ref. 2) to 396 p.p.m. in 2013 (ref. 3), due to emissions from fossil fuel combustion, cement production and land-use change. In the 1990s, the first international agreements to limit emissions of CO_2 were ratified, the United Nations Framework Convention on Climate Change (UNFCCC), in 1992, and subsequently the Kyoto Protocol, in 1997. Very recently at the Conference of the Parties (COP21) meeting in Paris 2015, it was agreed that further efforts are needed to limit global warming to a maximum of 2 °C. To achieve this, peak greenhouse gas emissions should be reached as soon as possible and by the second half of this century anthropogenic emissions should be balanced by sinks (UNFCCC, CP21, L.9). For such agreements to be effectual, accurate means for assessing the effectiveness of regional and national emission reduction policies are critical, and require better estimates of CO_2 emissions. Also, of upmost importance are improved estimates of the land and ocean biosphere sources and sinks and a better scientific understanding of the carbon cycle[4].

Asia has undergone rapid economic growth over the past two decades, which has been associated with large increases in fossil fuel emissions. Fossil fuel emissions increased between 1990 and 2010 by ~190% in India (to 0.55 PgC per year; 1 PgC $= 1 \times 10^{15}$ g carbon in CO_2) and ~240% in China (to 2.26 PgC per year) (Global Carbon Atlas, http://www.globalcarbonatlas.org) and in 2006 China surpassed the USA as the largest national emitter of CO_2 (ref. 5). Asia has also seen substantial changes in land-cover over the same time period. Between 1990 and 2010, forest extent increased in South Asia, by ~3% (2.1 Mha) and in East Asia by ~22% (45.4 Mha), whereas in Southeast Asia deforestation occurred with a loss in forest extent of ~13% (33.2 Mha)[6]. Deforestation rates are particularly extensive in Indonesia, where the rate of forest clearance (including forested peatlands) reached 0.84 Mha per year in 2012, surpassing that in Brazil[7]. These changes have important consequences for the regional and global carbon budget. Deforestation in Southeast Asia is a large source of CO_2, releasing an estimated average of 0.23 PgC per year for 1990–2007 (ref. 8), while, East Asian ecosystems appear to be a substantial sink of CO_2, taking up an estimated 0.16–0.33 PgC per year due to afforestation/reforestation and regional climate change, especially in Southern China[8–10]. The uncertainty associated with land-use change fluxes, however, is very large.

Top–down approaches, specifically atmospheric inversions, can help to reduce the uncertainty on land biosphere fluxes. Atmospheric inversions use observations of CO_2 concentrations with a model of atmospheric transport to constrain CO_2 fluxes between the land/ocean and atmosphere. Because the problem is under-determined, a prior estimate of the fluxes is needed, and this estimate is updated based on the atmospheric information to give a *posterior* estimate (see Methods). Although atmospheric inversions of CO_2 have been used since the 1980s[11,12], there have been few inversion studies focusing on Asia owing to the fact that until recently few atmospheric CO_2 observations were available for this region. Recent programmes, such as CARIBIC (Civil Aircraft for the Regular Investigation of the atmosphere. On the basis of an Instrument Container)[13], CONTRAIL (Comprehensive Observation Network for TRace gases by AIrLiner)[14,15] and Asian Global Atmospheric Watch (GAW) have helped ameliorate this. CARIBIC data have been utilized in an inversion for CO_2 fluxes in South Asia[16] and CONTRAIL data in inversions focusing on Asian CO_2 fluxes[17–19]. These inversions[16–18] consistently found a net sink for South Asia of −0.11 to −0.37 PgC per year (note that a negative flux is in the direction from the atmosphere to the land/ocean, that is, a sink). In contrast, for Southeast Asia, one recent study[17] found a source of ~0.45 PgC per year, while another[18] found a sink of ~−0.28 PgC per year. Differences among inversion results can arise from differences in modelled atmospheric transport, prior information and its assumed uncertainty, as well as from the observations used. Therefore, to establish confidence in the magnitude and sign of regional CO_2 fluxes, and to determine the range of uncertainty accounting not only for random errors, it is essential to examine ensembles of inversions.

In this study, we assess the carbon budget of Asia using an ensemble of seven atmospheric CO_2 inversions focusing on the regions of East, South and Southeast Asia (see Fig. 1). The inversions use different prior information, transport models and observation data (see Supplementary Table 1) to estimate the land biosphere CO_2 fluxes (including land-use change and fires) from 1996 to 2012. (Note that only four inversions are available for 1996 to 2000). All the inverse systems solve for the global CO_2 fluxes with varying resolutions over the Asian regions. By using global inverse systems it is possible to infer the Asian fluxes consistently with the global carbon budget and the atmospheric growth rate of CO_2 as well as assess how the optimization of the Asian regional fluxes influences the fluxes in the rest of the world. The inversion performance is examined by comparing the CO_2 concentration from the optimized models against independent observations from the CONTRAIL programme between 2008 and 2010. This study focuses on the temporal evolution of the fluxes and the quantification of trends and their uncertainty. This study is part of an effort examining the Asian greenhouse gas budget initiated by RECCAP (REgional Carbon Cycle Assessment and Processes[20]) as part of the international Global Carbon Project (www.globalcarbonproject.org/reccap/).

Results

Uncertainties from fossil fuel and cement inventories. Atmospheric inversions optimize the total land/ocean to atmosphere CO_2 flux (that is, the net biosphere plus the fossil fuel and cement (FFC) flux), therefore, the FFC fluxes must be subtracted *a posteriori* to obtain the land biosphere fluxes. However, the uncertainty in the FFC fluxes is not negligible. Although globally, the FFC flux is quite consistent among all inversions and with the CDIAC (Carbon Dioxide Information Analysis Center, 2013; http://cdiac.ornl.gov) inventory (within ±2%), this is not the case in Asia (Supplementary Fig. 1). In East Asia, the modelling groups assumed different mean values and trends in their FFC fluxes leading to disparities of up to ~1 PgC per year (33%) in 2012, while in South and Southeast Asia, the absolute disparities were smaller but still significant, ~0.2 PgC per year (33–50%) in 2012. The disparities arise from the different inventories on which their estimates were based, for example, CDIAC or EDGAR (Emission Database for Global Atmospheric Research, version 4.2; http://edgar.jrc.ec.europa.eu) and whether a static or evolving spatial distribution was used (see Supplementary Note 1).

Owing to the differences in prior FFC fluxes between inverse systems, it is recommended practice to subtract a standard FFC flux estimate from the total posterior fluxes when comparing land biosphere fluxes from different CO_2 inversions[21]. However, since the total posterior flux may still be biased, and depends to a certain degree on the prior information, biases in the prior FFC fluxes may be propagated into the posterior land biosphere fluxes. In this case, the bias in the land biosphere flux will be proportional to the difference between the prior FFC flux used

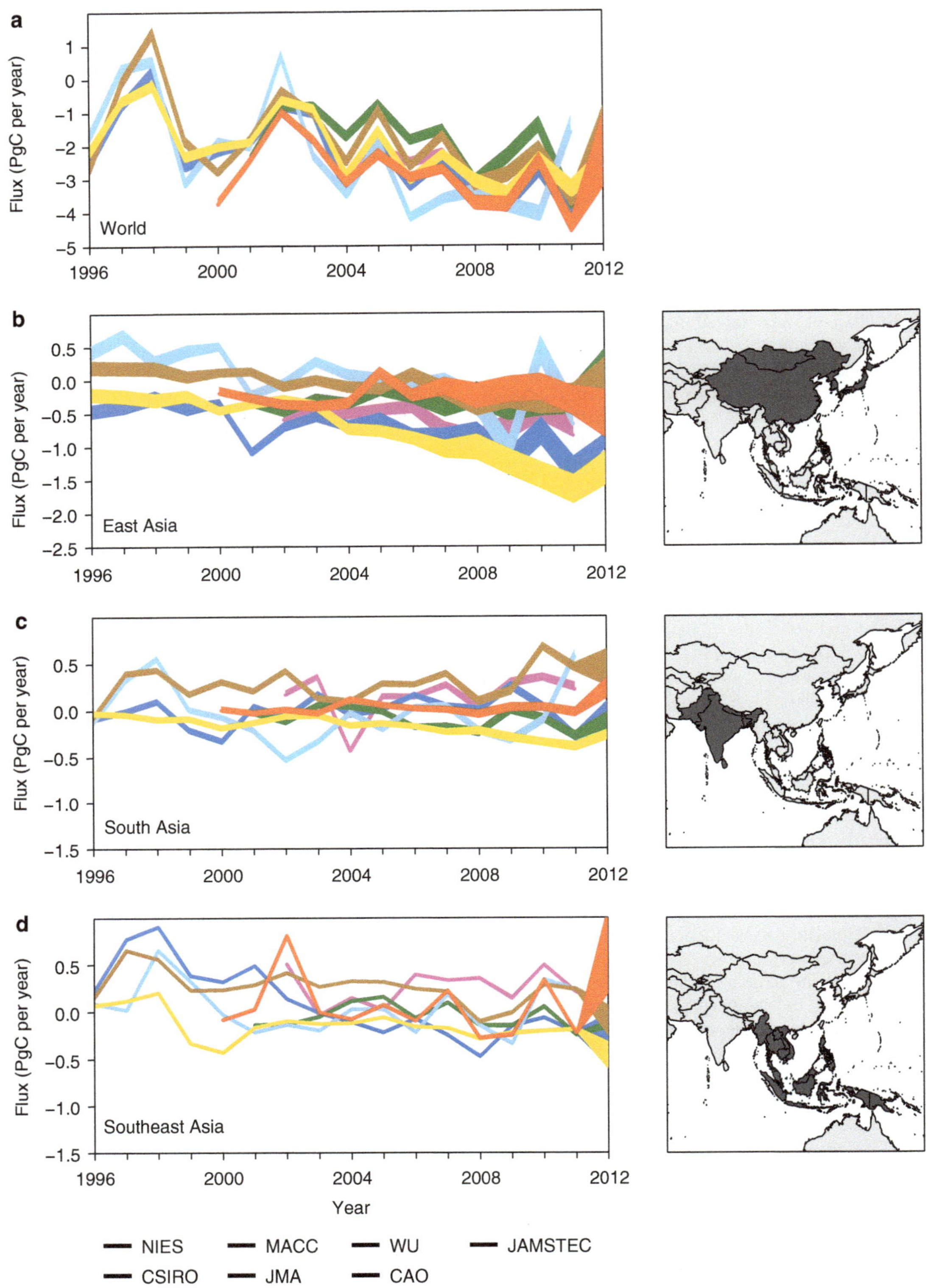

Figure 1 | Land biosphere fluxes estimated by the atmospheric CO_2 inversions. The posterior land biosphere (natural, land-use change and fire) fluxes (PgC per year) from each inversion. The width of each curve shows the range obtained by subtracting the FFC emissions estimates of the CDIAC, EDGAR and IEA inventories from the total posterior flux. Note that a negative flux indicates a carbon sink. The fluxes are shown globally (**a**) and for the regions: East Asia (**b**), South Asia (**c**) and Southeast Asia (**d**) as indicated in the adjacent maps. The inversions are named after the institute that performed them (NIES is the GELCA 64-region inversion system, MACC is the MACC-v13 inversion system used at LSCE, WU is CarbonTracker Europe, JAMSTEC is the ACTM-based inversion system, CSIRO is the CCAM-based inversion system, JMA is the JMA-CDTM-based inversion system, and CAO is the GELCA EOF-based inversion system, for details see Supplementary Table 1).

in each inversion, and the standard FFC flux (see Supplementary Fig. 2). An additional source of uncertainty comes from the choice of standard FFC flux that is subtracted; an overestimate of the standard FFC flux would lead to an overestimate of the land biosphere sink and vice versa. Commonly used FFC fluxes from inventories, i.e., CDIAC, EDGAR and IEA (International Energy Agency, 2014 edition; http://www.iea.org), differ substantially in the trend for East Asia from circa 2002 resulting in differences of ~0.5 PgC per year (~17%) in 2012, much higher than the uncertainty of the inventories of 8% (2σ) reported in ref. 22. (Note that IEA inventory does not include cement emissions so these were accounted for by using the cement emission estimates from CDIAC). Large inconsistencies have been previously reported in Chinese energy-use statistics and emission factors, leading to differences between the reported emissions of 14 to 18% in recent years[23,24], which highlights the considerable systematic uncertainty in the FFC fluxes, especially for East Asia.

Figure 1 shows the posterior (optimized using atmospheric CO_2 observations) land biosphere CO_2 fluxes from each of the inversions (see also Table 1). Note that the land biosphere fluxes include natural, land-use change, and fire fluxes. We show the range of posterior land biosphere fluxes for each inversion calculated using three standard FFC fluxes from the most commonly used inventories: CDIAC, EDGAR and IEA (see Supplementary Note 2). The uncertainty in the posterior land biosphere fluxes increases over time, especially in East Asia, owing to disparities in the FFC fluxes. For the analyses in the following sections, we use an ensemble of seven inversions with three different standard FFC fluxes.

Global land biosphere and ocean fluxes. Before discussing the Asian land biosphere fluxes, we first consider the global total as this provides a baseline for evaluating the inversion performance. The global posterior land fluxes from all inversions exhibit a high degree of consistency in the inter-annual variability and long-term trend (Fig. 1). All inversions show a reduction in the land sink (compared with the decadal mean) in 1997–1998 corresponding to a strong El Niño event (four inversions available) and in 2002–2003 corresponding to persistent El Niño conditions (7 inversions available). El Niño is associated with reduced precipitation, lower rates of net primary productivity (NPP), and increased fire disturbance in tropical and subtropical South America and Africa, Southeast Asia and Australia[25–27]. Six of the inversions also find an increased land sink in 2011, corresponding to a sustained La Niña period, which has been attributed to increased precipitation and net ecosystem uptake in semi-arid regions of South America, South Africa and Australia[28]. The 2011 land sink of -3.6 ± 0.5 (-4.6 to -1.1) PgC per year (median ± median absolute deviation, and full range of the ensemble) is the largest over 1996–2012 and is in good agreement with a recent process-based model estimate[28] with a land sink of -3.9 ± 1.3 PgC per year, which is greater than the 97.5th percentile over 1981–2012.

Over the period 1996–2007, we find a global land sink of -1.7 ± 0.3 (-2.5 to -1.2) PgC per year consistent with two other estimates, based on different methods, for the same time period. The first is from the Global Carbon Project (GCP) and is calculated as the residual between the FFC emissions and the change in atmospheric CO_2 plus the ocean flux based on ocean data[29]. GCP estimates a land sink (that is, the balance of land-use change and natural fluxes) of -1.00 ± 0.64 PgC per year. This, however, does not include the riverine 'loop' of the natural carbon cycle that existed before human perturbation and which is included in the inversion estimates. The riverine loop consists of a land biosphere uptake of CO_2 of 0.45 PgC per year, which is transported via rivers to the ocean where it is outgassed to the atmosphere[30]. Adding the land biosphere flux associated with the riverine loop to the GCP estimate gives a land sink of -1.45 ± 0.64 PgC per year, which is comparable to our estimate. The second estimate of the global land sink, based on the changes in atmospheric O_2 and CO_2 concentrations[31], is -1.1 ± 0.8 PgC per year, and again is not significantly different from our estimate owing to the large uncertainty. Furthermore, we find that the annual land biosphere sink increased by 1.7 ± 0.2 (1.4–2.0) PgC between 1996–2001 and 2008–2012 (based on the four inversions with data before 2000). In most inversions, the sink increased in tropical South America, eastern North America, Siberia and East Asia and, in three inversions, also in Europe. A trend of increasing land sink over the past circa 15 years has also been identified in previous studies[32,33]. At least part of the observed trend may be attributable to the change in occurrence frequency of El Niño and La Niña events between 1996–2004 and 2005–2012: in the second period there were less frequent and intense El Niño events and a sustained La Niña event. To ascertain whether the land sink is increasing also in response to land management changes, CO_2 and nitrogen fertilization, one would need a longer time series to filter-out climate oscillation effects, such as ENSO, and the influence of volcanic aerosols[34]. Process-based ecosystem models and forest inventories suggest that CO_2 fertilization, nitrogen deposition and forest regrowth are largely driving a long-term increase in the land biosphere sink[33].

The mean global ocean sink over 1996–2012 from all inversions is -1.6 ± 0.3 (-2.1 to -1.2) PgC per year and is not statistically different from the estimate from a recent synthesis study[35] for 1990–2009 of -1.55 ± 0.6 PgC per year (note this includes a correction of 0.45 PgC per year for the riverine transport of carbon from the land to the ocean[30], where it is eventually released to the atmosphere). We also find a small trend of increasing ocean sink (Mann Kendall test P value <0.1) with a mean annual increase of 0.01 PgC per year and is within the

Table 1 | Fossil fuel and cement emissions and land and ocean fluxes of CO_2.

Period	Air-borne fraction	Global FFC	Global Land	Global Ocean	East Asia FFC	East Asia Land	South Asia FFC	South Asia Land	Southeast Asia FFC	Southeast Asia Land
1996-2001	0.52 *0.54* (0.48-0.54)	6.69 (6.64-6.83)	−1.58 (−3.11 to −1.13)	−1.70 (−2.45 to −1.29)	1.48 (1.35-1.65)	−0.06 (−0.55 to 0.44)	0.35 (0.28-0.46)	−0.06 (−0.10 to 0.12)	0.22 (0.17-0.28)	0.26 (−0.09 to 0.53)
2002-2007	0.56 *0.57* (0.53-0.60)	7.86 (7.55-7.94)	−1.94 (−2.51 to −1.09)	−1.57 (−2.13 to −0.65)	2.03 (1.59-2.31)	−0.33 (−0.77 to 0.13)	0.45 (0.33-0.59)	−0.09 (−0.18 to 0.08)	0.27 (0.22-0.37)	0.02 (−0.12 to 0.31)
2008-2012	0.46 *0.46* (0.44-0.49)	9.12 (8.87-9.24)	−3.17 (−3.65 to −2.28)	−1.77 (−2.36 to −1.31)	2.85 (1.85-3.01)	−0.46 (−1.18 to −0.01)	0.64 (0.39-0.83)	−0.07 (−0.33 to 0.04)	0.34 (0.26-0.45)	−0.03 (−0.25 to 0.32)

FFC, fossil fuel and cement.
The median (and range) of the FFC emissions, land biosphere and ocean flux estimates from the inversions (PgC per year) are given. Note that negative numbers indicate a flux from the atmosphere to the land/ocean. The land biosphere fluxes (natural, land-use change and fire) are from the inversions and using the CDIAC, EDGAR, and IEA FFC emission estimates. For East Asia, JMA was excluded for the period 2008-2012. For South Asia, NIES, CSIRO and CAO were excluded. Also shown is the median (and range) of the global air-borne fraction of the FFC emissions from the inversions and calculated directly from atmospheric observations (in italics; data from the Global Carbon Project).

range of trends found by the synthesis study[35] for the past two decades of 0.01–0.05 PgC per year. However, the ocean sink is somewhat dependent on the prior ocean flux estimates used in the inversions, and thus the two estimates are not completely independent.

Assessment of the Asian fluxes. While the global fluxes from inversions are mass conserving, this is not the case on regional scales. Therefore, we evaluate the total posterior flux from each inversion by comparing atmospheric concentrations from forward simulations with independent observations (that is, not assimilated in the inversion) from CONTRAIL aircraft measurements[14]. The CONTRAIL measurements were made using an *in situ* analyser onboard commercial aircraft with regular flights from Japan to other Asian countries (for further details see Supplementary Note 3). We selected CONTRAIL data at seven locations where vertical profiles are available, that is, above the international airports at Bangkok, Singapore, Jakarta and Manila (used to evaluate the Southeast Asian fluxes), Seoul and Taipei (East Asian) and Delhi (South Asian) for 2008–2010. The models were sampled at the times and locations of the observations (for JMA and CSIRO only monthly mean model values were available).

To determine the importance of systematic errors in the inverse models for the regional flux estimates, we compared the annual mean posterior flux with the modelled and observed afternoon CO_2 vertical gradients and planetary boundary layer (PBL) concentrations (Fig. 2). The vertical gradient was calculated between 1 and 4 km, that is, from the PBL to the free troposphere. As well as the annual mean, the comparison was made for winter (December to February, DJF) and summer (June to August, JJA) vertical gradients and PBL concentrations (Supplementary Fig. 3). Systematic over-/underestimation of the CO_2 vertical gradient compared with that observed can indicate a bias in the posterior fluxes and may be due to too weak/strong vertical transport[36].

For South and Southeast Asia, the annual vertical CO_2 gradient and PBL concentration differences are positively correlated with the posterior fluxes, such that large modelled gradients and PBL concentrations correspond to large fluxes, and vice versa. This suggests that the model-observation differences are related to errors in the fluxes rather than to systematic biases in the vertical mixing between the PBL and free troposphere. In summer and winter, the model-observation differences for South Asia are much larger than for the other regions and may owe to the fact that the comparison is based on only one site, Delhi, which is continental and strongly influenced by local fluxes. For East Asia, there is no correlation between the annual (and winter) vertical gradients and the posterior fluxes, and the modelled gradients and PBL concentrations are close to those observed, suggesting that there is no significant bias in the annual mean (and winter) fluxes. In summer, however, the correlation is negative and the s.d. of the modelled and observed gradients is larger. This may result from the stronger influence from the southeasterly direction during this season, and hence the vertical gradients and PBL concentrations represent a combination of land and ocean influences.

Since the summer and winter model-observation differences are generally in the same direction for each model, that is, the summer and winter errors do not compensate, we use the annual mean model-observation differences to evaluate the posterior fluxes from each inversion. The annual means also contain the largest number of data points and are, therefore, more robust. For East and Southeast Asia, we do not find any strong bias in the gradients or PBL concentrations. Therefore, on this basis we do not omit any of the inversions from the ensemble for these two

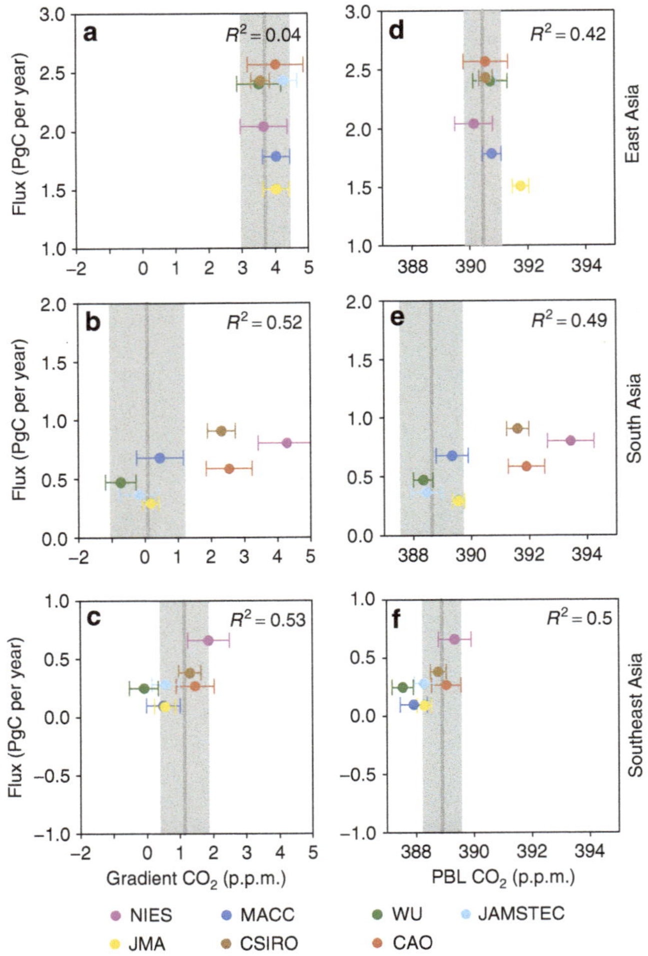

Figure 2 | Posterior total fluxes versus CO_2 vertical gradients and PBL concentrations for 2008-2010. The annual mean flux (PgC per year) versus CO_2 gradient (p.p.m.) between 1 and 4 km for each inversion is shown for East Asia (**a**), South Asia (**b**) and Southeast Asia (**c**). Also shown is the annual mean flux versus the PBL CO_2 concentration (p.p.m.) for East Asia (**d**), South Asia (**e**) and Southeast Asia (**f**). Each of the inversions was sampled at the times and locations of the CONTRAIL aircraft data and the error bars show the 1-σ s.d. of the inversion results. The grey vertical lines indicate the values of the observed gradients and PBL concentrations and the shading shows the 1-σ s.d. The correlation (R^2) between the fluxes and the gradients and PBL concentrations is given in each plot.

regions. However, for South Asia, the inversions NIES, CSIRO and CAO show large positive biases for the vertical gradient and PBL concentrations (>2 p.p.m.), therefore, we have omitted these results from the ensemble for this region.

East Asia. We find that the East Asian land biosphere was on average a sink of −0.29 (−0.81 to 0.18) PgC per year for 1996–2012. Some of the differences among the inversion results for East Asia can be related to differences in the prior information (Supplementary Figs 1, 2 and 4). The JMA inversion used FFC emission estimates with a fixed spatial distribution based on that in 1995 and only scaled the global total. Because the growth rate in East Asia for the 2000s was higher than the global mean, this resulted in lower FFC and total fluxes in the JMA inversion for East Asia; the total posterior flux in JMA for 2008–2012 was 36% lower than the mean of the other inversions. Thus, subtracting the standard FFC flux from the JMA posterior total results in a more negative land biosphere flux compared with the other inversions. As CONTRAIL data were not available for the latter part of this period, when the bias is largest, this could not be seen in the comparison with independent atmospheric observations. In any case, to avoid potentially biasing the results, we exclude the JMA inversion from the ensemble statistics for East Asia for 2008-2012.

The three inversions with data before 2001 (and excluding JMA), all indicate an increasing land sink (that is, the flux became more negative) over time (Mann Kendall test P value <0.01). From the ensemble, we estimate that the land sink increased from −0.06 (−0.55 to 0.44) PgC per year in 1996–2001 to −0.46 (−1.18 to −0.01) PgC per year in 2008–2012; with a median difference of 0.56 (0.30–0.81) PgC per year (Fig. 3). The three inversions (MACC, JAMSTEC and CSIRO) also show a small increase in the amplitude of the seasonal cycle in the East Asian land biosphere flux in 2008–2012 relative to 1996–2001 (Supplementary Fig. 5), while there was no trend in the amplitude of the seasonal cycle in the FFC fluxes used in the inversions.

Our land sink estimate for 2001–2005, of −0.33 (−0.77 to 0.10) PgC per year, is commensurate with a study using an ecosystem model[37], which found an average sink for China of approximately −0.27 PgC per year for the same time period. Our results are also consistent with a recent study[38] based on a combination of satellite measurements of greenness, process-based ecosystem models, and atmospheric inversions, which found a sink for East Asia of −0.22 to −0.29 PgC per year for 1990–2009, compared with our estimate of −0.25 (−0.74 to 0.18) PgC per year for 1996–2009. The increase in the land sink is supported by forest inventory data, which show an increase in the forest sink in China of ~34% between 1990–1999 and 2000–2007 owing to intensive national afforestation/reforestation programmes[8]. To investigate this further, we examine the Normalized Difference Vegetation Index (NDVI; from the Advanced Very High Resolution Radiometer from NOAA satellites, see Supplementary Note 4) and an independent estimate of Net Ecosystem Exchange (NEE) based on an up-scaling model of CO_2 eddy covariance data[39]. Figure 4 shows the median and range of the East Asian land biosphere flux given by the six inversions (excluding JMA) for the three standard FFC fluxes, as well as NDVI and the eddy-covariance-based NEE anomaly over East Asia. Although NDVI showed no significant trend, NEE became more negative (that is, a larger sink) over the same period by ~0.2 PgC. Thus, a sink increase is supported by

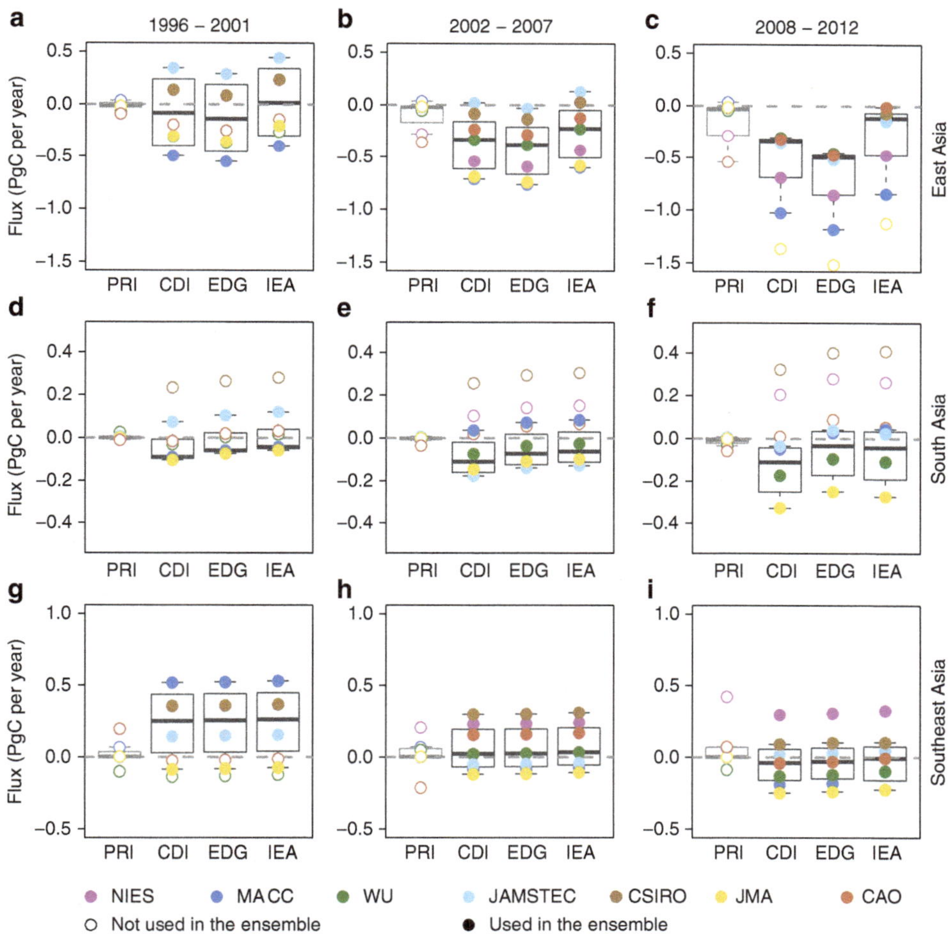

Figure 3 | Summary of the land biosphere fluxes based on the inversion ensemble. Prior and posterior land biosphere fluxes (PgC per year) for intervals 1996-2001 (**a**,**d**,**g**), 2002-2007 (**b**,**e**,**h**) and 2008-2012 (**c**,**f**,**i**). The land biosphere fluxes are calculated by subtracting CDIAC (CDI), EDGAR (EDG) and IEA FFC emission estimates, respectively, from the total posterior fluxes. The thick horizontal bar within the box shows the median, the box extent shows the upper and lower quartiles, and the whiskers extend to the furthest point within the range of the median plus/minus the interquartile range. The open circles indicate the prior (PRI) and the solid circles the posterior estimate from the individual inversions. For East Asia, JMA was excluded for 2008-2012, while for South Asia, NIES, CSIRO and CAO were excluded from the ensembles and are also shown as open circles. Inversions that only started in 2000 or 2001 were also excluded from the 1996-2001 ensembles. The grey dashed line indicates the level of a neutral land biosphere (that is, zero flux).

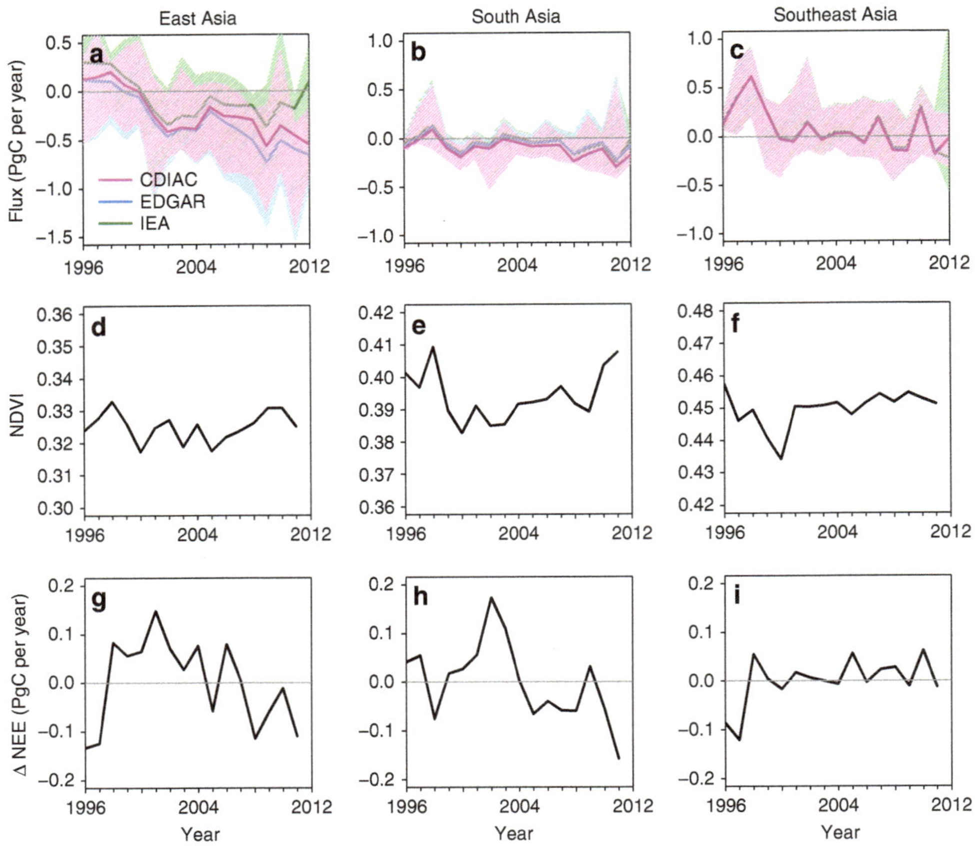

Figure 4 | Trends in the posterior land biosphere fluxes. The land biosphere fluxes (PgC per year), NDVI and eddy-covariance-based NEE anomalies (PgC per year) are shown as annual means for East Asia (**a,d,g**), South Asia (**b,e,h**) and Southeast Asia (**c,f,i**). The fluxes are calculated using FFC emission estimates from CDIAC (purple), EDGAR (blue) and IEA (green), the solid line shows the median and the shading indicates the range of the inversion ensemble. For East Asia, JMA was excluded, while for South Asia, NIES, CSIRO and CAO were excluded. (Negative NEE indicates a carbon sink.)

independent data and, in addition to afforestation/reforestation, may be driven by regional climate change; the temperature in East Asia increased by 0.4 °C between 1999 and 2009 (ref. 38).

The increase in the East Asian sink from the inversion ensemble is larger than suggested by either the eddy-covariance-based NEE or the forest inventory estimates. The larger sink increase in our estimates may be partly due to the uncertainty in the FFC fluxes. For example, using FFC fluxes with a low growth rate, such as IEA, the sink increase between 1996–2001 and 2008–2012 is 0.45 (0.30–0.59) PgC, while using a high growth rate estimate, such as EDGAR, it is 0.63 (0.56–0.81) PgC (Fig. 4). In summary, while atmospheric observations support an increase in the land biosphere sink in East Asia, the amount by which it has increased cannot be accurately determined until the uncertainty in FFC emissions is substantially reduced.

South Asia. From the inversion ensemble, we find that the land biosphere in South Asia was close to carbon neutral, − 0.05 (− 0.18 to 0.03) PgC per year for 1996–2012. Similarly to East Asia, some differences between the inversion results are apparent owing to the FFC flux estimates. The JMA inversion used an FFC flux with a low growth rate (lower than the other inversions and the three standard FFC fluxes) and correspondingly this inversion shows the largest increase in land biosphere sink. On the other hand, the CSIRO inversion used a high FFC flux and this inversion also shows the most positive land biosphere flux.

While we find that the South Asian land biosphere was nearly carbon neutral, previous inversion studies[16–18] found it to be a carbon sink of − 0.10 to − 0.37 PgC per year between 2002 and 2008. However, for the overlapping period of these studies, that is, 2007–2008, our estimate of − 0.13 (− 0.24–0.07) PgC per year is not significantly different. We also compare our results with a study based on a synthesis of 10 ecosystem models[40] for the period 2000–2009. We find a smaller sink of − 0.11 (− 0.19 to 0.06) PgC per year compared with the ecosystem model estimate of -0.19 ± 0.19 PgC per year.

Of the inversions with data before 2001, only one (JMA) show a significant trend in the land sink for South Asia and this may be due to the low growth rate in its FFC fluxes (Supplementary Fig. 2). Furthermore, we found no change in the amplitude of the seasonal cycle in land biosphere flux over the 17-year period, consistent with the findings of ecosystem models[40]. The overall absence of a trend is in contrast to the decrease in the eddy-covariance-based NEE estimate since 2002 (Fig. 4). Aside from possible errors in the eddy-covariance-based NEE (the discussion of which is outside the scope of our study) a possible explanation for the discrepancy is the poor atmospheric observation coverage for South Asia, and thus modest uncertainty reduction, 27% (9–37%, Supplementary Fig. 6), which may mean that the trend could not be detected. Alternatively, if the trend in FFC emissions were underestimated, this would also lead to an underestimation of the increase in the land sink from the inversions.

Southeast Asia. For 1996–2001, three of the four inversions with data throughout this time period found the Southeast Asian land biosphere to be a source, and the ensemble median and range was 0.26 (− 0.09 to 0.53) PgC per year. This period included the strong El Niño event of 1997–1998, which was associated with drought and lower rates of NPP, as well as increased fire disturbance[27,41]. For 2002–2012, however, we could not determine any difference from carbon neutrality for the land biosphere, − 0.02 (− 0.18 to 0.27) PgC per year. Two inversions (MACC and CSIRO) show a significantly increasing land sink in Southeast Asia over 1996–2012 (Mann Kendall test P value < 0.1), which mostly results from the large positive land biosphere flux (source) in 1997–1998 related to strong El Niño conditions. Only one inversion (NIES) estimates a significant source for 2002–2012. This can be explained by the use of a positive prior land flux, the influence of which can be seen in the posterior land fluxes.

Previous inversion studies have given contradicting results for Southeast Asia, that is, some estimate a source[17] and others a sink[18]. For the average over 1996–2012, we find a weak source for the Southeast Asian land biosphere of 0.06 (− 0.15 to 0.28) PgC per year, which is consistent with observed land-use changes (13% of forested area was lost in Southeast Asia between 1990 and 2010 (ref. 6)) and with estimates of above ground biomass loss of 0.09 PgC per year for the average over 1993–2012 based on satellite observations[42]. Our results, however, suggest that the Southeast Asian source reduced in 2002–2012 compared with 1996–2001, and all of the inversions with the data before 2001, show lower mean fluxes over the latter period. Furthermore, since 2001 the eddy-covariance-based NEE has been close to zero, while NDVI has been stable, supporting our finding of a near-neutral land biosphere for 2002–2012. The change to a neutral land biosphere may reflect the slowing rate of deforestation in Southeast Asia; according to FAO statistics the rate of deforestation slowed to 1.1 Mha per year between 2000 and 2010 compared with 2.7 Mha per year between 1990 and 2000 (ref. 6).

Discussion

We have used an ensemble of seven atmospheric CO_2 inversions and three global FFC flux estimates, based on the inventories of CDIAC, EDGAR and IEA, to infer the land biosphere fluxes for East, South and Southeast Asia, and to ascertain the robustness and overall uncertainty of the results. Asia as a whole emitted an average of 2.7–2.9 PgC per year over 1996–2012 from fossil fuel combustion and cement production with an average increase of ~5% per year. Of this, an average of 17% (0–26%) of the emitted carbon was sequestered by the land biosphere, with most of the sink located in East Asia (Supplementary Fig. 7). This is smaller than the mean fraction, 28% (20–38%) of global fossil fuel emissions sequestered by the global land biosphere.

The East Asian land biosphere was on average a carbon sink of − 0.29 (− 0.81 to 0.18) PgC per year or equivalently 14% (0–39%) of East Asia's total FFC emissions, over 1996–2012. Between 1996–2001 and 2008–2012, we find an increase in the annual sink of 0.56 (0.30–0.81) PgC, which explains 35% of the increase in the global land sink. However, the magnitude of this increase in the East Asian sink is contingent on the assumed increase in FFC emissions (the inventories differ by up to 17% for East Asia). The uncertainty in the growth rate of the FFC emissions in the inventories contributes 32% to the uncertainty in the land sink change; if China's emissions have grown commensurate with the low rate estimate of for example, IEA, then the land sink increase will be at the lower end of the range, and vice versa if they have followed the high rate estimate of for example, EDGAR. This highlights the need for better regional estimates of FFC emissions and trends for determining land biosphere fluxes from inverse systems. For South Asia, we find that on average the land biosphere was close to carbon neutral, − 0.05 (− 0.18 to 0.03) PgC per year over 1996–2012 and that there was no significant trend. During this time period, there was only a modest net increase, ~2 Mha, in forested area as afforestation/reforestation in India was nearly counter balanced by deforestation in Nepal and Pakistan[6]. For Southeast Asia, we find that the land biosphere was on average a carbon source between 1996 and 2001 of 0.26 (− 0.09 to 0.53) PgC per year but between 2002 and 2012, we cannot determine any difference from carbon neutrality with a flux of − 0.02 (− 0.18 to 0.27) PgC per year.

Our results contrast with the general findings of land surface models[33], which allocate the increase in the global land sink to tropical and southern regions with little change to even a decrease in the temperate land sink over the past two decades. This disparity is likely owing to land-use changes; for instance, the regrowth of forests is not accounted for in land surface models but probably contributes substantially to the increasing land sink in East Asia[8]. While in tropical Asia the land biosphere models likely overestimate the increase in the sink, again by not accounting for land-use and land-cover changes, and because few land surface models include a coupled carbon–nitrogen cycle and as a consequence may have a too strong CO_2 fertilization response.

Methods

Inversion methodology. All of the inversion systems used in this study have been previously described[16,21,43–48] and are summarized in Supplementary Table 1. Atmospheric CO_2 inversions optimize the land/ocean to atmosphere fluxes using spatiotemporal gradients in atmospheric CO_2 concentration from observations. The relationship between the fluxes, $\mathbf{x}$ (the quantity to be optimized), and CO_2 concentration, $\mathbf{y}$, is described by an atmospheric transport model, H, and can be written as:

$$\mathbf{y} = H(\mathbf{x}) + \boldsymbol{\varepsilon} \tag{1}$$

where $\boldsymbol{\varepsilon}$ is error associated with model representation (such as scale differences between the model and the observations), transport and measurements. Since there is insufficient information to solve equation (1) for the fluxes, statistical optimization methods are used, which require prior information. Using the Bayesian formalism, the problem can be expressed as the maximization of the probability density function of the fluxes given the prior information and observations and is equivalent to finding the minimum of the cost function:

$$J(\mathbf{x}) = \frac{1}{2}(\mathbf{x} - \mathbf{x}_b)^T \mathbf{B}^{-1}(\mathbf{x} - \mathbf{x}_b) + \frac{1}{2}(\mathbf{y} - H(\mathbf{x}))^T \mathbf{R}^{-1}(\mathbf{y} - H(\mathbf{x})) \tag{2}$$

where $\mathbf{B}$ and $\mathbf{R}$ are the error covariance matrices for the prior fluxes and the observations, respectively. The fluxes that minimize the cost function can be found by solving the first order derivative of equation (2), and requires the gradient of the transport operator. The inversion systems can be grouped into three basic categories according to the way they minimize the cost function: (1) inversions using a variational method and an adjoint model of transport and which resolve the fluxes at the spatial resolution of the transport model (MACC); (2) inversions using a Bayesian synthesis for a number of predefined regions (JMA, JAMSTEC and CSIRO); and (3) inversions using Kalman smoother methods (WU, NIES and CAO). In each framework, the total CO_2 flux is optimized and the posterior land biosphere fluxes are found by subtracting a standard FFC emissions estimate from the total. For access to the inversion and/or processing code, a request should be sent to the authors.

Prior information and observations. For the prior land biosphere fluxes, JAMSTEC, CSIRO and JMA used estimates based on the ecosystem model, CASA[49] while WU used the SiBCASA model[50], MACC used the ORCHIDEE model[51] and NIES and CAO used the VISIT model[52]. For the prior ocean fluxes, JAMSTEC, CSIRO, JMA and MACC used a pCO_2 observation-based estimate[53], WU used ocean fluxes estimated from an ocean-data inversion[30], while NIES and CAO used fluxes from the OTTM ocean-data inversion model[54]. The observation dataset used also differed among the inversions (Supplementary Fig. 8). NIES, CAO and WU used observations from the NOAA ESRL GMD Observation Package[55] and assimilated values at the time of sampling, or in the case on WU, as daily afternoon averages for sites below 1,000 m and daily morning averages for sites above 1,000 m. CSIRO and JAMSTEC used the GlobalView-CO_2 package[56] at monthly intervals. JMA also used monthly mean observations but used its own set of observations including 69 ground-based sites and 21 ocean-based sites as well as aircraft profiles and ship transects. MACC used 33 *in situ* sites and 106 flask sites and assimilated the data as daily afternoon averages for sites below 1,000 m and daily morning averages for sites above 1000 m.

References

1. Myhre, G. *et al.* in *Working Group I Contribution to the IPCC Fifth Assessment Report Climate Change 2013: The Physical Science Basis* (eds Jacob, D., Ravishankara, A. R. & Shine, K.) (2013).
2. Etheridge, D. M. *et al.* Natural and anthropogenic changes in atmospheric CO_2 over the last 1000 years from air in Antarctic ice and firn. *J. Geophys. Res.Atmos* **101,** 4115–4128 (1996).
3. WMO. The state of greenhouse gases in the atmosphere based on global observations through 2013. *WMO Greenhouse Gas Bull.* **10,** 8 (2014).
4. Ciais, P. *et al.* Current systematic carbon-cycle observations and the need for implementing a policy-relevant carbon observing system. *Biogeosciences* **11,** 3547–3602 (2014).
5. Gregg, J. S., Andres, R. J. & Marland, G. China: Emissions pattern of the world leader in CO_2 emissions from fossil fuel consumption and cement production. *Geophys. Res. Lett.* 35, L08806 (2008).
6. FAO. *Global Forest Resource Assessment* (Food and Agriculture Organization of the United Nations, 2010).
7. Margono, B. A., Potapov, P. V., Turubanova, S., Stolle, F. & Hansen, M. C. Primary forest cover loss in Indonesia over 2000-2012. *Nat. Clim. Change* **4,** 730–735 (2014).
8. Pan, Y. *et al.* A large and persistent carbon sink in the World's forests. *Science* **333,** 988–993 (2011).
9. Piao, S. *et al.* The carbon balance of terrestrial ecosystems in China. *Nature* **458,** 1009–1013 (2009).
10. Zhang, H. F. *et al.* Net terrestrial CO_2 exchange over China during 2001-2010 estimated with an ensemble data assimilation system for atmospheric CO_2. *J. Geophys. Res.Atmos* **119,** 3500–3515 (2014).
11. Enting, I. G. & Mansbridge, J. V. Seasonal sources and sinks of atmospheric CO_2 direct inversion of filtered data. *Tellus B* **41,** 111–126 (1989).
12. Tans, P. P., Conway, T. J. & Nakazawa, T. Latitudinal distribution of the sources and sinks of atmospheric carbon dioxide derived from surface observations and an atmospheric transport model. *J. Geophys. Res.Atmos* **94,** 5151–5172 (1989).
13. Brenninkmeijer, C. A. M. *et al.* Civil Aircraft for the regular investigation of the atmosphere based on an instrumented container: The new CARIBIC system. *Atmos. Chem. Phys.* **7,** 4953–4976 (2007).
14. Machida, T. *et al.* Worldwide measurements of atmospheric CO_2 and Other trace gas species using commercial airlines. *J. Atmos. Ocean. Technol.* **25,** 1744–1754 (2008).
15. Matsueda, H. *et al.* Evaluation of atmospheric CO_2 measurements from new flask air sampling of JAL airliner observations. *Pap. Meteor. Geophys.* **59,** 1–17 (2008).
16. Patra, P. K. *et al.* Carbon balance of South Asia constrained by passenger aircraft CO_2 measurements. *Atmos. Chem. Phys.* **11,** 4163–4175 (2011).
17. Niwa, Y. *et al.* Imposing strong constraints on tropical terrestrial CO_2 fluxes using passenger aircraft based measurements. *J. Geophys. Res. Atmos* **117,** D11303 (2012).
18. Jiang, F. *et al.* Carbon balance of China constrained by CONTRAIL aircraft CO_2 measurements. *Atmos. Chem. Phys.* **14,** 10133–10144 (2014).
19. Zhang, H. F. *et al.* Estimating Asian terrestrial carbon fluxes from CONTRAIL aircraft and surface CO_2 observations for the period 2006-2010. *Atmos. Chem. Phys.* **14,** 5807–5824 (2014).
20. Canadell, J. G. *et al.* An international effort to quantify regional carbon fluxes. *EOS Trans. Am. Geophys. Union* 92, 81–82, 2011).
21. Peylin, P. *et al.* Global atmospheric carbon budget: results from an ensemble of atmospheric CO_2 inversions. *Biogeosciences* **10,** 6699–6720 (2013).
22. Andres, R. J., Boden, T. A. & Higdon, D. A new evaluation of the uncertainty associated with CDIAC estimates of fossil fuel carbon dioxide emission. *Tellus B* **66,** 23616 (2014).
23. Guan, D., Liu, Z., Geng, Y., Lindner, S. & Hubacek, K. The gigatonne gap in China's carbon dioxide inventories. *Nat. Clim. Change* **2,** 672–675 (2012).
24. Liu, Z. *et al.* Reduced carbon emission estimates from fossil fuel combustion and cement production in China. *Nature* **524,** 335–338 (2015).
25. Myneni, R. B., Los, S. O. & Tucker, C. J. Satellite-based identification of linked vegetation index and sea surface temperature anomaly areas from 1982-1990 for Africa, Australia and South America. *Geophys. Res. Lett.* **23,** 729–732 (1996).
26. Woodward, F. I., Lomas, M. R. & Quaife, T. Global responses of terrestrial productivity to contemporary climatic oscillations. *Philos. Trans. R. Soc. Lond. B Biol. Sci.* **363,** 2779–2785 (2008).
27. Patra, P. K., Maksyutov, S. & Nakazawa, T. Analysis of atmospheric CO_2 growth rates at Mauna Loa using CO_2 fluxes derived from an inverse model. *Tellus B* **57,** 357–365 (2005).
28. Poulter, B. *et al.* Contribution of semi-arid ecosystems to interannual variability of the global carbon cycle. *Nature* **509,** 600–603 (2014).
29. Le Quéré, C. *et al.* Global carbon budget 2013. *Earth Syst. Sci. Data* **6,** 235–263 (2014).
30. Jacobson, A. R., Mikaloff Fletcher, S. E., Gruber, N., Sarmiento, J. L. & Gloor, M. A joint atmosphere-ocean inversion for surface fluxes of carbon dioxide: 1. methods and global-scale fluxes. *Global Biogeochem. Cycles* **21,** GB1019 (2007).
31. Keeling, R. F. & Manning, A. C. *Studies of Recent Changes in Atmospheric O_2 Content in Treatise on Geochemistry* 2nd Edn 385–404 (Elsevier, 2014).
32. Ballantyne, A. P., Alden, C. B., Miller, J. B., Tans, P. P. & White, J. W. C. Increase in observed net carbon dioxide uptake by land and oceans during the past 50 years. *Nature* **488,** 70–72 (2012).
33. Sitch, S. *et al.* Recent trends and drivers of regional sources and sinks of carbon dioxide. *Biogeosciences* **12,** 653–679 (2015).
34. Raupach, M. R., Canadell, J. G. & Le Quéré, C. Anthropogenic and biophysical contributions to increasing atmospheric CO_2 growth rate and airborne fraction. *Biogeosciences* **5,** 1601–1613 (2008).
35. Wanninkhof, R. *et al.* Global ocean carbon uptake: magnitude, variability and trends. *Biogeosciences* **10,** 1983–2000 (2013).
36. Stephens, B. B. *et al.* Weak northern and strong tropical land carbon uptake from vertical profiles of atmospheric CO_2. *Science* **316,** 1732–1735 (2007).
37. Tian, H. *et al.* Net exchanges of CO_2, CH_4, and N_2O between China's terrestrial ecosystems and the atmosphere and their contributions to global climate warming. *J. Geophys. Res.* **116,** G02011 (2011).
38. Piao, S. L. *et al.* The carbon budget of terrestrial ecosystems in East Asia over the last two decades. *Biogeosciences* **9,** 3571–3586 (2012).
39. Jung, M. *et al.* Global patterns of land-atmosphere fluxes of carbon dioxide, latent heat, and sensible heat derived from eddy covariance, satellite, and meteorological observations. *J. Geophys. Res. Biogeosci.* **116,** G00J07 (2011).
40. Patra, P. K. *et al.* The carbon budget of South Asia. *Biogeosciences* **10,** 513–527 (2013).
41. van der Werf, G. R. *et al.* Continental-scale partitioning of fire emissions during the 1997 to 2001 El Niño/La Niña period. *Science* **303,** 73–76 (2004).
42. Liu, Y. Y. *et al.* Recent reversal in loss of global terrestrial biomass. *Nat. Clim. Change* **5,** 470–474 (2015).
43. Chevallier, F. *et al.* Inferring CO_2 sources and sinks from satellite observations: Method and application to TOVS data. *J. Geophys. Res.* **110,** D24309 (2005).
44. Maki, T. *et al.* New technique to analyse global distributions of CO_2 concentrations and fluxes from non-processed observational data. *Tellus B* **62,** 797–809 (2010).
45. Rayner, P. J. *et al.* Interannual variability of the global carbon cycle (1992-2005) inferred by inversion of atmospheric CO_2 and $\delta^{13}CO_2$ measurements. *Global Biogeochem. Cycles* **22,** GB3008 (2008).
46. Peters, W. *et al.* Seven years of recent European net terrestrial carbon dioxide exchange constrained by atmospheric observations. *Global Change Biol.* **16,** 1317–1337 (2010).
47. Bruhwiler, L. M. P., Michalak, A. M., Peters, W., Baker, D. F. & Tans, P. An improved Kalman Smoother for atmospheric inversions. *Atmos. Chem. Phys.* **5,** 2691–2702 (2005).
48. Zhuravlev, R. V., Ganshin, A. V., Maksyutov, S. S., Oshchepkov, S. L. & Khattatov, B. V. Estimation of global CO_2 fluxes using ground-based and satellite (GOSAT) observation data with empirical orthogonal functions. *Atmos. Oceanic Opt.* **26,** 507–516 (2013).
49. Randerson, J. T., Thompson, M. V., Conway, T. J., Fung, I. Y. & Field, C. B. The contribution of terrestrial sources and sinks to trends in the seasonal cycle of atmospheric carbon dioxide. *Global Biogeochem. Cycles* **11,** 535–560 (1997).
50. van der Velde, I. R. *et al.* Terrestrial cycling of $^{13}CO_2$ by photosynthesis, respiration, and biomass burning in SiBCASA. *Biogeosciences* **11,** 6553–6571 (2014).
51. Krinner, G. *et al.* A dynamic global vegetation model for studies of the coupled atmosphere-biosphere system, Global Biogeochem. *Cycles* 19, GB1015 (2005).
52. Saito, M., Ito, A. & Maksyutov, S. Optimization of a prognostic biosphere model for terrestrial biomass and atmospheric CO_2 variability. *Geosci. Model Dev* **7,** 1829–1840 (2014).
53. Takahashi, T. *et al.* Climatological mean and decadal change in surface ocean pCO_2, and net sea-air CO_2 flux over the global oceans. *Deep Sea Res. II* **56,** 554–577 (2009).
54. Valsala, V. & Maksyutov, S. Simulation and assimilation of global ocean pCO_2 and air-sea CO_2 fluxes using ship observations of surface ocean pCO_2 in a simplified biogeochemical offline model. *Tellus B* **62,** 821–840 (2010).
55. Cooperative Global Atmospheric Data Integration Project, updated annually. Multi-laboratory compilation of atmospheric carbon dioxide data for the period 2000–2012 (obspack_co2_1_PROTOTYPE_v1.0.4_2013-11-25), NOAA Global Monitoring Division, Boulder, CO, USA.
56. Cooperative Global Atmospheric Data Integration Project. 2013, updated annually. Multi-laboratory compilation of synchronized and gap-filled atmospheric carbon dioxide records for the period 1979–2012 (obspack_co2_1_GLOBALVIEW-CO2_2013_v1.0.4_2013-12-23). Compiled by NOAA Global Monitoring Division: Boulder, CO, USA.

Acknowledgements

We thank the support of the Asia Pacific Network for Global Change for the grant 'Greenhouse gas budgets of South and Southeast Asia'. J.G.C. thanks the Australian Climate Change Science Program for their support. P.K.P., T.S., M.I. and S.M. acknowledge the support of the Environment Research and Technology Development Fund (2-1401) of the Ministry of the Environment, Japan. I.T.L.L. and W.P. received funding from the China Exchange Program of the Dutch Royal Academy of Research (KNAW) through travel grant no. 12CDP006 and a grant for computing time (SH-060-13) from the Netherlands Organization for Scientific Research (NWO). CSIRO inversions were undertaken on the NCI National Facility, supported by the Australian Commonwealth Government. We also gratefully acknowledge the following people who provided atmospheric observations that were used in the inversions: J. A. Morgui, E. Brunke, S. Hammer, M. Delmotte, M. Leuenberger, H. Meijer, H. Matsueda, E. Gloor, P. Krummel, M. van der Schoot, B. Stephens, M. Takahashi, M. Torn, K. Tsuboi, M. Ramonet, D. Worthy, S. Morimoto, A. J. Gomez-Pelaez, T. Nakazawa, L. V. Gatti, S. Aoki, K. Masarie, A. Andrews, R. Langenfelds, S. Biraud, C. Sweeney, M. Steinbacher, P. Steele, E. Dlugokencky, R. Keeling, E. Scheel, T. Aalto, H. Koide, S. Wofsy, M. Schmidt, L. Haszpra, I. Levin, P. Tans, K. Thoning, J. B. Miller, C. Labuschagne, J. Hatakka, S. Fang, M. Alemanno, M-K. Park, B. Wastine, L. Zhou, A. di Diodato, K. R. Kim, H. Mukai, J. Page, L. Lauria, H. Lee, N. Paramonova, G. Brailsford, L. Ciattaglia, Y. Fukuyama, F. Apadula, A. Lanza, M. Froehlich, E. Friedbacher, K. Uhse, C. Rafanelli, B. Noone, M. Yosuke, T. Conway, S. Piacentino, and T. Mkololo.

Author contributions

R.L.T. performed the data processing, analyses, synthesis of the results and prepared the manuscript; P.K.P assisted in the data processing, contributed inversion results and assisted in the preparing the manuscript; F.C. assisted in the data processing, contributed inversion results and assisted in the preparing the manuscript; S.M. assisted in data processing and in organizing the study; R.M.L contributed inversion results and assisted in the preparing the manuscript; T.Z. contributed inversion results and assisted in the preparing the manuscript; I.T.L.L contributed inversion results and assisted in the preparing the manuscript; W.P. contributed inversion results and assisted in the preparing the manuscript; A.G. contributed inversion results; R.Z. contributed inversion results; T.M. contributed inversion results; T.N. contributed inversion results; T.S. contributed inversion results; M.I. contributed inversion results; T.S. contributed inversion results; T.M. contributed observations for the inversion evaluation, B.P. contributed climate, remote-sensing and NEE data; J.G.C assisted in preparing the manuscript; P.C. assisted in preparing the manuscript.

Additional information

Competing financial interests: The authors declare no competing financial interests.

18

Massive subsurface ice formed by refreezing of ice-shelf melt ponds

Bryn Hubbard[1], Adrian Luckman[2], David W. Ashmore[1], Suzanne Bevan[2], Bernd Kulessa[2], Peter Kuipers Munneke[2], Morgane Philippe[3], Daniela Jansen[4], Adam Booth[5], Heidi Sevestre[6], Jean-Louis Tison[3], Martin O'Leary[2]& Ian Rutt[2]

Surface melt ponds form intermittently on several Antarctic ice shelves. Although implicated in ice-shelf break up, the consequences of such ponding for ice formation and ice-shelf structure have not been evaluated. Here we report the discovery of a massive subsurface ice layer, at least 16 km across, several kilometres long and tens of metres deep, located in an area of intense melting and intermittent ponding on Larsen C Ice Shelf, Antarctica. We combine borehole optical televiewer logging and radar measurements with remote sensing and firn modelling to investigate the layer, found to be $\sim$10 °C warmer and $\sim$170 kg m^{-3} denser than anticipated in the absence of ponding and hitherto used in models of ice-shelf fracture and flow. Surface ponding and ice layers such as the one we report are likely to form on a wider range of Antarctic ice shelves in response to climatic warming in forthcoming decades.

[1] Centre for Glaciology, Department of Geography and Earth Sciences, Aberystwyth University, Aberystwyth SY23 3DB, UK. [2] Glaciology Group, Department of Geography, Swansea University, Swansea SA2 8PP, UK. [3] Laboratoire de Glaciologie, Département Géosciences, Environnement et Société, Université Libre de Bruxelles, Bruxelles 1050, Belgium. [4] Alfred Wegener Institut, Helmholtz-Zentrum für Polar- und Meeresforschung D-27568, Bremerhaven, Germany. [5] School of Earth and Environment, University of Leeds, Leeds LS2 9JT, UK. [6] Department of Arctic Geology, University Centre in Svalbard, N-9171 Longyearbyen, Norway. Correspondence and requests for materials should be addressed to B.H. (email: byh@aber.ac.uk).

The ~49,000 km^2 Larsen C Ice Shelf (LCIS) is considered susceptible to future collapse because of its exposure to intense surface melting[1] across the northern sector of the Antarctic Peninsula (Fig. 1). Satellite and airborne radar data on the LCIS indicate low firn-air content[2], promoting the formation of melt ponds and the potential for hydrofracturing[3,4]. Analysis of remotely sensed images of Cabinet Inlet, located in the northwest sector of the LCIS, reveals the presence during some summer months of surface melt ponds that generally form in flow-parallel troughs that are some tens of kilometres long and hundreds of metres wide (Fig. 1). These ponds form in Cabinet Inlet as a result of melting by föhn winds that blow from the Graham Land mountains and eastwards through the shelf's northernmost-fringing inlets[5–7]. Although appearing to generate substantial surface melt, these winds do not persist for more than a few days even through the summer, and ponds appear intermittently on satellite images (Fig. 1). The additional stress and hydrofracturing potential of such ponded surface water has been proposed as a means of shelf destabilization[8–10] and has been implicated in the collapse of Larsen B Ice Shelf in 2002[11,12]. However, the implications of such ponding for the formation of new ice and its influence on ice-shelf structure have not been reported.

In this study, we report the presence of a massive ice layer, at least 16 km across, several kilometres long and tens of metres deep, present beneath an area of intermittent pond formation on LCIS, Antarctica. We combine field data with firn-modelling and remotely sensed data to investigate the layer's properties and formation. The layer is found to be composed of two units: an upper, solid ice unit formed largely from the continual refreezing of ponded water; and a lower, infiltration ice unit formed largely from the refreezing of meltwater that has percolated into very dense firn. The layer is found to be ~10 °C warmer and ~170 kg m^{-3} denser than that which would have been present in the absence of the influence of intense surface melting and pond formation. The implications of the layer's presence for ice-shelf thickness estimates, flow and stability are explored.

Results

Borehole drilling and logging. In early austral summer 2014/2015, we drilled a ~100-m long borehole into the flank of a Cabinet Inlet trough indicated by satellite imagery to have repeatedly hosted melt ponds over the past 15 years, but not since 2008/2009. Although the shelf was snow-covered at the time of drilling, inspection of the wall of a 2-m deep pit revealed the presence of both numerous ice layers within the snowpack and an unusually thick ice layer at a depth of 2.0 m that prevented continued excavation. The borehole was logged to a depth of ~97 m by optical televiewer (OPTV)[13], providing a geometrically accurate image of the complete borehole wall at a vertical and lateral resolution of ~1 mm. The resulting OPTV log (Fig. 2a) contrasts starkly with those retrieved from other accumulating

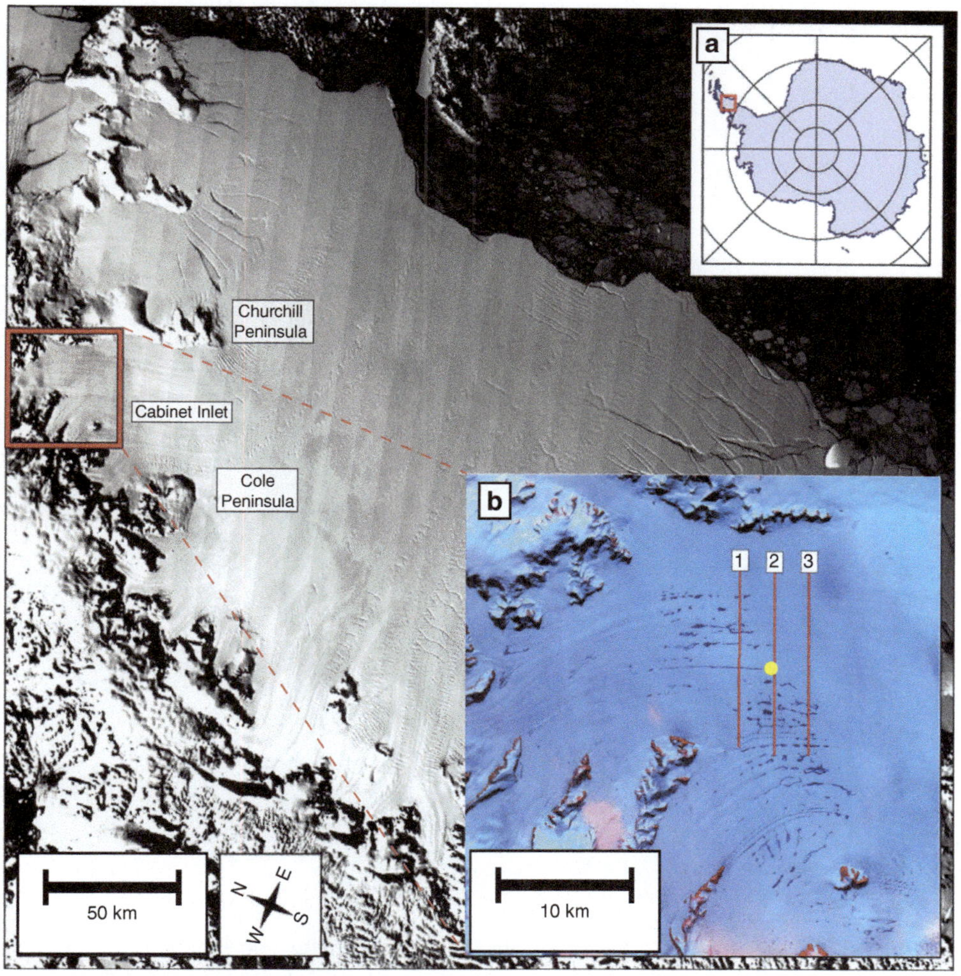

Figure 1 | Cabinet Inlet basemap and surface ponding. Location map of Cabinet Inlet study site on Larsen C Ice Shelf based on MODIS data from 3rd December 2014. (**a**) Main figure location in Antarctica (red box). (**b**) Landsat expansion of Cabinet Inlet from 31st December 2001, showing surface ponding (dark patches) and the locations of the borehole (yellow dot) and 200-MHz GPR transects (red lines) presented in Fig. 4.

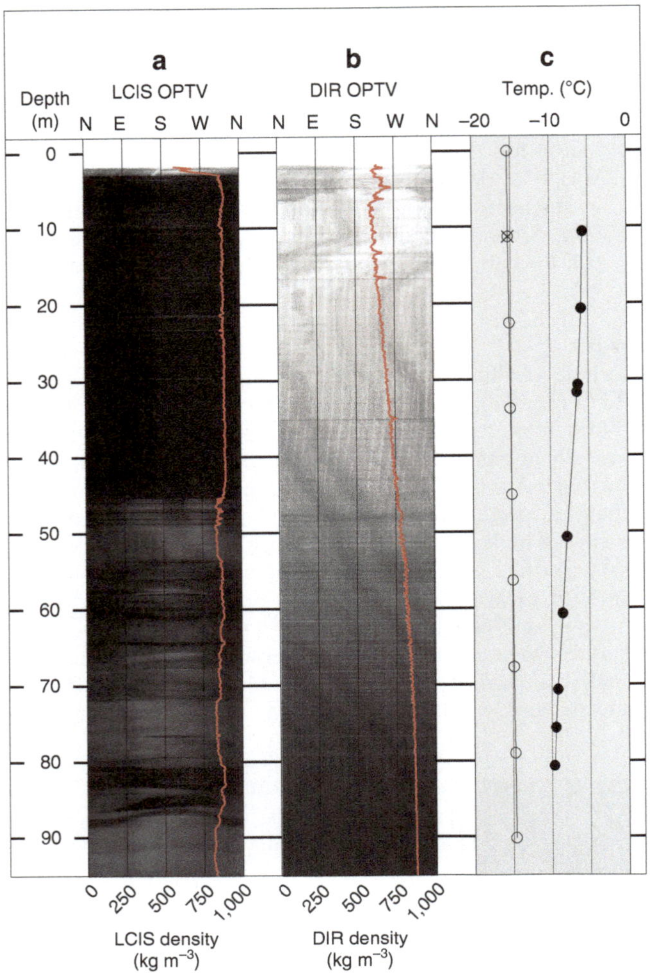

Figure 2 | Cabinet Inlet borehole data. (**a**) OPTV log of the LCIS borehole. This raw log is of the complete borehole wall, unrolled to progress north—east—south—west—north from left to right. The log is classified into two units, described and interpreted in the text: Unit 1 extends from a depth of ~2.9-45 m, and Unit 2 from a depth of ~45 m to the base. (**b**) OPTV log of a typical ice-shelf borehole unaffected by significant melting (Derwael Ice Rise, Roi Baudouin Ice Shelf, Antarctica) showing (**a**) the normal reduction in luminosity with depth, interpreted as progressive densification from snow, through firn to ice, and (**b**) the usual presence of regular horizontal planes closing up with depth, interpreted as annual layering. Density profiles derived from OPTV luminosity (described in Methods) are superimposed as red lines on both OPTV logs. (**c**) Englacial temperatures (**a**) measured by borehole thermistor string (solid dots), (**b**) that would normally be used in an ice-shelf model for this site (open circles), and (**c**) predicted by our firn model at a depth of 11 m (cross).

ice-shelf and ice-sheet locations (for example, Fig. 2b), including the interior of the Greenland Ice Sheet[14] and an East Antarctic ice shelf and ice rise[15,16]. In each of these other cases, OPTV log luminosity decreases gradually with depth as surface snow metamorphoses through firn to dense ice over depths of several tens of metres. In contrast, our Cabinet Inlet OPTV log (Fig. 2a) shows a sharp contact between high-luminosity surface snow or firn and low-luminosity ice at a depth of only 2.9 m below the ice-shelf surface. This ice extends to the 97-m deep base of the log. Converting the luminosity of this OPTV log to density reveals a mean density for this entire ice layer of 870 kg m^{-3}. Figure 2a also reveals a transition in ice type at a depth of ~45 m, with the overlying layer (named Unit 1) hosting more frequent horizontal layers and being more dense (mean = 888 kg m^{-3}) than the underlying ice layer (Unit 2; mean = 854 kg m^{-3}), which is principally composed of bubbly host ice containing coarse, irregularly dipping bubble-free ice layers. The proximity of this generally massive ice layer to the shelf surface, and the sharpness of its contact with the overlying snowpack, preclude formation by the usual process of compaction–metamorphism. Instead, we interpret this Unit 1 ice as having formed as a consequence of refreezing, following periods of intense surface melting and intermittent pond formation. The characteristics revealed by the OPTV image are consistent with this unit (2.9 to ~45 m) being largely bubble-free pond ice formed from the refreezing of surface water ponded during extended periods of intense, presumably summer, melting. Here the fine-scale horizontal layering apparent in Fig. 2a likely reflects the episodic nature of the process, which would involve four general stages: (a) snow accumulation, (b) surface melting, (c) meltwater infiltration into underlying snow, eventually resulting in its saturation and pond formation, and (d) the freezing of that meltwater layer to form pond ice, probably during early austral autumn. In contrast, Unit 2 (approximately 45–97-m depth) also contains layers of bubble-poor ice, but in this zone they are typically decimetres thick, contorted and isolated within host ice that is optically brighter (likely due to the presence of reflective bubbles; Fig. 2a). We interpret this lower unit as ice dominated by infiltration refreezing formed by meltwater percolating into underlying firn. This still involves the influence of intense surface melting but, in contrast to the Unit 1, of generally insufficient magnitude to form a continuous quasi-massive layer of largely bubble-free pond ice. These physical characteristics and depth ranges of the two units we identify are consistent with the upper (Unit 1) 'pond ice' forming within the Cabinet Inlet region of pond formation and the lower (Unit 2) 'infiltration ice' forming, and being inherited from, up-flow of the region of pond formation (Fig. 1).

Firn modelling and satellite image analysis. We evaluate this hypothesis for the recent past through a one-dimensional firn densification and hydrology model[17], driven by surface mass fluxes and temperature data from the RACMO2.3 regional climate model for Cabinet Inlet[18]. Model results (Fig. 3a,b) indicate the occurrence of intermittent, but substantial, surface melt events at the borehole location, consistent with Unit 1 forming from the refreezing of surface meltwater. As well as predicting the presence of such a layer, the model predicts (Fig. 3b) that the layer's upper surface was, in summer 2014–2015, expected to be 2.9-m below the snow surface, and that the overlying snowpack contains a substantial ice layer between ~1.9 and 2.2 m, agreeing with our direct observations from snow-pit digging, the OPTV log (Fig. 2a) and ground-penetrating radar (GPR) data (Fig. 4) addressed below. Further, this analysis is supported by a time series of moderate resolution imaging spectroradiometer (MODIS) satellite images (Fig. 3c), indicating that melt ponds formed annually between 2001 and 2009, while none appears in any of the images available between early 2009 and the time of fieldwork in late 2014. This reconstruction matches very closely the RACMO-based firn densification reconstruction (Fig. 3b) with 'pond years', coinciding with periods of intense near-surface firn densification (for example, 2005–2007).

Ground-penetrating radar. An approximation of the lateral extent of the massive subsurface ice layer we report is provided by the area known from satellite images to host melt ponds: ~60-km across-flow and ~20-km along-flow (Fig. 1). However, the precise degree to which this zone is underlain by massive

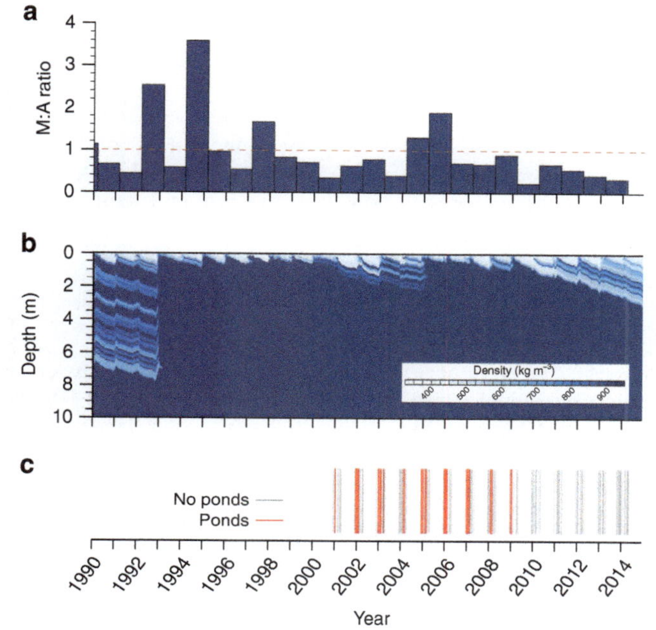

Figure 3 | Firn model and remotely sensed outputs for Cabinet Inlet for the period 1st January 1990 to the time of fieldwork on 1 st December 2014. (**a**) Ratio of melt to accumulation (M:A) predicted by the firn model for each 12-month period (1st March-28th February). The horizontal dashed line marks an M:A ratio of 1. (**b**) Predicted firn density plotted against depth. The predicted massive shallow ice layer (dark blue) is never more than ~8 m below the shelf surface and extends to within ~1 m of the surface in years 1993-2001 and 2005-2009, both containing individual years of high M:A ratio. (**c**) Classification of time series of summer MODIS satellite images (available since 2001) of Cabinet Inlet according to the presence (red bars) or absence (grey bars) of surface ponds.

pond and/or infiltration ice, and the depth of such layers where they are present beyond our borehole, cannot be determined from temporally discrete satellite images. To evaluate this we carried out GPR profiling at 200 MHz along three transects focused on the borehole location (Fig. 1). The resulting radargrams (Fig. 4) reveal the presence of numerous near-surface reflectors and one substantial reflector at a depth that varies between ~1 and ~3 m along the transects. Very little radar energy above the background noise level was received from below this reflection. This absence of signal return is consistent with the presence of a refrozen ice layer that, although characterized by minor density stratification, is physically and chemically uniform compared with firn layering unmodified by meltwater. This main reflector is ~2.9-m deep at the borehole location, coincident with field-based digging, drilling and OPTV logging, as well as with firn modelling (above). We therefore infer that this strong GPR reflection represents the upper surface of a spatially extensive ice layer that is both thick and relatively homogeneous. Since meltwater ponds are confined to surface troughs that persist in MODIS images for decades within this region of Cabinet Inlet, it is unlikely that this ice layer is ubiquitously composed of Unit 1 pond ice. Away from the troughs, this near-surface reflector therefore probably indicates the uppermost surface of an ice layer that is more similar to Unit 2 infiltration ice, that is, still influenced by intense surface melting and subsurface refreezing, but in these areas not actually forming standing ponds. In the absence of further direct investigation, such as by ice coring or borehole analysis, it is not yet possible to determine at high resolution the lateral variability of the two units we identify. Nonetheless, our OPTV and GPR data together indicate that a widespread ice layer extends at least

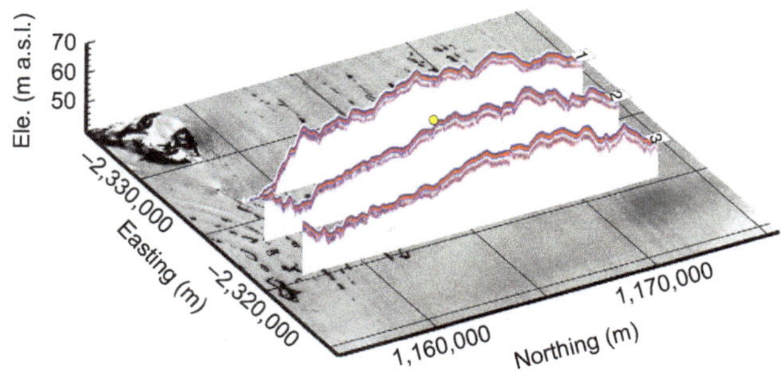

Figure 4 | Cabinet Inlet radargrams. 200 MHz GPR profiles along the three transects identified in Fig. 1, overlaid on a Landsat image of the region from 31st December 2001 when surface melt ponds were present. The borehole location is marked by the yellow dot.

across all of our ~16-km flow-orthogonal and ~6-km flow-parallel GPR study area. Our borehole OPTV log also indicates that Units 1 and 2 combined extend to a depth of at least 97 m at the location of our borehole. While it is highly likely that this thickness—defined by the depth of the base of Unit 2—varies spatially across Cabinet Inlet, it is not currently possible to specify this distribution without further borehole and/or GPR data.

Discussion

The presence of the refrozen ice layer we report above affects the physical properties of the LCIS in at least two ways. First, the layer's density is substantially higher than that of the snow and firn that would otherwise have formed over this depth range by standard compaction–metamorphism. For example, a recent LCIS model[19] used a density of ~700 kg m^{-3} for the shelf's uppermost ~100 m, based on inverting seismic data recorded in the shelf's southern sector. In contrast, our OPTV-derived densities indicate a measured density of ~870 kg m^{-3} over this depth range in the Cabinet Inlet area, 24% higher. Such a density enhancement influences calculations of the shelf's thickness based on the surface elevation data[20]. In this case, using an ice density of 917 kg m^{-3}, a sea water density of 1,026 kg m^{-3}, a surface elevation of 63.5 m (Fig. 4) and the assumed firn density of 700 kg m^{-3} for the uppermost 97 m of the shelf yields a total shelf thickness of 382 m. Repeating the calculation with our OPTV-reconstructed density of 870 kg m^{3} for the uppermost 97 m yields a total shelf thickness of 551 m. Although substantially thicker than that calculated from the 'standard firn' model, a thickness of 551 m is close to that of 564 m reconstructed for the area by the Bedmap2 consortium[21]. This close correspondence reflects the fact that the (altimetry-based) Bedmap reconstructions include a correction to account for enhanced densification, resulting from active surface melting on ice shelves[22]. Our OPTV-based density reconstruction also yields a firn-air content (the column-length equivalent accounted for by material with a density lower than that of bubble-free glacial ice) of 5.0 m. While this value is substantially lower than the 23.0 m that would result from firn of density 700 kg m^{-3}, it is far closer to the spatially distributed range of 0–4 m predicted for the area based on recent combined analyses of remotely sensed surface elevation and shelf thickness fields[9,10]. However, direct comparisons such as these are somewhat confounded by our firn-air content of 5.0 m being based on a single-point measurement (recorded, as noted above, on the limb of an elongate trough hosting an ephemeral surface pond), whereas the reconstructions based on remotely sensed data integrate data over a coarser spatial field.

Second, as well as altering its density, the ice column will be warmed by latent heat released by the freezing of ponded and percolating meltwater. This effect is quantified by a thermistor string installed into the borehole following OPTV logging (Fig. 2c). The mean annual temperature measured at a depth of 11 m was −5.9 °C, while the mean annual surface temperature at the site (which normally defines that at a depth of 10–15 m in the absence of significant refreezing) is −16.9 °C. Further, the entire ice profile recorded by our thermistor string (Fig. 2c) is warmer than would be expected without factoring in refreezing of ponded meltwater. This effect is confirmed by our firn model, which under-predicts englacial temperatures despite accounting for heat released by the refreezing of percolating meltwater. In this case, modelled pore-water refreezing contributes ~1.5 °C to the firn, yielding a predicted temperature at 11 m of −15.4 °C (Fig. 2c). This is still almost 10 °C colder than our measured firn temperature at that depth, demonstrating that the refreezing of ponded meltwater (not included in the model) provides significant, hitherto unconsidered, englacial heating in this region. As well as being denser, this ice is therefore also warmer than that which would otherwise be used in numerical models of the flow of the LCIS[19].

Replacing on the order of 10 × 10 × 0.1 km of standard firn with relatively warm and dense ice will exert some influence on ice shelf flow and stability. However, evaluating this influence at the ice-shelf scale is not straightforward. In the first instance, the warmer ice layer we identify will be less viscous than colder ice, and this may go some way to accounting for the anomalously high-rate factor that has in the past been necessary for numerical models of the flow of the LCIS, and in particular its northern sector, to match empirical data[23,24]. A consequence of such a temperature-induced acceleration would be a general reduction in back stress within confined embayments, such as Cabinet Inlet, potentially increasing shear stresses along the flow unit's lateral margins. This process is consistent with observations that the northern sector of Larsen B Ice Shelf experienced an increase in lateral rifting before its break up in 2002 (ref. 25). In contrast, the influence of the layer's presence on brittle deformation may be in the opposite direction, with enhanced ductile flow accommodating strain and the solid ice we report being more resistant to tensile fracture than lower-density and finer-grained firn. In such cases, the flow-parallel alignment of elongate solid ice bodies such as that we report herein might serve to resist the flow-orthogonal crevassing that commonly develops in response to longitudinal tensile stresses, approaching the marine limit of ice shelves. Finally, the spatial extent of this layer at the ice-shelf scale, and the way in which its deformation interacts with that of other material units, such as suture ice, and basal channels and crevasses, will also influence the way in which the layer's presence affects overall shelf stability. In particular, longitudinal troughs located on the surface of ice shelves have been related to spatially coincident basal channels[26], which have themselves been associated with shelf instability[27] through local thinning and crevassing[28,29]. Although it is not yet known whether the surface troughs on LCIS investigated herein are associated with such basal channels, the recent finding that material density is locally enhanced below similar surface troughs on the Roi Baudouin Ice Shelf, Antarctica[30], is consistent with the enhanced melting and massive ice formation reported herein. In the light of these complexities, identifying and exploring the net influence of pond ice formation on ice-shelf stability can only be achieved with confidence by including the full spatial extent and physical properties of the refrozen ice layer into a shelf-wide flow and fracture model, which may be regarded as a future research priority.

The surface ponding responsible for the subsurface ice layer we report herein is currently restricted to warmer regions of Antarctica's fringing ice shelves and in particular, but not exclusively, to the northern and western sectors of the Antarctic Peninsula. However, regional warming is predicted to spread southwards and intensify substantially over forthcoming decades[5,31], so ponding is expected also to become more widespread. Similarly, massive layers of warm, dense ice are therefore highly likely to form within ice shelves present across a substantial area of Antarctica, perhaps the entire continent if warming continues into the 22nd century[5], with important consequences for ice-shelf flow, hydrofracture and stability.

Methods

Borehole drilling and logging. The borehole was drilled by pressurized hot water and logged by OPTV[13]. The resulting OPTV log, analysed by WellCAD and BIFAT software[32], provides a geometrically accurate image, with a pixel size of ~1 mm, of the material composition of the complete borehole wall, as well as the structural geometry of layers and inclusions intersecting it[33]. Since the OPTV log records an image of reflected light, it also provides a proxy for the density of compacted snow, firn and ice due to the progressive decrease in reflectivity, as voids close and bubbles are occluded and collapsed[13,15]. Hubbard *et al.*[15] exploited this relationship and identified an exponential relationship between OPTV luminosity (L) and material density (D) on the basis of samples recovered from a core retrieved from a borehole logged by OPTV on the Roi Baudouin Ice Shelf, East Antarctica. However, due to equipment loss, the light-emitting diode brightness of the OPTV probe used in the current study was different from that of Hubbard *et al.*[15], and a new calibration was undertaken. This was based on the correlation of 40 core samples recovered from a logged borehole, again located on the Roi Baudouin Ice Shelf (Fig. 5). The new calibration yields a best-fit regression equation of $D = 950–40.1\ e^{(0.0101L)}$ ($R^2 = 0.82$), with root mean square values of the residuals of 40.4, 35.2 and 21.7 kg m^{-3} for the density ranges 600–700, 700–800 and 800–900 kg m^{-3}, respectively. While these values provide an error range for absolute densities derived from OPTV luminosity, the relative changes in density reported herein, that is, along a single borehole log, reduce to the precision of the method. This is approximated by the density range (910 kg m^{-3}) divided by the luminosity range (256), or ~3.5 kg m^{-3}.

Borehole temperatures (Fig. 2c) were recorded by negative temperature coefficient thermistors, recorded across a Wheatstone half-bridge by Campbell Scientific micro-loggers. Resistances were converted to temperature using a polynomial[34] fitted to the manufacturer's calibration curve refined via a second-stage calibration in a distilled water/ice bath. Once recalibrated, sensors were replaced into a new water/ice bath to determine temperature error, yielding a root mean square error of ±0.03 °C. Borehole temperatures were logged for at least 100 days, and the undisturbed ice temperature was calculated from an exponential function fitted to the cooling curve[35]. By this time, all temperatures recorded below the near-surface zone of seasonal thermal disturbance had stabilized to values that varied (over timescales of days to weeks) by <0.05 °C. Thus, the englacial temperatures we report herein were not influenced by the hot-water drilling process.

Firn modelling. The firn model used is IMAU-FDM v1.0, which takes into account firn compaction, meltwater percolation and refreezing[17,18]. At the surface, the firn model is forced with mass fluxes (snowfall, snowmelt, rain, sublimation, and snowdrift sublimation and erosion) and surface temperature from the regional

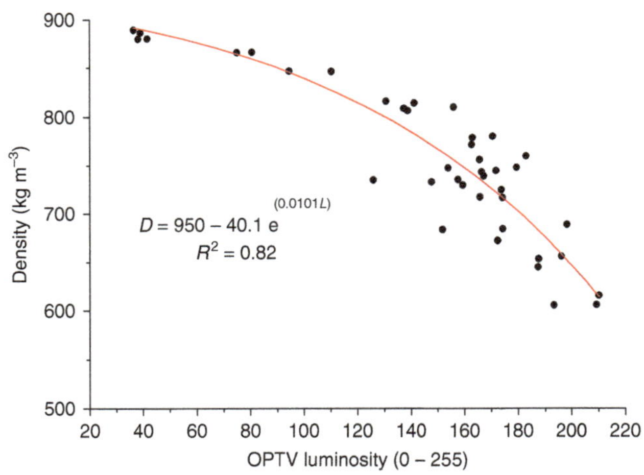

Figure 5 | Optical televiewer density-luminosity calibration. Material density, measured gravimetrically on core samples, plotted against equivalent OPTV luminosity, both measured along a borehole located on Roi Baudouin Ice Shelf, East Antarctica.

climate model RACMO2.3, run at a 5.5-km horizontal resolution in a domain over the Antarctic Peninsula.

Satellite image analysis. MODIS Terra and Aqua level 1 (250 m) data were ordered via the level 1 and atmosphere archive and distribution system, and used to check for the presence of melt ponds in Cabinet Inlet. A total of 577 cloud-free band 2 (848 nm) images between 1st December and 30th April from 2000 (Aqua) or 2002 (Terra) to 2015 were classified to generate the time series shown in Fig. 3c.

Ground-penetrating radar. GPR data (Fig. 4) were collected using a Sensors and Software Pulse Ekko Pro system operated in common offset mode with a 0.8-ns sample interval and a 4,000-ns sample window. The console was mounted on the skidoo and the 200 MHz antennae towed 15 m behind on a plastic sledge at ~10 km h^{-1} with an eight stack trace recorded every 3 m. A Leica VIVA GS10 GNSS rover unit was connected directly to the GPR console; the base station was located at the drill site. Surface elevation was referenced to the Earth Gravitational Model 1996 geoid, taken to represent sea level, and no correction was made for dynamic topography. GPR data were processed in Reflex-W. Processing steps included de-wow of 25-ns filter length, spherical divergence compensation, spectral whitening between 70 and 200 MHz, and a two-dimensional mean-averaging filter. The static correction was based on Leica Geo Office post-processed GNSS data, resulting in a vertical accuracy of ± 0.5 m. Radar wave propagation velocities used to convert travel times to the depths shown in Fig. 4 were 0.2 m ns^{-1} for the uppermost 3 m (based on a local 500 MHz common midpoint gather from the uppermost reflector of Unit 1) and a typical value for glacial ice[36] of 0.17 m ns^{-1} below that.

References

1. Holland, P. R. *et al.* Oceanic and atmospheric forcing of Larsen C Ice-Shelf thinning. *Cryosphere* **9**, 1005–1024 (2015).
2. Holland, P. R. *et al.* The air content of Larsen Ice Shelf. *Geophys. Res. Lett.* **38**, L10503 (2011).
3. Luckman, A. *et al.* Basal crevasses in Larsen C Ice Shelf and implications for their global abundance. *Cryosphere* **6**, 113–123 (2012).
4. McGrath, D. *et al.* Basal crevasses on the Larsen C Ice Shelf, Antarctica: Implications for meltwater ponding and hydrofracture. *Geophys. Res. Lett.* **39**, L16504 (2012).
5. Luckman, A. *et al.* Surface melt and ponding on Larsen C Ice Shelf and the impact of föhn winds. *Antarctic Sci.* **26**, 625–635 (2014).
6. Kuipers Munneke, P., van den Broeke, M. R., King, J. C., Gray, T. & Reijmer, C. H. Near-surface climate and surface energy budget of Larsen C ice shelf, Antarctic Peninsula. *Cryosphere* **6**, 353–363 (2012).
7. Cape, M. R. *et al.* Foehn winds link climate-driven warming to ice shelf evolution in Antarctica. *J. Geophys. Res.* **120**, 11037–11057 (2015).
8. van den Broeke, M. Strong surface melting preceded collapse of Antarctic Peninsula ice shelf. *Geophys. Res. Lett.* **32**, L12815 (2005).
9. Scambos, T. *et al.* Ice shelf disintegration by plate bending and hydro-fracture: satellite observations and model results of the 2008 Wilkins ice shelf break-ups. *Earth Planet Sci. Lett.* **280**, 51–60 (2009).
10. Kuipers Munneke, P., Ligtenberg, S. R. M., Van Den Broeke, M. R. & Vaughan, D. G. Firn air depletion as a precursor of Antarctic ice-shelf collapse. *J. Glaciol.* **60**, 205–214 (2014).
11. Glasser, N. F. & Scambos, T. A. A structural glaciological analysis of the 2002 Larsen B ice-shelf collapse. *J. Glaciol.* **54**, 3–16 (2008).
12. Banwell, A. F., MacAyeal, D. R. & Sergienko, O. V. Breakup of the Larsen B Ice Shelf triggered by chain reaction drainage of supraglacial lakes. *Geophys. Res. Lett.* **40**, 5872–5876 (2013).
13. Hubbard, B., Roberson, S., Samyn, D. & Merton-Lyn, D. Digital optical televiewing of ice boreholes. *J. Glaciol.* **54**, 1–8 (2008).
14. Hubbard, B. & Malone, T. Optical-televiewer-based logging of the uppermost 630 m of the NEEM deep ice borehole, Greenland. *Ann. Glaciol.* **54**, 83–89 (2013).
15. Hubbard, B. *et al.* Ice-shelf density reconstructed from optical-televiewer borehole logging. *Geophys. Res. Lett.* **2013**, GL058023 (2013).
16. Hubbard, B. *et al.* Optical-televiewer-based identification and characterization of material facies associated with an Antarctic ice-shelf rift. *Ann. Glaciol.* **53**, 137–146 (2012).
17. Ligtenberg, S. R. M., Helsen, M. M. & van den Broeke, M. R. An improved semi-empirical model for the densification of Antarctic firn. *Cryosphere* **5**, 809–819 (2011).
18. van Wessem, J. M. *et al.* The modelled surface mass balance of the Antarctic Peninsula at 5.5 km horizontal resolution. *Cryosphere* **10**, 271–285 (2016).
19. Jansen, D. *et al.* Present stability of the Larsen C ice shelf, Antarctic Peninsula. *J. Glaciol.* **56**, 593–600 (2010).
20. Bamber, J. & Bentley, C. A comparison of satellite-altimetry and ice-thickness measurements of the Ross Ice Shelf, Antarctica. *Ann. Glaciol.* **20**, 357–364 (1994).
21. Fretwell, P. *et al.* Bedmap2: improved ice bed, surface and thickness datasets for Antarctica. *Cryosphere* **7**, 375–393 (2013).
22. Griggs, J. A. & Bamber, J. L. Ice shelf thickness over Larsen C, Antarctica, derived from satellite altimetry. *Geophys. Res. Lett.* **36**, L19501 (2009).
23. Khazendar, A., Rignot, E. & Larour, E. Acceleration and spatial rheology of Larsen C Ice Shelf, Antarctic Peninsula. *Geophys. Res. Lett.* **38**, L09502 (2011).
24. Borstad, C. P., Rignot, E., Mouginot, J. & Schodlok, M. P. Creep deformation and buttressing capacity of damaged ice shelves: theory and application to Larsen C ice shelf. *Cryosphere* **7**, 1931–1947 (2013).
25. Rack, W. & Rott, H. Pattern of retreat and disintegration of Larsen B Ice Shelf, Antarctic Peninsula. *Ann. Glaciol.* **39**, 505–510 (2004).
26. Stanton, T. P. *et al.* Channelized ice melting in the ocean boundary layer beneath Pine Island Glacier, Antarctica. *Science* **341**, 1236–1239 (2013).
27. Alley, K. E., Scambos, T. A., Siegfried, M. R. & Fricker, H. A. Impacts of warm water on Antarctic ice shelf stability through basal channel formation. *Nat. Geosci.* **9**, 290–293 (2016).
28. Vaughan, D. G. *et al.* Subglacial melt channels and fracture in the floating part of Pine Island Glacier, Antarctica. *J. Geophys. Res.* **117**, F03012 (2012).
29. Sergienko, O. V. Basal channels on ice shelves. *J. Geophys. Res.* **118**, 1342–1355 (2013).
30. Drews, R. *et al.* Constraining variable density of ice shelves using wide-angle radar measurements. *Cryosphere* **10**, 811–823 (2016).
31. Ligtenberg, S. R. M., van de Berg, W. J., van den Broeke, M. R., Rae, J. G. L. & van Meijgaard, E. Future surface mass balance of the Antarctic ice sheet and its influence on sea level change, simulated by a regional atmospheric climate model. *Clim. Dyn.* **41**, 867–884 (2013).
32. Malone, T., Hubbard, B., Merton-Lyn, D., Worthington, P. & Zwiggelaar, R. Borehole and Ice Feature Annotation Tool (BIFAT): a program for the automatic and manual annotation of glacier borehole images. *Comput. Geosci.* **51**, 381–389 (2013).
33. Roberson, S. & Hubbard, B. Application of borehole optical televiewing to investigating the 3-D structure of glaciers: implications for the formation of longitudinal debris ridges, midre Lovenbreen, Svalbard. *J. Glaciol.* **56**, 143–156 (2010).
34. Steinhart, J. S. & Hart, S. R. Calibration curves for thermistors. *Deep Sea Res. Oceanogr. Abstr.* **15**, 497–503 (1968).
35. Humphrey, N. & Echelmeyer, K. Hot-water drilling and borehole closure in cold ice. *J. Glaciol.* **36**, 287–298 (1990).
36. Plewes, L. A. & Hubbard, B. A review of the use of radio-echo sounding in glaciology. *Progress Phys. Geogr.* **25**, 203–236 (2001).

Acknowledgements

This research was funded by the Natural Environment Research Council, grants NE/L005409/1 and NE/L006707/1, and Aberystwyth University's Capital Equipment fund. The Leica VIVA GS10 GNSS system and Sensors and Software Pulse Ekko Pro radar transmitter were loaned by the Natural Environment Research Council Geophysical Equipment Facility, loan number 1028. We thank the British Antarctic Survey for logistical support, and in particular the project's field assistants Ashly Fusiarski and Nicholas Gillett.

Author contributions

A.L. and B.H. led the overall project. Fieldwork was carried out by D.A., S.B., B.H., B.K., H.S., A.B. and A.L. D.J., M.O.L. and I.R. contributed to ice-shelf modelling, guiding the field programme and interpretations. S.B. and A.L. led the remote-sensing contributions. P.K.M. carried out the firn modelling. M.P., J.-L.T. and B.H. contributed to the OPTV calibration. All authors contributed to writing the manuscript, which was led by B.H.

Additional information

Competing financial interests: The authors declare no competing financial interests.

19

Tracking ocean heat uptake during the surface warming hiatus

Wei Liu[1], Shang-Ping Xie[1] & Jian Lu[2]

Ocean heat uptake is observed to penetrate deep into the Atlantic and Southern Oceans during the recent hiatus of global warming. Here we show that the deep heat penetration in these two basins is not unique to the hiatus but is characteristic of anthropogenic warming and merely reflects the depth of the mean meridional overturning circulation in the basin. We find, however, that heat redistribution in the upper 350 m between the Pacific and Indian Oceans is closely tied to the surface warming hiatus. The Indian Ocean shows an anomalous warming below 50 m during hiatus events due to an enhanced heat transport by the Indonesian throughflow in response to the intensified trade winds in the equatorial Pacific. Thus, the Pacific and Indian Oceans are the key regions to track ocean heat uptake during the surface warming hiatus.

[1]Climate, Atmospheric Sciences, and Physical Oceanography (CASPO), Scripps Institution of Oceanography, University of California, San Diego, 9500 Gilman Drive, La Jolla, California 92093, USA. [2]Pacific Northwest National Laboratory, 902 Battelle Blvd, Richland, Washington 99354, USA. Correspondence and requests for materials should be addressed to W.L. (email: wel109@ucsd.edu).

The global surface temperature increase has slowed over the past 16 years[1–3]. The warming hiatus is regarded as a result of natural variability against the centennial warming trend, while radiative forcing change also contributes[4–6]. Two leading theories have been proposed to explain the hiatus. One suggests that the hiatus is closely associated with the negative phase of the Interdecadal Pacific Oscillation (IPO)[7–9], manifested as a La Niña-like cooling and intensified easterly winds over the equatorial Pacific[10,11]. Consistent with this theory, the recent hiatus can be successfully simulated by nudging the sea surface temperature (SST) or trade winds to observations over the tropical Pacific[12–14]. The other idea is based on an energy view, focusing on the vertical energy redistribution in the ocean on the decadal timescale. The surface warming hiatus is accompanied by an excessive global deep ocean (>700 m) heat uptake[15], consistent with a positive radiative imbalance at the top of the atmosphere[16–20]. These two theories are not mutually exclusive. Indeed, the accelerated trade winds at the IPO-negative phase tilts the equatorial thermocline, causing an apparent vertical heat redistribution[13,21].

Recent studies[21–23] further attempted to infer mechanisms for the warming hiatus by tracking regional ocean heat uptake at different levels. Specifically, an earlier model study[21] found that surface warming hiatus is significantly correlated with accelerated warming in deep levels in all basins, below 750 m in the Atlantic and Southern Oceans and below 300 m in the Pacific and Indian Oceans. Observational data analyses, however, showed different results, from models and between data sets. In the Ishii data (Methods), the increase of ocean heat content (OHC) between 300 and 1,500 m accelerates in the Atlantic and Southern Oceans but changes little in the Pacific and Indian Oceans[22]. In the World Ocean Atlas (WOA) data (Methods), on the other hand, the recent surface warming hiatus is associated with a heat rearrangement within the upper 300 m between the Pacific and Indian Oceans[23], along with an increased heat transport from the Pacific to the Indian Ocean by the Indonesian throughflow (ITF)[24]. These debates highlight a gap between models and observations, large uncertainties in OHC observations[19,25–28], and a complex relationship between global mean surface temperature (GMST) and OHC[29].

In this study, we revisit the energy theory and combine multiple observational data sets and state-of-art model simulations to examine the relationship between the GMST and regional OHC on decadal timescales. We show that the deep ocean heat uptake in the Atlantic and Southern Oceans primarily reflects the downward penetration of anthropogenic heat via the deep meridional overturning circulation (MOC), while OHC in the upper 350 m in the Pacific and Indian Oceans is closely tied to the surface warming hiatus.

Results

Global and regional OHC changes. The observed GMST (defined as globally averaged SST in this study, not including land surface temperature) from the Ishii data[30] (Methods) shows a surface warming slowdown since 1998, with a near-zero warming trend during 2002–2012 (Fig. 1). Using the same data, ref. 22 shows that the OHC anomalies (from the 1970–2012 mean) increase during the hiatus period, and the warming penetrates deeper than 300 m only in the Atlantic and Southern Oceans (Supplementary Fig. 1). The large heat penetration was then taken as suggesting the importance of these basins for the hiatus. Here, instead of referencing to the 1970–2012 mean as in ref. 22, we compute ocean temperature anomalies as deviations from the 1970 values because referencing to 1970 facilitates an easy inspection of the OHC variation over the entire analysis period.

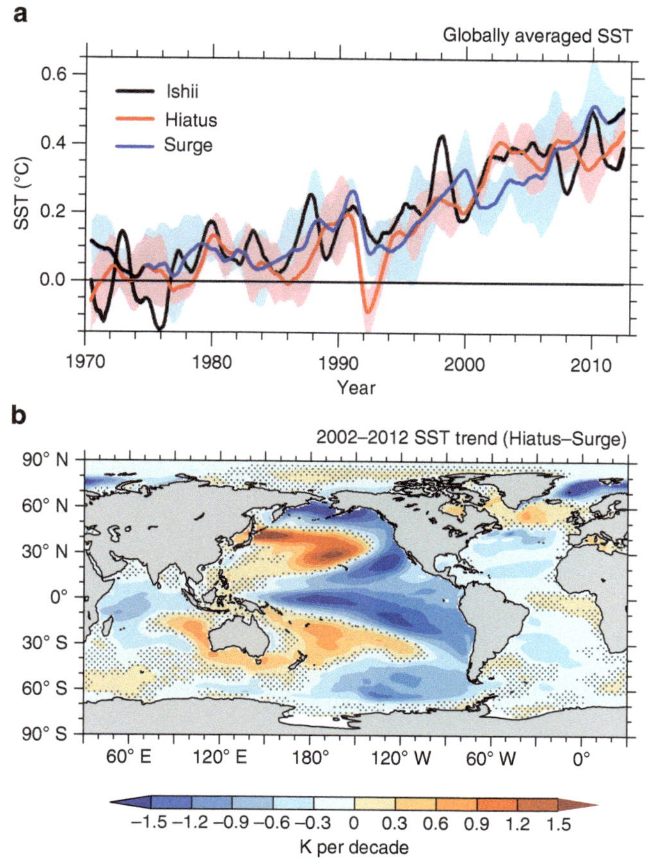

Figure 1 | Observed and simulated SST. (**a**) Monthly mean globally averaged SST from the Ishii data (black) and the CESM ensemble simulations. Two groups (named the Hiatus and Surge) of ensembles are shown, whose variations (one s.d.) are shaded in light pink and light blue and ensemble means are drawn in red and blue. All the curves are shown as a 12-month running mean by subtracting the annual mean value of the first year (year 1970). (**b**) The SST trend difference between the ensemble means of the CESM Hiatus and Surge groups during 2002-2012 (shading in K per decade). Stippling indicates region below 95% significance computed from a two-tailed *t*-test.

From our rendering in Fig. 2, the most prominent feature of historical OHC variation is the signal of anthropologic warming: both global and regional OHCs continuously increase during 1970–2012. The interbasin difference of OHC increase is not unique to the hiatus period but common to the entire analysis period. Between 300 and 1,500 m, the OHC in the Atlantic, Southern Ocean, Pacific and Indian Ocean (Supplementary Fig. 1) exhibits a warming trend of 0.19, 0.23, 0.05 and 0.09 × 10^{23} J per decade during 1998–2012, which accounts for 33.8%, 40.7%, 8.2% and 15.8% of the global OHC warming trend. Such interbasin partition holds when we extend the analysis period to 1970–2012 (and to other observational data sets, see Supplementary Fig. 2). The Atlantic, Southern Ocean, Pacific and Indian Ocean explain 30.7%, 41.3%, 13.5% and 5.4% of the global warming within 300–1,500 m, respectively. This result demonstrates the dominance of the Atlantic and Southern Oceans in the deep ocean sequestration of heat, and more importantly, indicates that the penetration of heat below 700 m is not uniquely tied to interdecadal modulations of the surface warming rate.

Model simulations are consistent with the above result. From the Community Earth System Model (CESM) large ensemble simulations[31] (Methods), four members are selected each for the

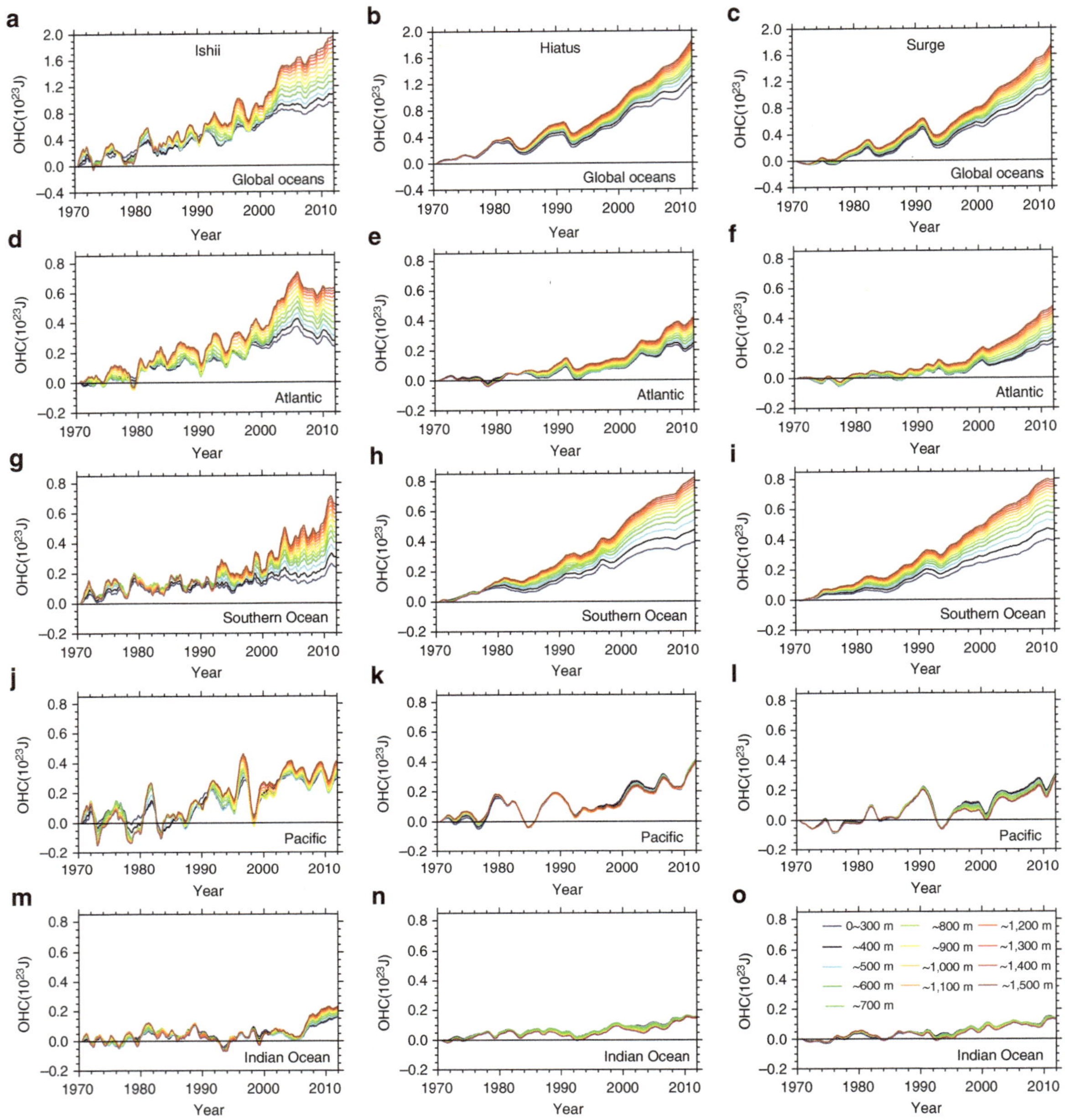

Figure 2 | Observed and simulated OHC. OHC integrated from the surface to indicated depths in global oceans, the Atlantic, Southern Ocean, Pacific and Indian Ocean from the Ishii data (**a**,**d**,**g**,**j**,**m**) and the ensemble means of the CESM Hiatus (**b**,**e**,**h**,**k**,**n**) and Surge (**c**,**f**,**i**,**l**,**o**) groups. All the curves are shown as a 12-month running mean by subtracting the annual mean value of the first year (year 1970).

Hiatus and Surge groups. The Hiatus group members are where the decadal trend of GMST is once negative during 2002–2012, whereas the Surge group are the runs corresponding to individual Hiatus members but with the largest warming trend during the same span (Supplementary Fig. 3; Supplementary Table 1). In Fig. 1a, the Hiatus group features a warming hiatus, with a near-zero warming rate since the early 2000s, while the Surge group shows a continuous surface warming, much like the ensemble mean of the Couple Model Intercomparison Project phase 5 (CMIP5) models[8,32]. The Hiatus and Surge groups share the same radiative forcing, but have different initial conditions. The difference between two groups is thus due to natural variability. The spatial pattern of the Hiatus minus Surge difference is characterized by negative SST trends in the tropical Pacific, and positive trends in the subtropical North and South Pacific (Fig. 1b), a pattern resembling the La Niña-like negative phase of the IPO.

Despite the differences in the GMST warming rate, subsurface OHCs (Methods) in the Ishii data, the Hiatus and Surge groups share a common evolution pattern. During 1970–2012, the global OHC keeps increasing, and the rate increases when it is integrated over a greater depth (Fig. 2). The Atlantic and Southern Oceans contribute most to OHC increase at depths >300 m, and this dominance in deep ocean contribution is not dependent on the GMST warming rate. During 1998–2012, the increase of 300–1,500 m OHC in the Atlantic and Southern Oceans accounts, respectively, for 30.7% and 50.1% of global OHC increase in the Hiatus group, and 43.2% and 46.3% in the Surge group. It is clear that results from either group are consistent with observations that show a deep ocean uptake in these basins. The deep heat storage takes place in the Atlantic and Southern Oceans over the entire 43 years, regardless whether the GMST warming slows down (in the Ishii data and the Hiatus group) or accelerates (in the Surge group). The deep heat penetration in the Atlantic and

Southern Oceans is the result of anthropogenic warming, not unique to the hiatus. Therefore, the deep warming in these two basins is not the basis to argue for their importance in driving the hiatus.

To further test the robustness of above results, we look for more warming hiatus and surge periods in the future events of the large ensembles. Running 10-year linear trends of GMST from 38-member simulations reveal another five hiatus events with negative decadal trends after 2012. There are nine hiatus decades in total. For each hiatus decade, we inspect the trends in the other 37 members and choose the one with the largest warming trend as the corresponding warming surge decade. The same span ensures the pair of hiatus and surge events share the identical radiative forcing. Thus, we expand the Hiatus and Surge groups to nine members each by including five pairs of hiatus (Supplementary Fig. 4) and surge decades (Supplementary Table 1) in the future simulations (the new groups are referred as the fHiatus and fSurge, see Methods). The decadal trend of composite GMST is -0.014 ± 0.023 K per decade in the fHiatus group and 0.376 ± 0.098 K per decade in the fSurge group, where the range ($\pm$ one s.d. of group $\times$ 1.86) represents the 95% range from a one-sided Student's *t*-test[21]. The trend spreads of two groups do not overlap, so these two groups are well separated regarding the surface warming rate.

Figure 3 compares the composite average trends of global and regional OHCs between the fHiatus and fSurge groups, where the error bar denotes one s.d. among members. To track the vertical heat redistribution over decades, we separate the ocean into four layers: 0–50 m, 50–350 m, 350–700 m and below 700 m. For the upper 350 m, the composite global OHC trend for nine hiatus decades is 0.39×10^{23}J per decade, a reduction of 31.6% compared with the composite trend in the fSurge group. This reduction is compensated in the deeper layers where the OHC trend for the hiatus decades is greater than the surge decades. For the 350–700 m layer, the composite global average in the fHiatus group is 32.0% larger (0.21 versus 0.15×10^{23}J per decade), and for the layer below 700 m, the composite global average in the fHiatus group is 5.6% larger (0.40 versus 0.37×10^{23}J per decade). The vertical heat redistribution is such that more heat is stored in the deep layers during hiatus decades[21].

Deep ocean below 700 m contributes ~38% of the whole ocean heat uptake. The difference in deep ocean heat uptake between the hiatus and surge groups is one order of magnitude smaller than their mean uptake (Fig. 3a). The s.d. within each group is also a few times smaller than the group mean. Thus, the deep ocean heat uptake observed during the hiatus decade reflects primarily anthropogenic warming, not the decadal variations that cause the hiatus. In support of observational results[23], our analyses of the large ensemble simulations show that statistically we do not expect to see a significant correspondence between decadal modulations of the GMST trend and global deep ocean heat uptake.

A clear pattern emerges from the regional ocean heat uptake below 700 m (Fig. 3b): it is large in the Atlantic and Southern Oceans (each making up ~36% of the global uptake) but small in the Pacific and Indian Oceans (~28% combined), a result consistent with observations[22]. Like the global mean, the deep ocean heat uptake in the Atlantic and Southern Oceans is dominated by anthropogenic warming, while the decadal

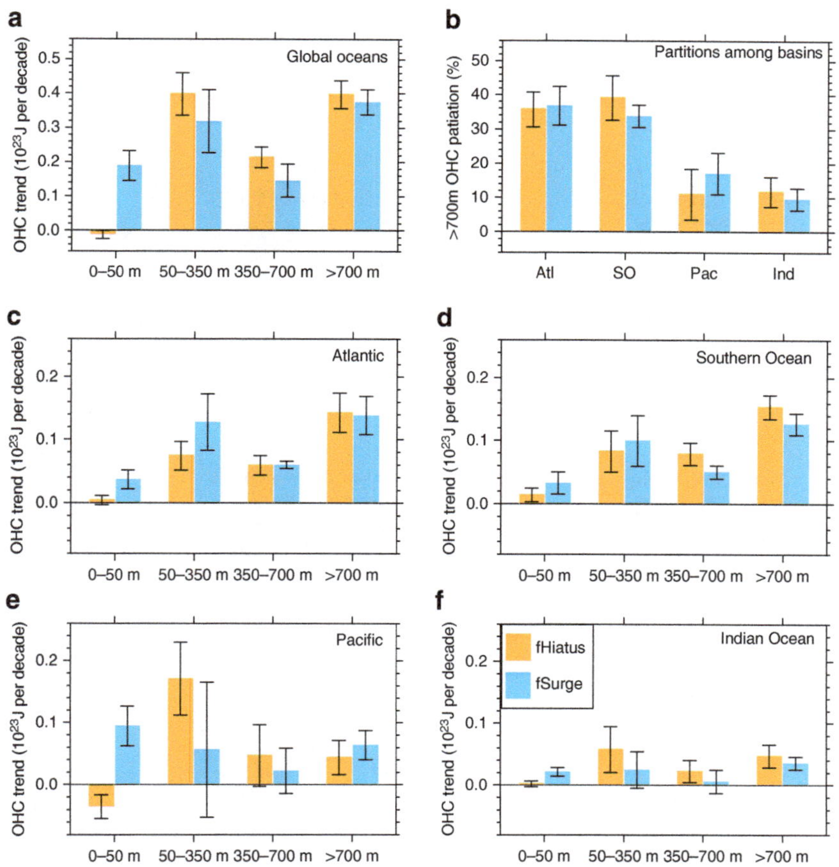

Figure 3 | OHC during hiatus and surge events. (**a**) Composite linear trends of global OHC during hiatus (orange bars) and surge (deep sky-blue bars) decades in the 21st century for four layers (surface to 50 m, 50-350 m, 350-700 m and below 700 m) with error bars showing one s.d. (**c**–**f**) Similar to **a** but for the Atlantic, Southern Ocean, Pacific and Indian Ocean. (**b**) The partition of individual basins for the OHC trend in oceans deeper than 700 m.

variability (including the fHiatus-fSurge difference) is an order of magnitude smaller. We can therefore apply the same conclusion made for global heat uptake below 700 m to these individual basins: the penetration of heat below 700 m in these basins does not make a major contribution to surface hiatus events or surges, either individually, or collectively.

Mechanisms for regional OHC variations. Physically, the disconnection between the decadal variations in GMST and OHC below 700 m reflects the fact that the variations of GMST and OHC below 700 m are governed by different mechanisms. GMST is affected by the atmosphere-ocean interaction and SST variability is organized into coherent patterns like the IPO[8,9,13]. Below 700 m, temperature variations are governed by distinct subsurface ocean dynamics, especially the MOCs in the Atlantic and Southern Oceans. MOCs prove to set the depth at which anthropogenic warming penetrates, which extend to great depths (> 1,500 m) in the Atlantic and Southern Oceans. The former is known as the Atlantic meridional overturning circulation (AMOC) (Fig. 4c) and the latter is a residual circulation resulting from wind-driven and eddy-mediating mechanisms (Fig. 4f). The sinking motions of the deep MOCs (40–80°N in the Atlantic and 40–50°S in the Southern Ocean) sequester anthropogenic heat in the deep ocean below 700 m in these two regions, causing a pronounced abyssal warming (Fig. 4a–f). MOCs in the Pacific and Indian Oceans, however, are limited to shallow depths (< ~300 m). They appear as symmetric cells about the equator in the former while as a single anticlockwise cell in the latter, trapping anthropogenic heat mainly within the upper ~300 m in these two basins (Fig. 4g–l).

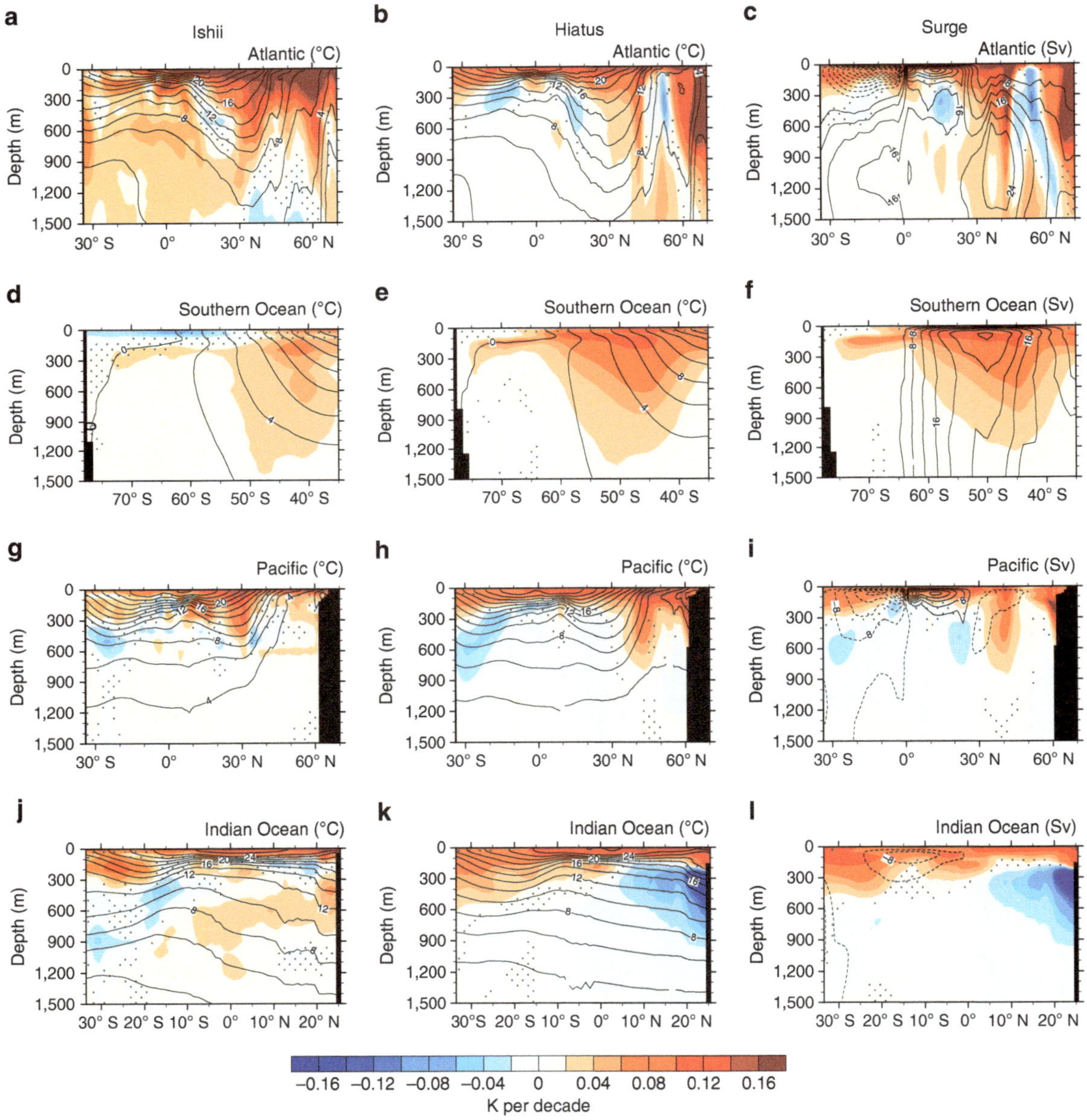

Figure 4 | Observed and simulated subsurface temperature trends. Zonal mean temperature trends during 1970–2012 (shading in *K* per decade) in the upper 1,500 m in global oceans, the Atlantic, Southern Ocean, Pacific and Indian Ocean from the Ishii data (**a**,**d**,**g**,**j**) and the ensemble means of the CESM Hiatus (**b**,**e**,**h**,**k**) and Surge (**c**,**f**,**i**,**l**) groups. Stippling indicates region below 95% significance computed from a two-tailed *t*-test. Contours show the 1970–2012 climatology of isotherm (°C) and meridional overturning stream-function (S_v, $1S_v = 1 \times 10^6\,m^3\,s^{-1}$). Since isotherms and meridional overturning stream-functions are highly similar between the Hiatus and Surge groups, for simplicity, the former is included in panels of the Ishii data and the hiatus group, while the latter is included in panels of the Surge group.

Heat uptake above 700 m, by contrast, shows certain correlation with GMST. Within the top three layers, differences in Indo-Pacific heat uptake between the fHiatus and fSurge groups are robust and on the same order of magnitude as their mean uptakes (Fig. 3e,f). During hiatus periods, anomalous cooling happens in the surface mixed layer (0–50 m) in all basins, especially in the Pacific. Over the Pacific and Indian Oceans, the 0–50 m cooling is compensated by warming within 50-350 m (Fig. 3e,f). This result is different from an earlier model study[21], which showed an overall cooling in the upper 300 m in the Indian Ocean but consistent with observations[23]. Here we include four observational data sets to compare with model results: the Ishii data, the EN4 data[33], the WOA data[34] and the latest European Centre for Medium-Range Weather Forecasts ocean reanalysis system 4 (ECMWF ORAS4) product[15,35] (Methods). All the observational data sets and CESM simulation consistently show an anomalous warming in the Indian Ocean below ~50 m (Supplementary Fig. 5e) and an accelerated OHC increase (Fig. 5f) during hiatus periods. This Indian Ocean OHC increase corresponds to a concurrent Pacific OHC decrease in the 0–100 m layer (Fig. 5e), indicating heat redistribution between these two basins. The distribution patterns vary among data sets due to data uncertainties, while the model result well lies within the range of observational uncertainties (Fig. 5b). Our model result shows that the total Indo-Pacific OHC change is close to zero within the upper 350 m (Fig. 5b), meaning that most of the hiatus-related cooling in the surface and mixed layer is compensated by warming in 50–350 m. Therefore, enhanced heat uptake below 350 m, as suggested by ref. 21, is not required in these two basins.

Figure 6b,c and Supplementary Fig. 6d further show that warming in the Indian Ocean mostly happens in the tropics at the thermocline depth (70–150 m). This warming pattern appears related to a shift towards a La Niña-like state[23] and a change of the ITF[24]. To further investigate the La Niña-like shift and related Indo-Pacific heat rearrangement, we consider the difference of climate trend between the Hiatus and Surge groups during 2002–2012. Our results show a pattern in the Pacific consistent with the La Niña-like negative phase of the IPO as described by ref. 13. Specially, anomalous high sea level pressure (SLP) centres in the mid-latitudes and intensified trade winds reflect the accelerated Walker and Hadley cells (Supplementary Fig. 7a). Strengthened surface winds accelerate equatorial surface currents (Supplementary Fig. 7c), the Equatorial Undercurrent (EUC) (Supplementary Fig. 7d) and the Pacific shallow MOC (Fig. 6a and Supplementary Fig. 7e,f). The tropical thermocline deepens in the central and western Pacific (Fig. 6c; Supplementary Fig. 7b) with an anomalous subsurface warming maximum due to

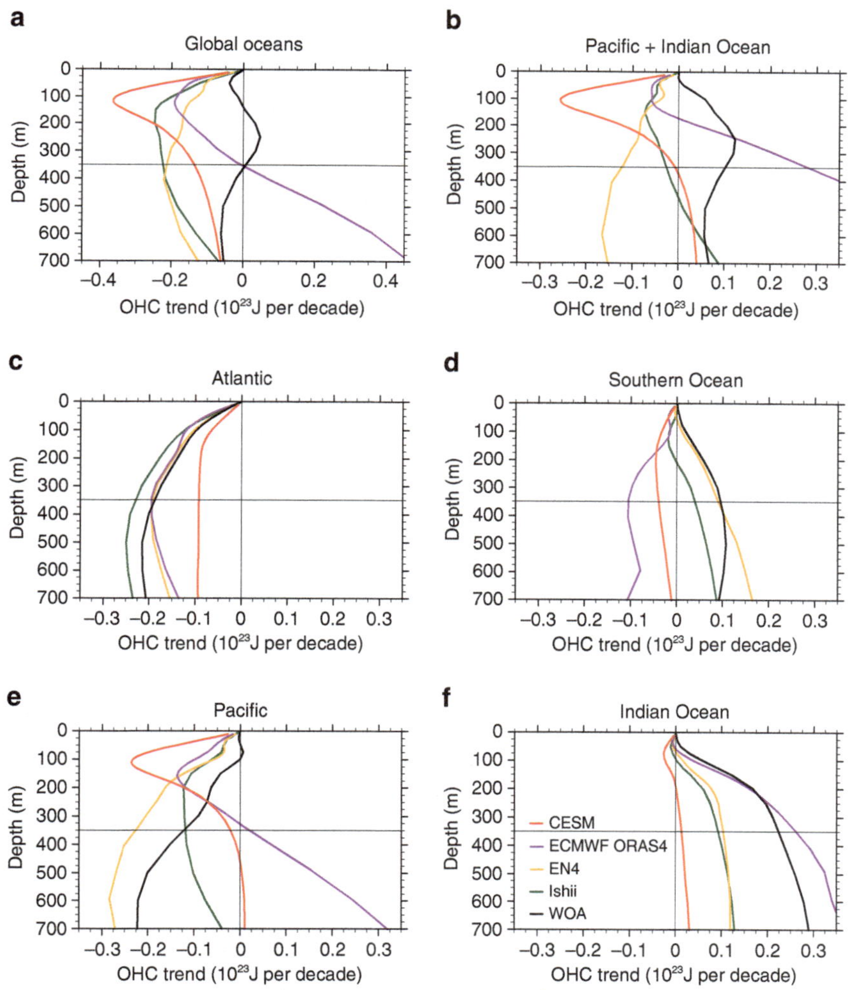

Figure 5 | Global and regional OHC trend differences as a function of depth between Hiatus and Surge events. Results in (**a**) global oceans, (**b**) the Pacific and Indian Oceans, (**c**) the Atlantic, (**d**) the Southern Ocean, (**e**) the Pacific and (**f**) the Indian Ocean are from observational data sets and the CESM simulations. OHC is integrated from the surface to indicated depths. Observations include the ECMWF ORAS4 reanalysis product (purple), the EN4 data (orange), the Ishii data (dark green) and the WOA data (blue). The CESM result (red) shows the OHC trend difference between the fHiatus and fSurge groups (the fHiatus ensemble mean minus the fSurge ensemble mean).

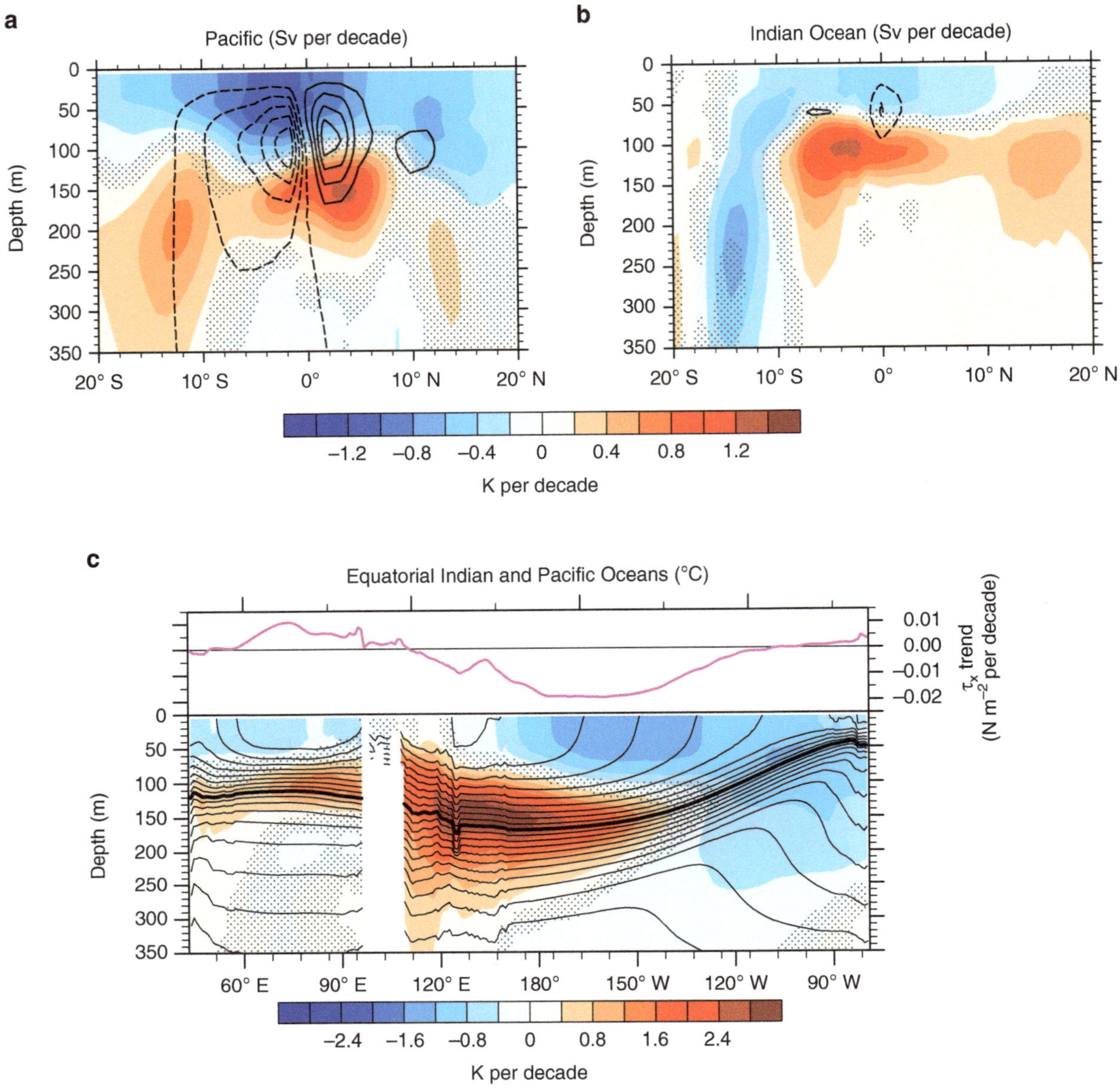

Figure 6 | Temperature trend differences in the tropical Pacific and Indian Oceans in the upper 350 m between the CESM Hiatus and Surge groups during 2002-2012. (**a**) The trend difference of zonal mean temperature (shading in *K* per decade) in the tropical Pacific, superposed by the trend difference of meridional overturning stream-function (contoured by $2S_v$ per decade, with zero contours omitted). (**b**) Similar to **a** but for the tropical Indian Ocean. (**c**) Trend differences of zonal wind stress (orchid) and temperature (shading in *K* per decade) along equatorial band (5° S-5° N) in the Indian and Pacific Oceans. The mean isotherms during 2002-2012 (38-ensemble mean) are also included as contours with an interval of 1 °C. The 20 °C contour is thickened to indicate the depth of thermocline.

increased equatorial pycnocline heat convergence, and shoals in the eastern Pacific with enhanced upwelling and surface cooling (Figs 6c; 1b).

The anomalous subsurface warming and accelerated OHC increase in the Indian Ocean is closely associated with the change in the Pacific. Anomalous warm water in the tropical western Pacific can be transported into the Indian Ocean via the Indonesian passages. During hiatus events, both the ITF volume transport and the ITF heat transport markedly increase[24], especially over the upper 350 m (Supplementary Fig. 8), which substantially contributes to the OHC increase in the Indian Ocean. Besides, local response in the Indian Ocean during the La Niña-like shift also facilitates the acceleration of the OHC increase in the Indian Ocean. Anomalous low SLP occurs in the subtropical southeastern Indian Ocean, with relaxed southeasterlies along Sumatra (Supplementary Fig. 7a). The weakened along-shore winds reduce the upwelling and deepen the thermocline off the coast. At the equator, anomalous westerlies pile up water eastwards and depress the thermocline in the eastern tropical Indian Ocean (Fig. 6c). Both processes contribute to the anomalous subsurface warming and OHC increase in the Indian Ocean.

Discussion

We show that the observed global OHC in the 300–1,500 m is dominated by a continuous warming trend over the past 43 years, and the penetration depth of ocean warming in individual basins is determined by the depth of the regional MOC. The deep (700–1,500 m) ocean heat uptake observed during the 21st century in the Atlantic and Southern Oceans primarily reflects anthropogenic heat entering the subsurface ocean via the deep

MOC. This deep penetration occurs at a similar rate during both hiatus and surge periods and is therefore not a major contributor to decadal changes in the warming rate of GMST. Our analyses of the CESM large ensemble simulations further show that the regional pattern of decadal ocean heat uptake below 700 m is indeed robust but corresponds poorly to decadal variations in global surface warming rate. OHC in the upper 350 m, particularly in the Pacific and Indian Oceans, shows robust features of decadal variations. During hiatus decades, the Indian Ocean shows an anomalous warming and accelerated OHC increase below 50 m, which is associated with a La Niña-like climate shift and an enhanced heat transport of the ITF. This Indian Ocean warming corresponds to a concurrent Pacific cooling in the 0–100 m layer, thus indicative of an Indo-Pacific heat rearrangement. We find that warming within 50–350 m mostly compensates the 0–50 m cooling in the Pacific and Indian Oceans so that enhanced deep-level (below 350 m) heat uptake is not required in these two basins. We therefore conclude that the ocean heat uptake in the upper 350 m in the Pacific and Indian Oceans is closely tied to the surface warming hiatus, whereas the regional pattern of deep ocean heat uptake (below 700 m) is not a valid basis to infer which basin drives the surface warming hiatus.

The vertical and interbasin heat redistribution in the ocean is not well understood. In addition to MOCs, the wind-driven decadal variability in ocean circulation is a plausible mechanism for heat redistribution[36]. For example, decadal variability of the Southern Ocean heat uptake is suggested to be caused by modulations in winds associated with the Southern Annual Mode[37]. The need to develop a rigorous energy view and test it for the hiatus[38] spotlights the important problem of three-dimensional redistribution of heat in the ocean.

Methods

The OHC calculation. In this study, OHC within a certain layer is calculated as

$$\mathrm{OHC} = \varrho_0 C_\mathrm{P} \int_{z_1}^{z_2} T(z)\mathrm{d}z \tag{1}$$

where ϱ_0 and C_P denote sea water density and specific heat capacity, respectively. z_1 and z_2 denote the upper and lower limits of the layer depth. $T(z)$ denotes temperature profile, a function of depth z. Because we focus on the variability over the decadal time scales, a 12-month running is applied to the OHC time series. We calculate the OHC anomalies to the first year (year 1970) to manifest the continuous subsurface warming (Fig. 2), and to the period of 1970–2012 to reproduce the results by ref. 22 (Supplementary Fig. 9).

The observational data sets. Four observational data sets are selected that cover the span of 1970–2012. They are composed of two categories: (1) objective analyses of *in-situ* observations (for example, expendable bathy-thermographs (XBTs), conductivity-temperature-depth (CTD) measurements from research ships and Argo floats) and (2) reanalysis products that assimilate a variety of available observations in an ocean model. The former includes the EN4 data[33], the WOA data[34] and the data from a research leading by Ishii[30] (called the Ishii data in this study). The latter includes the ECMWF ORAS4 product[15,35]. In particular, the EN4 data is monthly mean temperature at 42 levels extending to beyond 5,000 m from 1900 to the present. The WOA data is annual mean temperature at 16 levels in the upper 700 m during 1955–2012. The Ishii data is monthly mean temperature at 24 levels in the upper 1,500 m during 1945–2012. The ECMWF ORAS4 product is monthly temperature at 42 levels extending to beyond 5,000 m during 1958–2014. For the Ishii data, the annual mean anomalies for globally integrated upper 700 m OHC during 1950–2014 is available from Japan Meteorological Agency. Due to a sparse sampling and its induced large OHC error in the early era (Supplementary Fig. 10), the period of 1970–2012 is selected for the SST and OHC analyses. In this study, to examine the changes in the OHC (Fig. 5) and temperature (Supplementary Fig. 5) between recent hiatus event and prior surge event, we consider the global and regional OHC (temperature) trends for the periods from 2003 to 2012 and from 1992 to 2001, and calculate the difference between the two periods.

The CESM large ensemble simulations. The CESM community provides a production of a large ensemble (38 members) using a 1-degree CESM-CAM5 with biogeochemistry (BGC)[31]. Ensemble members go from 1920–2100 using historical forcing (1920–2005) and Representative Concentration Pathway 85 (RCP 8.5) forcing (2005–2100). We calculate the running 10-year linear trends of globally averaged SST from the 38-ensemble members and choose the members simulating the recent hiatus event by the criteria of negative decadal trend appearing within the period of 2002–2012. As shown in Supplementary Fig. 3, four members (M09, M16, M19 and M31) happen to capture the recent hiatus event, while four (M13, M29 M37 and M07) simulate a warming surge during the same span. Thus we collect the former four as the 'Hiatus' group and the latter four as the 'Surge' group. For each group, the ensemble mean SST and OHC are calculated and compared in Figs 1 and 2. The model drift is estimated from a 500-year pre-industrial control run and removed in the SST and OHC analyses. Moreover, to test the robustness of the OHC difference between the hiatus and surge events, we expand the Hiatus and Surge groups by including five pairs of hiatus and surge decades (Supplementary Table 1) in the future simulations and name the new groups are as the fHiatus and fSurge. To calculate the ensemble mean or standard deviation of the OHC (Figs 3 and 5) and temperature (Supplementary Fig. 5) of the fHiatus and fSurge, values in future hiatus and surge events are scaled to 2002–2011 by multiplying by a ratio of 38-ensmble mean global OHC trend during 2002–2012 to 38-ensmble mean global OHC trend during the event period.

References

1. Easterling, D. R. & Wehner, M. F. Is the climate warming or cooling? *Geophys. Res. Lett.* **36,** L08706 (2009).
2. Katsman, C. A. & van Oldenborgh, G. J. Tracing the upper ocean's "missing heat". *Geophys. Res. Lett.* **38,** L14610 (2011).
3. Foster, G. & Rahmstorf, S. Global temperature evolution 1979–2010. *Environ. Res. Lett.* **6,** 044022 (2011).
4. Solomon, S. *et al.* Contributions of Stratospheric Water Vapor to Decadal Changes in the Rate of Global Warming. *Science* **32,** 1219–1223 (2010).
5. Kaufmann, R. K., Heikki, K., Michael, L. M. & James, H. S. Reconciling anthropogenic climate change with observed temperature 1998–2008. *Proc. Natl Acad. Sci. USA* **108,** 11790–11793 (2011).
6. Fyfe, J. C., von Salzen, K., Cole, J. N. S., Gillett, N. P. & Vernier, J.-P. Surface response to stratospheric aerosol changes in a coupled atmosphere–ocean model. *Geophys. Res. Lett.* **40,** 584–588 (2013).
7. Meehl, G. A., Hu, A., Arblaster, J. M., Fasullo, J. & Trenberth, K. E. Externally forced and internally generated decadal climate variability associated with the Interdecadal Pacific Oscillation. *J. Clim.* **26,** 7298–7310 (2013).
8. Meehl, G. A., Teng, H. & Arblaster, J. M. Climate model simulations of the observed early-2000s hiatus of global warming. *Nat. Clim. Change* **4,** 892–902 (2014).
9. Trenberth, K. E., Fasullo, J. T., Branstator, G. & Phillips, A. S. Seasonal aspects of the recent pause in surface warming. *Nat. Clim. Change* **4,** 911–916 (2014).
10. Zhang, Y., Wallace, J. M. & Battisti, D. S. ENSO-like interdecadal variability: 1900-93. *J. Clim.* **10,** 1004–1020 (1997).
11. Power, S., Casey, T., Folland, C., Colman, A. & Mehta, V. Inter-decadal modulation of the impact of ENSO on Australia. *Clim. Dyn.* **15,** 319–324 (1999).
12. Kosaka, Y. & Xie, S.-P. Recent global-warming hiatus tied to equatorial Pacific surface cooling. *Nature* **501,** 403–407 (2013).
13. England, M. H. *et al.* Recent intensification of wind-driven circulation in the Pacific and the ongoing warming hiatus. *Nat. Clim. Change* **4,** 222–227 (2014).
14. Watanabe, M. *et al.* Contribution of natural decadal variability to global warming acceleration and hiatus. *Nat. Clim. Change* **4,** 893–897 (2014).
15. Balmaseda, M. A., Trenberth, K. E. & Källén, E. Distinctive climate signals in reanalysis of global ocean heat content. *Geophys. Res. Lett.* **40,** 1754–1759 (2013).
16. Hansen, J. *et al.* Earth's energy imbalance: Confirmation and implications. *Science* **308,** 1431–1435 (2005).
17. Hansen, J., Sato, M., Kharecha, P. & von Schuckmann, K. Earth's energy imbalance and implications. *Atmos. Chem. Phys.* **11,** 13421–13449 (2011).
18. Trenberth, K. E., Fasullo, J. T. & Balmaseda, M. A. Earth's Energy Imbalance. *J. Clim.* **27,** 3129–3144 (2014).
19. Trenberth, K. E., Fasullo, J. T. & Kiehl, J. Earth's global energy budget. *Bull. Am. Meteorol. Soc.* **90,** 311–323 (2009).
20. Loeb, N. G. *et al.* Observed changes in top-of-the-atmosphere radiation and upper-ocean heating consistent within uncertainty. *Nat. Geosci* **5,** 110–113 (2012).
21. Meehl, G. A., Arblaster, J. M., Fasullo, J. T., Hu, A. & Trenberth, K. E. Model-based evidence of deep-ocean heat uptake during surface-temperature hiatus periods. *Nat. Clim. Change* **1,** 360–364 (2011).
22. Chen, X. & Tung, K. K. Varying planetary heat sink led to global-warming slowdown and acceleration. *Science* **345,** 897–903 (2014).
23. Nieves, V., Willis, J. K. & Patzert, W. C. Recent hiatus caused by decadal shift in Indo-Pacific heating. *Science* **349,** 532–535 (2015).
24. Lee, S.-K. *et al.* Pacific origin of the abrupt increase in Indian Ocean heat content during the warming hiatus. *Nat. Geosci* **8,** 445–449 (2015).
25. Lyman, J. M. *et al.* Robust warming of the global upper ocean. *Nature* **465,** 334–337 (2010).

26. Durack, P. J., Gleckler, P. J., Landerer, F. W. & Taylor, K. E. Quantifying underestimates of long-term upper-ocean warming. *Nat. Clim. Change* **4,** 999–1005 (2014).
27. Smith, D. M. *et al.* Earth's energy imbalance since 1960 in observations and CMIP5 models. *Geophys. Res. Lett.* **42,** 1205–1213 (2015).
28. Roemmich, D. *et al.* Unabated planetary warming and its ocean structure since 2006. *Nat. Clim. Change* **5,** 240–245 (2015).
29. Palmer, M. D., McNeall, D. J. & Dunstone, N. J. Importance of the deep ocean for estimating decadal changes in Earth's radiation balance. *Geophys. Res. Lett.* **38,** L13707 (2011).
30. Ishii, M., Shouji, A., Sugimoto, S. & Matsumoto, T. Objective analyses of SST and marine meteorological variables for the 20th century using COADS and the Kobe Collection. *Int. J. Climatol.* **25,** 865–879 (2005).
31. Kay, J. E. *et al.* The Community Earth System Model (CESM) Large Ensemble Project: A Community Resource for Studying Climate Change in the Presence of Internal Climate Variability. *Bull. Amer. Meteor. Soc.* **96,** 1333–1349 (2015).
32. Fyfe, J. C., Gillett, N. P. & Zwiers, F. W. Overestimated global warming over the past 20 years. *Nat. Clim. Change* **3,** 767–769 (2013).
33. Good, S. A., Martin, M. J. & Rayner, N. A. EN4: quality controlled ocean temperature and salinity profiles and monthly objective analyses with uncertainty estimates. *J. Geophys. Res.* **118,** 6704–6716 (2013).
34. Levitus, S. *et al.* World ocean heat content and thermosteric sea level change (0–2000 m), 1955–2010. *Geophys. Res. Lett.* **39,** L10603 (2012).
35. Balmaseda, M. A., Mogensen, K. & Weaver, A. T. Evaluation of the ECMWF ocean reanalysis system ORAS4. *Q.J.R. Meteorol. Soc* **139,** 1132–1161 (2013).
36. Huang, R. X. Heaving modes in the world oceans. *Clim. Dyn.* **45,** 3563–3591 (2015).
37. Drijfhout, S. S. *et al.* Surface warming hiatus caused by increased heat uptake across multiple ocean basins. *Geophys. Res. Lett.* **41,** 7868–7874 (2014).
38. Xie, S.-P., Kosaka, Y. & Okumura, Y. M. Distinct energy budgets for anthropogenic and natural changes during global warming hiatus. *Nat. Geosci* **9,** 29–33 (2016).

Acknowledgements

This work is supported by NSF AGS-1249145. Jian Lu was supported by the U.S. Department of Energy Office of Science Biological and Environmental Research (BER) as part of the Regional and Global Climate Modeling program. The CESM model output data are publicly available on the Earth System Grid (ESG). For observational and reanalysis data, the EN4 data are publicly available at http://hadobs.metoffice.com/en4/, the Ishii data are publicly available at http://rda.ucar.edu/data sets/ds285.3/ and http://www.data.jma.go.jp/gmd/kaiyou/english/ohc/ohc_data_en.html, the WOA data are publicly available at https://www.nodc.noaa.gov/OC5/3M_HEAT_CONTENT/ and the ECMWF ORAS4 data are publicly available at https://climatedataguide.ucar.edu/climate-data/oras4-ecmwf-oceanreanalysis-and-derived-ocean-heat-content and Integrated Climate Data Center -ICDC.

Author contributions

W.L. performed the analyses. W.L. and S.-P.X. wrote the paper. J.L. contributed to the interpretation of results.

Additional information

Competing financial interests: The authors declare no competing financial interests.

20

Timescales of methane seepage on the Norwegian margin following collapse of the Scandinavian Ice Sheet

Antoine Crémière[1,2], Aivo Lepland[1,2,3,4], Shyam Chand[1,2], Diana Sahy[5], Daniel J. Condon[5], Stephen R. Noble[5], Tõnu Martma[3], Terje Thorsnes[1,2], Simone Sauer[1,2] & Harald Brunstad[6]

Gas hydrates stored on continental shelves are susceptible to dissociation triggered by environmental changes. Knowledge of the timescales of gas hydrate dissociation and subsequent methane release are critical in understanding the impact of marine gas hydrates on the ocean-atmosphere system. Here we report a methane efflux chronology from five sites, at depths of 220–400 m, in the southwest Barents and Norwegian seas where grounded ice sheets led to thickening of the gas hydrate stability zone during the last glaciation. The onset of methane release was coincident with deglaciation-induced pressure release and thinning of the hydrate stability zone. Methane efflux continued for 7–10 kyr, tracking hydrate stability changes controlled by relative sea-level rise, bottom water warming and fluid pathway evolution in response to changing stress fields. The protracted nature of seafloor methane emissions probably attenuated the impact of hydrate dissociation on the climate system.

[1] Marine Geology, Geological Survey of Norway, Postal box 6315 Sluppen, 7491 Trondheim, Norway. [2] CAGE—Centre for Arctic Gas Hydrate, Environment and Climate, Department of Geology, UiT the Arctic University of Norway, 9037 Tromsø, Norway. [3] Institute of Geology, Tallinn University of Technology, 19086 Tallinn, Estonia. [4] Department of Geology, University of Tartu, Ravila 14A, 50411 Tartu, Estonia. [5] British Geological Survey, Keyworth, Nottingham NG12 5GG, UK. [6] Lundin Norway AS, 1366 Lysaker, Norway. Correspondence and requests for materials should be addressed to A.C. (email: antoine.cremiere@ngu.no) or to A.L. (email: aivo.lepland@ngu.no).

Large quantities of methane exist as free gas, dissolved in pore fluids and solid gas hydrates within marine sediments[1]. Gas hydrates are found in a variety of geological settings[2] but are most common on continental margins where the combination of appropriate pressure–temperature conditions and the presence of organic-rich sediments are conducive to abundant methane generation. The stability of the marine gas hydrate reservoir is primarily affected by changes in temperature and hydrostatic pressure. Consequently, changes in the thickness of the gas hydrate stability zone (GHSZ) are likely to occur along glaciated continental margins where pressure regimes are governed by alternating glacial loading and unloading cycles, and gas hydrate dissociation may be amplified by the influx of warmer waters following ice retreat[2–5]. If bottom water warming and/or hydrostatic pressure decrease is large enough, gas hydrate dissociation and release of a significant amount of methane are possible[4,6–10]. The transfer of this methane to the water column has the potential to amplify ocean acidification and de-oxygenation and, possibly, atmospheric greenhouse gas concentrations[11,12]. The rates and timescales over which gas hydrates dissociate control the impact of hydrate-derived methane on the global climate system.

Methane released by the dissociation of marine gas hydrates has been inferred to be the cause of a number of negative $\delta^{13}C$ excursions in the geological record[6,7]. Large magnitude gas hydrate dissociation events are also proposed to have occurred in the more recent past, as evidenced by major fluid flow structures in sedimentary basins, such as pockmarks[4,13]. However, the role of marine gas hydrate releases in influencing atmospheric methane concentrations during the Quaternary remains uncertain. For example, the radiocarbon and hydrogen isotopic composition of methane from Greenland ice cores suggests that atmospheric methane rise at 11.6 and 38 ka can be predominantly accounted for through wetland emissions[14,15]. Although it appears that Quaternary marine gas hydrate dissociation events did not have a significant impact on the climate system, recent estimates of the gas hydrate reservoirs stored beneath the grounded Antarctic ice sheet suggest that methane release resulting from accelerated ice wastage has the potential to act as a positive feedback on future climate warming[16].

Methane is consumed at a variety of stages during transport through the sediment and overlying water column. On a global scale, >80% of methane migrating within the sedimentary package can be microbially consumed at the sedimentary sulphate–methane transition zone by the anaerobic oxidation of methane (AOM)[11,17]. However, where gas supply is high and flow through the sediment is advective, the capacity of this benthic microbial filter can be outstripped, resulting in release of methane into the overlying water column[18]. A proportion (up to 100%) of methane within the water column is consumed via dissolution of free gas bubbles and methanotrophic microbial activity[19,20] before it reaches the ocean surface and escapes into the atmosphere. Consequently, the short residence time of methane in the oceans and atmosphere (~10 years), combined with the processes that control consumption in the sediment and water column, requires a rapid (decadal) release and transfer of large volumes of methane to achieve a significant impact on ocean/atmosphere chemistry.

The processes that influence gas hydrate stability in continental margins act on timescales ranging from near instantaneous (for example, slope failure) to 10^2–10^3 years (ice sheet collapse), whereas significant sea-level and bottom water temperature variations operate over 10^3 years[21,22]. The transfer of methane from the seabed into the water column is typically detected by water column acoustic gas flares, with observations restricted to timescales of less than a few decades[20,23]. However, assessments of the timescales over which gas hydrate systems respond to the processes that drive changes in pressure and temperature in the subsurface are limited. Studies from GHSZ pinch outs indicate rapid response to temperature change in the immediate subsurface over a few decades[8,22]. Modelling results show that heat transfer and temperature re-equilibration of ~100 m of marine sediments would take 10^2–10^3 years[8,24], and that gas propagated from such depth may take an additional thousand years to vent at the seafloor[24]. However, empirical constraints on methane efflux triggered by hydrate dissociation resulting from glacial unloading of a continental margin, both past and present are, as yet, lacking.

Direct information on the timing of past methane release events can be obtained through U-Th dating of methane-derived authigenic carbonates[25,26], which form at the sedimentary sulphate–methane transition zone, due to elevated carbonate alkalinity resulting from AOM in areas experiencing relatively intense methane fluxes over 10^2–10^3 years[25]. To date, U-Th dating of methane-derived carbonates has mostly focused on samples from mid-latitudes and was used to assess the periodicity of fluid migration along fault systems[27,28] and the effects of long-term changes in sea-level and bottom water temperature on methane efflux and/or gas hydrate dissociation[26,29,30]. Conversely, U-Th dating of carbonates from methane seeps off Svalbard indicates that the stability of gas hydrates at the updip of the GHSZ at high-latitude sites is sensitive to seasonal bottom water temperature fluctuations[21].

The shelf areas of the Norwegian and Barents Seas offer ideal settings in which to assess the processes governing gas hydrate stability on glaciated continental margins with grounded ice and hence can serve as a palaeo-analogue of the Antarctic gas hydrate reservoir[16]. Here we report on the timing of methane release in the Barents and Norwegian Seas, where a drop in seabed pressure associated with the retreat of the grounded Scandinavian Ice Sheet (SIS) has been inferred as the driving force behind the dissociation of a significant amount of gas hydrates[3,13,31] (Fig. 1). We use U-Th geochronology, stable carbon and oxygen isotope analyses and petrographic observations on authigenic carbonate crusts associated with methane seeps to provide constraints on the timescales of past methane fluxes from five locations for that region. Our new U-Th data set, combined with model estimates of the amount of methane potentially stored as gas hydrates before and after the collapse of the SIS offers an opportunity to study the timescales of methane efflux across the seafloor in response to the environmental changes associated with the last deglaciation.

Results

Studied areas. The southwest Barents and Norwegian Sea are active provinces for oil and gas exploration, with extensively documented, focused, hydrocarbon-rich fluid migration through regional fault networks[3,32,33]. Mega-scale glacial lineations (for example, in the Ingøydjupet depression) indicate that the shelf areas of those seas were covered by the grounded SIS during the last glaciation[34–36]. The pressure exerted by grounded ice on the underlying sediments was associated with a ~500 m thickening of the GHSZ[3,37] and reactivation of faults and fractures as fluid pathways. Age constraints on grounded ice wedges indicate that ice persisted on the southwest Barents Sea shelf at least until 17 ka[38,39], and that ice streams extended to the edge of the shelf at least twice[34,36]. In this area the seafloor is marked by ubiquitous pockmarks, circular depressions resulting from the expulsion of fine-grained sediments in response to fluid seepage[3,13,31,40,41]. Although some pockmarks coincide with observed water-column gas flares, indicating active seepage,

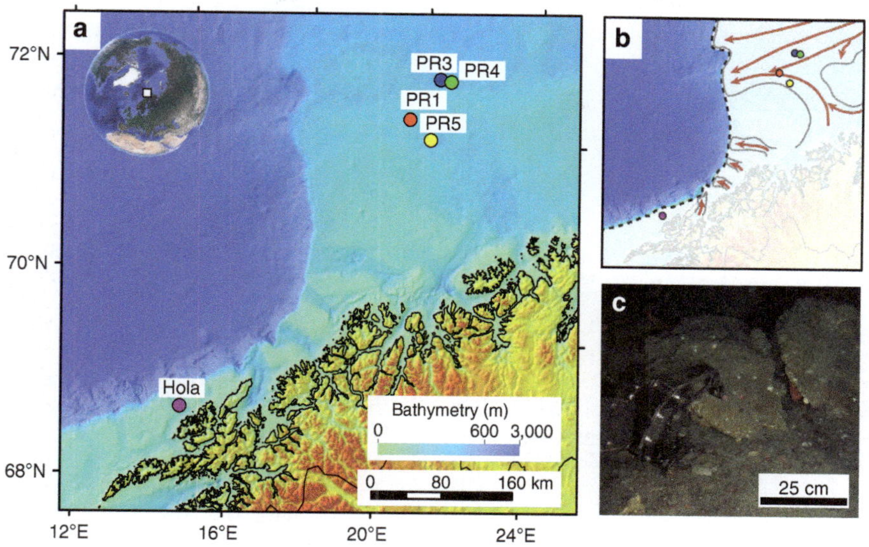

Figure 1 | Bathymetric map of the study area and remotely operated underwater vehicle (ROV) recovery of authigenic carbonate. (**a**) The five carbonate sampling sites (PR1, PR3, PR4, PR5 and Hola) are shown as circles. (**b**) The white shaded area represents the extent of the Scandinavian Ice-sheet during the LGM[51] with main ice streams depicted by red arrows[35,39]. (**c**) Carbonate crust sampled at the seafloor.

most are inactive and constitute records of widespread fluid escape at the seafloor related to abrupt gas hydrate dissociation after the last deglaciation[3,13,31]. The cross-cutting relationships between pockmarks and surrounding sediments, coupled with radiocarbon dating, indicate that pockmark formation post-dated the deposition of glaciomarine sediments at ~15 ka[31].

An autonomous underwater vehicle was used to map the seabed at high spatial resolution (<10 cm) in some pockmark areas and areas of known seepage[3,42], and revealed the presence of numerous carbonate crusts. Guided by the autonomous underwater vehicle mapping results, carbonate crusts were sampled by a remotely operated underwater vehicle from four sites in the southwest Barents Sea at depths ranging between 320 and 400 m (PR1, PR3, PR4 and PR5; Fig. 1 and Table 1). Additional samples were collected from the Hola site (depth 218 m) off the Vesterålen islands in the Norwegian Sea. Active gas seepage, inferred from gas flares in the water column, occurs at all sites except at PR5.

Carbonate crust characterization. The typical thickness of studied crust samples varies between 10 and 30 cm (Fig. 2). Authigenic carbonates occur as impure early-generation cements, lithifying mud to gravel-size detrital glacial/glaciomarine sediments and as pure detritus-free later-generation botryoidal-layered fillings of centimetre-scale cavities formed within the early-generation carbonate-cemented sediments (Fig. 2b,c). Such cavities are interpreted as conduits generated by a focused fluid flow, which washed out weakly cemented sediments. Microcrystalline and radial fibrous aragonite cements are the dominant authigenic carbonate phases, but minor micritic Mg-calcite (13–15% mol $MgCO_3$) is also present (Supplementary Fig. 1 and Supplementary Table 1). Both aragonite and high Mg-calcite cements contain disseminated framboidal pyrite indicative of active microbial sulfate reduction during carbonate precipitation[43]. The dominance of aragonite among diagenetic carbonate phases probably indicates precipitation under methane fluxes capable of sustaining high rates of AOM activity in sediments just beneath the seafloor[44].

Carbon and oxygen isotope compositions of the sampled carbonate crusts were measured on powders hand-drilled on a millimetre scale, to spatially resolve distinctive carbonate generations. Carbonate $\delta^{13}C$ values range from −43.1 to −13.0‰ Vienna Pee Dee Belemnite (VPDB) (Fig. 3 and Table 1, average −32.3 ± 4.3‰ VPDB), indicating that microbial oxidation of methane-rich fluids is the dominant source of carbon. The absence of very ^{13}C-depleted ($\delta^{13}C < -45$‰ VPDB) carbonates in the studied crusts further suggests that the principal source of methane is thermogenic, which is typically less ^{13}C depleted than microbial methane[45]. A predominantly thermogenic origin is also consistent with the stable isotope composition of methane gas currently seeping from the seabed ($\delta^{13}C$-CH_4 of −48‰ and δD-CH_4 of −186‰ Vienna Standard Mean Ocean Water (VSMOW); Supplementary Fig. 2).

The $\delta^{18}O$ values of the authigenic carbonates provide information about the temperature and oxygen isotope composition of the ambient porewater in which carbonates precipitate. The oxygen isotope composition of carbonates precipitating in equilibrium with the bottom water after deglaciation, that is, at 15 ka ($\delta^{18}O = 1$‰ VSMOW[46] and 1 °C) would have a $\delta^{18}O$ value of 4.7‰ VPDB for aragonite[47] and 5.6‰ VPDB for Mg-calcite with 15% mol $MgCO_3$ (refs 48,49). The high $\delta^{18}O$ values (that is, up to 6.4‰ VPDB; Fig. 3) measured in the early-generation Mg-calcite-cemented sediments cannot be explained by precipitation of Mg-calcite in equilibrium with the bottom waters during the early deglaciation period and probably indicate the influence of ^{18}O-rich water potentially released during gas hydrate dissociation[50].

U-Th geochronology. A total of 14 methane-derived authigenic carbonate crusts from the five sites were selected for U-Th dating; the sampling targeted both early-stage carbonate cements and late-stage cavity fills (Supplementary Data 1). The latter are characterized by low detrital content ($[^{230}Th/^{232}Th]_{AR}$ ranging between 5 and 427) and give dates with a precision on the order of ±0.5–1% (2σ). Early-stage cements prove to be more difficult to date effectively, owing to the presence of a high proportion of detrital material (Fig. 2c), with uncertainties magnified by the impact of the corrections required to account for the presence of initial detrital Th (Fig. 2b and Supplementary Data 2). We obtain 59 reliable dates (defined as $[^{230}Th/^{232}Th]_{AR} > 2$, see Fig. 4 and Supplementary Note), mostly from late-stage cavity fills, which show that 12 of the analysed crusts formed from 17 to 7 ka.

Table 1 | U-Th isochron ages, stable carbon and oxygen isotopic composition, coordinate and water depth for carbonate samples.

Area	Sample	U-Th age (ka ± 2σ)			$\delta^{13}C$	$\delta^{18}O$	Latitude N	Longitude W	Water depth (m)
		Min.	Max.	*n*	(‰ VPDB)	(‰ VPDB)			
PR1									
	P1201001	9.2 ± 0.5	13.7 ± 0.1	8	− 30.5 to − 18.1	4.0 to 5.3	72° 09′ 28.5″	19° 43′ 38.5″	319
	P1210002	11.3 ± 0.6	13.0 ± 0.1	7	− 31.4 to − 13.0	3.4 to 5.4	72° 09′ 28.1″	19° 43′ 37.7″	320
	P1210004	8.9 ± 0.1	11.5 ± 0.5	3	− 35.5 to − 23.0	2.8 to 5.9	72° 09′ 27.8″	19° 43′ 38.0″	320
	P1210006	2.38 ± 0.03	5.4 ± 0.1	6	− 43.1 to − 33.4	3.0 to 4.2	72° 09′ 28.0″	19° 43′ 38.7″	320
	P1210007	8.2 ± 0.1	10.9 ± 0.1	2	− 33.8 to − 19.5	3.7 to 5.4	72° 09′ 27.8″	19° 43′ 37.9″	320
PR3									
	P1210010	11.6 ± 0.1	15.1 ± 1.4	3	− 32.6 to − 27.6	4.4 to 5.3	72° 35′ 20.4″	20° 35′ 10.8″	403
	P1210011	11.8 ± 0.1	11.9 ± 0.1	3	− 36.0 to − 31.1	4.3 to 5.0	72° 35′ 18.9″	20° 35′ 11.0″	403
	P1210012	11.8 ± 0.1		1	− 34.6 to − 30.2	4.6 to 5.3	72° 35′ 20.4″	20° 35′ 10.8″	403
	P1210014	10.9 ± 0.1	17.5 ± 0.7	3	− 33.2 to − 26.8	4.6 to 5.6	72° 35′ 18.9″	20° 35′ 11.1″	403
PR4									
	P1201017	9.61 ± 0.04	12.3 ± 0.2	4	− 41.3 to − 26.2	3.6 to 6.4	72° 34′ 03.8″	20° 52′ 21.0″	391
	P1210018	12.9 ± 1.0	14.4 ± 0.6	3	− 38.6 to − 28.7	4.3 to 6.4	72° 34′ 02.2″	20° 52′ 09.3″	391
PR5									
	P1210032	3.3 ± 0.4		1	− 40.0 to − 33.4	3.6 to 4.1	71° 59′ 12.4″	20° 28′ 40.5″	393
	P1210036	7.1 ± 0.1	15.9 ± 4.2	4	− 40.3 to − 33.5	4.1 to 4.3	71° 59′ 12.6″	20° 28′ 41.3″	393
Hola									
	Hola	6.1 ± 1.6	11.3 ± 0.6	11	− 38.0 to − 26.7	2.5 to 3.7	68° 55′ 05.8″	14° 17′ 02.6″	218

VPDB, Vienna Pee Dee Belemnite.

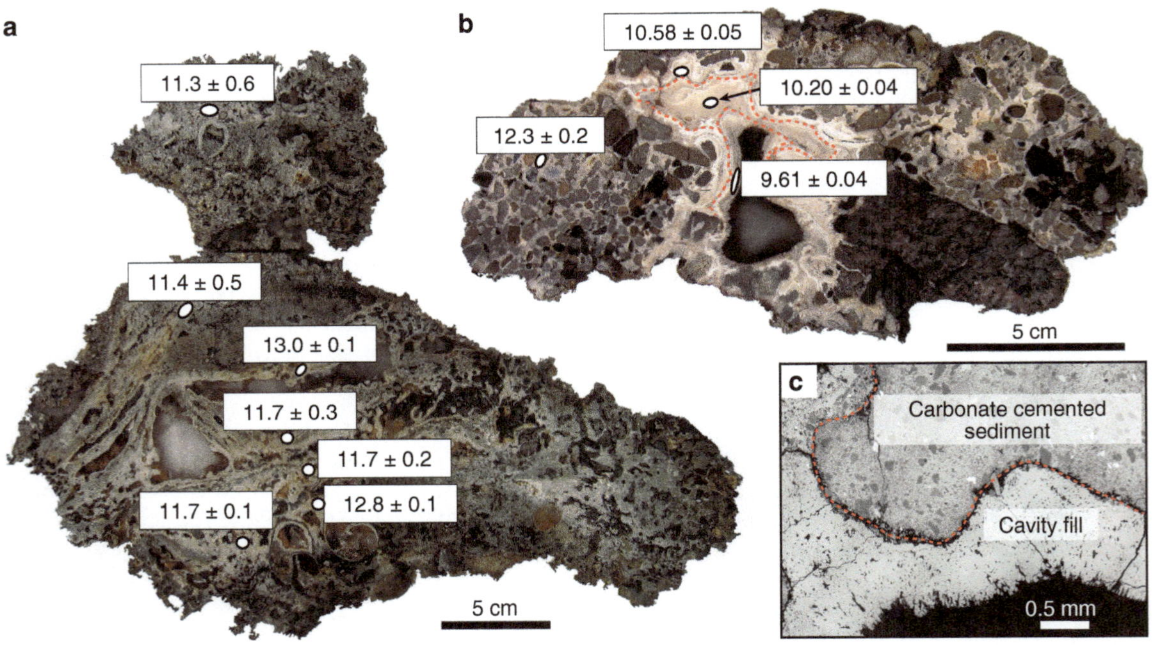

Figure 2 | Authigenic carbonate crusts. Detailed U-Th ages (in ka ± 2σ) on representative cross-sections of (**a**) sample P1210002 and (**b**) sample P1210017. (**c**) Scanning electron microscope image showing early-stage cements containing detrital material and detritus-free layer of radial fibrous aragonite cavity fill.

Younger dates, between 5 and 2 ka, are obtained from two crusts, one each from the PR1 and PR5 sites (Fig. 4). Owing to the preponderance of late-stage aragonite cavity fills, the U-Th data set provides minimum age estimates for the onset of emissions, which is recorded by the less well-dated early-stage cements. Both the older (17–7 ka) and younger (5–2 ka) crusts span several 100–5,000 years (Fig. 4) and constrain the minimum duration and absolute timing of methane efflux at each site. Data from multiple subsamples collected from the same crust are consistent with the observed textural relationships (that is, cavity fills are younger than early-stage cements; Fig. 2b).

Gas hydrate stability modelling. Gas hydrate stability was modelled at steady state based on the presence of an ~1,100-m-thick grounded ice sheet[51], the regional average geothermal gradient of 31 °C km^{-1} (ref. 52) and a gas composed of 96% methane, 3% ethane and 1% propane. Such a gas composition is consistent with the base of the GHSZ corresponding to the observed bottom-simulating seismic reflectors in the Barents Sea[37,53]. The model estimates the thickness of the GHSZ at up to 600 m below the seafloor during the Last Glacial Maximum (LGM), thinning gradually towards the shelf edge due to decreasing ice load (Fig. 5a). Following the retreat of grounded ice, which triggered isostatic rebound, sea-level rise and bottom water warming, the GHSZ thinned to *ca.* 100 m in most of the southwest Barents Sea, including the study area. A thicker GHSZ of up to 400 m persists in the deeper, central part of the southwest Barents Sea (Fig. 5b). Based on the volume change of gas hydrates between the LGM (Fig. 5a) and the present day (Fig. 5b), we estimate that 5–43 Gt of methane was released in the studied area (4×10^5 km^2).

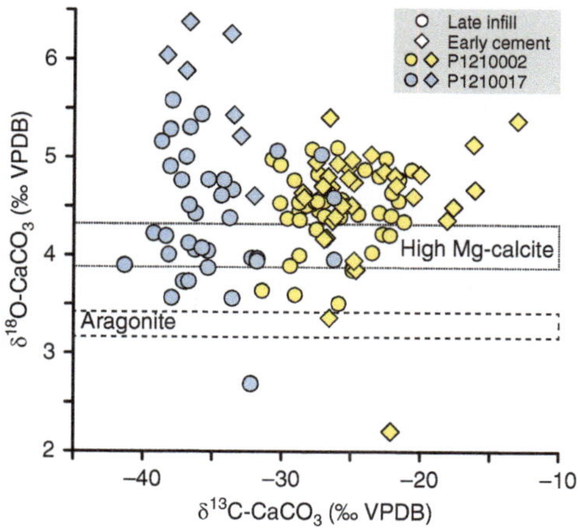

Figure 3 | Stable C and O isotopes. Isotope compositions of early generation cements and late generation cavity infills of samples hand-drilled from two representative carbonate crusts. The boxes deliniated by dashed lines indicate the $\delta^{18}O$ range of carbonate (aragonite[47] and Mg-calcite with 13-15% mol $MgCO_3$ (refs 48,49)) precipitating in equilibrium with present-day bottom water (temperature of 2-3 °C and seawater $\delta^{18}O$ of 0‰ VSMOW).

Discussion

The clustering of U-Th dates indicates that the main crust-forming methane flux episode took place between 17 and 7 ka (Fig. 4), which is consistent with the estimated timing of pockmark development in the Barents Sea[31] and North Sea[54], suggesting that fluid release activity was controlled by a regional process. Although the onset of crust growth is less well constrained, the oldest U-Th age (17.5 ± 0.7 ka) coincides with the deglaciation of the southwest Barents Sea (~18–16 ka) (Fig. 4), which is constrained by ^{14}C dating of foraminifera associated with glaciomarine sediments[39]. The U-Th ages and the presence of isotopically heavy oxygen (>5‰ VPDB) support a model in which methane is discharged from gas hydrate dissociation, triggered by pressure changes on the continental shelf due to collapse and retreat of the SIS[3,31]. The U-Th data further indicate that substantial methane efflux continued along the ice-free northern Norwegian margin for another *ca.* 10 kyr. Although the carbonate crust U-Th data are limited in coverage, the clustering of U-Th dates (Fig. 4) does suggest that methane efflux may have been pulsed. U-Th ages of distinct layers within cavity infills provide evidence for sustained methane fluid flow through individual conduits over 900 years (Fig. 2b). Such focused fluid flow events may be analogous to the observed active

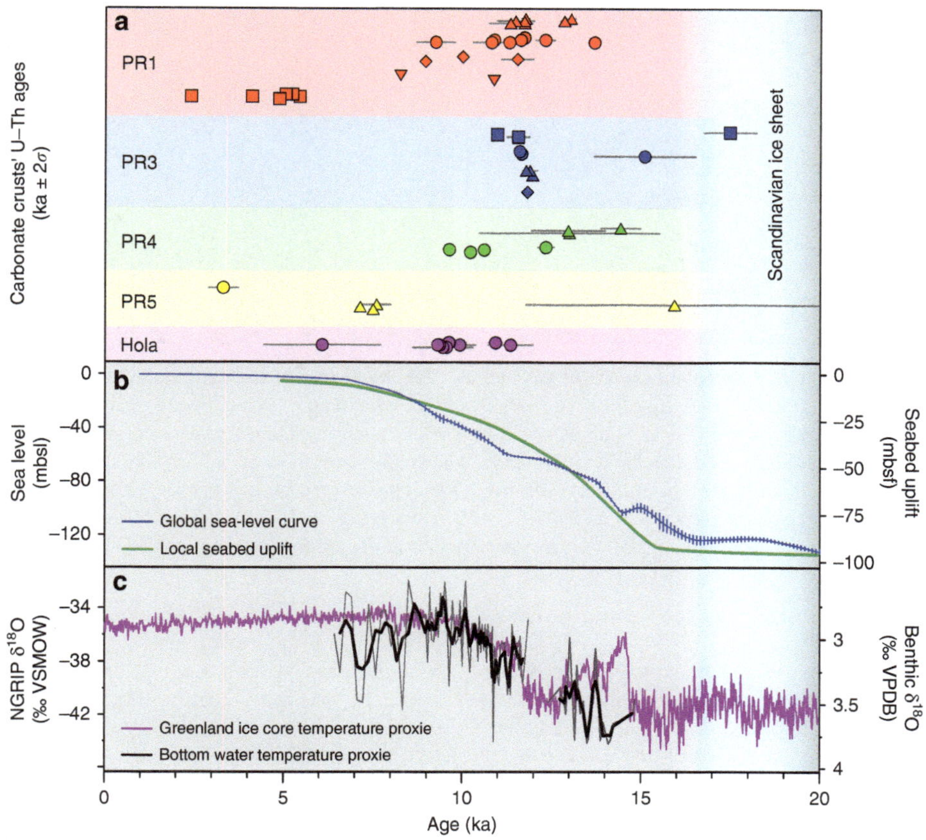

Figure 4 | U-Th carbonate crust dating results compared with paleoclimate and sea-level records since the LGM. (**a**) U-Th ages (error bars are 2σ) on 14 studied carbonate crusts (up to 11 ages per crust) grouped by site (see Fig. 1 for locations). (**b**) Global sea-level curve (error bars are 1σ (ref. 61)) and seabed uplift curve (extracted from ref. 57). (**c**) Stable oxygen isotope records of ice core, North Greenland Ice Core Project (NGRIP)[75] and of benthic foraminifera from a sediment core from the southwest Barents Sea[76]. The blue vertical shaded area depicts the period when grounded ice-sheet prevailed across the study area[39].

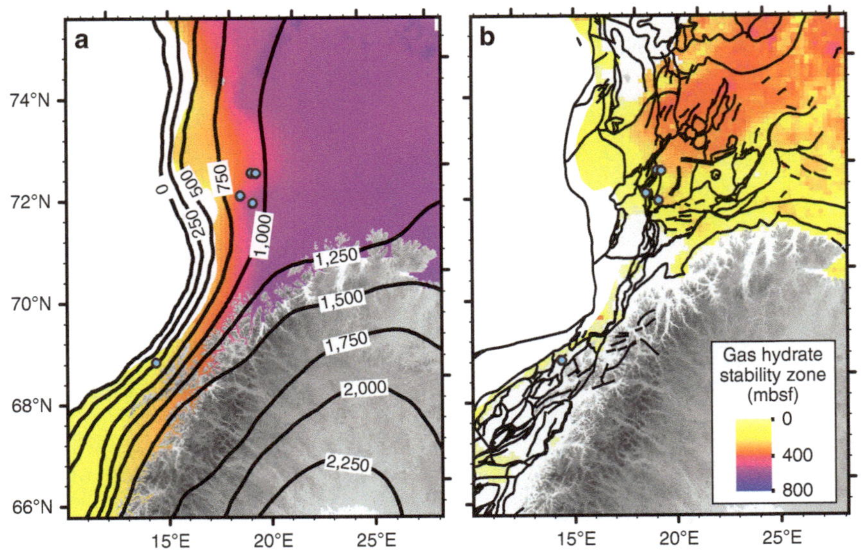

Figure 5 | Steady-state modelled thickness of the GHSZ. Thickness of GHSZ during the LGM (**a**) and at present-day conditions (**b**). Black contours denote (**a**) the ice-sheet thickness in metres during the LGM[51] and (**b**) broad scale faulting system[77]. The average value of the geothermal gradient used is 31 °C km^{-1} (ref. 52). It is noteworthy that the abyssal plain has not been taken into account in our model.

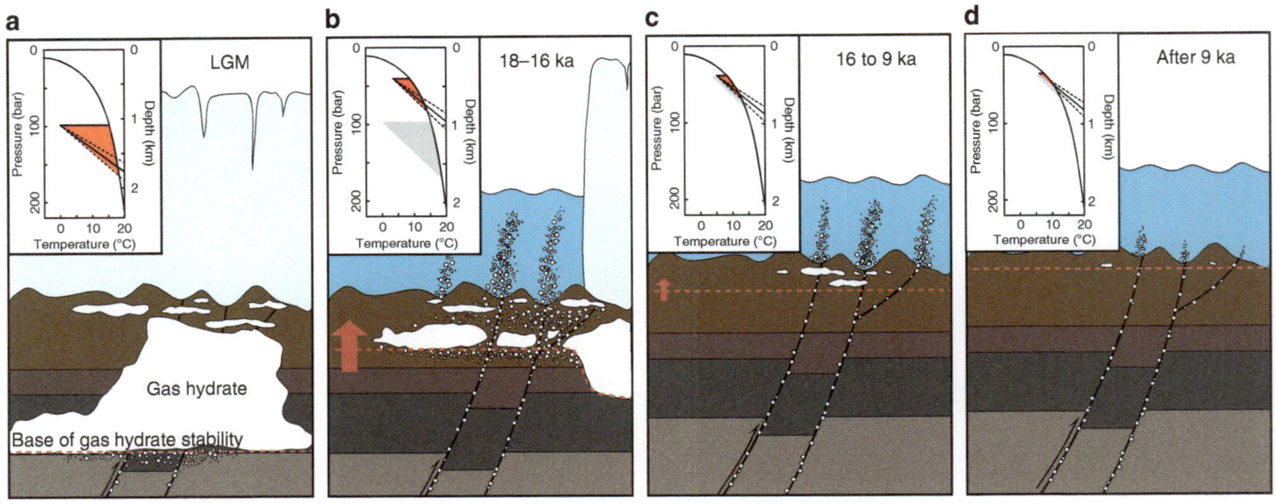

Figure 6 | Schematic sketch illustrating different snapshots of gas hydrate stability at steady state and fluid flow dynamics through time in the southwest Barents Sea shelf. (**a**) During the LGM, gas hydrate stability shown with the red area in the top-left corner was extending up to 600 m below the seabed. (**b**) Methane migrates through fractures and porous media as a result of gas hydrate dissociation triggered by grounded ice sheet retreat 18-16 ka. (**c**) Gas hydrate dissociation continues during the isostatic rebound and bottom water warming from ~16 to ~9 ka. (**d**) After ~9 ka to present, gas plumes occur locally connected to open deep-seated faults. The average geothermal gradient and associated 2σ uncertainties (31 ± 6 °C km^{-1} (ref. 52)) are shown by solid and dashed lines, respectively, at the base of gas hydrate stability fields (red areas). The red arrow depicts relative change of the base of the GHSZ (red dashed line). Temperature and pressure constraints used for assessing change in GHSZ are in Supplementary Table 3.

water column acoustic gas flares and as such could provide a means to extrapolate observed fluxes.

The integrated geochemistry and geochronology of methane-derived authigenic carbonates, coupled with gas hydrate modelling, provides a conceptual and temporal framework for methane seepage in the southwest Barents Sea. During the LGM, the SIS covered most of Northern Europe, including the Norwegian continental shelf and the entire Barents Sea[55,56]. Glacial loading by about 1,100 m of grounded ice[51] created a 150- to 200-m isostatic depression of the lithosphere[57] and the resulting increase of hydrostatic pressure in the underlying sediments extended the GHSZ to up to 600 m below the seafloor throughout the Barents Sea and the Norwegian shelf (Figs 5a and 6). Ice loading probably reactivated widespread basement-penetrating fault systems enhancing the migration of gas originating in Triassic and Jurassic source rocks and hydrocarbon reservoirs[58]. The accumulation of thermogenic gas in the upper part of the sediment column, corresponding to the GHSZ, would have enabled widespread gas hydrate formation beneath the ice cap during the LGM (Fig. 6a).

The retreat of the SIS from the shelf margin began at 19 ka and comprised several episodes of ice retreat and advance with grounding zone wedge formation[35,39]. The western Barents Sea, including our study area, was deglaciated over a 2-kyr period, with ice-free conditions reached at ~16 ka[39]. The pressure drop associated with ice sheet unloading led to thinning of the GHSZ by as much as 400 m, resulting in gas hydrate dissociation and thereby increased pore pressure, triggering methane advection

(Fig. 6b). Assuming a Darcy linear flow model, the dissipation of pore pressure from the seafloor to the base of the LGM GHSZ (600 m below seafloor) during glacial retreat is expected to have taken between 0.005 and 4.6 kyr, to reach the steady state (permeability of 0.04 mD to 1.51 D[59] and water viscosity of 2.04×10^{-3} Pa·s). In addition, the rate of gas hydrate dissociation, position within the sedimentary column and the fluid flow velocity would potentially affect the time over which the fluids reach the seafloor. As carbonate crusts are forming at fluid flow velocities between 20 and 60 cm per year[60], the methane migration at such velocities from the base of the LGM GHSZ (600 m) to the seafloor would take between 1 and 3 kyr.

Following the SIS collapse, isostatic rebound caused a *ca.* 90 m uplift of the seafloor between 16 and 6 ka (Fig. 4), and at the same time sea-level rose *ca.* 120 m (ref. 61), resulting in a minor net increase in hydrostatic pressure. Changes in stress over this time interval associated with glacial unloading and isostatic rebound resulted in reactivation of pre-existing faults in certain orientations relative to the evolving stress field[62] and would have facilitated fluid migration from shallow gas reservoirs generated by gas hydrate dissociation and/or from deeper petroleum systems. Gas hydrate dissociation was further promoted by the inflow of warm Atlantic water into the Barents Sea shelf after deglaciation, as bottom water temperatures rose from 4 °C at ~15 ka to 6 °C at ~10 ka with 1–2 °C fluctuations on 100-year timescales (Fig. 4), remaining relatively stable during the Holocene. The heat transfer would take several hundred years to propagate through the sediments and reach the base of the GHSZ[24]. Consequently, post-glacial gas hydrate destabilization possibly continued until ~9–7 ka under the combined influence of seabed uplift and temperature increase (Fig. 6c). At present, gas seepage as evidenced by flares in the water column in the Barents Sea is mainly observed above deep-seated faults (Fig. 6d) with methane sourced from petroleum reservoirs.

We estimate that $10 \times 10^3 - 88 \times 10^3$ Tg of methane was stored as gas hydrate in the southwest Barents Sea shelf during the LGM, which is equivalent to 0.6–4.9% of the current total oceanic gas hydrate reservoir[63]. As our study area only represents *ca.* 30% of the glaciated Barents Sea shelf and *ca.* 5% of all glaciated shelves with grounded ice sheets (that is, Norwegian margin, Antarctica and Greenland), it is likely that significant amounts of methane were stored in such shelf settings during the LGM. However, the impact of oceanic gas hydrate dissociation on the climate system is limited by the atmospheric transfer rate of methane, which is a function of timescales of dissociation and release rates at the seafloor, the efficiency of methane oxidizers and bubble dissolution in the water column, and the thickness of the water column. Modelling estimates show that under normal oceanographic conditions, methane released at water depths >200 m would be almost entirely (>80%) consumed during transit towards the sea surface[20], and that a catastrophic bubble release is required, to drive transport of methane from deeper waters (>100 m) to the atmosphere[20]. As water depth in the investigated areas of the Norwegian and Barents Sea was on the order of 200–400 m after ice sheet collapse, the resulting impact on atmospheric methane concentrations was probably muted. Furthermore, methane release at the sediment–water interface was also probably modulated by the availability of excess methane and fluid pathways to the seafloor, with slower transit through the sediment column increasing the magnitude of microbial methane consumption within the sedimentary sulphate methane transition zone. Assuming that 90% of the methane released by gas hydrate dissociation is consumed by AOM[11], and that 1–5% of it would pass through the water column and reach the atmosphere, the mean integrated flux to the atmosphere in the southwest Barents Sea over 10 kyr would be about 0.0005–0.0172 Tg · per year (minimum and maximum estimates) of methane, which is relatively small when compared with the ~200 Tg · per year methane flux from all natural sources[64]. Such a scenario of protracted, relatively low-intensity methane efflux in response to abrupt environmental changes has rarely been evaluated by studies that postulate linkages between gas hydrate dissociation and transfer of methane to the marine realm/atmosphere. However, on shallow continental shelf regions (that is, <150–200 m water depth) that have also experienced glacial–interglacial cycles, it is plausible that a higher proportion of the methane released by gas hydrate dissociation may have passed through the water column and reached the atmosphere.

The main episode of carbonate crust formation in the Barents Sea after the collapse of the SIS broadly overlaps with elevated atmospheric methane concentrations as recorded by ice cores from Greenland and Antarctica[65]. The potential importance of gas hydrate destabilization due to warming of upper oceanic water masses as a cause for elevated atmospheric methane has been debated in several publications[4,10]. The hydrogen isotope and ^{14}C characteristics of methane trapped in ice cores suggest insignificant emissions from marine gas hydrates during times of high atmospheric methane after deglaciation and stability of gas hydrates[14,66]. However, the general inference of gas hydrate stability within continental margin sediments through glacial–interglacial cycles is not globally applicable, as it is dependent on local changes in temperature and hydrostatic pressure. This is particularly important in case of glaciated continental margins where gas hydrate destabilization was triggered by local reduction in pressure effects of collapsing grounded ice sheets. Reliable assessment of the influence of hydrate-released methane from glaciated margins on the climate system after the LGM requires global quantification of methane storage, release and consumption budgets, constraints of timescales of hydrate dissociation, and temporal and volume estimates of the dynamics of ice sheets. Such global data are not currently available. Abrupt, globally synchronous methane release over a timescale of 10^2 years from deglaciated shelf areas with grounded ice appears unlikely given the protracted nature of hydrate-derived methane efflux after the LGM, as our findings from the southwest Barents Sea indicate, as well as the asynchronous deglaciation of different shelf areas.

The analysis of methane-derived authigenic carbonate through the integration of U-Th geochronology and geochemical proxies, combined with gas hydrate modelling provides a means for evaluating past methane release from glaciated continental margins where gas hydrate dynamics are governed by glacial–interglacial cycles. Modelling results demonstrate that gas hydrate accumulation beneath grounded ice-sheets on the Norwegian margin, in a setting analogous to the present day Antarctic shelf[16], generates potentially significant methane reservoirs the stability of which is sensitive to environmental changes affecting local pressure and temperature regimes. Although significant amounts of methane could have been released at the seafloor and transferred to the water column, U-Th geochronology suggests methane release over a *ca.* 10 kyr interval implying that flux rates were modulated by the second-order processes controlling both the rate of dissociation (isostatic rebound and bottom water warming), fluid transport (changes in stress fields leading to fault reactivation under the control of isostatic rebound) and consumption (via efficiency of microbial oxidation). Overall, the protracted nature of methane release on the Norwegian margin and its minimal impact on atmospheric methane concentrations highlight the complexity of the gas hydrate system and the importance of mechanisms mediating gas hydrate dissociation and fluid advection in response to abrupt climatic change.

Methods

Mineralogy. Petrographical characterization of carbonates was achieved by a combination of optical and scanning electron microscopy techniques. Mineralogy was determined at the Geological Survey of Norway by X-ray diffraction (Philips X'pert apparatus, Cu K_α radiation in 2–70° 2θ range) on bulk carbonate powder. Identification of the main carbonate phases was done using EVA software (Supplementary Table 1). The d_{104} displacement was used to estimate the $MgCO_3$ mol% in calcite[67].

Stable isotope compositions. Stable carbon and oxygen isotopes of hand-drilled carbonate were analysed with a GasBench II preparation line connected to a Thermo Scientific Delta V Advantage IRMS (Thermo Fisher Scientific). Reproducibility is better than ± 0.2‰ for both $\delta^{13}C$ and $\delta^{18}O$. The stable carbon and hydrogen isotopes of methane (Supplementary Fig. 2) were analysed by a Delta plus XP IRMS (Thermo Fisher Scientific) with analytical reproducibility better than ± 1‰ for $\delta^{13}C$ values and better than ± 10‰ for δD values, respectively. Stable isotope compositions are reported in conventional delta (δ) units relative to the VPDB reference for carbon and oxygen, and relative to the VSMOW reference for hydrogen isotopes.

U-Th dating. U-Th dating was carried out on 76 microdrilled carbonate samples weighing between 2–80 mg (Supplementary Data 1 and Supplementary Fig. 3). A further seven unconsolidated sediment samples (three to six replicates from each, weighing 30–130 mg) were also analysed, to constrain the impact of incorporated detrital material on the U-Th systematics of the carbonate crusts. Analytical protocols for the separation and purification of U and Th from carbonate crust and detritus samples are similar to those of Douarin *et al.*[68], with modifications aimed at ensuring the following: (i) complete dissolution of the detrital material incorporated in the crusts and (ii) oxidation of organic material liable to produce isobaric interferences during measurements of Th isotope ratios[69]. All evaporation steps took place in a closed EvapoClean device, to minimize cross-contamination and reduce fall-in blanks. Samples were leached in ~8 M HNO_3 for 1 h and centrifuged to separate soluble carbonate-rich fractions from less-soluble detrital material. Detritus fractions were dissolved over 1–7 days in a mixture of $HClO_4$:HF:HNO_3 (1:2:2.5), using ~50 µl HF per estimated mg of material, dried down, re-dissolved in 8 M HCl and dried down again, to ensure the conversion of residual fluorides to chlorides. Detritus fractions were then dissolved in 8 M HNO_3, recombined with their respective carbonate fractions, spiked with a mixed ^{229}Th-^{236}U tracer, left to equilibrate overnight and dried down. All samples went through two overnight oxidation steps in a mixture of 2 ml 16 M HNO_3 and 0.2 ml 30% H_2O_2 followed by evaporation to dryness. Pre-concentration of U and Th through Fe co-precipitation and initial separation on 0.6 ml columns using AG-1 × 8 anion exchange resin were done following the procedure of Edwards *et al.*[70]. Th fractions were further purified using a second pass through AG-1 × 8 resin and were filtered using 0.22-µm pore-size syringe filters to remove resin particles. Both U and Th fractions were oxidized twice in 2 ml 16 M HNO_3 and 0.2 ml 30% H_2O_2, and dissolved in 1 ml 0.1 M HCl and 0.035 M HF. Before mass spectrometry analyses, all samples were filtered to remove particles originating from the perfluoroalkoxy alkane (FEP) beakers used for sample preparation.

Isotope ratio measurements were made on a Thermo Neptune Plus multi-collector ICP-MS, with samples introduced via an Aridus II desolvating nebulizer using an ESI PFA nebulizer tip with an ~50 µl min^{-1} uptake rate. Typical operating conditions included the addition of 4–8 ml min^{-1} high-purity N_2 to the sample-carrying Ar stream, to minimize U and Th oxide formation in the plasma.

U measurements were made using a normal sample cone and X-skimmer cone, and a static multicollector data collection protocol with ^{234}U measured on an axial secondary electron multiplier, and ^{233}U, ^{235}U, ^{236}U and ^{238}U measured on Faraday cups equipped with $10^{11}\,\Omega$ resistors. Blocks of five samples were bracketed by analyses of CRM 112a U and CRM 112a U + IRMM 3636 spike. Exponential mass bias corrections were based on the measured values of the $^{233}U/^{236}U$ ratio of the IRMM 3636 spike normalized to a value of 1.01906. Secondary electron multiplier/ Faraday gain corrections were based on the $^{234}U/^{235}U$ ratio of bracketing unspiked CRM 112a analyses. Hydride formation and tailing were monitored at the beginning of each analytical session, with measurements made at mass 237 and 239, while aspirating an unspiked CRM 112a solution, and were corrected during offline data reduction.

Owing to the young age of the samples, Th measurements required better sensitivity and were predominantly made using a Jet sample cone and X-skimmer cone. ^{230}Th was measured on the axial secondary electron multiplier, with ^{229}Th and ^{232}Th measured on Faraday cups equipped with $10^{11}\,\Omega$ resistors. Blocks of ten samples were bracketed by blocks of five analyses of an in-house ^{229}Th-^{230}Th-^{232}Th reference solution calibrated against CRM 112a, which was used to monitor mass bias and secondary electron multiplier/Faraday gain.

U-Th age calculations were performed in Isoplot v. 3.75 (ref. 71). The impact of measured detritus $^{232}Th/^{238}U$ activity ratio variability on the geological interpretation of U-Th dates within the scope of this study was found to be negligible for samples with ($^{230}U/^{232}Th$) activity ratios above ~2 (see Supplementary Note). Corrected U-Th dates from samples with ($^{230}Th/^{232}Th$) < 2 (n = 16) were either negative or statistically equivalent to zero (at the 2σ level) and were consequently excluded from the data set (Supplementary Data 1 and 2). The remaining analyses were corrected using a calculated average composition of carbonate-free detritus samples (Option 3, Supplementary Table 2).

Gas hydrate modelling. The gas hydrate stability modelling was carried out assuming steady state using CSMHYD software[72]. Consequently, this approach illustrates only changes in gas hydrate stability independent of the transient behaviour of the system. We used the bathymetry grids of the southwest Barents Sea shelf (https://www.ngdc.noaa.gov/mgg/global/global.html) and World Ocean Database 2005 bottom water temperature (http://www.nodc.noaa.gov/OC5/ indprod.html) from NOAA and a feeding thermogenic gas composition of 96% methane, 3% ethane and 1% propane[37]. As the underlying overturned sediments have varying consolidation with a range of different thermal conductivity and hence heat flow values[59], the geothermal gradient in Barents Sea can vary from 22.8 to 69.3 °C km^{-1}, but is mostly confined around 31 ± 6 °C km^{-1} (2σ)[52]. For the Last Glacial Maximum (LGM), ice thickness models predicted for the Barents Sea varies in coverage and thicknesses, ranging from 750 to 2,000 m in the study area[51,56,73]. For predicting the gas hydrate stability during the LGM model, we used the data from ref. 56, which predicts an ice-sheet thickness of ~1,100 m for the southwest Barents Sea falling within the most accepted range and we applied a bottom ice temperature of − 1.5 °C.

The model of gas hydrate stability was used to estimate the volume of the GHSZ ($12.7 \times 10^4\,km^3$) in the southwest Barents Sea shelf during the LGM. Considering a sediment porosity[58] of 0.1–0.3 conservatively filled by 1–2% of gas hydrate in which methane occupies 70–100% of the cages[74], the weight of methane stored in gas hydrate would range between 10 and 88 Gt, calculated with a volumetric ratio between methane and gas hydrate of 164 and a methane density of 0.70805 kg m^{-3} at standard conditions for temperature and pressure (0 °C and 1 bar).

References

1. Milkov, A. V. Global estimates of hydrate-bound gas in marine sediments: how much is really out there? *Earth Sci. Rev.* **66,** 183–197 (2004).
2. Kvenvolden, K. A. Gas hydrates—geological perspective and global change. *Rev. Geophys.* **31,** 173–187 (1993).
3. Chand, S. *et al.* Multiple episodes of fluid flow in the SW Barents Sea (Loppa High) evidenced by gas flares, pockmarks and gas hydrate accumulation. *Earth Planet. Sci. Lett.* **331-332,** 305–314 (2012).
4. Kennett, J. P. *Methane Hydrates in Quaternary Climate Change: The Clathrate Gun Hypothesis* (American Geophysical Union, 2003).
5. Portnov, A., Vadakkepuliyambatta, S., Mienert, J. & Hubbard, A. Ice-sheet-driven methane storage and release in the Arctic. *Nat. Commun.* **7,** 10314 (2016).
6. Dickens, G. R., O'Neil, J. R., Rea, D. K. & Owen, R. M. Dissociation of oceanic methane hydrate as a cause of the carbon isotope excursion at the end of the Paleocene. *Paleoceanography* **10,** 965–971 (1995).
7. Hesselbo, S. P. *et al.* Massive dissociation of gas hydrate during a Jurassic oceanic anoxic event. *Nature* **406,** 392–395 (2000).
8. Marín-Moreno, H., Minshull, T. A., Westbrook, G. K., Sinha, B. & Sarkar, S. The response of methane hydrate beneath the seabed offshore Svalbard to ocean warming during the next three centuries. *Geophys. Res. Lett.* **40,** 5159–5163 (2013).
9. Reagan, M. T., Moridis, G. J., Elliott, S. M. & Maltrud, M. Contribution of oceanic gas hydrate dissociation to the formation of Arctic Ocean methane plumes. *J. Geophys. Res. Oceans* **116,** C09014 (2011).
10. Nisbet, E. G. Have sudden large releases of methane from geological reservoirs occurred since the Last Glacial Maximum, and could such releases occur again? *Philos. Trans. Math. Phys. Eng. Sci.* **360,** 581–607 (2002).
11. Reeburgh, W. S. Oceanic methane biogeochemistry. *Chem. Rev.* **107,** 486–513 (2007).
12. Valentine, D. L., Blanton, D. C., Reeburgh, W. S. & Kastner, M. Water column methane oxidation adjacent to an area of active hydrate dissociation, Eel river Basin. *Geochim. Cosmochim. Acta* **65,** 2633–2640 (2001).
13. Rise, L., Bellec, V. K., Chand, S. & Bøe, R. Pockmarks in the southwestern Barents Sea and Finnmark fjords. *Nor. J. Geol.* **94,** 263–282 (2014).
14. Petrenko, V. V. *et al.* $^{14}CH_4$ measurements in Greenland Ice: investigating Last Glacial termination CH_4 sources. *Science* **324,** 506–508 (2009).
15. Bock, M. *et al.* Hydrogen isotopes preclude marine hydrate CH_4 emissions at the onset of Dansgaard-Oeschger events. *Science* **328,** 1686–1689 (2010).
16. Wadham, J. L. *et al.* Potential methane reservoirs beneath Antarctica. *Nature* **488,** 633–637 (2012).
17. Hinrichs, K. & Boetius, A. *The Anaerobic Oxidation of Methane: New Insights in Microbial Ecology and Biogeochemistry* (2002).
18. Wallmann, K., Drews, M., Aloisi, G. & Bohrmann, G. Methane discharge into the Black Sea and the global ocean via fluid flow through submarine mud volcanoes. *Earth Planet. Sci. Lett.* **248,** 545–560 (2006).
19. Boetius, A. & Wenzhöfer, F. Seafloor oxygen consumption fuelled by methane from cold seeps. *Nat. Geosci.* **6,** 725–734 (2013).
20. McGinnis, D. F., Greinert, J., Artemov, Y., Beaubien, S. E. & Wüest, A. Fate of rising methane bubbles in stratified waters: How much methane reaches the atmosphere? *J. Geophys. Res. Oceans* **111,** C09007 (2006).

21. Berndt, C. *et al.* Temporal constraints on hydrate-controlled methane seepage off Svalbard. *Science* **343,** 284–287 (2014).
22. Phrampus, B. J. & Hornbach, M. J. Recent changes to the Gulf Stream causing widespread gas hydrate destabilization. *Nature* **490,** 527–530 (2012).
23. Skarke, A., Ruppel, C., Kodis, M., Brothers, D. & Lobecker, E. Widespread methane leakage from the sea floor on the northern US Atlantic margin. *Nat. Geosci.* **7,** 657–661 (2014).
24. Darnell, K. N. & Flemings, P. B. Transient seafloor venting on continental slopes from warming-induced methane hydrate dissociation. *Geophys. Res. Lett.* **42,** 10765–10772 (2015).
25. Bayon, G., Henderson, G. M. & Bohn, M. U–Th stratigraphy of a cold seep carbonate crust. *Chem. Geol.* **260,** 47–56 (2009).
26. Teichert, B. M. A. *et al.* U/Th systematics and ages of authigenic carbonates from Hydrate Ridge, Cascadia Margin: recorders of fluid flow variations. *Geochim. Cosmochim. Acta* **67,** 3845–3857 (2003).
27. Crémière, A., Bayon, G., Ponzevera, E. & Pierre, C. Paleo-environmental controls on cold seep carbonate authigenesis in the Sea of Marmara. *Earth Planet. Sci. Lett.* **376,** 200–211 (2013).
28. Liebetrau, V., Eisenhauer, A. & Linke, P. Cold seep carbonates and associated cold-water corals at the Hikurangi Margin, New Zealand: new insights into fluid pathways, growth structures and geochronology. *Mar. Geol.* **272,** 307–318 (2010).
29. Feng, D. *et al.* U/Th dating of cold-seep carbonates: an initial comparison. *Deep Sea Res. Part II Top. Stud. Oceanogr.* **57,** 2055–2060 (2010).
30. Watanabe, Y., Nakai, S., Hiruta, A., Matsumoto, R. & Yoshida, K. U–Th dating of carbonate nodules from methane seeps off Joetsu, Eastern Margin of Japan Sea. *Earth Planet. Sci. Lett.* **272,** 89–96 (2008).
31. Pau, M., Hammer, Ø. & Chand, S. Constraints on the dynamics of pockmarks in the SW Barents Sea: evidence from gravity coring and high-resolution, shallow seismic profiles. *Mar. Geol.* **355,** 330–345 (2014).
32. Mienert, J. & Posewang, J. Evidence of shallow- and deep-water gas hydrate destabilizations in North Atlantic polar continental margin sediments. *Geo-Mar. Lett.* **19,** 143–149 (1999).
33. Vogt, P. R., Gardner, J. & Crane, K. The Norwegian–Barents–Svalbard (NBS) continental margin: introducing a natural laboratory of mass wasting, hydrates, and ascent of sediment, pore water, and methane. *Geo-Mar. Lett.* **19,** 2–21 (1999).
34. Laberg, J. S. & Vorren, T. O. Late Weichselian submarine debris flow deposits on the Bear Island Trough Mouth Fan. *Mar. Geol.* **127,** 45–72 (1995).
35. Ottesen, D., Dowdeswell, J. A. & Rise, L. Submarine landforms and the reconstruction of fast-flowing ice streams within a large Quaternary ice sheet: the 2500-km-long Norwegian-Svalbard margin (57°–80°N). *Geol. Soc. Am. Bull.* **117,** 1033–1050 (2005).
36. Sættem, J., Poole, D. A. R., Ellingsen, L. & Sejrup, H. P. Glacial geology of outer Bjørnøyrenna, southwestern Barents Sea. *Mar. Geol.* **103,** 15–51 (1992).
37. Chand, S. *et al.* Gas hydrate stability zone modelling in areas of salt tectonics and pockmarks of the Barents Sea suggests an active hydrocarbon venting system. *Mar. Pet. Geol.* **25,** 625–636 (2008).
38. Rüther, D. C., Mattingsdal, R., Andreassen, K., Forwick, M. & Husum, K. Seismic architecture and sedimentology of a major grounding zone system deposited by the Bjørnøyrenna Ice Stream during Late Weichselian deglaciation. *Quat. Sci. Rev.* **30,** 2776–2792 (2011).
39. Winsborrow, M. C. M., Andreassen, K., Corner, G. D. & Laberg, J. S. Deglaciation of a marine-based ice sheet: Late Weichselian palaeo-ice dynamics and retreat in the southern Barents Sea reconstructed from onshore and offshore glacial geomorphology. *Quat. Sci. Rev.* **29,** 424–442 (2010).
40. Solheim, A. & Elverhøi, A. Gas-related sea floor craters in the Barents Sea. *Geo-Mar. Lett.* **13,** 235–243 (1993).
41. Jung, W.-Y. & Vogt, P. R. Effects of bottom water warming and sea level rise on Holocene hydrate dissociation and mass wasting along the Norwegian-Barents Continental Margin. *J. Geophys. Res. Solid Earth* **109,** B06104 (2004).
42. Chand, S. *et al.* Active venting system offshore Northern Norway. *Eos Trans. Am. Geophys. Union* **89,** 261–262 (2008).
43. Peckmann, J. *et al.* Methane-derived carbonates and authigenic pyrite from the northwestern Black Sea. *Mar. Geol.* **177,** 129–150 (2001).
44. Aloisi, G. *et al.* CH_4-consuming microorganisms and the formation of carbonate crusts at cold seeps. *Earth Planet. Sci. Lett.* **203,** 195–203 (2002).
45. Schoell, M. The hydrogen and carbon isotopic composition of methane from natural gases of various origins. *Geochim. Cosmochim. Acta* **44,** 649–661 (1980).
46. Fairbanks, R. G. A 17,000-year glacio-eustatic sea level record: influence of glacial melting rates on the Younger Dryas event and deep-ocean circulation. *Nature* **342,** 637–642 (1989).
47. Kim, S.-T., O'Neil, J. R., Hillaire-Marcel, C. & Mucci, A. Oxygen isotope fractionation between synthetic aragonite and water: Influence of temperature and Mg2 + concentration. *Geochim. Cosmochim. Acta* **71,** 4704–4715 (2007).
48. Tarutani, T., Clayton, R. & Mayeda, T. Effect of polymorphism and magnesium substitution on oxygen isotope fractionation between calcium carbonate and water. *Geochim. Cosmochim. Acta* **33,** 987–98 (1969).
49. Friedman, I. & O'Neil, J. R. *Compilation of Stable Isotope Fractionation Factors of Geochemical Interest* (U.S. G.P.O., 1977).
50. Bohrmann, G., Greinert, J., Suess, E. & Torres, M. Authigenic carbonates from the Cascadia subduction zone and their relation to gas hydrate stability. *Geology* **26,** 647–650 (1998).
51. Siegert, M. J., Dowdeswell, J. A., Hald, M. & Svendsen, J.-I. Modelling the Eurasian Ice Sheet through a full (Weichselian) glacial cycle. *Glob. Planet. Change* **31,** 367–385 (2001).
52. Vadakkepuliyambatta, S., Hornbach, M. J., Bünz, S. & Phrampus, B. J. Controls on gas hydrate system evolution in a region of active fluid flow in the SW Barents Sea. *Mar. Pet. Geol.* **66**(Part 4): 861–872 (2015).
53. Laberg, J. S. & Andreassen, K. Gas hydrate and free gas indications within the Cenozoic succession of the Bjørnøya Basin, western Barents Sea. *Mar. Pet. Geol.* **13,** 921–940 (1996).
54. Forsberg, C. F. *et al. Formation Of Pockmarks In The Norwegian Channel* (Society of Underwater Technology, 2007).
55. Landvik, J. Y. *et al.* The last glacial maximum of Svalbard and the Barents Sea area: ice sheet extent and configuration. *Quat. Sci. Rev.* **17,** 43–75 (1998).
56. Svendsen, J. I. *et al.* Late Quaternary ice sheet history of northern Eurasia. *Quat. Sci. Rev.* **23,** 1229–1271 (2004).
57. Fjeldskaar, W. The amplitude and decay of the glacial forebulge in Fennoscandia. *Nor. Geol. Tidsskr.* **74,** 2–8 (1994).
58. Henriksen, E. *et al.* Chapter 17 Uplift and erosion of the greater Barents Sea: impact on prospectivity and petroleum systems. *Geol. Soc. Lond. Mem.* **35,** 271–281 (2011).
59. Bugge, T. *et al.* Shallow stratigraphic drilling applied in hydrocarbon exploration of the Nordkapp Basin, Barents Sea. *Mar. Pet. Geol.* **19,** 13–37 (2002).
60. Luff, R., Wallmann, K. & Aloisi, G. Numerical modeling of carbonate crust formation at cold vent sites: significance for fluid and methane budgets and chemosynthetic biological communities. *Earth Planet. Sci. Lett.* **221,** 337–353 (2004).
61. Lambeck, K., Rouby, H., Purcell, A., Sun, Y. & Sambridge, M. Sea level and global ice volumes from the Last Glacial Maximum to the Holocene. *Proc. Natl. Acad. Sci.* **111,** 15296–15303 (2014).
62. Brandes, C., Steffen, H., Steffen, R. & Wu, P. Intraplate seismicity in northern Central Europe is induced by the last glaciation. *Geology* **G36710,** 1 (2015).
63. Boswell, R. & Collett, T. S. Current perspectives on gas hydrate resources. *Energy Environ. Sci.* **4,** 1206–1215 (2011).
64. Ciais, P. *et al.*in *Climate Change 2013: The Physical Science Basis. Contribution of Working Group I to the Fifth Assessment Report of the Intergovernmental Panel on Climate Change* 465–570 (Cambridge University Press, 2013).
65. Bender, M. *et al.* Climate correlations between Greenland and Antarctica during the past 100,000 years. *Nature* **372,** 663–666 (1994).
66. Sowers, T. Late quaternary atmospheric CH_4 isotope record suggests marine clathrates are stable. *Science* **311,** 838–840 (2006).
67. Goldsmith, J. & Graf, D. Relation between lattice constants and composition of the Ca-Mg carbonates. *Am. Mineral.* **43,** 84–101 (1958).
68. Douarin, M. *et al.* Growth of north-east Atlantic cold-water coral reefs and mounds during the Holocene: a high resolution U-series and ^{14}C chronology. *Earth Planet. Sci. Lett.* **375,** 176–187 (2013).
69. Shen, C.-C. *et al.* Uranium and thorium isotopic and concentration measurements by magnetic sector inductively coupled plasma mass spectrometry. *Chem. Geol.* **185,** 165–178 (2002).
70. Edwards, R. L., Chen, J. H. & Wasserburg, G. J. U-238 U-234-TH-230-TH-232 systematics and the precise measurement of time over the past 50 000 years. *Earth Planet. Sci. Lett.* **81,** 175–192 (1987).
71. Ludwig, K. R. *User's Manual for Isoplot 3.75: A Geochronological Toolkit for Microsoft Excel* **5** (Kenneth R. Ludwig, 2003).
72. Sloan, E. D. & Koh, C. A. *Clathrate Hydrates of Natural Gases* (CRC Press, 2008).
73. Siegert, M. J. & Dowdeswell, J. A. Numerical reconstructions of the Eurasian Ice Sheet and climate during the Late Weichselian. *Quat. Sci. Rev.* **23,** 1273–1283 (2004).
74. Lorenson, T. D. & Collett, T. S. in *Proceedings of the Ocean Drilling Program, Scientific Results* **164,** 37–46 (2000).
75. Andersen, K. K. *et al.* High-resolution record of Northern Hemisphere climate extending into the last interglacial period. *Nature* **431,** 147–151 (2004).
76. Aagaard-Sørensen, S., Husum, K., Hald, M. & Knies, J. Paleoceanographic development in the SW Barents Sea during the Late Weichselian–Early Holocene transition. *Quat. Sci. Rev.* **29,** 3442–3456 (2010).
77. Gabrielsen, R. H., Færseth, R. B., Jensen, L. N., Kalheim, J. E. & Riis, F. *Structural Elements of the Norwegian Continental Shelf, Part I: The Barents Sea Region* 47 NPD Bulletin, 1990).

Acknowledgements

This study benefited from reviews from H. Marín Moreno and two anonymous reviewers. We are grateful to G. Bayon, C. Pierre, A. Prave and B. Schoene for helpful discussions. We acknowledge J. Schönenberger for her help with XRD analyses. We thank Lundin Norway AS and the Geological Survey of Norway for permission to publish the data. This research is part of the Centre of Excellence: Arctic Gas hydrate, Environment and Climate (CAGE) funded by the Norwegian Research Council (grant No. 223259). T.M. acknowledges the Estonian Research Council (Grant PUT611).

Author contributions

A.C., A.L., D.J.C., S.C. and D.S. designed the study and wrote the manuscript. D.S. produced U-Th data with input from A.C., A.L., D.J.C. and S.R.N. T.M. performed stable isotope analyses. S.C. and A.C. carried out the gas hydrate stability modelling. A.L., T.T., S.C. and H.B. carried out site surveys and sample collection. S.S. helped with sample preparation. All authors contributed to the final editing of the manuscript.

Additional information

Competing financial interests: The authors declare no competing financial interests.

Initialized decadal prediction for transition to positive phase of the Interdecadal Pacific Oscillation

Gerald A. Meehl[1], Aixue Hu[1] & Haiyan Teng[1]

The negative phase of the Interdecadal Pacific Oscillation (IPO), a dominant mode of multi-decadal variability of sea surface temperatures (SSTs) in the Pacific, contributed to the reduced rate of global surface temperature warming in the early 2000s. A proposed mechanism for IPO multidecadal variability indicates that the presence of decadal timescale upper ocean heat content in the off-equatorial western tropical Pacific can provide conditions for an interannual El Niño/Southern Oscillation event to trigger a transition of tropical Pacific SSTs to the opposite IPO phase. Here we show that a decadal prediction initialized in 2013 simulates predicted Niño3.4 SSTs that have qualitatively tracked the observations through 2015. The year three to seven average prediction (2015-2019) from the 2013 initial state shows a transition to the positive phase of the IPO from the previous negative phase and a resumption of larger rates of global warming over the 2013-2022 period consistent with a positive IPO phase.

[1]National Center for Atmospheric Research, 3090 Center Green Drive, Boulder, Colorado 80301, USA. Correspondence and requests for materials should be addressed to G.A.M. (email: meehl@ucar.edu).

Retrospective hindcasts have shown skill in simulating the two phase changes of the Interdecadal Pacific Oscillation (IPO) since 1960, the first from negative to positive in the 1970s and the second from positive to negative in the late 1990s[1–5]. A proposed mechanism for IPO multidecadal variability indicates that for such decadal transitions to occur, there should be a build-up of upper ocean heat content in the off-equatorial western tropical Pacific such that an El Niño/Southern Oscillation (ENSO) event can then trigger a transition of tropical Pacific SSTs to the opposite IPO phase[5–8].

This mechanism is a variant of the delayed-action oscillator[9] that depends on off-equatorial wind forcing[10,11], which could be produced by tropical-midlatitude interaction. Convective heating anomalies in the tropical Pacific, associated with precipitation and SST anomalies there, produce anomalous Rossby waves in the atmosphere and sea level pressure signals in the mid-latitude North and South Pacific. These sea level pressure anomalies are associated with surface wind stress curl anomalies that force anomalous ocean Rossby waves near 25°N and 25°S. These slow-moving ocean Rossby waves subsequently produce, on decadal timescales, off-equatorial ocean heat content anomalies in the western tropical Pacific. The build-up of off-equatorial heat content that is necessary for an IPO transition contributes to a 15- to 20-year timescale for the IPO[6]. The interannual variability associated with ENSO, combined with the off-equatorial ocean heat content anomalies at the western boundary, then produce equatorial ocean Kelvin waves that change the equatorial thermocline depth in such a way as to reverse the sign of equatorial Pacific SST anomalies to the opposite phase of the IPO. The mechanism, as originally formulated[6], has been independently confirmed[7] and generalized in the context of the subtropical gyre and the Pacific subtropical cells (STCs)[12]. This is a variant of a previous mechanism for Pacific decadal variability that postulated IPO-like variability could be associated with amplitude modulation of ENSO events, thus also connecting interannual and decadal timescale variability in the Pacific[13].

Transitions in the sign of the IPO have been shown to be simulated in single and multi-model initialized decadal hindcasts and verified both through epoch differences and empirical orthogonal function metrics[1–5]. For the negative IPO from roughly 2000–2013, there was a slowdown or reduced rate of global surface warming (also sometimes referred to as the early-2000s hiatus[14–17]). The previous positive phase of the IPO from the 1970s to late 1990s was associated with an accelerated rate of global warming and has been documented in climate models[17–19].

Here we connect these transitions to the mechanism described above[6,7] and use those results to provide the context for a decadal prediction initialized in 2013, which shows a transition of the IPO from its recent negative state back to positive. Although there have been earlier attempts to predict a recent IPO transition from negative to positive[20], here we provide a process-based mechanism on which to base the credibility of the prediction of the transition to the positive phase of the IPO. We choose the CCSM4, because it is the only current climate model with an extensive documentation with regards to processes and initialized predictions involved with IPO transitions[1,2,17], as well as initialized predictions of decadal climate variability in the North Atlantic[21] (see Methods). There are ten ensemble members run for each initial year.

Results

Off-equatorial heat content. As noted above, there are two crucial elements of the IPO mechanism outlined above[6]: a build-up of off-equatorial ocean heat content anomalies in the western tropical Pacific over at least a 15-year period before a transition and an interannual ENSO event that triggers a transition (El Niño for transition from negative to positive and La Niña for transition from positive to negative). For the IPO transitions during the period of the CMIP5 decadal hindcasts from 1960 to the early 2000s[22], the ocean states used to initialize the hindcasts for CCSM4 are shown in Fig. 1. For the tropical northwestern Pacific region (5°N–30°N, 125°E–180°E), the vertically integrated heat content to the depth of the 20 °C isotherm in the ocean initial state is $\sim 2.7 \times 10^{24}$ J just before the 1976–77 El Niño event, with roughly 2.2×10^{24} J in the tropical southwestern Pacific region (5°S–30°S, 150°E–160°W) (Fig. 1a). The 1972–73 El Niño, evidenced by an increase in vertically integrated heat content to the depth of the 18 °C isotherm in the eastern equatorial Pacific (5°S–5°N, 160°W–80°W) to values near 1.5×10^{24} J (Fig. 1b) is followed by a subsequent decrease over the next few years in off-equatorial southwestern tropical Pacific heat content by nearly 0.1×10^{24} J (Fig. 1a). However, the subsequent 1976–77 El Niño, with an increase in vertically integrated heat content in the eastern equatorial Pacific of $\sim 1.4 \times 10^{24}$ J (Fig. 1b), is associated with a subsequent drop in off-equatorial northwestern and

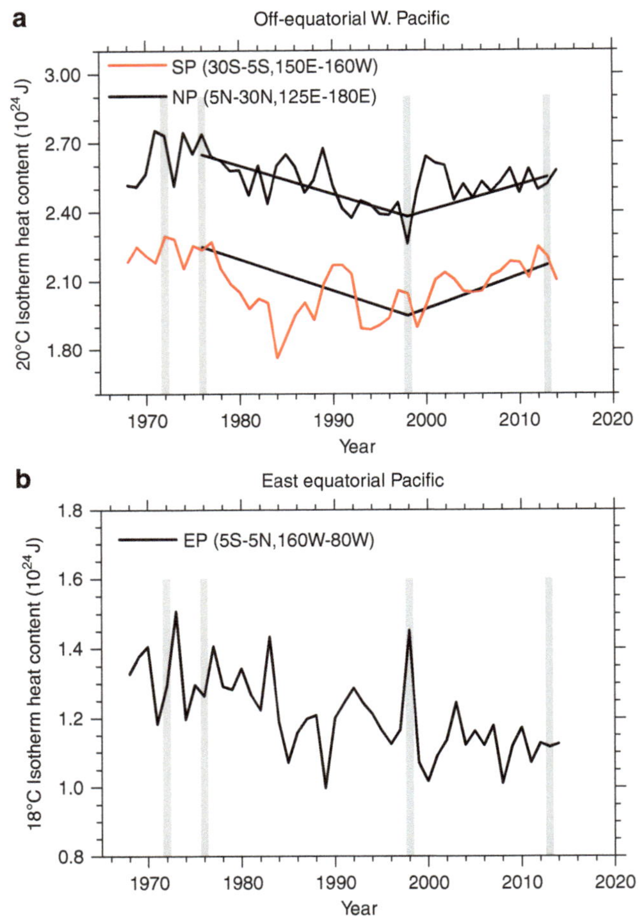

Figure 1 | Off-equatorial western Pacific and east equatorial Pacific heat content. (**a**) Annual mean time series of vertically integrated ocean heat content ($\times 10^{24}$ J) to the depth of the 20 °C isotherm for regions in the off-equatorial western Pacific from the ocean initial states used for the initialized model hindcasts; blue line for an area in the northwest tropical Pacific (5°N-30°N, 125°E-180°E); red line for an area in the southwest tropical Pacific (5°S-30°S, 150°E-160°W); shaded grey vertical lines indicate start points for the 1972-73 El Niño, 1976-77 El Niño, 1998-2000 La Niña and 2014-2015 El Niño, respectively; solid lines are linear trends from 1976 to 1998 and from 1998 to 2013; (**b**) same as in **a**, except for vertically integrated ocean heat content down to the depth of the 18 °C isotherm for an area in the equatorial eastern Pacific (5°N-5°S, 160°W-80°W).

southwestern tropical Pacific heat content for the next decade and longer (Fig. 1a). With the caveat that the initial ocean state values are less certain before the satellite era that began in the late 1970s (see Methods), this sequence suggests that the 1972–73 El Niño started the discharge of off-equatorial western Pacific heat content that was then given a boost by the 1976–77 El Niño, which closely followed, and thus contributed to the onset of the positive phase of the IPO in the mid-1970s[19].

Values for the southwestern Pacific region bottom out after the 1982–83 El Niño, which occurred while the IPO was still in a positive phase. The weak La Niña in 1984–85 contributed to a subsequent increase in off-equatorial heat content in the southwestern area of $\sim 0.3 \times 10^{24}$ J, but the northwestern area heat content continued to decline. Thus, the changes in the southwestern area were necessary but not sufficient to produce an IPO transition and the positive IPO phase continued until the late 1990s.

After the partial recovery around 1990, the southwestern Pacific region heat content declined again until a minimum value of $\sim 2.0 \times 10^{24}$ J was reached in 1997, whereas there was a minimum value of roughly 2.3×10^{24} J in the northwestern Pacific region around 1998, coincident with the large 1997–98 El Niño (Fig. 1a). These low heat content values in the off-equatorial western Pacific in the ocean initial states were then associated with the 2-year La Niña event (1998–2000) with low eastern equatorial Pacific heat content values down to near 1.0×10^{24} J (Fig. 1b) and a transition to the negative phase of the IPO. Next, off-equatorial western tropical Pacific vertically integrated heat content values started to climb again until the initial-state heat content in the northwestern tropical Pacific region in 2013 was nearly 2.6×10^{24} J, close to its value in the 1976 initial state when the IPO previously transitioned to positive. The southwestern Pacific region heat content in 2012 and 2013 was around 2.1–2.2×10^{24} J, also its highest value since the mid-1970s. Indications from these ocean heat content values from the initial states for the initialized hindcasts suggest that if even a moderate El Niño were to occur shortly after 2013, there could be an IPO transition from negative to positive.

It would be tempting to speculate, based on this small sample, that threshold values amounting to around 2.6×10^{24} J in the northwestern area and 2.2×10^{24} J in the southwestern area would be necessary for an El Niño event to trigger a transition to positive IPO. Conversely, heat content values of roughly 2.4×10^{24} and 2.0×10^{24} J in the northwestern and southwestern areas, respectively, could be necessary for a La Niña event to trigger a transition to the negative phase of the IPO. However, based on the small sample size and the limitations of the ocean initialization scheme (see Methods), these values can only be considered as instructive at this time, pending further research with more models.

Hindcast skill. Credibility for an initialized prediction in the future must be evaluated in the context of past predictions or hindcasts. Thus, it is necessary to quantify IPO hindcast skill in CCSM4 to see whether there are false IPO transitions that appear in initialized hindcasts with previous El Niño or La Niña events. Figure 2 shows hindcast verification in terms of pattern correlation for Pacific SSTs (Pacific basin, ocean grid points 40°S–70°N, 100°E–80°W) for CCSM4 for the ensemble average of ten ensemble members for each initial year starting in 1960 (that is, the correlation is calculated using the ensemble average[4]). This is a version of a similar figure shown for initialized hindcasts for all the CMIP5 models[4]. Hindcast skill for the year three to seven predictions (5 year average over years 3–7, hereafter referred to as '3- to 7-year' hindcasts or predictions) is defined as a pattern correlation for Pacific SSTs above the 95% significance level

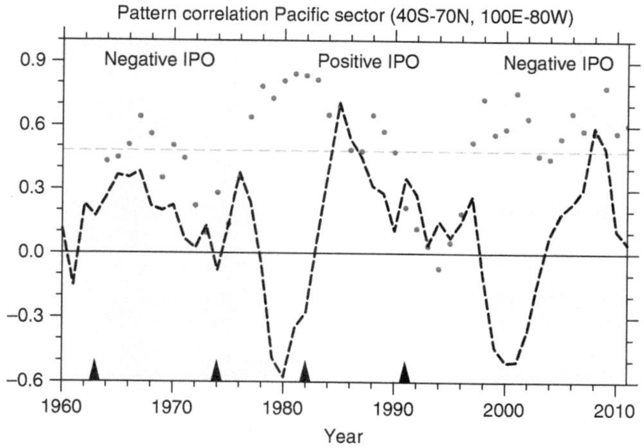

Figure 2 | Hindcast skill for the IPO. Anomaly pattern correlation for the Pacific SSTs in hindcasts initialized every year from 1960 for CCSM4 (ocean points only, 40°S-70°N, 100°E-80°W; same as in Fig. 2e in ref. 4, but just for CCSM4 see Methods). Horizontal dashed line is 95% significance. Blue dashed line is persistence prediction defined as the average of years three to seven in the future having the same SST anomalies as the 5-year average before the start of the prediction. Each red dot represents the central year of the 3- to 7-year average hindcast; thus, the value for the initial year 1960 for the prediction averaged for 1962-66 is plotted for 1964 and so on. Phase changes of the IPO from negative to positive in the late 1970s and from positive to negative in the late 1990s are denoted by shading. Hindcast skill notably drops below 95% when influenced by the volcanic eruptions (denoted by triangles above *x* axis) of Fuego in the early 1970s and Pinatubo in the early 1990s, but not Agung in the early 1960s or El Chichon in the early 1980s[23].

(horizontal dashed line, see Methods). Most 3- to 7-year hindcasts are skillful by that measure and exceed persistence (defined as the 5-year average anomalies before the initial year and persisted to forecast years three to seven) except for two periods, one in the mid-1970s and the other in the mid-1990s. These two periods have been shown to have reduced hindcast skill due to the effects of the Fuego and Pinatubo volcanic eruptions, respectively[23], when the post-eruption sequence of tropical Pacific SSTs does not match the model ensemble-average forced response. The previously documented skill in IPO transitions from negative to positive in the late 1970s and from positive to negative in the late 1990s is denoted by red dots above 95% significance for changes in shading in Fig. 2. Model skill in predicting transitions is indicated by the combination of having a good pattern correlation and poor performance of a persistence forecast (blue dashed line). It is noteworthy that the persistence predictions have no skill in predicting across major decadal transitions (actually having negative values in Fig. 2 at the time of the transitions), while the initialized predictions have significant skill across the transitions. In addition, the CCSM4 shows skill in predicting the correct state of the IPO even for years when there were significant El Niño (for example, no IPO transition for the 1982–83 El Niño) or La Niña (for example, no IPO transition for the 1988–89 La Niña) events.

Figure 2 also addresses the concern that an initialized hindcast could either trend to model climatology and always 'transition' away from its initial state or simply persist the initial state and never 'transition' to another IPO state. The performance of the model in the hindcasts shows that, with the exception of the volcano-related drops in skill noted above (with the caveat that the volcanoes affect 11 out of the 48 five-year average prediction periods in Fig. 2), the model can capture IPO transitions as well as persist existing initialized IPO states.

ENSO as a trigger for decadal transitions. As the previous two IPO transitions were simulated with some success as noted above and if ENSO events act as a trigger for the western tropical Pacific off-equatorial heat content anomalies in the ocean initial states noted in Fig. 1 as in previous studies[6], then ENSO evolution for those IPO transitions should have been simulated as well. There have been previous indications that ENSO events can be predicted with initialized climate models for lead times of at least 2 years[24]. Here we show the time evolution of the predictions for all ensemble members and the ensemble average for the first 5 years after initialization. Clearly, there will be spread in the predictions, but the hindcast verification in Fig. 2 uses the ensemble average and that is of primary interest here.

Owing to the limitations of the ocean initialization scheme that affect ocean initial states more in the pre-satellite (before the late-1970s) than post-satellite era (see Methods), we follow the conservative approach previously used to analyse predictions of the mid-1970s climate shift[1] and show results for the model initialized in January 1976. The ensemble average for the Niño3.4 index in the eastern equatorial Pacific (5°N–5°S, 170°W–120°W, a standard ENSO index) tracks the observations for the onset of the El Niño in mid-1976 and then the ensemble average remains anomalously warm through 1978, in fact longer than what was observed (Fig. 3a). Although there is spread in the initialized predictions after 1977, the 1976–1977 period of above-normal eastern equatorial SSTs is captured by all ten ensemble members and was associated with the IPO transition from negative to positive for 3- to 7-year predictions during this time period[1].

For the IPO transition from positive to negative in the late 1990s, there is greater confidence in the ocean initial states (see Methods); thus, we follow a previous analysis[4] and show the predictions initialized in January 1996 that have more ensemble spread than for the 1976–77 case, but the ensemble average shows above-normal Niño3.4 SSTs for the 1997–98 event of lower amplitude than what was observed. Of more significance for the IPO transition, the ensemble average model prediction shows below-normal Niño3.4 SSTs for the observed 2-year La Niña from 1998 to 2000, with most ensemble members having negative values for that time period in qualitative agreement with the observations. This La Niña simulation was associated with the

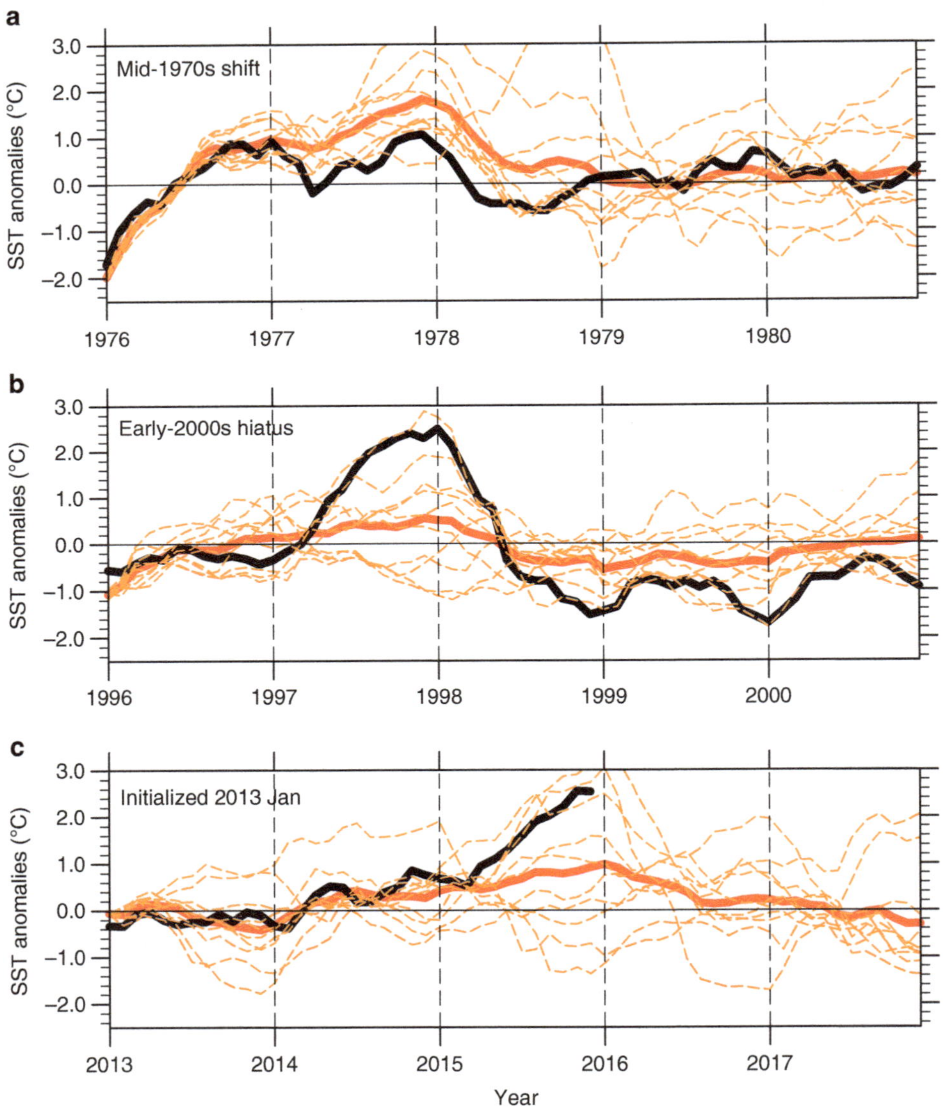

Figure 3 | Initialized predictions of eastern equatorial Pacific sea surface temperatures. Monthly mean time series of initialized predictions for the Niño3.4 area (5°S-5°N, 120°W-170°W), calculated as anomalies from a 1975 to 2014 climatology; observations (black lines[34]), individual model ensemble members are dashed orange lines; ensemble average is solid red line; (**a**) prediction for the mid-1970s shift, initialized 1 January 1976 (ref. 1); (**b**) prediction for early-2000s hiatus, initialized 1 January 1996 (ref. 4); (**c**) prediction initialized 1 January 2013.

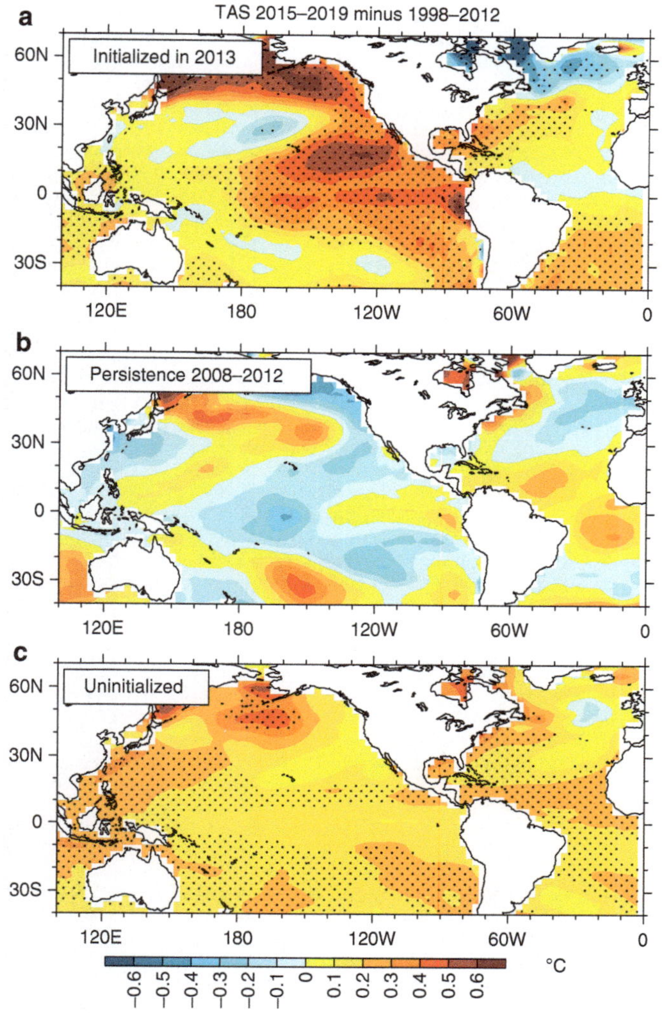

Figure 4 | Predicted patterns of surface temperature. (**a**) Global surface air temperature anomaly patterns (°C) for prediction initialized on 1 January 2013 for the 5-year average 2015–2019 minus observed reference period 1998–2012 from NCEP/NCAR reanalyses[34] (see Methods); (**b**) persistence prediction defined as the average of years three to seven in the future having the same SST anomalies as the 5-year average before the start of the prediction as in Fig. 2, for the 2015–2019 average having the same SST anomalies as the 2008–2012 average, and (**c**) uninitialized prediction for years 2015–2019 minus the model data averaged over the period 1998–2012; stippling indicates 95% significance level from one-sided *t*-test.

prediction of the transition of the IPO from positive to negative in the model for the 3- to 7-year predictions initialized in 1996 (ref. 1).

Prediction for 2015–2019. Thus, there are indications of some model skill in simulating ENSO events that could act as triggers for the off-equatorial ocean heat content in the western tropical Pacific to contribute to IPO transitions (Fig. 1). Therefore, it is of interest to investigate simulations initialized in 2013 for the 3- to 7-year predictions (2015–2019) of the IPO and globally averaged surface air temperature. We choose the initial year 2013 so that we have 2 years of subsequent observations to see whether there has been any skill for predicted Niño3.4 SSTs that could indicate a role in an IPO transition. As noted above, for a credible prediction of an IPO transition, there must not only be heat content build-up in the off-equatorial western Pacific over a period of ~15 years, but also a reasonable simulation of ENSO for the first few years after initialization, for the 3- to 7-year prediction to have skill.

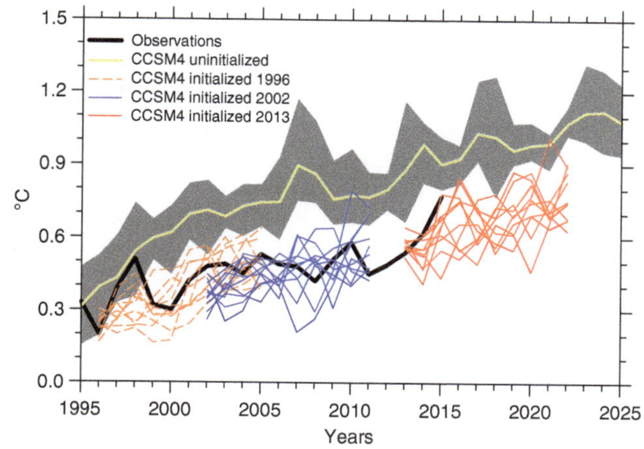

Figure 5 | Time series of global surface air temperature anomalies. Annual mean surface temperature anomalies (°C) relative to observed climatology of 1960–1990 for the CCSM4 uninitialized six member ensemble average (solid yellow line) and the envelope of individual ensemble simulations (grey shading); annual mean values from observations is solid black line[25]; individual ensemble member time series for 10-year predictions initialized 1 January 1996 (dashed orange lines) to capture the late-1990s transition from positive to negative IPO[4], an example of predictions for the hiatus initialized 1 January 2002 (blue lines) and predictions initialized on 1 January 2013 (red lines), the latter showing larger predicted warming rates than the former during the hiatus.

For predictions initialized in January 2013 (Fig. 3c), the observations started to warm in 2014 and the model-simulated Niño3.4 SSTs also warmed. In 2015, the observed Niño3.4 SSTs continued to increase, to approach the magnitude of the warmest predicted ensemble members (Fig. 3c). The ensemble average predictions from the model show above-normal Niño3.4 SSTs for 2015 continuing through 2016. The associated SST anomalies for the 3- to 7-year prediction (2015–2019 minus a 15-year average before the initial year, 1998–2012, used as a reference climatology with which to compare the initialized prediction) initialized in 2013 show a positive phase of the IPO, with above-normal SSTs predicted over the eastern tropical Pacific extending into the northeast Pacific, with low amplitude-negative SST anomalies in the northwest and southwest Pacific (Fig. 4a). This pattern is nearly opposite to the persistence prediction (persisting 2008–2012 average anomalies forward to 2015–2019; Fig. 4b) and is also considerably different from the uninitialized simulation that shows warming almost everywhere (Fig. 4c).

As noted above, there is a strong association between the IPO tendency and globally averaged surface air temperature trends. CCSM4 hindcasts initialized in 1996 simulated the onset of the reduced rate of global surface warming with the transition of the IPO from positive to negative in the late 1990s[4]. The predicted linear trend from 1996 to 2005 of $+0.30 \pm 0.09$ °C per decade (error bars are ± 1 s.d.) is smaller than the uninitialized value of $+0.39 \pm 0.08$ °C per decade and closer to the observed value of $+0.24$ °C per decade (Fig. 5). Meanwhile, a hindcast initialized in 2002 during the hiatus has a predicted linear trend from 2002 to 2011 of $+0.20 \pm 0.11$ °C per decade, roughly 30% smaller than the prediction initialized in 1996 before the hiatus.

Prediction of 2013–2022 global surface temperature trend. With a transition of the IPO from negative to positive in a prediction initialized in 2013, it could be expected from previous studies[14–18] that there should be a resumption of higher rates of globally averaged surface temperature increase. For the prediction initialized in 2013, the linear trend from 2013 to 2022 is $+0.22 \pm 0.13$ °C per decade, nearly 60% larger than the uninitialized projection for that time period of $+0.14 \pm 0.12$ °C per decade (Fig. 5) and nearly three times larger than the observed trend[25] of $+0.08$ °C per decade over the period 2001 to 2014 during the early twenty-first century slowdown[26]. This indicates an acceleration of the global surface temperature warming trend over the prediction period from 2013 to 2022 associated with the positive phase of the IPO, compared with that from external forcing alone and compared with the observed warming trend during previous negative phase of the IPO that was associated with a slowdown of the rate of increase of global surface temperature[25].

Discussion

The 3- to 7-year hindcasts with CCSM4 initialized every year show skill in simulating patterns of SST in the Pacific associated with the IPO, except for periods when there are disruptions from volcanic eruptions with a post-eruption sequence of Pacific SST patterns that do not agree with the ensemble average forced response in the model[23]. Other studies have shown less skill for the smaller North Pacific area defined for the Pacific Decadal Oscillation (PDO)[27]. Although the PDO and IPO are closely related[28], the smaller area strictly defined for the PDO in the North Pacific has more spatial noise and less predictive skill for that small area than the much larger Pacific-wide SST pattern defined for the IPO. In addition, the STC in the Southern Hemisphere shows larger variability in some models, affecting the STC transport convergence anomaly that could affect the North Pacific STC and thus the PDO index, which is defined for the North Pacific.

A proposed mechanism for the IPO indicates that a sustained build-up of off-equatorial ocean heat content in the western tropical Pacific for at least 15 years, due to tropical–mid-latitude interaction and ocean Rossby waves associated with the phase of the IPO, is required for an ENSO event to then trigger a transition of the IPO to its opposite phase. Initial state heat content used for the hindcasts with CCSM4 indicates that there was indeed such a build-up before the simulated IPO transition in the mid-1970s that was associated with the prediction of the 1976–77 El Niño event, with possible contributions from the 1972 to 1973 El Niño event. For the late-1990s IPO transition from positive to negative, a heat content deficit in the off-equatorial western tropical Pacific was associated with the simulated 1998–2000 La Niña event. Other El Niño or La Niña events did not trigger an IPO transition, owing to the requisite time for heat content to build up in the off-equatorial western Pacific. Since the late 1990s, heat content has built up in the initial state used for the predictions, thus suggesting that even a moderate-sized El Niño event could trigger a transition to the positive phase of the IPO. Predictions initialized in 2013 show that the simulated Niño3.4 SSTs have tracked the observations with low-amplitude warming in 2014 and larger warming in 2015. Such qualitative success in Niño3.4 simulations for past IPO transitions suggests that an IPO transition probably started in 2014. Indeed, the 3- to 7-year predictions for 2015–2019 that were initialized in 2013 indicate such an IPO transition has occurred, with a resumption of accelerated rates of global warming above those in the uninitialized model simulations. If such predictions were to become operational, the next step (beyond the scope of this study) would be to provide probabilistic climate change information for various time frames in the near-term future.

Methods

Initialization procedure. As described in a previous study[21], the hindcasts in CCSM4 are initialized using a forced ocean–sea ice simulation designed to reproduce the evolution of the ocean and sea ice states from the start of 1948 through the end of 2013. The CCSM4 ocean and sea ice models are coupled and forced at the surface with Coordinated Ocean-ice Reference Experiment (CORE-II) version 2-observed historical atmospheric data[29,30]. The forcing data include a complete set of air–sea momentum, heat and freshwater fluxes based on the National Centers for Environmental Prediction/National Center for Atmospheric Research (NCEP/NCAR) reanalyses. There is a weak restoring of model surface salinity to observed climatology and no assimilation of subsurface observations. An integration is run through four consecutive 66-year cycles of 1948–2013 forcing. The ocean and sea ice models in hindcast experiments are initialized with 1 January restart files for a particular year from the last (fourth) cycle of the spin-up simulation. Initial conditions for the atmosphere and land surface are taken from corresponding years of a six-member ensemble of uninitialized twentieth-century runs. The ten-member ensembles run for each initial year starting in 1960 are generated by randomly selecting atmosphere and land initial conditions from different uninitialized twentieth-century runs[31] and/or from different days in the month of January. Results from these uninitialized twentieth-century experiments[31] are shown in Figs 4c and 5. The initialized hindcasts and the uninitialized twentieth-century experiments include the same external forcings of solar irradiance, greenhouse gases, aerosols and volcanic activity.

As atmospheric model reanalysis data are used in the CORE-II forcing of the spin-up integrations used for the ocean initial states in the hindcasts, it is likely to be that the quality of the forcing data becomes better in the more recent period, whereas there are issues with reanalysis quality earlier in the record, in particular with precipitation and radiation fluxes over oceans (for example, before the 1970s)[32,33]. Therefore, most of the emphasis in this study is on initialized hindcasts after 1975.

Bias adjustment procedure. The initialized hindcast experiments are bias adjusted to remove model systematic errors[1,21]. Climatological monthly mean differences from observations (surface air temperature from the NCEP/NCAR reanalyses)[34] for the 10-year hindcasts are computed and composited by month following the initial dates of all the hindcast simulations. Next, these average time-evolving monthly mean differences are subtracted from each member of each 10-year hindcast to remove the model systematic error, leaving the model signal from external forcing and internally generated decadal climate variability.

Assessing hindcast skill. Anomalies for the 3- to 7-year bias-adjusted predictions are calculated as the year 3–7 average minus the average of the observations for the previous 15 years before the initial year of the hindcast[1,4]. Other metrics show similar results[2]. To test whether the pattern correlation coefficient between the prediction and observations is distinguishable from chance associations in the large-scale pattern, a Monte Carlo test is performed that consists of 10,000 randomly constructed patterns based on detrended twentieth-century simulations from a multi-model ensemble of CMIP5 models[4]. The 95th percentile of the pattern correlation coefficient of the random pattern is 0.47 in Fig. 2. Persistence is defined as the observed 5-year average before the initial year of the prediction minus a 15-year average for 6–20 years before the initial year[4].

Observations. The Nino3.4 index is available from http://www.esrl.noaa.gov/psd/gcos_wgsp/Timeseries/. Observed global mean surface temperature data available from http://www.ncdc.noaa.gov/cag/time-series/global. The NCEP/NCAR reanalyses are available from http://www.esrl.noaa.gov/psd/data/gridded/data.ncep.reanalysis.html.

References

1. Meehl, G. A. & Teng, H. Case studies for initialized decadal hindcasts and predictions for the Pacific region. *Geophys. Res. Lett.* **39,** L22705 (2012).
2. Meehl, G. A. & Teng, H. Regional precipitation simulations for the mid-1970s shift and early-2000s hiatus. *Geophys. Res. Lett.* **41,** 7658–7665 (2014).
3. Meehl, G. A. & Teng, H. CMIP5 multi-model initialized decadal hindcasts for the mid-1970s shift and early-2000s hiatus and predictions for 2016–2035. *Geophys. Res. Lett.* **41,** 1711–1716 (2014).
4. Meehl, G. A., Teng, H. & Arblaster, J. M. Climate model simulations of the observed early-2000s hiatus of global warming. *Nat. Clim. Change* **4,** 898–902 (2014).
5. Ding, H., Greatbatch, R. J., Latif, M., Park, W. & Gerdes, R. Hindcast of the 1976/77 and 1998/99 climate shifts in the Pacific. *J. Climate* **26,** 7650–7661 (2013).
6. Meehl, G. A. & Hu, A. Megadroughts in the Indian monsoon region and southwest North America and a mechanism for associated

multi-decadal Pacific sea surface temperature anomalies. *J. Climate* **19,** 1605–1623 (2006).
7. Farneti, R., Molteni, F. & Kucharski, F. Pacific interdecadal variability driven by tropical-extratropical interactions. *Clim. Dyn.* **42,** 3337–3355 (2014).
8. Giese, B. S., Urizar, S. C. & Fučkar, N. S. Southern Hemisphere origins of the 1976 climate shift. *Geophys. Res. Lett.* **29,** 1-1-1-4 (2002).
9. White, W. B., Tourre, Y. M., Barlow, M. & Dettinger, M. A delayed action oscillator shared by biennial, interannual, and decadal signals in the Pacific basin. *J. Geophys. Res.* **108,** 3070 (2003).
10. Wang, X., Jin, F.-F. & Wang, Y. A tropical ocean recharge mechanism for climate variability. Part I: equatorial heat content changes induced by the off-equatorial wind. *J. Climate* **16,** 3585–3598 (2003).
11. Wang, X., Jin, F.-F. & Wang, Y. A tropical ocean recharge mechanism for climate variability. Part II: a unified theory for decadal and ENSO modes. *J. Climate* **16,** 3599–3616 (2003).
12. Farneti, R., Dwivedi, S., Kucharski, F., Molteni, F. & Griffies, S. M. On Pacific subtropical cell variability over the second half of the twentieth century. *J. Climate* **27,** 7102–7112 (2014).
13. Jin, F. Low frequency modes of the tropical ocean dynamics. *J. Climate* **14,** 3872–3881 (2001).
14. Meehl, G. A. *et al.* Model-based evidence of deep-ocean heat uptake during surface-temperature hiatus periods. *Nat. Clim. Change* **1,** 360–364 (2011).
15. Kosaka, Y. & Xie, S.-P. Recent global-warming hiatus tied to equatorial Pacific surface cooling. *Nature* **501,** 403–407 (2013).
16. England, M. H. *et al.* Slowdown of surface greenhouse warming due to recent Pacific trade wind acceleration. *Nat. Clim. Change* **4,** 222–227 (2014).
17. Watanabe, M. *et al.* Contribution of natural decadal variability to global warming acceleration and hiatus. *Nat. Clim. Change* **4,** 893–897 (2014).
18. Meehl, G. A., Hu, A., Arblaster, J. M., Fasullo, J. & Trenberth, K. E. Externally forced and internally generated decadal climate variability associated with the Interdecadal Pacific Oscillation. *J. Climate* **26,** 7298–7310 (2013).
19. Meehl, G. A., Hu, A. & Santer, B. D. The mid-1970s climate shift in the Pacific and the relative roles of forced versus inherent decadal variability. *J. Climate* **22,** 780–792 (2009).
20. Thoma, M., Greatbatch, R. J., Kadow, C. & Gerdes, R. Decadal hindcasts initialized using observed surface wind stress: evaluation and prediction out to 2024. *Geophys. Res. Lett.* **42,** 6454–6461 (2015).
21. Yeager, S., Karspeck, A., Danabasoglu, G., Tribbia, J. & Teng H., H. A decadal prediction case study: late twentieth-century North Atlantic ocean heat content. *J. Climate* **25,** 5173–5189 (2012).
22. Meehl, G. A. *et al.* Decadal climate prediction: an update from the trenches. *Bull. Am. Meteorol. Soc.* **95,** 243–267 (2014).
23. Meehl, G. A., Teng, H., Maher, N. & England, M. H. Effects of the Mount Pinatubo eruption on decadal climate prediction skill of Pacific sea surface temperatures. *Geophys, Res. Lett.* **42,** 10,840–10,846 (2015).
24. Luo, J.-J., Masson, S., Behera, S. & Yamagata, T. Extended ENSO predictions using a fully coupled ocean-atmosphere model. *J. Climate* **21,** 84–93 (2008).
25. Karl, T. R. *et al.* Possible artifacts of data biases in the recent global surface warming hiatus. *Science.* **348,** 1469–1472 (2015).
26. Fyfe, J. C. *et al.* Making sense of the early-2000s warming slowdown. *Nat. Clim. Change* **6,** 224–228 (2016).
27. Kim, H.-M., Webster, P. J. & Curry, J. A. Evaluation of short-term climate change prediction in multi-model CMIP5 decadal hindcasts. *Geophys. Res. Lett.* **39,** L10701 (2012).
28. Han, W. *et al.* Intensification of decadal and multi-decadal sea level variability in the western tropical Pacific during recent decades. *Clim. Dyn.* **43,** 1357–1379 (2014).
29. Large, W. G. & Yeager, S. G. The global climatology of an interannually varying air-sea flux data set. *Clim. Dyn.* **33,** 341–364 (2008).
30. Griffies, S. M. *et al.* Coordinated ocean-ice reference experiments (COREs). *Ocean Model.* **26,** 1–46 (2009).
31. Meehl, G. A. *et al.* Climate system response to external forcings and climate change projections in CCSM4. *J. Climate* **25,** 3661–3683 (2012).
32. Trenberth, K. E. & Guillemot, C. J. Evaluation of the atmospheric moisture and hydrological cycle in the NCEP/NCAR reanalyses. *Clim. Dyn.* **14,** 213–231 (1998).
33. Trenberth, K. E., Fasullo, J. T. & Mackaro, J. Atmospheric moisture transports from ocean to land and global energy flows in reanalyses. *J. Climate* **24,** 4907–4924 (2011).
34. Kalnay, E. *et al.* The NCEP/NCAR 40-year reanalysis project. *Bull. Amer. Meteorol. Soc.* **77,** 437–471 (1996).

Acknowledgements

The authors thank Mark Cane for encouraging us to publish this work. Portions of this study were supported by the Regional and Global Climate Modeling Program (RGCM) of the U.S. Department of Energy's Office of Biological and Environmental Research (BER) Cooperative Agreement DE-FC02-97ER62402 and the National Science Foundation. The National Center for Atmospheric Research is sponsored by the National Science Foundation. The CCSM4 decadal prediction runs used resources of the National Energy Research Scientific Computing Center, which is supported by the Office of Science of the U.S. Department of Energy under contract DE-AC02-05CH11231, and the Oak Ridge Leadership Computing Facility located in the Oak Ridge National Laboratory, which is supported by the Office of Science of the Department of Energy under contract DE-AC05-00OR22725. An award of computer time was provided by the Innovative and Novel Computational Impact on Theory and Experiment (INCITE) program.

Author contributions

G.M. directed and wrote this work with contributions from all authors. G.M., A.H. and H.T. performed the analyses. H.T. ran the model ensemble hindcasts and predictions, and with A.H. conducted analyses. All the authors discussed the results.

Additional information

Competing financial interests: The authors declare no competing financial interests.

22

Summer rainfall over the southwestern Tibetan Plateau controlled by deep convection over the Indian subcontinent

Wenhao Dong[1], Yanluan Lin[1], Jonathon S. Wright[1], Yi Ming[2], Yuanyu Xie[1], Bin Wang[1,3], Yong Luo[1], Wenyu Huang[1], Jianbin Huang[1], Lei Wang[4], Lide Tian[4], Yiran Peng[1] & Fanghua Xu[1]

Despite the importance of precipitation and moisture transport over the Tibetan Plateau for glacier mass balance, river runoff and local ecology, changes in these quantities remain highly uncertain and poorly understood. Here we use observational data and model simulations to explore the close relationship between summer rainfall variability over the southwestern Tibetan Plateau (SWTP) and that over central-eastern India (CEI), which exists despite the separation of these two regions by the Himalayas. We show that this relationship is maintained primarily by 'up-and-over' moisture transport, in which hydrometeors and moisture are lifted by convective storms over CEI and the Himalayan foothills and then swept over the SWTP by the mid-tropospheric circulation, rather than by upslope flow over the Himalayas. Sensitivity simulations confirm the importance of up-and-over transport at event scales, and an objective storm classification indicates that this pathway accounts for approximately half of total summer rainfall over the SWTP.

[1]Ministry of Education Key Laboratory for Earth System Modeling, Center for Earth System Science, and Joint Center for Global Change Studies (JCGCS), Tsinghua University, Room S813, MengMinwei Science Building, Qinghuayuan 1, Beijing 100084, China. [2]Geophysical Fluid Dynamics Laboratory, Princeton/NOAA, Princeton, New Jersey 08540-6649, USA. [3]State Key Laboratory of Numerical Modeling for Atmospheric Science and Geophysical Fluid Dynamics, Institute of Atmospheric Physics, Chinese Academy of Sciences, Beijing 100029, China. [4]Key Laboratory of Tibetan Environment Changes and Land Surface Processes, Institute of Tibetan Plateau Research, Chinese Academy of Sciences (CAS), and the CAS Center for Excellence in Tibetan Plateau Earth Sciences, Beijing 100101, China. Correspondence and requests for materials should be addressed to Y.Lin (email: yanluan@tsinghua.edu.cn).

The Tibetan Plateau (TP), sometimes referred to as 'the world's water tower'[1], is a critical region for water resources in Asia. Nearly one sixth of the global population depends on rivers with headwaters on the TP, and changes in precipitation over the TP have great impacts on glaciers, river discharge, plant phenology and natural hazards[1–5]. Summer (June–September) precipitation, which occurs mainly in convective storms (CSs), accounts for 60–90% of total annual precipitation over the TP and surrounding areas[6–8]. Trends and variability in summertime precipitation over the TP are therefore likely to exert strong influences on climate and socio-economic conditions throughout many parts of Asia.

The spatiotemporal variation of summer precipitation over the TP is complex, and is mainly regulated by a combination of summer monsoons and regional moisture recycling[2,3,9,10]. Rainfall over the eastern TP has been extensively studied and well-documented, with a relative abundance of ground measurements, and is closely linked to the South Asian monsoon via a moisture corridor that extends along the Brahmaputra River from the Bay of Bengal[9,11,12]. However, despite recent sharp declines in glacier extent[2,3], the southwestern TP (SWTP) has received considerably less attention. The SWTP is a semiarid region with a vulnerable ecosystem, and precipitation systems over the SWTP often serve as precursors of synoptic systems downstream[4,6,11]. Some convective instabilities that develop over the SWTP cause extreme rainfall and severe flooding in East Asia as they develop and move eastward[9,13].

The factors that control precipitation variations over the SWTP remain poorly understood. Here we establish that summer rainfall over the SWTP and CEI is closely related via a previously unreported 'up-and-over' moisture transport path, in which moist, cloud-laden air lifted by convective systems over CEI is swept over the SWTP by the mid-tropospheric circulation. This relationship is evident in both observational data sets and global climate model simulations, and depends on the coincident occurrence of convective systems over CEI and southwesterlies in the middle and upper troposphere over the Himalayas. Several lines of evidence confirm both the existence of the up-and-over moisture transport path and its importance in the water budget of the SWTP, including satellite measurements, isotopic ratios in precipitation, numerical sensitivity simulations and backward trajectory calculations. The results consistently indicate that up-and-over transport accounts for at least half of total summer rainfall over the SWTP, and show that this pathway dominates moisture supply in this region. The SWTP will be substantially drier without the up-and-over transport associated with convective systems over CEI.

Results

Linkage of summer rainfall between SWTP and CEI. Evidence of a strong connection between rainfall over the SWTP and rainfall over the Indian subcontinent emerges from historical records of precipitation (Fig. 1; see Methods for details). In particular, despite a large difference in rainfall magnitude across the Himalayas (Fig. 1a), significant decreasing trends have been observed over both the SWTP and central-eastern India (CEI; Fig. 1b). Decreases in rainfall over the Indian subcontinent over the past six decades have been attributed both to natural variability[7,14] and to anthropogenic forcing[15–17]. Summer rainfall variability over the SWTP is highly correlated with summer rainfall variability over CEI (Fig. 1c). Indeed, summer precipitation over CEI is a reliable predictor for summer precipitation over the SWTP (Fig. 2a), with correlations of $r=0.68$ for APHRODITE[18] data (based on rain gauge measurements during 1951–2007; $P<0.001$) and $r=0.64$ for TRMM[19] data (during 1998–2013; $P<0.01$). The correlation between TRMM and APHRODITE estimates of precipitation exceeds 0.85 during their period of overlap (1998–2007; $P<0.01$) with a difference of $<8\%$ between their mean magnitudes. Strong correlations between precipitation variability over these two regions are likewise evident in two other observational data sets, as well as in simulations of historical climate and scenario-based projections of future climate by many global climate models (Supplementary Fig. 1). These results indicate a close and robust relationship between summer rainfall over the SWTP and summer rainfall over CEI that transcends data sources and model configurations. Quantitative differences in the correlation coefficients may be tied to differences in model physics, dynamics or resolutions, and are not explored here.

The strong relationship between summer rainfall over the SWTP and summer rainfall over CEI is somewhat surprising, as these two regions are separated by the Himalayan mountain range. The Himalayas are on average more than 5,000 m above mean sea level between 75 and 95° E, and act as a natural barrier for northward water vapour transport into the SWTP. Northward flow impinging on the Himalayas is mechanically lifted by the orography, with the southern foothills of the Himalayas among the most prominent regions of orographic precipitation on Earth[20,21]. This situation creates a strong precipitation gradient in the lee of the Himalayas. The sharp decrease in precipitation on the leeside of the mountain range suggests that the Himalayas strictly limit upslope moisture transport under stably stratified conditions. In other words, these mountains are sufficiently high that condensation during upslope flow 'wrings out' most of the water, leaving little moisture available for precipitation in the interior of the SWTP. However, the interior of the SWTP experiences more precipitation than other similar places blocked by high mountain ranges, such as desert areas in the southwestern United States and South America. Furthermore, numerical modelling experiments[9] and analyses of the isotopic composition of rainfall[22–27] suggest that much of the rainfall over the SWTP during boreal summer can be traced back to the nearby ocean. The strong connection between the two regions must therefore be associated with alternative routes for moisture transport, rather than direct upslope flow. These results call for a re-evaluation of moisture transport into the SWTP and a careful analysis of how the relationship between the SWTP and CEI is established.

An up-and-over route for moisture transport into SWTP. Summertime rainfall over the Indian subcontinent is dominated by deep CSs[10,21,28–30]. These CSs originate over the surrounding oceans, and are then observed to move northward or northwestward in company with strong onshore moisture transport[31]. Individual CSs may organize and develop into monsoon depressions, which in turn contribute significantly to monsoon rainfall over the northern part of the Indian subcontinent[32,33]. CS development preferentially occurs when monsoon southwesterlies and high-latitude northwesterlies converge and ascend along the southwestern flank of the Himalayas[13].

Almost every large precipitation event over the SWTP occurs in tandem with a corresponding event over CEI. For example, precipitation over these two regions during summer 2002 (Fig. 2b) exhibits similar variations at both daily and weekly time scales ($r=0.72$, $P<0.01$). Variations at these time scales correspond to episodes of strong rainfall; for example, between June and September 2002, the same 34 days accounted for $>60\%$ of the total rainfall over both regions. The co-variability of precipitation over the two regions is thus primarily determined by the correspondence of individual convective precipitation events.

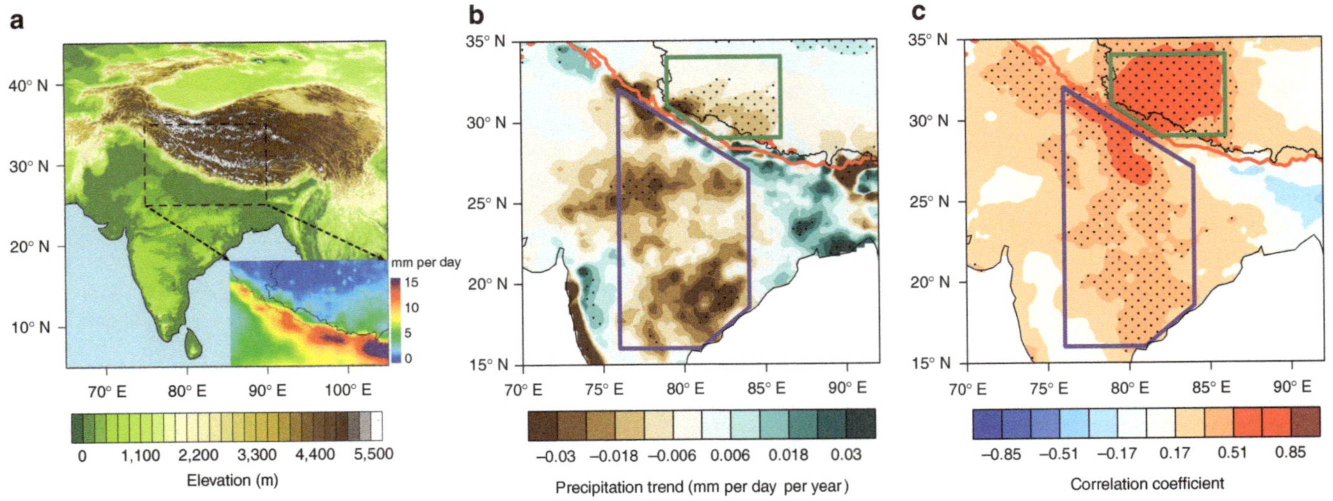

Figure 1 | June-September precipitation characteristics. (**a**) Topography of the Tibetan Plateau and surrounding areas. The inset shows summer precipitation averaged over 1998-2013 using TRMM. (**b**) Precipitation linear trends during 1951-2007 using APHRODITE. (**c**) Correlation coefficient between area-averaged precipitation over the southwestern Tibetan Plateau (SWTP; green polygon) and precipitation over the entire study region (1951-2007, APHRODITE). Dots in **b**,**c** indicate statistically significant areas ($P<0.01$, using a two-sided t-test). Purple polygon in **b**,**c** indicates central-eastern India (CEI). The red isoline in **b**,**c** is the 2,500 m elevation contour locating the Tibetan Plateau.

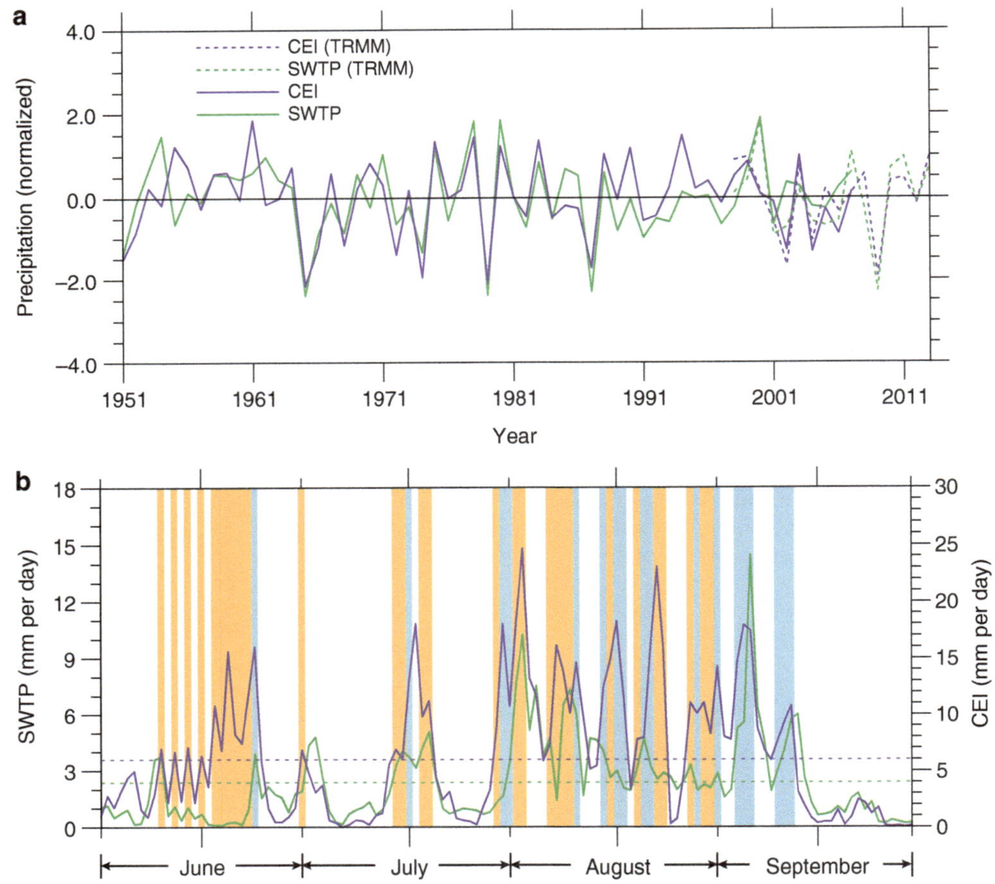

Figure 2 | Summer precipitation variations. (**a**) Time series of standardized anomalies in summer mean precipitation over the southwestern Tibetan Plateau (SWTP; green) and central-eastern India (CEI; purple) using APHRODITE (1951-2007) and TRMM (1998-2013). (**b**) As in **a** but for daily precipitation from TRMM during June-September 2002. The light blue shading marks days with intrusive CSs, while the light orange shading marks days with non-intrusive CSs. Green dashed lines indicate the seasonal mean precipitation over the SWTP and purple for 6 mm per day over CEI.

We propose that this link manifests as strong transport of moisture and hydrometeors at middle-to-upper levels of the troposphere: moist, cloud-laden air is elevated by deep CSs over the Indian subcontinent and then advected over the Himalayas into the SWTP by the mid-tropospheric circulation. We term this transport pathway the 'up-and-over' route.

Intrusive and non-intrusive deep CSs. Not all precipitation events over CEI are accompanied by strong precipitation over the SWTP. In some cases, convective precipitation over CEI is confined to the south side of the Himalayas. We therefore categorize CSs over CEI into two types based on the characteristics of the mid-tropospheric flow near the Himalayas (see Methods). We define each CS as intrusive if the mid-tropospheric circulation permits it to influence the SWTP, and non-intrusive if its effects are confined to the south side of the Himalayas. Supplementary Figure 2 shows examples of typical intrusive and non-intrusive CS events. Both examples were characterized by a closed low-level (850 hPa) centre of low pressure south of the Himalayas; however, strong rainfall extended northward into the SWTP for the intrusive CS and did not for the non-intrusive CS. The northward extension of rainfall in the intrusive case was associated with strong cross-barrier (southwesterly) winds of up to 10 m s^{-1} at 500 hPa and a region of persistent large-scale ascent that extended from the Indian subcontinent to the SWTP. At 300 hPa, strong southwesterlies ahead of an approaching upper level trough pushed the upper tropospheric high southeastward. These middle and upper tropospheric circulation patterns promoted the advection of convectively lifted moisture and hydrometeors from the Indian subcontinent northeastward over the SWTP, in stark contrast to the circulation patterns observed during the non-intrusive CS. In the non-intrusive case, the low-pressure system extended upward from the surface to 500 hPa, with predominant easterly flow along the mountain range. Meanwhile, the centre of high pressure at 300 hPa straddled the mountain range, with northwesterly flow towards the SWTP and northeasterly flow towards the Indian Subcontinent. The absence of cross-barrier flow in the middle and upper troposphere (500 and 300 hPa) effectively prevented the northward transport of moisture and hydrometeors, so that rainfall was mostly confined to the Indian subcontinent. The moist static energy at 300 hPa was likewise much larger during the intrusive case than during the non-intrusive case, indicating that upper levels of the troposphere experienced greater moistening and diabatic heating during the intrusive case.

WRF simulations and backward trajectory calculations. It is difficult to distinguish the relative contributions of up-and-over transport, upslope transport and local processes using observations alone. For example, the pattern of precipitation we attribute to up-and-over transport might instead be generated by two separate rainfall events over the SWTP and CEI. We therefore conduct a series of sensitivity simulations using the Weather Research and Forecasting (WRF) model (see Methods), in which we estimate the quantitative contributions of up-and-over transport, upslope transport and local recycling processes to the precipitation over the SWTP by reducing the advection of moisture and hydrometeors across the southern boundary into the SWTP within certain layers (Supplementary Fig. 3). A sensitivity simulation with hydrometeor advection entirely disallowed gives negligible precipitation change over the SWTP as compared with the control simulation (Figure not shown); however, this does not necessarily indicate that hydrometeor transport is unimportant given the large uncertainties in model simulated hydrometeor fields. In light of these uncertainties, we focus on the moisture transport within different layers for the following sensitivity simulations. We assume that water vapour advection across the southern boundary of the Plateau over CEI initially located more than 5 km above ground level (AGL) is mainly related to up-and-over transport, while advection of CEI water vapour initially located below 2.5 km AGL is mostly associated with upslope transport. Advection of CEI water vapour

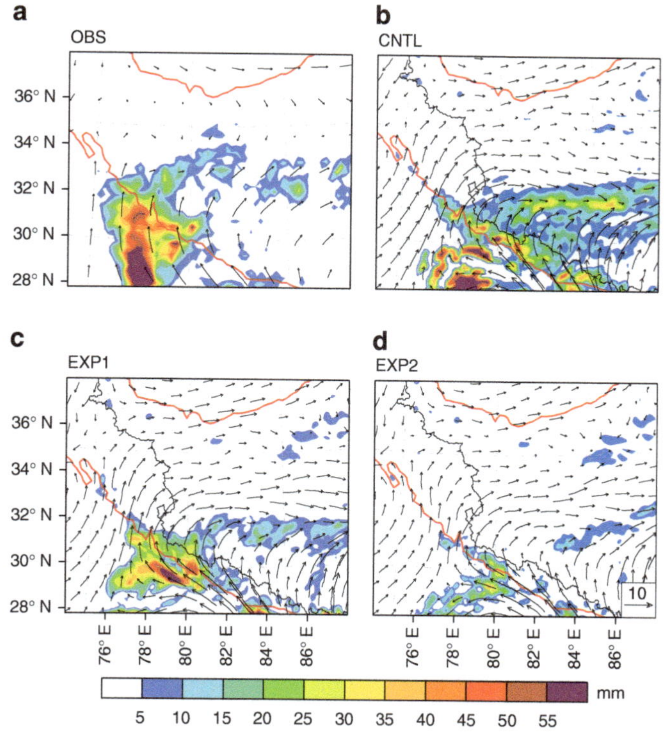

Figure 3 | Sensitivity simulations using WRF. The distribution of precipitation overlaid with 500 hPa winds from **(a)** observations and reanalyses (precipitation from TRMM and wind fields from ERA-Interim), **(b)** the control simulation of the intrusive rain event on 12 September 2002, **(c)** a simulation of the same case but with advection of moisture above 5,000 m AGL reduced by 50% and **(d)** a simulation of the same case but with advection of moisture above 2,500 m AGL altitude reduced by 50%. The red isoline is the 2,500 m elevation contour.

between 2.5 and 5 km AGL is assumed to be a mixture of the up-and-over and upslope transport. We use transport in this layer to represent uncertainties associated with partitioning between the two pathways.

The distribution of precipitation and the upper level circulation produced by the control simulation agree well with TRMM observations and ERA-Interim reanalyses (Fig. 3). Simulated precipitation over the SWTP decreases by ~56% when advection of upper layer (>5 km AGL) water vapour is reduced by half. Halving the transport of water vapour at levels >2.5 km, AGL further reduces the precipitation, with simulated rainfall only ~18% of that produced in the control simulation. Simulated precipitation over the SWTP essentially disappears (~2% of the control simulation) when water vapour transport across the southern boundary is halved through the entire vertical column. These results indicate that the vast majority of the precipitation in this case (~98%) depends on transport of moisture from the Indian subcontinent, with large contributions (56–82%) attributable to up-and-over transport. Additional simulations for other intrusive CSs show similar results (Supplementary Fig. 4), indicating that these contributions are robust.

The topographic features of the Himalayas, including both valleys and ridges, are well-represented by the 9-km grid used for the WRF simulations (Supplementary Fig. 5). Moisture transport can be found both in the valleys and at middle-to-upper levels, indicating that the upslope and up-and-over routes coexist. However, the reduction of moisture transport in the sensitivity simulations is most pronounced when advection of upper layer moisture is eliminated, indicating that moisture supply during these events depended most critically on up-and-over transport.

Our conclusions regarding the importance of up-and-over transport during the 12 September 2002 intrusive case are further supported by backward trajectory calculations using the FLEXPART particle dispersion model[34] (Supplementary Fig. 6). Trajectory-derived transport pathways preceding the rain event are qualitatively consistent with the results of the WRF simulations described above. The quantitative partitioning of 'up-and-over' and 'local convection' trajectories is very sensitive to the criteria used to define these subsets; however, reasonable criteria assign 50–75% of these particles to the up-and-over route, indicating that transport via the up-and-over route was at the very least comparable to local convection (including that dependent on upslope moisture transport) during this event. Overall, the evidence consistently supports both the existence and the critical importance of the up-and-over transport to SWTP precipitation at event scales.

Moisture transport and isotopic observations. Both satellite measurements and reanalyses confirm the existence of mid-tropospheric moisture corridors extending from CEI into the SWTP associated with intrusive CSs (for example, Supplementary Figs 7 and 8). By contrast, the same moisture fields show distinct breaks near the Himalayas for non-intrusive cases (not shown). The consistency between AIRS retrievals and ERA-Interim estimates of water vapour in this case indicates that the reanalysis is able to adequately capture the circulation and moisture fields over this area, as AIRS radiances were not assimilated by ERA-Interim until April 2003 (ref. 35). The vertical structure of the reanalysis moisture fields and circulation along the 80° E longitude transect (Supplementary Fig. 8) during typical intrusive and non-intrusive cases shows significant differences in the vicinity of the Himalayas and the SWTP. During the intrusive case, moist air and condensate lofted by strong upward motion were advected northward to the SWTP in the 400–600 hPa layer. By contrast, advection at these levels was directed southward during the non-intrusive case. Upslope transport also tends to be enhanced during intrusive events (Supplementary Fig. 8), but the sensitivity simulations described above indicate that rainfall during this event depended primarily on simultaneous increases in up-and-over transport. Contrasts in upper level circulation and rainfall between these two cases are also reproduced by the high-Asia refined analysis (Supplementary Fig. 9).

Additional evidence for the moisture source associated with intrusive CSs over the SWTP can be gleaned from the stable oxygen isotope ratio ($\delta^{18}O$) in precipitation. This tracer has been widely used to infer key characteristics of precipitation, including moisture source regions, transport history and the degree of rainout[24–27]. We use event-based observations of $\delta^{18}O$ in precipitation from four stations in the Tibetan Network for Isotopes in Precipitation (TNIP) and one station operated by the Institute of TP Research (ITP) during 1998–2007 to evaluate the contributions of different moisture sources and transport pathways to summertime rainfall over the SWTP. A total of 98 summertime rainfall events (59 intrusive cases and 39 non-intrusive cases) were selected corresponding to the identified CSs (see Supplementary Table 3 for details). As shown by previous studies, precipitation amount and $\delta^{18}O$ are inversely related in this region (Fig. 4), with a lower ratio of heavy isotopes during events with higher precipitation amounts[27]. Moreover, we find that intrusive CSs were consistently associated with larger precipitation amounts and lower values of $\delta^{18}O$ than non-intrusive CSs. This difference is consistent with differences in moisture source and transport effects, but is also potentially consistent with the well-known amount effect, which also manifests as an inverse relationship between $\delta^{18}O$ and

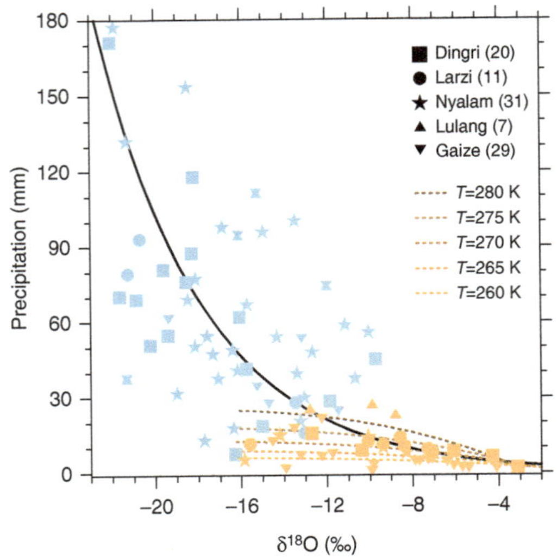

Figure 4 | Isotopic content of precipitation. Scatter plot of summertime precipitation amount and $\delta^{18}O$ in precipitation for 98 rainfall events at four TNIP observation stations and one ITP observation station during 1998–2007. Blue markers indicate values observed during intrusive CSs (59 events), while orange markers indicate value observed during non-intrusive CSs (39 events). The black solid line indicates the exponential trend line for all the markers. The coloured dashed lines indicate theoretical estimates of the amount effect in local convection with different initial temperature (ranges from 260 to 280 K) calculated by isotope theoretical model.

precipitation amount[36,37]. To distinguish between these two possibilities, we generate theoretical estimates of the amount effect in local convection (dashed curves in Fig. 4; see Methods for details) using a model intended to represent rain events due to either upslope transport or local recycling (which both deliver moisture to the atmospheric surface layer above the SWTP). The theoretical estimates are consistent with the observations collected during non-intrusive cases (when local convection is expected to dominate), but diverge sharply from the observations collected during intrusive cases. Ratios of $\delta^{18}O$ in summer precipitation over the SWTP for intrusive cases are much larger than expected based on the theoretical curves. Previous studies have likewise shown that water vapour detrained from tropical convective clouds is more enriched in heavy isotopes than expected based on theoretical estimates[38]. The isotopic observations collected during intrusive cases are therefore consistent with efficient moisture transport from the tropical ocean via the up-and-over pathway, and inconsistent with heavy convective precipitation that depends on moisture sources that deliver water vapour primarily to the surface layer (that is, upslope flow and local evapotranspiration).

Composite and mean precipitation profiles. To extend our analysis beyond event scales, we construct composites for all intrusive and non-intrusive CSs during the period 1998–2005 (Fig. 5). Intrusive CSs account for approximately half of total summertime precipitation over both CEI (46%) and the SWTP (47%), with small interannual fluctuations. Precipitation occurs over the SWTP during 90% of the intrusive CSs, while rainfall is mostly confined to the Indian subcontinent during non-intrusive CSs. Non-intrusive CSs account for ~35% of total summer rainfall over CEI, but with relatively little rainfall over the SWTP (17% of total summertime precipitation). Similar to the example cases discussed above, both composites have a low-level cyclonic circulation over CEI. CSs are more intense and extend further northwestward in the intrusive composite (Fig. 5a), while CSs and

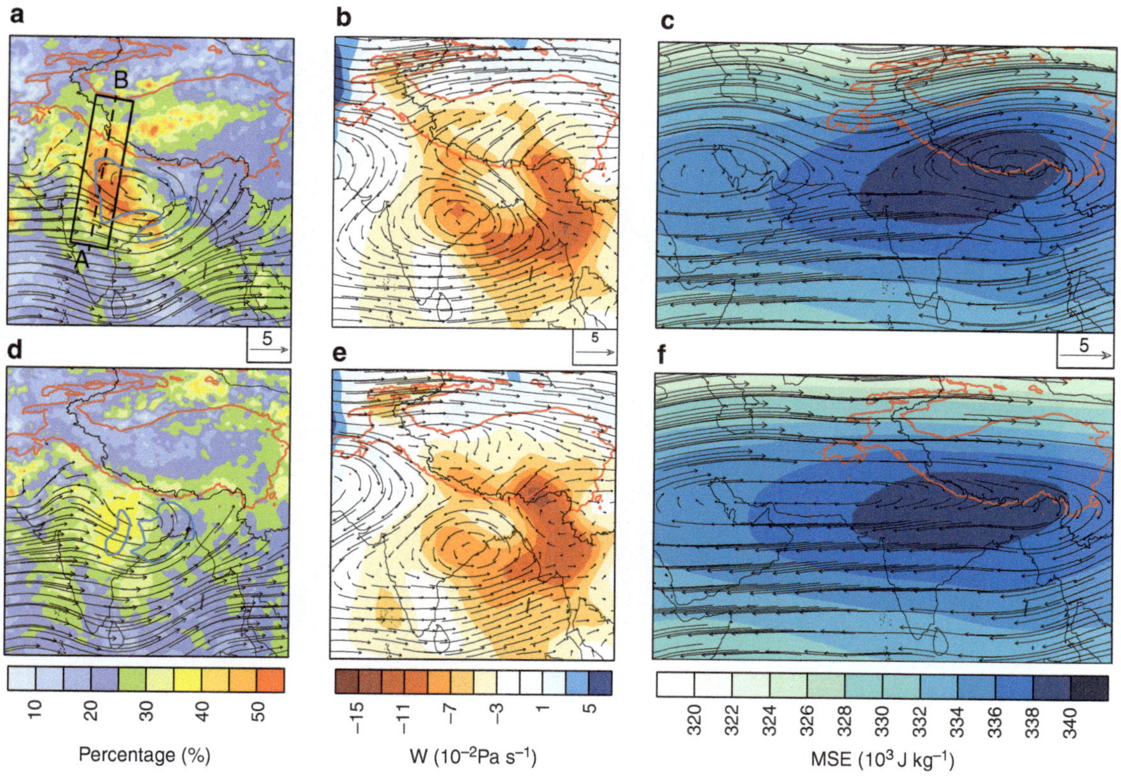

Figure 5 | Composite precipitation and circulation characteristics. (**a**) Rainfall percentage overlaid with 850 hPa winds, (**b**) vertical velocity and winds at 500 hPa and (**c**) moist static energy and winds at 300 hPa for all intrusive CSs (227 events) during 1998–2005. (**d–f**) As in **a–c** but for non-intrusive CSs (229 events). The region is extended westward in **c**,**f** to show the Iran High. Blue contours in **a**,**d** indicate regions with >10 storm centres. The red isoline is the 2,500 m elevation contour. The black rectangle in **a** is used in subsequent transect calculations.

the monsoon trough are typically shifted slightly eastward towards the northwestern Bay of Bengal in the non-intrusive composite (Fig. 5d). The area of heavy rainfall during intrusive CSs extends towards the northeast in a band that stretches from CEI to the SWTP, but the area of heavy rainfall during non-intrusive CSs is smaller and mostly confined to CEI. Northward cross-barrier flow from CEI to the SWTP at 500 hPa is apparent in the intrusive composite (Fig. 5b), and contrasts with southward flow from the SWTP to CEI in the non-intrusive composite (Fig. 5e). An eastward centre of high pressure at 300 hPa straddles the southern periphery of the Plateau in the intrusive composite (Fig. 5c), while anticyclonic winds associated with the Iran High are much stronger and extend further eastward in the non-intrusive composite (Fig. 5f), suggesting that intrusive and non-intrusive CSs preferentially occur during the TP and Iranian Plateau modes of the upper tropospheric anticyclone, respectively[39]. This difference means that the upper level circulation over the SWTP during non-intrusive CSs is dominated by flow from the mid-latitudes rather than by flow from the Indian subcontinent. Moist static energy, which reflects the effects of convective heating and moistening, is likewise larger in the intrusive composite, with a southwest–northeast tilt that extends further over the SWTP[30,40].

We apply the same method to create composite distributions of moisture anomalies in the middle and upper troposphere using twice-daily AIRS retrievals during 2002–2005 (Fig. 6). The intrusive composite features a tongue of positive moisture anomalies that extends northeastward from CEI to the SWTP in the middle and upper troposphere (Fig. 6a–c), consistent with the prevailing cross-barrier (southwesterly) flow. By contrast, positive moisture anomalies are confined mainly to the southern side of the Himalayas during non-intrusive CSs (Fig. 6d–f), consistent with the anomalous eastward extension of the Iran High and the associated anticyclonic flow. In summary, strong southwesterly flow promotes moisture transport over the SWTP during intrusive CSs, while strong northeasterly flow inhibits moisture transport over the SWTP during non-intrusive CSs.

Figure 7 shows mean precipitation along a transect perpendicular to the Plateau boundary for the intrusive and non-intrusive CS composites. Intrusive and non-intrusive CSs both result in large amounts of precipitation over the Indian subcontinent, but intrusive CSs produce much more precipitation over the SWTP than non-intrusive CSs. In particular, while orographic precipitation along the south slope of the plateau is apparent during both intrusive and non-intrusive CSs, precipitation extends northward into the Plateau during intrusive CSs but drops sharply at altitudes >2.5 km during non-intrusive CSs. This difference highlights the importance of up-and-over transport to precipitation over the SWTP. The up-and-over route is open during intrusive CSs, and enhances rainfall by delivering moisture and condensate to the inner part of the SWTP. By contrast, the up-and-over route is effectively closed during non-intrusive CSs, and precipitation rarely penetrates beyond the flank of the plateau.

Discussion

Our results emphasize that mid-tropospheric advection of convectively lofted hydrometeors and moisture from CSs in CEI northward over the Himalayas (that is, 'up-and-over' transport) is critically important for summertime precipitation variability over the SWTP. Relative to conventional upslope moisture transport, which is frequent but inefficient, up-and-over transport is more efficient but less frequent. Up-and-over

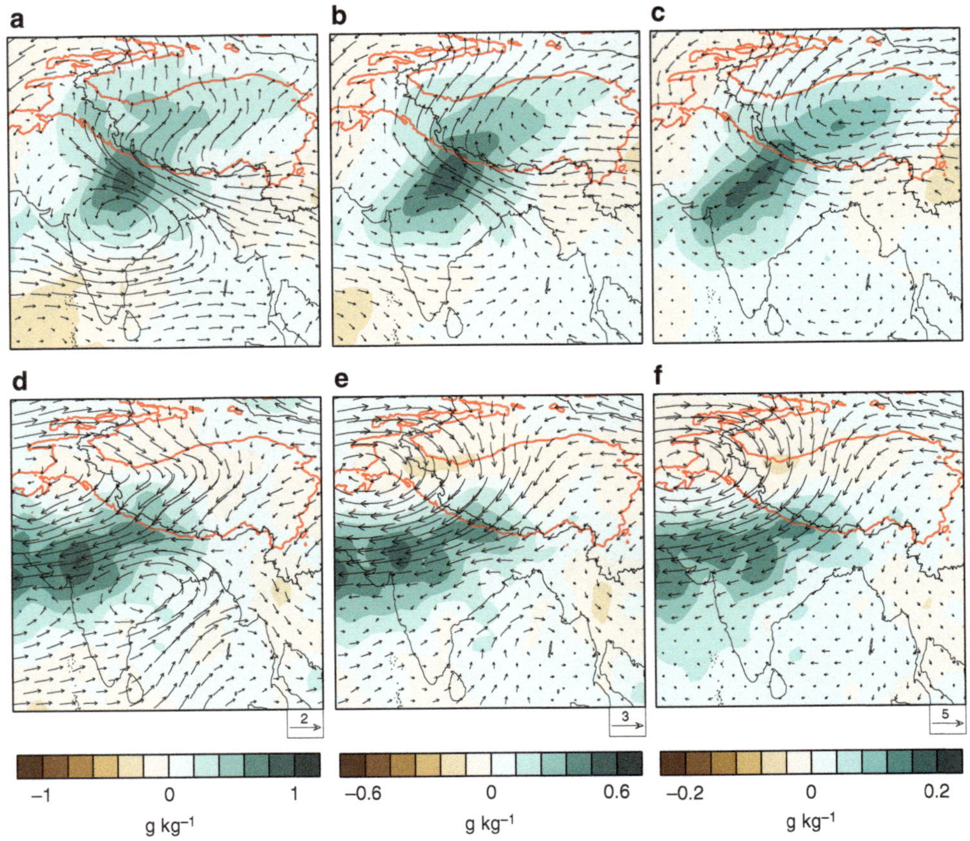

Figure 6 | Composite moisture and circulation characteristics. Mid- and upper level moisture anomalies from AIRS overlaid with wind anomalies from ERA-Interim at (**a**) 500 hPa, (**b**) 400 hPa and (**c**) 300 hPa for all intrusive CSs (103 events) during 2002–2005. (**d–f**) As in **a–c** but for non-intrusive CSs (90 events). The red isoline is the 2,500 m elevation contour.

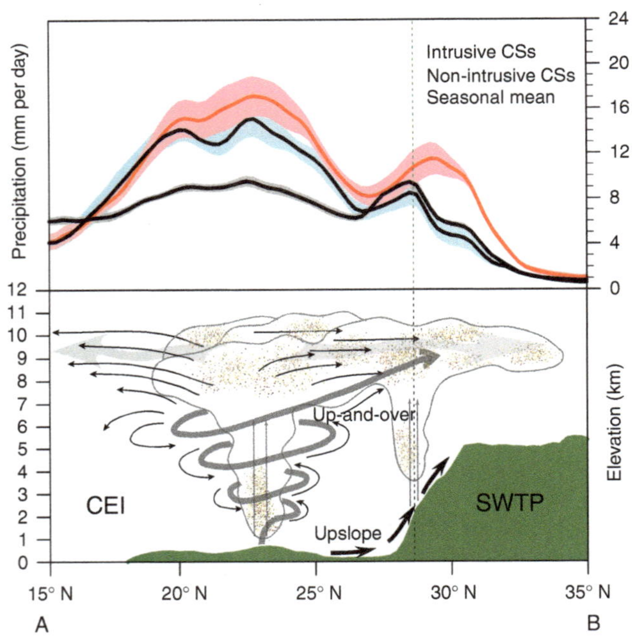

Figure 7 | Precipitation profiles and a schematic of the up-and-over transport. Mean precipitation averaged over the black box marked in Fig. 5a for the intrusive CSs (red), non-intrusive CSs (blue) and total summer rainfall over 1998–2005 based on TRMM data. The shaded regions represent 1 s.e.m. The dashed line indicates the periphery of the southwestern Tibetan Plateau (SWTP), defined as topography 2,500 m above mean sea level.

transport occurs on ~23% of days during summer and accounts for approximately half of total summer rainfall over the SWTP. Local convection also occurs frequently over the SWTP in summer, and follows a clear diurnal cycle with an afternoon peak. Episodes of local convection often follow intrusive CSs (Fig. 2b), when up-and-over transport is particularly strong. This lag may be explained by the importance of precipitation recycling processes[13,20] in these episodes. Up-and-over transport, complemented by upslope transport, and associated rainfall represent the ultimate source of much of the moisture that participates in these recycling processes, and therefore contribute to total precipitation over the SWTP even on days when this transport is weak or non-existent. In this sense, up-and-over transport associated with intrusive CSs over India fundamentally determines the overall wetness of the SWTP: without this transport, the SWTP would be substantially drier. Although most precipitation over the SWTP occurs during summer, future studies should investigate the role of precipitation variability in other seasons (especially that associated with mid-latitude systems during winter) in the regional water budget.

We present strong evidence that intrusive CSs over the Indian subcontinent directly and significantly influence rainfall over the SWTP at both seasonal and event scales. This result has important implications for understanding and evaluating projections of future changes in regional precipitation to climate forcings, and for assessing the resilience of the biosphere, hydrosphere and cryosphere in this region. Confidence in projected changes in regional rainfall is notoriously difficult to achieve, but is vitally important for planning adaptation strategies, managing water resources and establishing environmental protections. Summer precipitation is projected to increase over both CEI and the SWTP under a warmer climate[41], which would benefit 'summer accumulation' type glacier growth in this region[3]. Although changes in both regions remain uncertain, the importance and robust nature of the relationship between these two regions establishes a framework for evaluating model performance in this region, eliminating unrealistic simulations and ultimately bolstering confidence in projections of future change.

Methods

Core data. APHRODITE[18] is a long-term continental-scale daily gridded precipitation data set over Asia. These data have previously been used in studies of precipitation characteristics over both India[7] and the TP. In addition to APHRODITE, we use observational estimates of precipitation from the TRMM Multi-satellite Precipitation Analysis[19], GPCP[42], CMAP[43] and model output from CMIP5 historical simulations and future projections. Moisture fields are based on satellite retrievals from AIRS[44,45] and reanalysis output from ERA-Interim[35]. ERA-Interim reanalysis and forecast fields are also used for dynamic and thermodynamic conditions in both the case study and the construction of composites. Additional details are provided in Supplementary Tables 1 and 2.

Time series correlation coefficient. We calculate correlation coefficients using normalized anomalies (anomalies from the mean over the entire record divided by the s.d. over the entire record) of summer rainfall over CEI and the SWTP.

CS identification and categorization. We identify and classify CSs using 3-h CLAUS[46] and TRMM[19] data. First, CSs over CEI are detected based on a conventional definition[13] as contiguous areas larger than 10,000 km^2 with CLAUS brightness temperatures (T_b) colder than 219 K. Second, large rainfall events[8] are identified as days with rainfall amounts larger than 6 mm per day. CS days are defined as days that satisfy both of the above criteria over CEI. We then further classify CS days into two types based on mid-tropospheric (600–400 hPa) wind directions within a region between 78 and 82° E longitude and 26 and 32° N latitude. CS days with northward flow stronger than seasonal mean (1 m s^{-1}) in the mid-troposphere over the mountains are classified as intrusive, while CS days with southward flow in the mid-troposphere over the mountains are classified as non-intrusive. Finally, we use 850 hPa relative vorticity to track the location and intensity of each CS.

WRF simulations. WRF model simulations are conducted using one-way nesting approach via the ndown approach (Detailed model setup is listed in Supplementary Table 4). A high-resolution (9 km) inner domain is nested within a coarse-resolution (27 km) domain (Supplementary Fig. 3), with the lateral boundary conditions of the inner domain obtained from hourly outputs from the outer domain simulations. This approach enables us to alter hydrometeor and water vapour fluxes at the boundaries of the inner domain. In addition to the control simulation, three other sensitivity simulations are conducted to differentiate and quantify the contributions of moisture fluxes in different vertical layers to the rainfall amount over the SWTP. Advection of water vapour is reduced by 50% in the whole air column, above 5 km AGL and above 2.5 km AGL, respectively, in these three sensitivity simulations. Radiosonde measurements indicate that the middle and upper troposphere over CEI are nearly saturated (relative humidities of 70–100%) on CS days, whereas non-convective days are rather dry with relative humidities of 20–50% above the boundary layer (Supplementary Fig. 10). For simplicity, we reduce the specific humidity above 5 km AGL by 50% to mimic the situation without convection. This simulation is intended to quantify the contribution of up-and-over transport. The second simulation, in which we reduce moisture advection by 50% above 2.5 km AGL, is intended to quantify the contribution of upslope transport. The final simulation, in which we reduce moisture advection throughout the column by 50%, is intended to quantify the total contribution of transport across the southern boundary.

Isotope observations. The TNIP includes more than 30 observational stations on the TP, with the longest records dating back to the 1960s and the most recent records collected in 2007 (ref. 27). We select event-based precipitation data from four TNIP stations and one ITP station that measured $\delta^{18}O$ ratios in precipitation over the southern or SWTP during 1998–2007. We select the station at Lulang, located in the southeastern TP, to compare differences in moisture sources for rainfall over the SWTP and the southeastern TP. A total of 98 summer rainfall events corresponding to our categorized CSs were observed at the five selected stations (Supplementary Table 3). Additional details regarding these measurements have been provided by Yao *et al*[27].

Isotope theoretical model. The theoretical amount effect is modelled by treating local convection as a Rayleigh distillation process operating on a single parcel, with condensate immediately removed as precipitation after it forms[36]. Isotopic fractionation is assumed to occur at thermodynamic equilibrium. The liquid–ice transition is neglected for simplicity; its inclusion does not affect the qualitative nature of the results. The model includes five free parameters, the surface pressure (set to be 560 hPa) and the initial temperature (ranges from 260 to 280 K), relative humidity (set to 80%), vapour isotopic composition ($\delta^{18}O_v = -15‰$) and mass of the parcel. In the absence of suitable constraints, the last two parameters are used to fit the model to the observations. With the exception of $\delta^{18}O_v$, which acts to shift the model output left and right to lower or higher values of $\delta^{18}O$, variation of these parameters only affects the quantitative end points and does not affect the qualitative output of the model. The shapes of the theoretical curves, which are the basis for our conclusion that moisture source and transport effects govern the differences between intrusive and non-intrusive cases, are unaffected by these variations. The most important source of uncertainty in this model and its interpretation is the role of post-condensation exchange, as equilibration and partial evaporation of raindrops falling through an unsaturated cloud layer can have important impacts on the isotopic composition of precipitation at the ground[37]. These effects are dependent on raindrop size, relative humidity below the cloud base and the depth of the unsaturated layer, and are a key component of the overall amount effect. Reanalysis data indicate that the relative humidity of near-surface air over the SWTP is generally higher during intrusive cases than during non-intrusive cases. Given similar raindrop sizes, post-condensation exchange will therefore tend to increase $\delta^{18}O$ in precipitation more during non-intrusive cases than during intrusive cases; however, the opposite would need to be true for post-condensation exchange to explain the deviation of the intrusive observations from the theoretical curve. Although the effects of post-condensation exchange likely cause us to overestimate $\delta^{18}O_v$ relative to its actual value, they are unlikely to affect our main conclusion.

Backward trajectory analysis. Trajectories are calculated using version 9.02 of the FLEXPART particle dispersion model[34] driven by ERA-Interim reanalysis and forecast fields[35] at 1° × 1° horizontal resolution and 3-h temporal resolution. About 50,000 particles filling the domain 30–33° N, 80–84° E and 100–8,000 m AGL are initialized at 18UTC 12 September 2002 and tracked backward in time until 12UTC 8 September 2002 (102 h), with particle locations output hourly. To properly account for the complex topography and relative frequency of moist convection in this region, we conduct these simulations with parametrizations of moist convection, sub-grid scale turbulence and mesoscale velocity fluctuations enabled.

References

1. Xu, X., Lu, C., Shi, X. & Gao, S. World water tower: An atmospheric perspective. *Geophys. Res. Lett.* **35,** L20815 (2008).
2. Yao, T. *et al.* Different glacier status with atmospheric circulations in Tibetan Plateau and surroundings. *Nat. Clim. Change* **2,** 663–667 (2012).
3. Mölg, T., Maussion, F. & Scherer, D. Mid-latitude westerlies as a driver of glacier variability in monsoonal High Asia. *Nat. Clim. Change* **4,** 68–79 (2014).
4. Tao, S.-Y. & Ding, Y.-H. Observational evidence of the influence of the Qinghai-Xizang (Tibet) Plateau on the occurrence of heavy rain and severe convective storms in China. *Bull. Am. Meterol. Soc.* **62,** 23–30 (1981).
5. Dorji, T. *et al.* Plant functional traits mediate reproductive phenology and success in response to experimental warming and snow addition in Tibet. *Global Change Biol.* **19,** 459–472 (2013).
6. Maussion, F. *et al.* Precipitation seasonality and variability over the Tibetan Plateau as resolved by the High Asia Reanalysis. *J. Clim.* **27,** 1910–1927 (2014).
7. Turner, A. G. & Annamalai, H. Climate change and the South Asian summer monsoon. *Nat. Clim. Change* **2,** 587–595 (2012).
8. Singh, D., Tsiang, M., Rajaratnam, B. & Diffenbaugh, N. S. Observed changes in extreme wet and dry spells during the South Asian summer monsoon season. *Nat. Clim. Change* **4,** 456–461 (2014).
9. Chen, B., Xu, X., Yang, S. & Zhang, W. On the origin and destination ofatmospheric moisture and air mass over the Tibetan Plateau. *Theoret. Appl. Climatol.* **110,** 423–435 (2012).
10. Annamalai, H. & Slingo, J. M. Active/break cycles: diagnosis of the intraseasonal variability of the asian summer monsoon. *Clim. Dynam.* **18,** 85–102 (2001).
11. Wu, G. & Zhang, Y. Tibetan Plateau forcing and the timing of the monsoon onset over South Asia and the South China Sea. *Mon. Weather Rev.* **126,** 913–927 (1998).
12. Yang, K. *et al.* Recent climate change over the Tibetan Plateau and their impacts on energy and water cycle: A review. *Glob. Planet. Change* **112,** 79–91 (2014).
13. Sugimoto, S. & Ueno, K. Formation of mesoscale convective systems over the eastern Tibetan Plateau affected by plateau-scale heating contrast. *J. Geophys. Res.* **115,** D16105 (2010).

14. Wang, B. *et al.* Northern hemisphere summer monsoon intensified by mega-El nino/southern oscillation and atlantic multidecadal oscillation. *Proc. Natl Acad. Sci. USA* **110,** 5347–5352 (2013).
15. Bollasina, M. A., Ming, Y. & Ramaswamy, V. Anthropogenic aerosols and the weakening of the South Asian summer monsoon. *Science* **334,** 502–505 (2011).
16. Lau, K. M., Kim, M. K. & Kim, K. M. Asian summer monsoon anomalies induced by aerosol direct forcing: the role of the Tibetan Plateau. *Clim. Dynam.* **26,** 855–864 (2006).
17. Ramanathan, V. *et al.* Atmospheric brown clouds: Impacts on South Asian climate and hydrological cycle. *Proc. Natl Acad. Sci. USA* **102,** 5326–5333 (2005).
18. Yatagai, A. *et al.* APHRODITE: constructing a long-term daily gridded precipitation data set for Asia based on a dense network of rain gauges. *Bull. Am. Meteorol. Soc.* **93,** 1401–1415 (2012).
19. Huffman, G. J. *et al.* The TRMM Multisatellite Precipitation Analysis (TMPA): Quasi-global, multiyear, combined-sensor precipitation estimates at fine scales. *J. Hydrometeorol.* **8,** 38–55 (2007).
20. Liu, X., Bai, A. & Liu, C. Diurnal variations of summertime precipitation over the Tibetan Plateau in relation to orographically-induced regional circulations. *Environ. Res. Lett.* **4,** 045203 (2009).
21. Xie, S., Xu, H., Saji, N. H., Wang, Y. & Liu, W. Role of narrow mountains in large-scale organization of Asian monsoon convection. *J. Clim.* **19,** 3420–3429 (2005).
22. Numaguti, A. Origin and recycling processes of precipitating water over the Eurasian continent: experiments using an atmospheric general circulation model. *J. Geophys. Res.* **104,** 1957–1972 (1999).
23. Yang, M., Yao, T., Wang, H., Gou, X. & Tian, L. Estimating the criterion for determining water vapour sources of summer precipitation on the northern Tibetan Plateau. *Hydrol. Process.* **20,** 505–513 (2005).
24. Tian, L. *et al.* Oxygen-18 concentrations in recent precipitation and ice cores on the Tibetan Plateau. *J. Geophys. Res.* **108,** 4293 (2003).
25. Tian, L. *et al.* Stable isotopic variations in west China: A consideration of moisture sources. *J. Geophys. Res.* **112,** D10112 (2007).
26. Kurita, N. & Yamada, H. The role of local recycling evaluated using stable isotope data from over the middle of the Tibetan Plateau during the monsoon season. *J. Hydrometeor.* **9,** 760–765 (2013).
27. Yao, T. *et al.* A review of climatic controls on $\delta^{18}O$ in precipitation over the Tibetan Plateau: Observations and simulations. *Rev. Geophys.* **51,** 525–548 (2013).
28. Gadgil, S. The Indian monsoon and its variability. *Annu. Rev. Earth Pl. Sc.* **31,** 429–467 (2003).
29. Romatschke, U. & Houze, Jr R. A. Characteristics of precipitating convective systems in the South Asian monsoon. *J. Hydrometeor.* **12,** 3–26 (2011).
30. Chen, T. & Weng, S. Interannual and intraseasonal variations in monsoon depressions and their westward-propagating predecessors. *Mon. Wea. Rev.* **127,** 1005–1020 (1999).
31. Krishnamurthy, V. & Shukla, J. Intraseasonal and seasonally persisting patterns of Indian monsoon rainfall. *J. Clim.* **20,** 3–20 (2007).
32. Prajeesh, A. G., Ashok, K. & Rao, D. V. B. Falling monsoon depression frequency: A Gray-Sikka conditions perspective. *Sci. Rep.* **3,** 2989 (2013).
33. Molnar, P., Boos, W. R. & Battisti, D. S. Orographic controls on climate and paleoclimate of Asia: thermal and mechanical roles for the Tibetan Plateau. *Annu. Rev. Earth Planet. Sci.* **38,** 77–102 (2010).
34. Stohl, A. *et al.* Technical note: the Lagrangian particle dispersion model FLEXPART version 6.2. *Atmos. Chem. Phys.* **5,** 2461–2474 (2005).
35. Dee, D. P. *et al.* The ERA-Interim reanalysis: configuration and performance of the data assimilation system. *Quart. J. R. Meteorol. Soc.* **137,** 553–597 (2011).
36. Dansgaard, W. Stable isotopes in precipitation. *Tellus* **16,** 436–468 (1964).
37. Lee, J.-E. & Fung, I. 'Amount effect' of water isotopes and quantitative analysis of post-condensation processes. *Hydrol. Process.* **22,** 1–8 (2007).
38. Webster, C. R. & Heymsfield, A. J. Water isotope ratios D/H, 18O/16O, 17O/16O in and out of clouds map dehydration pathways. *Science* **302,** 1742–1745 (2003).
39. Zhang, Q., Wu, G. & Qian, Y. The bimodality of the 100 hPa South Asia high and its relationship to the climate anomaly over East Asia in summer. *J. Meteorol. Soc. Japan. Ser. II* **80,** 733–744.
40. Boos, W. R. & Kuang, Z. Dominant control of the South Asian monsoon by orographic insulation versus plateau heating. *Nature* **463,** 218–223 (2010).
41. Collins, M. *et al.* in: *Climate Change 2013: The Physical Science Basis* (eds Stocker, T. F. *et al.*) (Cambridge Univ. Press, 2013).
42. Adler, R. F. *et al.* The version 2 Global Precipitation Climatology Project (GPCP) monthly precipitation analysis (1979-Present). *J. Hydrometeor.* **4,** 1147–1167 (2003).
43. Xie, P. & Arkin, P. A. Global precipitation: a 17-year monthly analysis based on gauge observations, satellite estimates, and numerical model outputs. *Bull. Am. Meteor. Soc.* **78,** 2539–2558 (1997).
44. Aumann, H. H. *et al.* AIRS/AMSU/HSB on the Aqua mission: Design, science objectives, data products, and processing systems. *IEEE T. Geosci. Remote* **41,** 253–264 (2003).
45. Tian, B. *et al.* AIRS/AMSU/HSB version 6 level 3 product user guide. tech. rep., jet propulsion laboratory, Pasadena. Available at http://disc.sci.gsfc.nasa.gov/AIRS/documentation/v6_docs/v6releasedocs-1/V6_L3_User_Guide.pdf (2013).
46. Hodges, K. I., Chappell, D. W., Robinson, G. J. & Yang, G. An improved algorithm for generating global window brightness temperatures from multiple satellite infrared imagery. *J. Atmos. Oceanic Technol.* **17,** 1296–1312 (2000).

Acknowledgements

We gratefully acknowledge the APHRODITE, TRMM, GPCP, CMAP, CLAUS, AIRS and ERA-Interim projects for preparing the precipitation, brightness temperature and other atmospheric data, and the associated data repositories (listed in Supplementary Table 2) for providing public access to these data. We further thank the climate modelling groups for producing the model output and making it available. This work was supported by the Ministry of Science and Technology of China (Grant 2013CBA01805 and 2014CB441303). L.W. is supported by the National Natural Science Foundation of China (Grant 41322001) and the Hundred Talents Program of Chinese Academy of Sciences. L.T. is supported by the National Natural Science Foundation of China (Grant 41530748).

Author contributions

W.D. and Y.Lin. developed the essential research idea. W.D. performed the analysis and wrote the initial draft of the paper. Y.Lin., Y.X. and J.S.W. contributed editorial suggestions and supporting analyses. L.T. and L.W. provided the isotope measurements and their interpretation. Y.M., B.W., Y.Luo, W.H., J.H., Y.P., and F.X. contributed to the interpretation of the results and the preparation of the manuscript.

Additional information

Competing financial interests: The authors declare no competing financial interests.

23

Ecological responses to experimental glacier-runoff reduction in alpine rivers

Sophie Cauvy-Fraunié[1,2,3], Patricio Andino[2], Rodrigo Espinosa[2], Roger Calvez[4], Dean Jacobsen[2,5] & Olivier Dangles[1,2]

Glacier retreat is a worldwide phenomenon with important consequences for the hydrological cycle and downstream ecosystem structure and functioning. To determine the effects of glacier retreat on aquatic communities, we conducted a 4-year flow manipulation in a tropical glacier-fed stream. Compared with an adjacent reference stream, meltwater flow reduction induces significant changes in benthic fauna community composition in less than 2 weeks. Also, both algal and herbivore biomass significantly increase in the manipulated stream as a response to flow reduction. After the flow reduction ceased, the system requires 14-16 months to return to its pre-perturbation state. These results are supported by a multi-stream survey of sites varying in glacial influence, showing an abrupt increase in algal and herbivore biomass below 11% glacier cover in the catchment. This study shows that flow reduction strongly affects glacier-fed stream biota, prefiguring profound ecological effects of ongoing glacier retreat on aquatic systems.

[1] Institut de Recherche pour le Développement (IRD), UMR EGCE, IRD-247 CNRS-UP Sud-9191, Avenue de la Terrasse, Bâtiment 13, 91198 Gif-sur Yvette, France. [2] Pontificia Universidad Católica del Ecuador, Escuela de Ciencias Biológicas, Avenida 12 de Octubre, Quito 170150, Ecuador. [3] IRSTEA, UR MALY, 5 rue de la Doua, Villeurbanne 69100, France. [4] Institut de Recherche pour le Développement (IRD), UMR G-EAU, 361 Rue Jean-François Breton, BP 5095, 34196 Montpellier, France. [5] Freshwater Biological Laboratory, Department of Biology, University of Copenhagen, Universitetsparken 4, 2100 Copenhagen, Denmark. Correspondence and requests for materials should be addressed to S.C.-F. (email: sophie.cauvy@gmail.com) or to O.D. (email: olivier.dangles@ird.fr).

Worldwide recession of glaciers foreshadows global-scale shifts of alpine river flow regimes[1,2]. While most glaciers have been shrinking since the end of the Little Ice Age, around 160 years ago, shrinking rates have accelerated over the last 3–4 decades[3], leading to an increased threat to water resources[1], alpine biodiversity and associated ecosystem services[4–6]. Indeed, glacial contributions to river flow create a diversity of habitats that are important for a large number of plants and animals[7], and they support economically important wetlands and fisheries[6]. Moreover, climate-driven fluctuations in discharge following glacier melting are expected to profoundly affect ecological processes in alpine rivers, which are strongly influenced by hydrological and thermal regimes[8,9].

A major issue in elucidating the ecological effects of shrinking glaciers relates to predicting biological community responses to modifications in glacier runoff (for example, lower volumes of ice melt water) and variability (for example, more prolonged and frequent low flows[10]). After a critical threshold of reduction in ice volume, annual glacial runoff should decrease, inducing an increase in low-flow period frequency and intensity followed by continuous low flows until the complete loss of glacier outflow when the glacier disappears[2]. Pronounced decline in glacial discharge observed worldwide is one of the most concerning symptoms of glacier shrinking with unclear consequence on biodiversity[4]. While several studies have shown that many species may become endangered or extinct if glacial meltwater inputs are reduced or entirely lost[4,6,7], we still know little about mechanisms driving the response of aquatic biota to glacier retreat so that we can better evaluate their response to global warming[5,6]. A whole-ecosystem manipulation of meltwater reduction, simulating the effect of glacier retreat on glacier-fed streams, in which both disturbance effects and return times to pre-disturbance conditions are determined, should provide insights into the ecological response to forthcoming glacial meltwater alteration. Yet such an ecosystem manipulation has never been performed in alpine rivers.

Here we experimentally simulate temporary flow reduction in a glacier-fed stream of the Ecuadorian Andes, a region where a decrease in glacier runoff is expected to be one of the most pronounced worldwide[2]. Using a Before-After-Control-Impacted design[11], we divert one-third of the glacial meltwater of a glacier-fed stream (manipulated) while an adjacent similar stream (reference) remains undisturbed (see Supplementary Fig. 1 for the field manipulation design). The stream manipulation consists of three time intervals: (1) establishment of baseline conditions under unaltered stream flow ($\sim$1 year); (2) diversion of water flow, inducing a water flow reduction in the downstream reach ($\sim$1 year; the upper reach was kept undisturbed); and (3) reset to initial flow conditions to evaluate post-drought recovery trajectories ($\sim$2 years). This experimental design allows us to assess the response of benthic algal and faunal communities to flow reduction, and evaluate the ecosystem resilience, defined as the capacity of the benthic communities to return to their initial configuration after flow disturbance cease[12]. To determine whether our experimental results would be corroborated across a spatial gradient in glacial cover, we analyse the spatial variability in benthic algae and fauna across 33 stream sites, in the same study area. Corroboration could help predict future temporal ecological variability under reduction in glacial influence[13].We show that meltwater flow reduction induces changes in the benthic community, characterized by an increase in biomass of benthic algae and macroinvertebrate herbivores. Changes in benthic community composition after flow reduction is 30 times faster than the time required to return to pre-disturbance state after flow recovery. We discuss the ecosystem capacity to recover from meltwater flow reduction and emphasize the importance of understanding how glacier retreat will affect whole alpine freshwater ecosystems in a rapidly changing world.

Results

Abiotic and biotic changes under flow alteration. Our experimental water diversion significantly reduced the mean daily flow by 31% in the manipulated stream and the flow coefficient of variation by 20% (Fig. 1a and Supplementary Table 1). The most marked abiotic changes in the manipulated reach were an

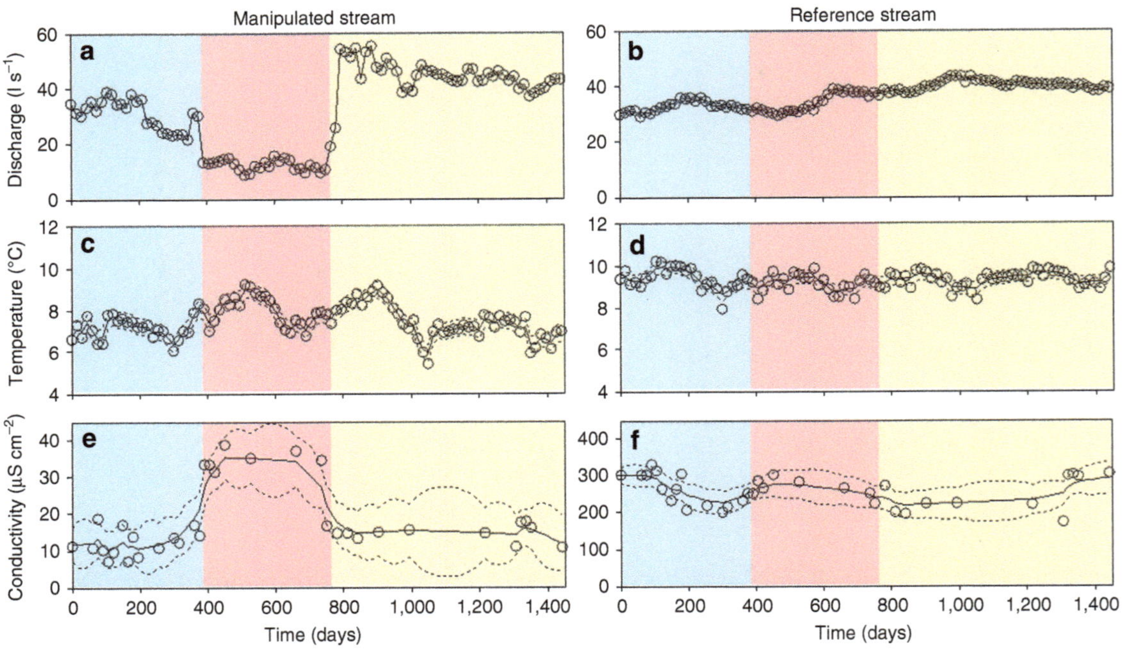

Figure 1 | Environmental time series. Time series (open dots), smoothed-state estimates (solid lines) and state 95% confidence level (dotted lines) of discharge (**a**,**b**), water temperature (**c**,**d**), and conductivity (**e**,**f**) for the downstream reach of the manipulated stream and the reference stream. Blue, red and yellow regions indicate base flow, flow alteration and flow recovery periods, respectively.

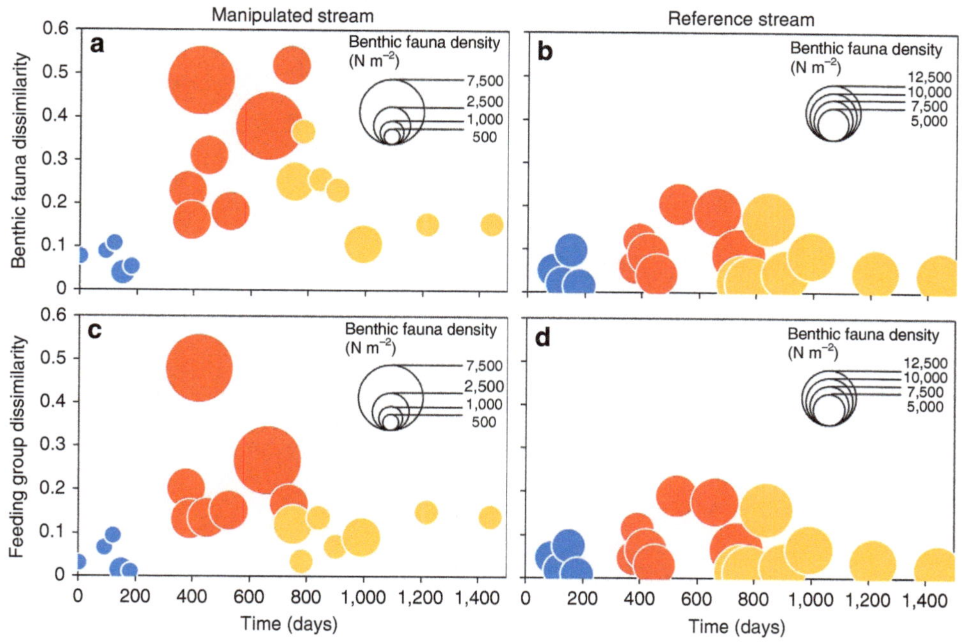

Figure 2 | Temporal variability in benthic fauna and feeding group dissimilarity between reaches of the manipulated and reference stream. Benthic fauna and feeding group dissimilarity between downstream and upstream reaches in the manipulated stream (**a**,**c**) and between subsequent sampling dates in the reference stream (**b**,**d**). The three experimental phases are shown in different colours (blue: base flow; red: flow alteration; yellow: flow recovery). Dot size refers to benthic fauna density. Note difference in dot size in the reference and manipulated streams due to much higher overall density in the reference stream.

increase of 1.4 °C in mean water temperature (Fig. 1c) and a 65% increase in mean conductivity (from 14.4 to 32.2 μS cm^{-2}) (Fig. 1e). These abiotic changes were not observed in the reference stream (Fig. 1b,d,f) or in the upstream reach of the manipulated stream (Supplementary Fig. 2). In terms of biota, flow reduction had a strong impact on benthic fauna community composition: dissimilarity between lower and upper stream reach communities rapidly (<15 days) increased following the flow reduction (Fig. 2a). Benthic communities remained relatively stable (that is, dissimilarity below 0.2) in the reference stream, although slight changes in the abundance of a few groups (for example, Hyalellidae, Baetidae and Simuliidae) occurred during the period of low flow in the manipulated stream (for example, increase in Hyalellidae and Simuliidae at day 553, and in Hyalellidae and Baetidae at day 664) and recovery phase (for example, decrease in Hyalellidae and increase in Baetidae at day 864; Fig. 2b). Flow reduction increased benthic fauna density by 6.5 times on average (± 1.8) but had no effect on richness. After flow returned to pre-disturbance values, it took the faunal community 14–16 months to return to its original composition, that is, within the confidence interval limits of the natural variability in taxon assemblages (Fig. 2a,b). Changes in benthic faunal assemblages during flow alteration were mainly due to increased abundances of herbivorous/detritivorous Diptera (Orthocladiinae and Ceratopogonidae), Ephemeroptera (Baetidae) and Coleoptera (Elmidae), which were the most significant indicator taxa of post-flow diversion samples (Simper analysis). Likewise, dissimilarity in feeding trait composition between upstream and downstream reaches of the manipulated stream was higher during flow alteration (Fig. 2c), while the feeding trait composition remained constant in the reference stream (Fig. 2d). In the manipulated stream, herbivores had by far the greatest contribution to dissimilarity in feeding trait composition among experimental phases, with higher densities during the low-flow period. Collector-gather and predator densities were also higher during flow alteration but they had an overall small contribution to the observed trophic changes (Table 1).

Algal and herbivore biomass shifts along flow manipulation. One of the major trophic impacts of the decrease in glacier runoff was the strong increase in the biomass of benthic primary producers (algae) and herbivorous invertebrates (Fig. 3a,c). State-space analyses revealed three significant shifts (large standardized smoothed-state residuals, red stars in Fig. 3a,c) in algal biomass in the manipulated reach: algal biomass increased significantly 2 weeks after flow reduction, decreased 40 days after flow recovery and then remained at intermediate levels up to a final decrease to the initial level, 20 months after flow recovery. Three shifts occurred in herbivore biomass: it increased significantly after 2 weeks of water diversion, reached maximum values at the end of the flow reduction period and decreased to intermediate levels 1 week after flow recovery. Herbivore biomass then slowly decreased to initial levels 14 months after flow recovery. We found no significant shifts in algal and herbivore biomass in the reference stream (Fig. 3b,d). Multivariate autoregressive state-space analysis on algal and herbivore biomass showed significant top-down and bottom-up effects between algae and herbivores in the downstream reach of the manipulated stream (negative effect of herbivores on algae, $B_{h \rightarrow a} = -0.044$ and positive effect of algae on herbivores $B_{a \rightarrow h} = 0.014$, corrected Akaike Information Criterion = 83.0), but not in the reference stream (see Supplementary Notes Part 2: 4 for details).

Ecological shifts across a gradient of glacial influence. These experimental results were supported by our complementary field survey of 33 stream sites covering a gradient in glacial influence. Our data revealed that, between 93.2 and 11.3% of glacier cover in the catchment, both algal and herbivore biomass remained at relatively constant low levels while below that threshold, both

Table 1 | Dissimilarity in feeding trait composition.

Functional feeding groups	Contribution		Mean density		
	BF→FA	FA→FR	BF	FA	FR
A. Manipulated stream					
Herbivores	57.82	28.08	647 (231)	4,325 (2,022)	1,878 (348)
Collector-gatherers	6.26	4.0	43 (25)	447 (322)	169 (83)
Predators	5.71	4.25	60 (43)	451 (357)	244 (246)
Collector-filterers	0.83	0.58	60 (42)	51 (47)	31 (29)
Shredders	0.72	0.49	4 (5)	46 (47)	23 (13)
Overall dissimilarity	71.35	37.4			
B. Reference stream					
Shredders	15.73	14.07	4,969 (245)	7,943 (2,875)	9,847 (1,803)
Herbivores	2.44	2.23	1,203 (235)	1,513 (451)	1830 (245)
Collector-gatherers	1.12	2.73	403 (75)	231 (117)	815 (618)
Predators	0.92	1.01	244 (98)	321 (152)	110 (74)
Collector-filterers	0.67	0.59	111 (23)	150 (220)	87 (71)
Overall dissimilarity	20.88	20.6			

Relative contribution (%) of the five functional feeding groups to the Bray-Curtis dissimilarity (SIMPER analysis) and overall dissimilarity (%) between base flow and flow alteration stages (BF→FA), and flow alteration and flow recovery stages (FA→FR) in the manipulated (A) and reference (B) reach. Mean density (ind m^{-2}) and s.d. of the five functional feeding groups for base flow (BF), flow alteration stages (FA), and flow recovery stages (FR) in the manipulated (A) and reference (B) reach.

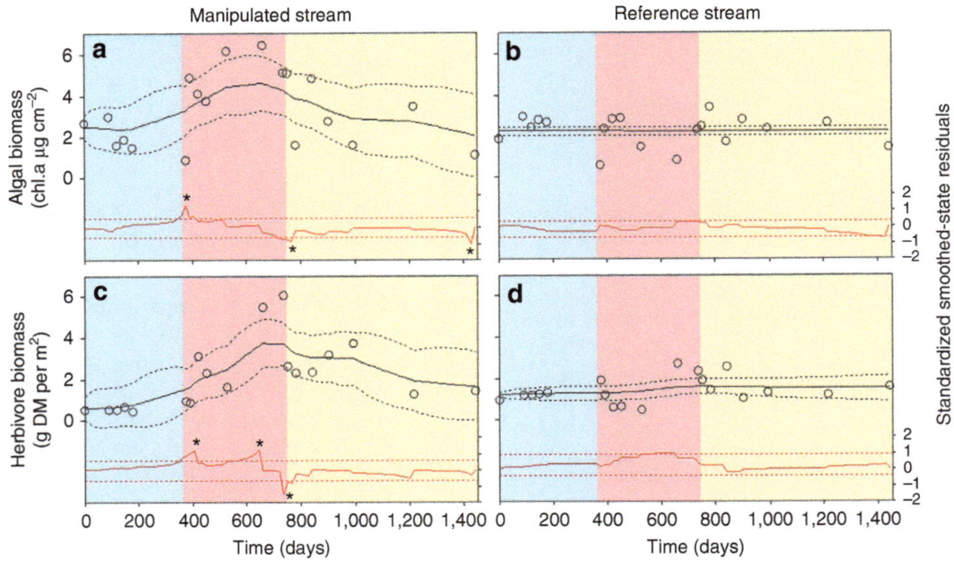

Figure 3 | Ecological time series. Time series (open dots), smoothed-state estimates (solid lines) and state 95% confidence level (dotted lines) of algal biomass (chl.a, chlorophyll a; **a**,**b**) and herbivore biomass (DM = dry mass; **c**,**d**) for the downstream reach of the manipulated stream and the reference stream. Red curves represent standardized smoothed-state residuals from state-space models for algae and herbivores time series. Dashed red lines are the 95% confidence intervals for a *t*-distribution. Red stars indicate when standardized smoothed-state residuals are beyond the dashed level lines. Blue, red and yellow regions indicate base flow, flow alteration and flow recovery periods, respectively.

algal and herbivore biomass significantly increased (Fig. 4). When plotting our experimental data on Fig. 4 (blue and red dots and squares), we found that algal and herbivore biomass values before and during experimental flow reduction closely matched the pattern observed with the multi-locality survey data.

Discussion

Previous studies on benthic fauna response to reduced flow have commonly found significant changes in the community structure and function characterized by change in density (either increases or decreases[14]), reduction in taxon richness[15] and reorganization of the feeding trait assemblage[16,17]. The impacts are however variable among studied systems (for example, natural flow regime, water quality and the features of the disturbance imposed[18–20]). Yet, manipulative experiments in glacier-fed systems are rare, especially for whole ecosystems. Our experimental design allows us to evaluate potential shifts in biota in response to meltwater reduction and assess ecosystem resilience after flow recovery to identify the ecosystem capacity to recover from periodic low flows. In glacier-fed stream, we expect an increase in both density and taxon richness of benthic fauna, as the reduction in glacier runoff should reduce the environmental harshness characterizing glacial meltwater, thereby allowing more species to establish[4,10,21]. We also expect trophic consequences of these structural changes due to the strong connectance in glacier-fed stream foodwebs[22–24].

While previous surveys have examined spatial and temporal patterns of stream community response to a range of glacial influence[4,6,7,10] our relatively long-term experiment offers fresh

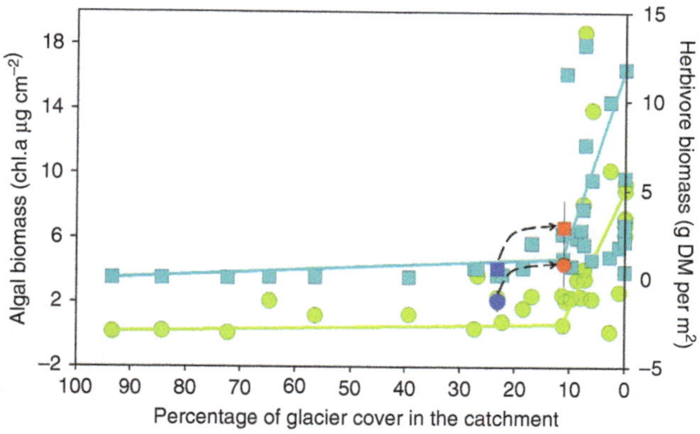

Figure 4 | Spatial variability in algae and herbivores along a glacier influence gradient. Relationship between algal (chl.a, chlorophyll a; green dots) and herbivore biomass (blue squares) as a function of the percentage of glacier cover in the catchment for 33 streams in the Antisana reserve (Ecuador). Douglas-Peucker (DP) models are fitted to algal and herbivore biomass. Data (mean 95% confidence interval) of algal (dots) and herbivore (squares) biomass obtained in our flow-reduction experiment are shown in blue (base flow) and red (low flow). The dashed line represents the shift from low to high level of algal and herbivore biomass between base flow and flow alteration stages.

viewpoints for the prediction of aquatic biota response to glacier retreat. Indeed, our study shows that reduction in glacier runoff would have complex consequences for aquatic biota structure and trophic organization via alterations in resources and trophic interactions. Our combined experimental study and survey supports that a reduction in glacier runoff, and associated glacial influence in mountain streams, beyond a threshold of about 11% of glacier cover in the catchment, would induce an abrupt shift in aquatic biota. As additional field evidence, we observed high algal biomass in natural glacier-fed streams during exceptionally low glacial-shrinking periods (for example, La Niña event, S. Cauvy-Fraunié pers. Obs.), a pattern also reported in temperate glacier-fed rivers[25]. Reductions in both flow and glacial influence imply higher water temperature, conductivity and lower turbidity[26], thereby increasing the overall productivity of stream systems[14,25].

Our experimental flow manipulation revealed that reduced meltwater discharge significantly influenced the structure of benthic communities. In particular, it enhanced herbivore densities, a response that is likely a result of an increase in benthic algal biomass[14]. Flow reduction had no effect on taxon richness, but rather induced benthic taxon replacement characterized by (1) the gain of new taxa (for example, Ceratopogonidae and Empididae) less adapted to the harsh glacial meltwaters[26,27], (2) the loss of taxa (for example, Blephariceridae and Scirtidae) caused by modifications in environmental conditions (for example, loss of fast flow habitat and higher conductivity)[14,27] and an increase in predation and competition pressures[10,14,28]. Specifically, lowered collector–filterer densities, such as Simuliidae, were probably attributable to the loss of fast-flow microhabitats as water velocity controls the rate of food delivery for filter feeders[29]. Our study therefore strongly suggests that the relationship between disturbance, species composition and species interactions will be a central issue in understanding climate change effects on glacier-fed aquatic systems, and probably in a wide variety of glacially influenced ecosystems[30,31].

Finally, another key finding of our work is that the recovery of benthic fauna communities affected by flow reduction was rather slow compared with previous studies with similar or stronger percentage of flow reduction[19,32]. Resilience can however vary greatly according to studied systems and species (for example, their dispersal capacity), the distance to species pools and the magnitude, duration and timing of flow disturbance[33–35]. Nevertheless, compared with the recurrence interval of low-flow events, the time required for recovery might become insufficient for ecosystem resilience between low flows. Indeed, increasingly variable melting flow regimes[2,36] would induce an increase in the frequency of minimum discharge, in terms of both intensity and duration[37]. If low meltwater discharges were rare, benthic communities would have more time to fully recover between each event. In contrast, more frequent reductions in meltwater discharge should prevent the ecosystem from fully recovering between consecutive flow disturbances, with potentially severe but unknown consequences on the aquatic communities[16]. In addition, after a certain amount of glacier volume lost, glacier runoff would undeniably decrease, thereby inducing shifts in benthic communities[27].

Our study provides unique insights into how glacially influenced rivers can be expected to respond to glacier retreat under global warming and therefore urges to further increase our understanding on how climate change will affect whole ecosystems. We found that reduction in glacier runoff rapidly induces a reorganization of benthic taxon assemblage, which requires a 30 times longer period to return to initial state. Resilience delay may be prolonged by competition exclusion by dominant herbivore taxa that established during low-flow conditions. Benthic communities constitute the bulk of primary and secondary production of alpine stream ecosystems[22] and have been widely used to indicate changes in environmental conditions not only in systems receiving meltwater contributions[10,38] but in other aquatic systems[39] as well. Our study also suggests that alpine river biota could be used as key indicators to identify early signals of ecosystem damaging, such as slowing return rates from perturbation. Such shifts are most relevant to concerns of biodiversity conservation and ecosystem services of aquatic systems in both temperate and tropical mountains where many types of human-induced perturbations affect resilience.

Methods

Experimental study streams. We studied two alpine streams, one reference and one manipulated, located at 4,100 m above sea level in the Antisana Reserve, Ecuador (0° 29′ 06″ S, 78° 08′ 31″ W). The two streams were located within the catchment of 'Los Crespos' glacier, which covered an area of about 1.82 km^2 in 2010, with an average cumulative mass balance trend of − 0.6 meters water equivalent per year (m w.e.yr^{-1}) (ref. 3). Both streams shared similar morphological characteristics in terms of slope (0.7–1.2%), width (0.9–1.2 m) and depth (20–50 cm). We monitored abiotic and biotic variables (see below) in two reaches (about 20 m) in the manipulated stream and in one reach in the reference stream. The monitoring started in October 2009 to establish baseline conditions over 1 year. In November 2010 (after 380 days) we built a stone dam on the manipulated stream to divert about one-third of the water, a perturbation that mimics predicted reduction of glacier runoff (2–30% of baseline flows) following glacier shrinkage[2]. Stream flow disturbance effects on biota were assessed over 1 year, and in November 2011 the dam was removed to evaluate stream biota recovery (until October 2013, Supplementary Fig. 1). In the manipulated stream, benthic fauna displacements over the ~200 m separating the two reaches may render them partly dependent. However, unpublished data on upstream migration traps left for 48 h indicated a low level of upstream migration, and we therefore considered the two reaches statistically independent. Our experimental design based on both manipulated and reference streams and upstream–downstream comparisons maximizes the likelihood that the observed patterns are causally related to flow reduction rather than external drivers[11]. The experiment is conducted over 4 years, the equivalent of ca. 20–40 generations of the studied organisms[40].

Abiotic and biotic parameter monitoring. Water pressure loggers (Hobo, Onset Computer Corp., USA) were installed over the study period to record water pressure and temperature every 30 min. One more logger was placed outside the streams at 4,100 m a.s.l. to correct for variation in atmospheric pressure. Pressure data were transformed into discharge values as detailed in a previous study[10]. Conductivity (at 25 °C), pH and dissolved oxygen were measured with portable meters (WTW, Germany) on several visits to the study reaches. At 19 dates spread over the study period (at $t = 0$, 87, 132, 160, 181, 388, 397, 419, 448, 533, 664, 759, 766, 798, 864, 923, 1,011, 1,238 and 1,471 days), algal biomass was quantified by collecting 9 small pebbles (2–4 cm) at random and extracting chlorophyll a in 96% ethanol for 1–3 days in the dark until further analysis in the laboratory (see ref. 38 for details). Also, five quantitative Surber samples (0.05 m^2; mesh size 200 μm) were collected randomly from pebble–cobble substratum. All samples were collected between 09:00 and 14:00 and preserved in the field in 70% ethanol. Invertebrates were counted and identified to morphospecies, genus or (sub) family. Functional feeding groups were assigned to each taxon following[41,42]. For the most abundant taxa (representing more than 95% of the relative abundance in the three stream sites), functional feeding groups were refined based on the dominant food items identified via gut content analyses of 10 individuals per taxon[43,44]. Macroinvertebrate guts were mounted on to microscope slides by removing the foregut following dissection of macroinvertebrates. Gut contents were placed in a drop of water on a microscope slide and secured with a coverslip. Five randomly selected fields within the slide were viewed at × 200 magnification. In each field on the slide, the approximate percentage by area of the five food types identified (diatoms, filamentous algae, fine detritus, coarse detritus and animal matter) was estimated. A percentage of herbivory was assigned to each dominant taxon according to the relative abundance of diatoms and filamentous algae observed in their gut contents (mean for the 10 individuals)[44]. For each sampling date, herbivore biomass was calculated based on the most dominant herbivore taxa density multiplied by their dry mass and their percentage of herbivory assigned. Organisms' dry mass (DM) was calculated from the measured body length (L) of each taxon, using the equation $DM = aL^b$ and using the parameters a and b provided by ref. 45 for each invertebrate family. As we had no data on the rate of algae and invertebrate reproduction, we could not calculate the productivity of our system. We thus used biomass values of both algae and herbivores as a surrogate of standing stocks.

Data analyses. Dissimilarity in macroinvertebrate assemblage and feeding trait composition between the upstream and downstream reaches of the manipulated stream was calculated for each sampling date using Bray–Curtis dissimilarity index[46] to assess how flow reduction affected the overall organization of benthic fauna community structure and trophic composition. An indicator species analysis (SIMPER procedure) was used to determine whether particular taxa were indicative of low-flow or recovery conditions, and which traits accounted for the greatest dissimilarity across the three experimental phases in the manipulated reach. The statistical significance of each indicator species was tested using Monte Carlo randomization with 1,000 runs. For comparison, similar analyses were performed on the reference stream to evaluate the natural variability in benthic fauna community structure and trophic composition.

To estimate trends in abiotic and biotic time series, we used state-space analyses, which describe the observed time series by an underlying state process evolving through time[47–48]. For each stream site, univariate autoregressive state-space models with Gaussian errors were fitted independently to discharge, water temperature, conductivity, algal biomass and herbivore biomass time series (15 days time step). Smoothed-state estimates (the expected values of the state process based on the maximum-likelihood values of the model parameters) were then computed using the Kalman smoother, and plotted to examine trends in each time series (Supplementary Notes Part 1: 1–3)[47–49]. To identify potential sudden level changes in algal and herbivore biomass time series, standardized smoothed-state residuals (the difference between the smoothed-state estimate at time t and $t-1$ standardized by its s.d.) were computed and compared with 95% confidence interval for a t-distribution[47,50] (Supplementary Notes Part 1: 3 and 4). To estimate interaction strengths between algae and herbivores, multivariate autoregressive state-space models were fitted to algal and herbivore biomass time series with discharge, water temperature and conductivity time series as environmental covariates[51–52]. In those models, the state processes described changes in algal and herbivore biomass based on interactions between the two trophic levels and environmental covariates, while the observed time series were described by the state processes and observation errors associated with incomplete sampling (Supplementary Notes Part 2). All analyses were performed using the *MARSS* package in R (R Development Core Team, 2013, version 3.0.2)[49]. R codes are available in the Supplementary Notes.

Complementary field survey. Previously collected data on algal and herbivore biomass in 33 stream sites with different glacial influence[53] (per cent of glacier cover in the catchment (%GCC) ranged from 0 to 93.2) were used to address the relevance of our experimental results. To compare algal and herbivore biomass in the manipulated stream with the 33 natural stream sites, we calculated the corresponding %GCC of the manipulated stream during flow alteration based on the relationship between glacial meltwater discharge and the %GCC of stream sites located along the study glacierized catchment (4,200–4,800 m). Both algal and herbivore biomass values of the 33 stream sites and of the manipulated stream before and during experimental flow reduction were plotted against %GCC. Breaks in the slope of the relationship between algal and herbivore biomass and glacial influence were identified using the Douglas–Peucker algorithm (Matlab version R2010a; The Mathworks Inc., Natick, MA, USA).

References

1. Sorg, A., Bolch, T., Stoffel, M., Solomina, O. & Beniston, M. Climate change impacts on glaciers and runoff in Tien Shan (Central Asia). *Nat. Clim. Change* **2,** 725–731 (2012).
2. Baraer, M., Bryan, G., McKenzie, J. M., Condom, T. & Rathay, S. Glacier recession and water resources in Peru's Cordillera Blanca. *J. Glaciol.* **58,** 134–150 (2012).
3. Rabatel, A. *et al.* Current state of glaciers in the tropical Andes: a multi-century perspective on glacier evolution and climate change. *Cryosphere* **7,** 81–102 (2013).
4. Jacobsen, D., Milner, A. M., Brown, L. E. & Dangles, O. Biodiversity under threat in glacier-fed river systems. *Nat. Clim. Change* **2,** 361–364 (2012).
5. Hannah, D. M. *et al.* Integrating climate-hydrology-ecology for alpine river systems. *Aquat. Conserv.* **17,** 636–656 (2007).
6. Milner, A. M., Brown, L. E. & Hannah, D. M. Hydroecological response of river systems to shrinking glaciers. *Hydrol. Process.* **23,** 62–77 (2009).
7. Finn, D. S., Khamis, K. & Milner, A. M. Loss of small glaciers will diminish beta diversity in Pyrenean streams at two levels of biological organization. *Global Ecol. Biogeogr.* **22,** 40–51 (2013).
8. Poff, N. L., Tokar, S. & Johnson, P. Stream hydrological and ecological responses to climate change assessed with an artificial neural network. *Limnol. Oceanogr.* **41,** 857–863 (1996).
9. Poff, N. L. & Zimmerman, J. K. Ecological responses to altered flow regimes: a literature review to inform the science and management of environmental flows. *Freshwater Biol.* **55,** 194–205 (2010).
10. Cauvy-Fraunié, S., Espinosa, R., Andino, P., Dangles, O. & Jacobsen, D. Relationships between stream macroinvertebrate communities and new flood-based indices of glacial influence. *Freshwater Biol.* **59,** 1916–1925 (2014).
11. Petraitis, P. *Multiple Stable States in Natural Ecosystems* (Oxford Univ. Press, 2013).
12. Nimmo, D., Mac Nally, R., Cunningham, S., Haslem, A. & Bennett, A. Vive la résistance: reviving resistance for 21st century conservation. *Trends Ecol. Evol.* **30,** 516–523 (2015).
13. Blois, J. L., Williams, J. W., Fitzpatrick, M. C., Jackson, S. T. & Ferrier, S. Space can substitute for time in predicting climate-change effects on biodiversity. *Proc. Natl Acad. Sci. USA* **110,** 9374–9379 (2013).
14. Dewson, Z. S., James, A. B. & Death, R. G. A review of the consequences of decreased flow for instream habitat and macroinvertebrates. *J. North Am. Benthol. Soc.* **26,** 401–415 (2007).
15. Dewson, Z. S., James, A. B. & Death, R. G. Invertebrate community responses to experimentally reduced discharge in small streams of different water quality. *J. North Am. Benthol. Soc.* **26,** 754–766 (2007).
16. Ledger, M. E., Brown, L. E., Edwards, F. K., Milner, A. M. & Woodward, G. Drought alters the structure and functioning of complex food webs. *Nat. Clim. Change* **3,** 223–227 (2013).
17. Walters, A. W. & Post, D. M. How low can you go? Impacts of a low-flow disturbance on aquatic insect communities. *Ecol. Appl.* **21,** 163–174 (2011).
18. Rolls, R. J., Leigh, C. & Sheldon, F. Mechanistic effects of low-flow hydrology on riverine ecosystems: ecological principles and consequences of alteration. *Freshwater Sci.* **31,** 1163–1186 (2012).
19. James, A. B. & Suren, A. M. The response of invertebrates to a gradient of flow reduction-an instream channel study in a New Zealand lowland river. *Freshwater Biol.* **54,** 2225–2242 (2009).
20. Vander Vorste, R., Corti, R., Sagouis, A. & Datry, T. Invertebrate communities in gravel-bed, braided rivers are highly resilient to flow intermittence. *Freshwater Sci.* **10,** 683274 (2015).
21. Brown, L. E., Hannah, D. M. & Milner, A. M. Vulnerability of alpine stream biodiversity to shrinking glaciers and snowpacks. *Global Change Biol.* **13,** 958–966 (2007).
22. Clitherow, L. R., Carrivick, J. L. & Brown, L. E. Food web structure in a harsh glacier-fed river. *PLoS ONE* **8,** e60899 (2013).
23. Füreder, L., Welter, C. & Jackson, J. K. Dietary and stable isotope (δ 13C, δ 15N) analyses in alpine stream insects. *Int. Rev. Hydrobiol.* **88,** 314–331 (2003).
24. Robinson, C. & Jolidon, C. Leaf breakdown and the ecosystem functioning of alpine streams. *J. North Am. Benthol. Soc.* **24,** 495–507 (2005).
25. Uehlinger, U., Robinson, C., Hieber, M. & Zah, R. The physico-chemical habitat template for periphyton in alpine glacial streams under a changing climate. *Hydrobiologia* **657,** 107–121 (2010).
26. Jacobsen, D. & Dangles, O. Environmental harshness and global richness patterns in glacier-fed streams. *Global Ecol. Biogeogr.* **21,** 647–656 (2012).

27. Cauvy-Fraunié, S., Espinosa, R., Andino, P., Jacobsen, D. & Dangles, O. Invertebrate metacommunity structure and dynamics in an andean glacial stream network facing climate change. *PLoS ONE* **10,** e0136793 (2015).
28. Brown, L. E. & Milner, A. M. Rapid loss of glacial ice reveals stream community assembly processes. *Global Change Biol.* **18,** 2195–2204 (2012).
29. James, A., Dewson, Z. & Death, R. The effect of experimental flow reductions on macroinvertebrate drift in natural and streamside channels. *River Res. Appl.* **24,** 22–35 (2008).
30. Loreau, M. *From Populations to Ecosystems: Theoretical Foundations for a New Ecological Synthesis (MPB-46)* (Princeton Univ. Press, 2010).
31. Hooper, D. U. *et al.* A global synthesis reveals biodiversity loss as a major driver of ecosystem change. *Nature* **486,** 105–108 (2012).
32. Rader, R. B. & Belish, T. A. Influence of mild to severe flow alterations on invertebrates in three mountain streams. *Regul. River* **15,** 353–363 (1999).
33. Lake, P. Ecological effects of perturbation by drought in flowing waters. *Freshwater Biol.* **48,** 1161–1172 (2003).
34. Bogan, M. T., Boersma, K. S. & Lytle, D. A. Resistance and resilience of invertebrate communities to seasonal and supraseasonal drought in arid-land headwater streams. *Freshwater Biol* **60,** 2547–2558 (2014).
35. Walters, A. W. Resistance of aquatic insects to a low-flow disturbance: exploring a trait-based approach. *J. North Am. Benthol. Soc.* **30,** 346–356 (2011).
36. Finn, D. S., Raesaenen, K. & Robinson, C. T. Physical and biological changes to a lengthening stream gradient following a decade of rapid glacial recession. *Global Change Biol.* **16,** 3314–3326 (2010).
37. Botter, G., Basso, S., Rodriguez-Iturbe, I. & Rinaldo, A. Resilience of river flow regimes. *Proc. Natl Acad. Sci. USA* **110,** 12925–12930 (2013).
38. Jacobsen, D. *et al.* Temporal variability in discharge and benthic macroinvertebrate assemblages in a tropical glacier-fed stream. *Freshwater Sci.* **33,** 32–45 (2014).
39. Wallace, J. B. & Webster, J. R. The role of macroinvertebrates in stream ecosystem function. *Annu. Rev. Entomol.* **41,** 115–139 (1996).
40. Rios-Touma, B. *Comunidades de macroinvertebrados en un rio altoandino: importancia del microhabitat, dinamica de la deriva, papel de la materia organica y relevancia de la ovoposicion.* PhD thesisUniv. Barcelona (2008).
41. Tomanova, S., Goitia, E. & Helešic, J. Trophic levels and functional feeding groups of macroinvertebrates in neotropical streams. *Hydrobiologia* **556,** 251–264 (2006).
42. Poff, N. L. *et al.* Functional trait niches of North American lotic insects: traits-based ecological applications in light of phylogenetic relationships. *J. North Am. Benthol. Soc.* **25,** 730–755 (2006).
43. Cummins, K. W. & Klug, M. J. Feeding ecology of stream invertebrates. *Annu. Rev. Ecol. Syst.* **10,** 147–172 (1979).
44. Dangles, O. Functional plasticity of benthic macroinvertebrates: implications for trophic dynamics in acid streams. *Can. J. Fish. Aquat. Sci.* **59,** 1563–1573 (2002).
45. Benke, A. C., Huryn, A. D., Smock, L. A. & Wallace, J. B. Length-mass relationships for freshwater macroinvertebrates in North America with particular reference to the southeastern United States. *J. North Am. Benthol. Soc.* **18,** 308–343 (1999).
46. Baselga, A. Separating the two components of abundance-based dissimilarity: balanced changes in abundance versus abundance gradients. *Methods Ecol. Evol.* **4,** 552–557 (2013).
47. Durbin, J. & Koopman, S. J. *Time Series Analysis by State Space Methods* (Oxford Univ. Press, 2012).
48. Commandeur, J. J. & Koopman, S. J. *An Introduction to State Space Time Series Analysis* (Oxford Univ. Press, 2007).
49. Holmes, E. E., Ward, E. J. & Wills, K. Marss: multivariate autoregressive state-space models for analyzing time-series data. *R J.* **4,** 11–19 (2012).
50. Harvey, A. C. & Koopman, S. J. Diagnostic checking of unobserved-components time series models. *J. Bus. Econ. Stat.* **10,** 377–389 (1992).
51. Ives, A., Dennis, B., Cottingham, K. & Carpenter, S. Estimating community stability and ecological interactions from time-series data. *Ecol. Monogr.* **73,** 301–330 (2003).
52. Hampton, S. E. *et al.* Quantifying effects of abiotic and biotic drivers on community dynamics with multivariate autoregressive (MAR) models. *Ecology* **94,** 2663–2669 (2013).
53. Cauvy-Fraunié, S. Hydroecology of Invertebrate Communities in Equatorial Glacier-fed Streams. Doctoral dissertation, Pierre Marie Curie Univ. (2014).

Acknowledgements

We warmly thank Jérôme Casas, Debra Finn, Sonia Kefi, LeRoy Poff, Scott Tiegs, Ross Vander Vorste and the three anonymous reviewers for their constructive suggestions on a previous version of the manuscript. We thank Cristina Ibarra for laboratory work. Part of this work was funded by the Fond Français pour l'Environnement Mondial (FFEM) and the Fondation pour la Recherche sur la Biodiversité (FRB) within the project 'Modeling BIOdiversity and land use interactions under changing glacial water availability in Tropical High Andean Wetlands' (BIOTHAW, AAP-SCEN-2011-II).

Author contributions

O.D., R.C. and D.J. conceived the study; S.C.-F., P.A. and R.E. collected the Ecuadorian data; S.C.-F. and O.D. analysed the data. The writing of the paper was led by S.C.-F. and O.D. with participation from D.J. S.C.-F. and O.D. contributed equally to the study.

Additional information

Competing financial interests: The authors declare no competing financial interests.

24

Incorrect interpretation of carbon mass balance biases global vegetation fire emission estimates

N.C. Surawski[1,†], A.L. Sullivan[2], S.H. Roxburgh[2], C.P. Mick Meyer[3] & P.J. Polglase[2,†]

Vegetation fires are a complex phenomenon in the Earth system with many global impacts, including influences on global climate. Estimating carbon emissions from vegetation fires relies on a carbon mass balance technique that has evolved with two different interpretations. Databases of global vegetation fire emissions use an approach based on 'consumed biomass', which is an approximation to the biogeochemically correct 'burnt carbon' approach. Here we show that applying the 'consumed biomass' approach to global emissions from vegetation fires leads to annual overestimates of carbon emitted to the atmosphere by 4.0% or 100 Tg compared with the 'burnt carbon' approach. The required correction is significant and represents ~9% of the net global forest carbon sink estimated annually. Vegetation fire emission studies should use the 'burnt carbon' approach to quantify and understand the role of this burnt carbon, which is not emitted to the atmosphere, as a sink enriched in carbon.

[1] CSIRO Agriculture, GPO Box 1700, Canberra, Acton 2601, Australian Capital Territory, Australia. [2] CSIRO Land and Water, GPO Box 1700, Canberra, Acton 2601, Australian Capital Territory, Australia. [3] CSIRO Oceans and Atmosphere, 107-121 Station Street, Aspendale, Victoria 3195, Australia. †Present addresses: Energy, Environment and Water Research Center (EEWRC), The Cyprus Institute, PO Box 27456, Nicosia 2121, Cyprus (N.C.S.); Fenner School of Environment & Society, Australian National University, Linnaeus Way, Acton, Australian Capital Territory 2601, Australia (P.J.P.). Correspondence and requests for materials should be addressed to N.C.S. (email: n.surawski@cyi.ac.cy).

Vegetation fires have existed in the Earth system since shortly after the appearance of terrestrial vegetation ~420 million years ago[1,2] and are a necessary disturbance for maintaining some ecosystems[3]. They also have a range of anthropogenically deleterious consequences such as damage to assets and infrastructure, loss of life, as well as degrading ambient air quality leading to negative impacts on human health[4]. Vegetation fires also perturb global biogeochemical cycles, with impacts on climate[4–6]. It is estimated that over the period 2001–2010, global vegetation fires emitted an average of 2.5 Pg per year ($1\,Pg = 1 \times 10^{15}\,g$) of carbon to the atmosphere[7], equivalent to 30% of annual-averaged anthropogenic emissions from fossil fuel combustion over 2004–2013 (ref. 8). Changes in global vegetation fire regimes, such as an increased fire frequency, have been predicted from modelling studies undertaken in the northwest of the United States[9], as well as Mediterranean-type ecosystems under a warmer and wetter future climate[10]. Furthermore, charcoal records from boreal forests suggest that current vegetation fire frequencies are at their highest level in the last 3,000 years[11] and predictions of a future increase in fire intensity have been made[12]. Given the global impact of vegetation fires on ecosystems and climate, accurate accounting methods are required to estimate their greenhouse gas emissions, to characterize their climate forcing and to properly design carbon mitigation strategies involving the management of vegetation.

Using the Intergovernmental Panel on Climate Change (IPCC) methodology[13], total emissions from fire are calculated using the product of activity and an emissions factor (EF). Activity refers to a process occurring in the environment that leads to emissions (for example, fire) and an EF is thus defined as emissions per unit activity.

When calculating an EF from fire, the principle of a carbon mass balance (CMB) is invoked, with the technique ultimately being attributed to either Ward *et al.*[14] or Radke *et al.*[15]. While the partitioning of the chemical species of interest and total emitted carbon into reference-gas-normalized emission ratios is consistently applied in vegetation fire emission studies[16–18], there have been inconsistencies in the treatment of vegetation exposed to fire but not emitted to the atmosphere (that is, post-burn combustion residue).

In this article, we show that neglecting changes in the post-burn carbon content (CC_{post}, %) leads to an upward bias in emission estimates for the 'consumed biomass' CMB interpretation compared with the 'burnt carbon' approach. We therefore recommend that vegetation fire studies employ the 'burnt carbon' approach to more accurately estimate emissions, by taking into account the change in carbon content from post-fire residues.

Results

Definitional preliminaries. Following the definitions from Surawski *et al.*[16], we define 'burnt' as fuel that has been thermally altered as a result of exposure to fire and either emitted to the atmosphere or left in the post-fire residue; and we define 'consumed' as that component of the fuel that is emitted to the atmosphere as a result of exposure to fire. We follow the definition of Keane *et al.*[19] that 'fuels are the dead and live biomass available for fire ignition and combustion'.

The basis of biased emission estimates. Since the publication of the CMB technique[15], the method has been interpreted and applied in two different ways in the vegetation fire emission literature[17,20]. The first is the 'consumed biomass' approach (Fig. 1, top pathway), which is based on the simplifying assumption that all burnt carbon is volatilized and emitted. We demonstrate below that this assumption is incorrect and leads to biased results. The second method, the 'burnt carbon' approach, additionally takes into account changes between the pre- and post-fire carbon content of the fuel and is unbiased (Fig. 1, bottom pathway).

Both approaches have the same starting point, which is the amount of biomass fuel before burning (B_{pre}, $t\,ha^{-1}$ on a dry matter basis) and its carbon content (CC_{pre}, %). In the absence of measurements of fuel carbon content, a carbon content of 50% is commonly assumed[18], although a value of 45% has also been suggested[21]. The pre-burning carbon load ($t\,ha^{-1}$ on a dry matter basis) can then be calculated from estimates of both B_{pre} and CC_{pre} ($C_{pre} = B_{pre} \times CC_{pre}$). Both interpretations of the CMB technique rely on estimates of biomass fuel after burning (B_{post}) to enable calculation of a combustion factor ($CF = (B_{pre} - B_{post})/B_{pre}$) (ref. 13). It is the subsequent analysis where the two interpretations of the CMB technique diverge and bias is introduced.

The 'consumed biomass' approach is more commonly applied[21–25] (and references therein) and does not require any further measurements after B_{post} has been calculated. In this case, estimation of an EF is based on consumed biomass only and reports EFs as the mass of a chemical species emitted per kilogram of dry fuel consumed. The 'burnt carbon' approach is far less commonly applied[16,17,26,27] and requires an estimate of the post-burning carbon content for burnt fuel that is not emitted to the atmosphere. Estimating CC_{post} enables the post-fire carbon load ($C_{post} = B_{post} \times CC_{post}$) to be calculated. This in turn enables burnt carbon to be partitioned between that emitted to the atmosphere ($\Sigma C_{emit} = C_{pre} - C_{post}$) and that remaining as post-burn combustion residue (C_{post}, for example, charred material enriched in carbon; Fig. 1). Consequently, the second interpretation of the CMB technique is based on 'burnt carbon' involving either atmospheric emission or resulting in an ashed or charred residue after exposure to fire (C_{post} in Fig. 1) with both of these ultimate fates being represented as a percentage of burnt carbon.

The explicit and incorrect assumption of the 'consumed biomass' CMB method (top pathway in Fig. 1) is that all burnt carbon is volatilized and emitted to the atmosphere[20,25,28]. In contrast, the 'burnt carbon' CMB method (bottom pathway in Fig. 1) correctly accounts for burnt carbon remaining as a residue after a vegetation fire. Many vegetation fire research studies have shown that not all burnt fuel carbon is volatilized and emitted to the atmosphere[17,26,29–34]. The need to take into account the full-carbon budget of vegetation fires has been raised previously[21,35,36], yet vegetation fire emission studies generally do not measure CC_{post} (with $CC_{post} > CC_{pre}$ for charred post-fire material[30,31,34]), which precludes use of the 'burnt carbon' CMB method and this leads to an overestimation bias of emissions for the 'consumed biomass' method relative to the 'burnt carbon' method.

To avoid this bias with the 'consumed biomass' CMB approach, it is necessary to multiply its EF by the fraction of burnt fuel carbon emitted to the atmosphere, $\Sigma C_{emit}/C_{fuel}$ (ref. 16), where ΣC_{emit} is the mass of emitted carbon and C_{fuel} is the mass of fuel carbon burnt (Supplementary Table 1). Only if all burnt fuel carbon is emitted to the atmosphere (that is, $\Sigma C_{emit}/C_{fuel} = 1$, which is very unlikely to be achieved in a real fire), would these two methods agree without inclusion of the correction factor. Otherwise, estimates of emissions using the 'consumed biomass' approach will be overestimated (that is, relative to the 'burnt' carbon approach), represented by the difference between complete volatilization and emission of carbon to the atmosphere, and the true fraction of carbon volatilized and emitted to the atmosphere ($\varepsilon = 1 - \Sigma C_{emit}/C_{fuel}$).

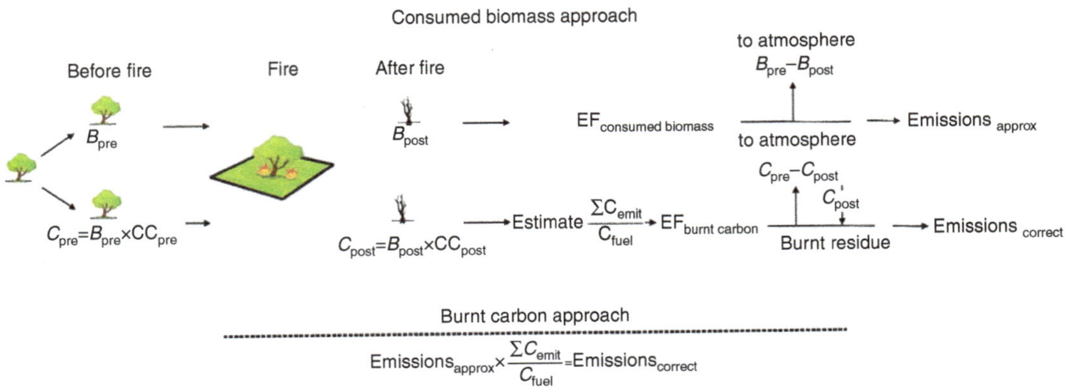

Figure 1 | A diagrammatic illustration of how biased emission estimates from fire occur. A schematic of the two different ways the carbon mass balance technique has been interpreted and applied in vegetation fire emission studies. C_{pre} and B_{pre} denote pre-fire carbon and biomass fuel, CC_{pre} and CC_{post} represent pre- and post-fire carbon content, and C_{post} and B_{post} denote post-fire carbon and biomass fuel, respectively. $\Sigma C_{emit}/C_{fuel}$ represents the correction factor required for the burnt carbon accounting framework. $Emissions_{approx}$ represents approximate emissions estimated using the consumed biomass approach and $Emissions_{correct}$ represents emission estimates using the burnt carbon approach. EF denotes emissions factor with the estimate depending on what accounting framework is used (that is, consumed biomass or burnt carbon).

Quantification of the correction factor. To quantify the magnitude of this overestimation, we obtained empirical estimates from the literature[17,29–34] (Supplementary Table 1) of the fraction of burnt fuel carbon emitted to the atmosphere or the fraction of burnt carbon retained in post-burn residues (and in one case both emissions and residues), to determine the likely values of ε in various vegetation types around the globe. ε values range from a minimum of 0.4% in a savanna ecosystem[34] to a maximum of 50% in a temperate forest[34]. On the basis of 40 records (Fig. 2), the median ε value is 4.0% with the interquartile range (25th–75th percentiles) spanning the interval 2.0–12.0%.

Discussion

In terms of estimating global emissions from vegetation fires, the widely used Global Fire Emissions Database[22], and others such as the Fire INventory from NCAR[23], and IPCC good practice guidelines for national greenhouse gas inventories[13] rely on the 'consumed biomass' approach for reporting EFs. If we use the median ε value of 4.0% and apply this to the globally averaged annual carbon emissions from 2001 to 2010 of 2.5 Pg (ref. 7), the overestimation of emissions (that is, for the 'consumed biomass' approach compared with the 'burnt carbon' approach) amounts to $\sim$100 Tg (1 Tg $= 1\times10^{12}$ g) of carbon per year, with an uncertainty range (25th–75th percentiles) of 50–300 Tg. Furthermore, the 'consumed biomass' approach (which assumes 100% volatilization of burnt carbon) precludes the ability to quantify the efficacy of any direct management of fire for post-fire residues for mitigating net vegetation fire emissions, such as the preferential firing of vegetation fuels for enhanced charred post-fire residues to increase sequestration in the pedosphere.

Altogether, the potential for a 50–300-Tg error in annual global vegetation fire emission estimates highlights that vegetation fire research should be conducted in a more coordinated manner[1,37], to ensure the complete biogeochemistry of fire events are adequately represented. Despite recommendations in the literature[21], measurements of CC_{post} are not routinely made in vegetation fire emission studies. This prevents $\Sigma C_{emit}/C_{fuel}$ from being calculated, leading to an overestimation of emissions for the 'consumed biomass' approach relative to the 'burnt carbon' approach. Furthermore, this overestimation means that the amount of carbon in burnt residues needs to be adjusted to maintain global carbon mass balance from vegetation fires. Apart from improving global vegetation fire emission estimates, it is envisaged that global studies of CC_{post} will broaden our understanding of ecosystem carbon fluxes due to fire, including the potential for managing burnt carbon residues to enhance ecosystem carbon storage and to spawn research efforts exploring the post-fire fate of carbon[34–36,38]. In particular, such efforts should estimate the global ε value with different fire behaviours and ecosystem types, as only a coarse estimate was obtained using current estimates in the literature. In conducting such research, however, it is important to realize that greater uncertainties are present for burnt areas, fuel load and combustion factors from global vegetation fires[39].

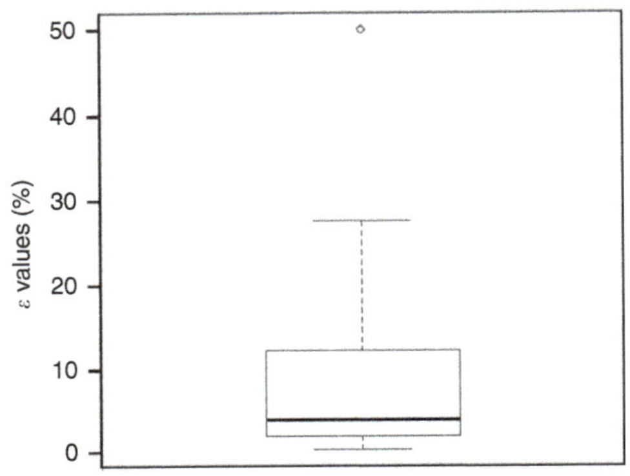

Figure 2 | A box plot of the global ε value distribution based on our literature review. The ε value is the percentage of carbon exposed to fire (that is, burnt) that remains as a post-fire residue, based on 40 records[17,29–34] (Supplementary Table 1). The five-number summary for ε is: minimum $=$ 0.4%, 25th percentile $=$ 1.98%, median $=$ 3.95%, 75th percentile $=$ 12.0%, maximum $=$ 50% (the single outlier is shown as an open circle). The interquartile range (IQR) $=$ 10.0.

Methods

EF reporting methods. Examples of studies that have adopted the approximate 'consumed biomass' approach to emissions accounting from vegetation fires are ubiquitous[21–25] (and references therein).

Examples of studies exploring the more complete 'burnt carbon' approach are far less common[16,17,26,27].

EF reporting fundamentals. Estimating emissions from vegetation fires involves multiplying the amount of fuel consumed[6] by an EF (ref. 24). The amount of fuel consumed is obtained by multiplying the area burnt (*A*) by the fuel load (*B*) and a combustion factor (CF). The basis of the CMB technique is that for the dominant carbonaceous species emitted in a plume (usually CO_2, CO and CH_4), excess mixing ratios (that is, molar concentrations above background) are measured.

Then, emission ratios are calculated by normalizing (that is, dividing) the excess mixing ratio of the species of interest to the excess mixing ratio of a reference gas, or in some cases the sum of the excess mixing ratios of total emitted carbon[40]. An EF is then obtained by dividing the emission ratio for the species of interest by that of the emitted and detected carbon

Conversion of the consumed biomass method to burnt carbon. Using the 'consumed biomass' interpretation of an EF (an approximation that assumes that all fuel carbon is volatilized and emitted to the atmosphere), emissions for the Xth carbon-based species (E_X) can be calculated by multiplying the area burnt, the fuel load and the combustion and EFs together via:

$$E_X = A \times B \times \mathrm{CF} \times \mathrm{EF}_X \qquad (1)$$

In equation (1), EF_X, reported on a kg species X per kg consumed basis, is defined by[18]:

$$\mathrm{EF}_X = C_{\mathrm{pre}} \times U_C \times \frac{C_X}{C_T}, \qquad (2)$$

where U_C is a conversion factor used to convert between atomic and molar masses, C_X is the number of moles of species X and C_T is the total number of carbon moles emitted to the atmosphere.

Assuming that CO_2 is used as a reference gas for the normalization of emission ratios (that is, other choices include CO or CH_4 (ref. 16) or total emitted carbon[40]), C_X/C_T is given by:

$$\frac{C_X}{C_T} = \frac{\frac{\Delta C_X}{\Delta CO_2}}{\sum_{j=1}^{N} \mathrm{NC}_j \times \frac{\Delta C_j}{\Delta CO_2}}, \qquad (3)$$

where ΔC_X represents the difference between the plume and ambient mixing ratios for species X, and NC_j represents the number of carbon atoms in species j, with summation being performed across the total number of carbon species N.

Alternatively, using the 'burnt carbon' interpretation, emissions for the Xth (that is, carbon based) species are given by:

$$E_X = A \times B \times \mathrm{CF} \times C_{\mathrm{pre}} \times U_C \times \mathrm{EF}_X. \qquad (4)$$

In equation (4), assuming again that CO_2 is used as a reference gas, EF_X (defined as the percentage of burnt carbon emitted as species X) is defined by[17]:

$$\mathrm{EF}_X = \frac{\frac{\sum C_{\mathrm{emit}}}{C_{\mathrm{fuel}}} \times \frac{\Delta C_X}{\Delta CO_2}}{\sum_{j=1}^{N} \mathrm{NC}_j \times \frac{\Delta C_j}{\Delta CO_2}}, \qquad (5)$$

where ΣC_{emit} is the mass of fuel carbon emitted to the atmosphere from j different carbon species, and C_{fuel} is the mass of fuel carbon burnt.

Note that:

$$\frac{\sum C_{\mathrm{emit}}}{C_{\mathrm{fuel}}} = \frac{C_{\mathrm{pre}} - C_{\mathrm{post}}}{C_{\mathrm{pre}}} = \frac{B_{\mathrm{pre}}\mathrm{CC}_{\mathrm{pre}} - B_{\mathrm{post}}\mathrm{CC}_{\mathrm{post}}}{B_{\mathrm{pre}}\mathrm{CC}_{\mathrm{pre}}} = 1 - \frac{B_{\mathrm{post}}\mathrm{CC}_{\mathrm{post}}}{B_{\mathrm{pre}}\mathrm{CC}_{\mathrm{pre}}}. \qquad (6)$$

Consequently, the only way that the equation (1) can equal equation (4) is if equation (2) is multiplied by $\Sigma C_{\mathrm{emit}}/C_{\mathrm{fuel}}$ (ref. 16). The presentation here focused on carbon-based species, but similar calculations can be performed for nitrogen-based species (such as N_2O).

Global estimates of the correction factor. A broad literature search was conducted using the Web of Science, Scopus and Google Scholar databases to parameterize this quantity. Literature searches were motivated by listings of major field-based vegetation fire emission campaigns commencing in the 1980s and ending with the Fire Lab at Missoula Experiments (Supplementary Table 1). On the basis of our literature search, in Supplementary Table 1, we parameterise the global ε value for different ecosystems and fuel types, as well as for different fire types, fuel sampling methods, carbon residue sampling methods and carbon residue quantification methods. A statistical summary of our ε value database is provided in Fig. 2 and its caption.

References

1. Bowman, D. M. J. S. *et al.* Fire in the Earth system. *Science* **324,** 481–484 (2009).
2. Scott, A. C. & Glasspool, I. J. The diversification of Paleozoic fire systems and fluctuations in atmospheric oxygen concentration. *Proc. Natl Acad. Sci. USA* **103,** 10861–10865 (2006).
3. Bowman, D. M. J. S. *et al.* The human dimension of fire regimes on Earth. *J. Biogeogr.* **38,** 2223–2236 (2011).
4. Moritz, M. A. *et al.* Learning to coexist with wildfire. *Nature* **515,** 58–66 (2014).
5. Crutzen, P. J., Heidt, L. E., Krasnec, J. P., Pollock, W. H. & Seiler, W. Biomass burning as a source of atmospheric gases CO, H2, N2O, NO, CH3Cl and COS. *Nature* **282,** 253–256 (1979).
6. Seiler, W. & Crutzen, P. J. Estimates of gross and net fluxes of carbon between the biosphere and the atmosphere from biomass burning. *Clim. Change* **2,** 207–247 (1980).
7. Randerson, J. T., Chen, Y., van der Werf, G. R., Rogers, B. M. & Morton, D. C. Global burned area and biomass burning emissions from small fires. *J. Geophys. Res.* **117,** G04012 (2012).
8. Le Quéré, C. *et al.* Global carbon budget 2014. *Earth Syst. Sci. Data* **7,** 47–85 (2015).
9. Westerling, A. L., Turner, M. G., Smithwick, E. A. H., Romme, W. H. & Ryan, M. G. Continued warming could transform Greater Yellowstone fire regimes by mid-21st century. *Proc. Natl Acad. Sci. USA* **108,** 13165–13170 (2011).
10. Batllori, E., Parisien, M.-A., Krawchuk, M. A. & Moritz, M. A. Climate change-induced shifts in fire for Mediterranean ecosystems. *Global Ecol. Biogeogr.* **22,** 1118–1129 (2013).
11. Kelly, R. *et al.* Recent burning of boreal forests exceeds fire regime limits of the past 10,000 years. *Proc. Natl Acad. Sci. USA* **110,** 13055–13060 (2013).
12. Flannigan, M. *et al.* Global wildland fire season severity in the 21st century. *For. Ecol. Manage.* **294,** 54–61 (2013).
13. IPCC. 2006 *IPCC Guidelines for National Greenhouse Gas Inventories* Vol. 4, Ch. 2 (IGES, 2006).
14. Ward, D. E. *et al.* in *75th Annual Meeting of the Air Pollution Control Association* (New Orleans, Louisiana,, 1982).
15. Radke, L. F., Hegg, D. A., Lyons, J. H., Brock, C. A. & Hobbs, P. V. in *Aerosols and Climate* (eds Hobbs, P. V. & Patrick McCormick, M.) 411–422 (A. Deepak Publishing, 1988).
16. Surawski, N. C., Sullivan, A. L., Meyer, C. P., Roxburgh, S. H. & Polglase, P. J. Greenhouse gas emissions from laboratory-scale fires in wildland fuels depend on fire spread mode and phase of combustion. *Atmos. Chem. Phys.* **15,** 5259–5273 (2015).
17. Hurst, D. F., Griffith, D. W. T. & Cook, G. D. Trace gas emissions from biomass burning in tropical Australian savannas. *J. Geophys. Res.* **99,** 16441–16456 (1994).
18. Yokelson, R. J. *et al.* Emissions of formaldehyde, acetic acid, methanol, and other trace gases from biomass fires in North Carolina measured by airborne Fourier transform infrared spectroscopy. *J. Geophys. Res.* **104,** 30109–30125 (1999).
19. Keane, R. E. Describing wildland surface fuel loading for fire management: a review of approaches, methods and systems. *Int. J. Wildland Fire* **22,** 51–62 (2013).
20. Nance, J. D., Hobbs, P. V. & Radke, L. F. Airborne measurements of gases and particles from an Alaskan wildfire. *J. Geophys. Res.* **98,** 14873–14882 (1993).
21. Andreae, M. O. & Merlet, P. Emission of trace gases and aerosols from biomass burning. *Global Biogeochem. Cycles* **15,** 955–966 (2001).
22. van der Werf, G. R. *et al.* Global fire emissions and the contribution of deforestation, savanna, forest, agricultural, and peat fires (1997–2009). *Atmos. Chem. Phys.* **10,** 11707–11735 (2010).
23. Wiedinmyer, C. *et al.* The Fire INventory from NCAR (FINN): a high resolution global model to estimate the emissions from open burning. *Geosci. Model Dev.* **4,** 625–641 (2011).
24. Ward, D. E. *et al.* Smoke and fire characteristics for cerrado and deforestation burns in Brazil - BASE-B experiment. *J. Geophys. Res.* **97,** 14601–14619 (1992).
25. Akagi, S. K. *et al.* Emission factors for open and domestic biomass burning for use in atmospheric models. *Atmos. Chem. Phys.* **11,** 4039–4072 (2011).
26. Lobert, J. M. *et al.* in *Global Biomass Burning: Atmospheric, Climatic, and Biospheric Implications* (ed. Levine, J. S.) 289–304 (The MIT Press, 1991).
27. Lobert, J. M., Scharffe, D. H., Hao, W. M. & Crutzen, P. J. Importance of biomass burning in the atmospheric budgets of nitrogen-containing gases. *Nature* **346,** 552–554 (1990).
28. Bertschi, I. T., Yokelson, R. J., Ward, D. E., Christian, T. J. & Hao, W. M. Trace gas emissions from the production and use of domestic biofuels in Zambia measured by open-path Fourier transform infrared spectroscopy. *J. Geophys. Res.* **108,** 8469 (2003).
29. Kuhlbusch, T. A. J. & Crutzen, P. J. Toward a global estimate of black carbon in residues of vegetation fires representing a sink of atmospheric CO2 and a source of O2. *Global Biogeochem. Cycles* **9,** 491–501 (1995).
30. Kuhlbusch, T. A. J. *et al.* Black carbon formation by savanna fires: Measurements and implications for the global carbon cycle. *J. Geophys. Res.* **101,** 23651–23655 (1996).
31. Forbes, M. S., Raison, R. J. & Skjemstad, J. O. Formation, transformation and transport of black carbon (charcoal) in terrestrial and aquatic ecosystems. *Sci. Total Enviro.* **370,** 190–206 (2006).
32. Russell-Smith, J. *et al.* Improving estimates of savanna burning emissions for greenhouse accounting in northern Australia: limitations, challenges, applications. *Int. J. Wildland Fire* **18,** 1–18 (2009).
33. Saiz, G. *et al.* Pyrogenic carbon from tropical savanna burning: production and stable isotope composition. *Biogeosciences* **12,** 1849–1863 (2015).
34. Santín, C., Doerr, S. H., Preston, C. M. & Gonzalez-Rodriguez, G. Pyrogenic organic matter production from wildfires: a missing sink in the global carbon cycle. *Global Change Biol.* **21,** 1621–1633 (2015).

35. Bird, M. I., Wynn, J. G., Saiz, G., Wurster, C. M. & McBeath, A. The pyrogenic carbon cycle. *Ann. Rev. Earth Planet. Sci.* **43,** 9.1–9.26 (2015).
36. Santín, C. *et al.* Towards a global assessment of pyrogenic carbon from vegetation fires. *Global Change Biol.* **21,** 1–16 (2015a).
37. Pyne, S. J. Problems, paradoxes, paradigms: triangulating fire research. *Int. J. Wildland Fire* **16,** 271–276 (2007).
38. Ball, R. Regulation of atmospheric carbon dioxide by vegetation fires. *Clim. Res.* **59,** 125–133 (2014).
39. Bond, T. C. Bounding the role of black carbon in the climate system: a scientific assessment. *J. Geophys. Res.* **118,** 5380–5552 (2013).
40. Meyer, C. P. *et al.* Direct measurements of the seasonality of emission factors from savanna fires in northern Australia. *J. Geophys. Res.* **117,** D20305 (2012).

Acknowledgements

We thank M. Battaglia and S. Eady for their support of this project. We thank Dale Hurst and Pep Canadell for reviewing a previous draft. We dedicate this paper to the memory of Professor Mike Raupach who made seminal contributions to carbon cycle research.

Author contributions

A.L.S. suggested undertaking this project of highlighting different interpretations of the CMB technique. S.H.R. suggested a graphical representation of different interpretations of the CMB technique. N.C.S. prepared Figs 1 and 2 and Supplementary Table 1, and wrote the manuscript with contributions from all co-authors.

Additional information

Competing financial interests: The authors declare no competing financial interests.

25

Glacial legacies on interglacial vegetation at the Pliocene-Pleistocene transition in NE Asia

Ulrike Herzschuh[1,2], H. John B. Birks[3,4,5], Thomas Laepple[1], Andrei Andreev[6,7], Martin Melles[6] & Julie Brigham-Grette[8]

Broad-scale climate control of vegetation is widely assumed. Vegetation-climate lags are generally thought to have lasted no more than a few centuries. Here our palaeoecological study challenges this concept over glacial-interglacial timescales. Through multivariate analyses of pollen assemblages from Lake El'gygytgyn, Russian Far East and other data we show that interglacial vegetation during the Plio-Pleistocene transition mainly reflects conditions of the preceding glacial instead of contemporary interglacial climate. Vegetation-climate disequilibrium may persist for several millennia, related to the combined effects of permafrost persistence, distant glacial refugia and fire. In contrast, no effects from the preceding interglacial on glacial vegetation are detected. We propose that disequilibrium was stronger during the Plio-Pleistocene transition than during the Mid-Pliocene Warm Period when, in addition to climate, herbivory was important. By analogy to the past, we suggest today's widespread larch ecosystem on permafrost is not in climate equilibrium. Vegetation-based reconstructions of interglacial climates used to assess atmospheric CO_2-temperature relationships may thus yield misleading simulations of past global climate sensitivity.

[1] Periglacial Research Section, Alfred Wegener Institute, Helmholtz Centre for Polar and Marine Research, Telegraphenberg A43, 14473 Potsdam, Germany. [2] Institute of Earth and Environmental Sciences, Faculty of Sciences, University Potsdam, 14479 Potsdam-Golm, Germany. [3] Department of Biology, University of Bergen, Postboks 7803, 5020 Bergen, Norway. [4] Bjerknes Centre for Climate Research, Bergen, Norway. [5] Environmental Change Research Centre, University College London, London WC1E 6BT, UK. [6] Institute of Geology and Mineralogy, Faculty of Mathematics and Natural Sciences, University of Cologne, 50674 Köln (Cologne), Germany. [7] Institute of Geology and Petroleum Technologies, Faculty of Natural Sciences, Kazan Federal University, Kazan 420008, Russia. [8] Department of Geosciences, University of Massachusetts Amherst, Amherst, Massachusetts 01003-9297, USA. Correspondence and requests for materials should be addressed to U.H. (email: Ulrike.Herzschuh@awi.de).

Natural vegetation patterns in space and time are assumed to reflect mainly climate variability on all environmentally relevant temporal and spatial scales[1,2]. For example, it has been shown that vegetation turnover in space can substitute for turnover in time on millennial timescales[3], which fits with the findings that the climatic niches of individual plant taxa are generally stable over even longer timescales[4]. Although lag times between vegetation and climate changes have been extensively discussed with respect to tree migration patterns in Europe and North America following the last glacial stage[5–7], most proxy studies do not suggest major vegetation–climate disequilibria[8]. Relevant model evidence is lacking. Patterns of Late Pleistocene interglacial vegetation histories (marine isotope stage (MIS) 11, 9, 7 and 5) as inferred from long-term pollen records from Greece[9] and the Massif Central[10] are, in addition to their established relationship with interglacial climate, hypothesized to reflect the different locations of glacial refugia, creating unique interglacial migrational patterns. This accords with the suggestion that in addition to climate, post-glacial migration limitations might also explain the current distribution of many European plant species[11]. Furthermore, an analysis integrating modern plant community composition, their climatic niches and climate change since the Last Glacial Maximum implies that North and South American forests are still responding to last glacial climate change[12]. However, systematic investigations of changes in vegetation–climate relationships over glacial-interglacial timescales are lacking and thus the concept of 'glacial legacies' on interglacial vegetation has not been demonstrated using proxy data.

High-northern latitudes are ideally suited to detect former non-equilibrium patterns as climate change is particularly strong due to polar amplification[13]. Furthermore, the southern expansion of permafrost soils during glacial climate states[14] may have increased the distance between glacial refugia and potential interglacial refugial areas. However, non-climatic vegetation drivers such as herbivory[15,16], fire[17] or soil disturbance[18] should also be considered. Finally, tectonic impacts and their climatic effects may represent a hidden driver of long-term vegetation change in the Arctic[19].

The only high-latitude pollen record continuously spanning several glacial-interglacial cycles is from Lake El'gygytgyn in Chukotka, Russian Far East ranging from 3,530 to 2,150 kyr ago BP (refs 20,21), covering both the Mid-Pliocene Warm Period (MPWP: 3,530–2,900 kyr ago) and the Plio-Pleistocene transition (PPT: 2,900–2,150 kyr ago). The identification of vegetation–climate disequilibrium requires proxy records that reliably reflect regional climate change and are independent of the Lake El'gygytgyn pollen data and of sufficient temporal length and resolution. Unfortunately, no independent climate record from Lake El'gygytgyn or any other arctic terrestrial site of sufficient data quality exists. Accordingly, only marine records can be used. Here we assume and later confirm that the globally integrating benthic isotope stack (LR04)[22] and nearest sea-surface temperature (SST)[23] records from the Pacific can be used to approximate the climate conditions at our study site. While all climate records indicate that the relative intensities of glacials and interglacials vary over time, a behaviour likely related to differences in orbital configuration[24] and internal climate dynamics, these intensities appear to be largely globally coherent.

To assess the impact of herbivory, fire and soil disturbance on vegetation, common non-pollen palynomorphs in lake sediments can be used including *Sporormiella*[25] (a coprophilous fungal spore), *Gelasinospora*[26] (a fungal spore characteristic of burned soils) and *Glomus*[27,28] (an endomycorrhizal fungal spore characteristic of disturbed soils), respectively.

Here through multivariate analyses we show that variation among interglacial pollen assemblages at Lake El'gygytgyn through the PPT is best explained by long-term drivers (that is, proxies for climate and the vegetation condition of the preceding glacial), which we interpret to indicate major vegetation–climate disequilibrium. In contrast, short-term drivers (that is, proxies for contemporaneous climate, soil disturbance/erosion and/or herbivory) best explain statistically the variations in the pollen data of the PPT glacial and MPWP interglacials and glacials. From these results we conclude that the severe glacial climate and related extensive permafrost during some PPT led to distant glacial refugia, particularly of evergreen tree taxa, that hindered their subsequent interglacial establishment and the development of forests attaining equilibrium with climate. Our results imply that the current widespread larch ecosystem actually represents a transitional vegetation type reflecting severe last glacial (MIS2) conditions rather than contemporaneous Holocene interglacial (MIS1) climate.

Results

Vegetation change on glacial–interglacial timescales. A principal curve of the Lake El'gygytgyn pollen record, which summarizes complex multivariate data in one dimension, shows strong temporal variations (Fig. 1 and Supplementary Figs 1 and 2, 47% variation explained). The most obvious turnover at the glacial–interglacial timescale has already been described and related to 41 kyr obliquity insolation cyclicity[20]. Furthermore, the principal curve is consistent with a landscape openness curve based on the pollen-based biome-reconstruction technique of the same pollen record[29]. This suggests that millennial-scale vegetation variability generally reflects global climate shifts between glacial and interglacial states known from many proxy-based reconstructions[22,23,30]. The sequence of pollen types along the principal curve (Fig. 2) generally reflects their present-day relative occurrences in vegetation ranging from arctic steppe-tundra to dark needle-leaf forests[31]. As we focus on variability at glacial–interglacial timescales, similar principal curve analyses were performed with pollen spectra averaged for each marine isotope stage (see Methods). They show a very similar trend (Fig. 1, 75% variation explained: Supplementary Fig. 3). Inferred glacial–interglacial vegetation variability generally confirms SST reconstructions[23], suggesting that the glacial–interglacial amplitude during the MPWP was rather low compared with that of the PPT. Furthermore, the principal curve suggests that differences in pollen composition among glacials and among interglacials are in the same range as glacial–interglacial variability, particularly during the PPT. As expected, the pollen spectra-specific contribution to compositional turnover (= Local-Contribution-to-Beta-Diversity[32]), here used as a measure of the uniqueness of the pollen composition in a marine isotope stage, is generally high for stages with extreme principal curve values.

A globally coherent pattern of stage intensities. To test our assumption that we can approximate the climate conditions at our site with remote marine temperature and isotope records, we first investigate the global pattern of glacial and interglacial intensities. We compare the time series of glacial peak temperatures of a global set of SST records (Supplementary Table 1 and Supplementary Fig. 4) with the glacial intensities recorded in the LR04 stack (see Methods). The same analysis is also performed on the interglacial intensities. To ensure that most of the records overlap, we analyse the correlation during the PPT (2,150–2,900 kyr ago). The glacial intensities in all records cores show a significant ($P<0.05$) positive correlation to the inverted

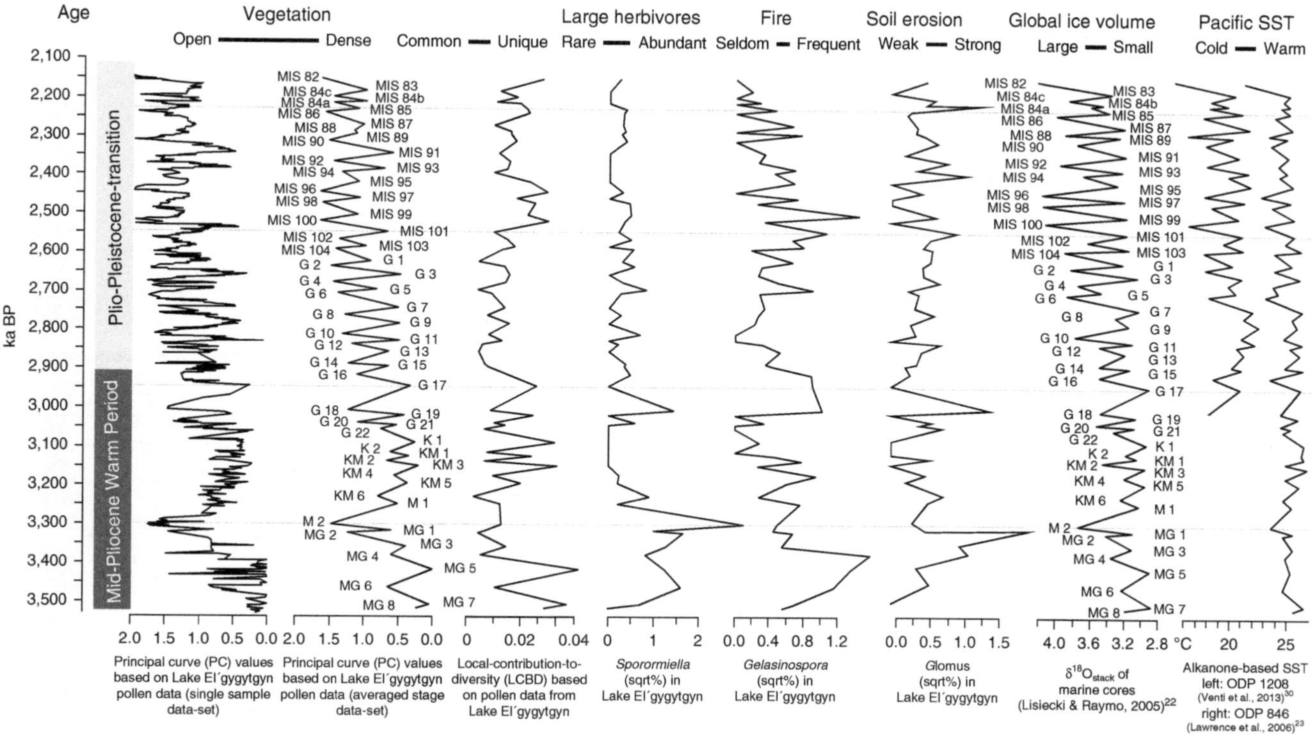

Figure 1 | Proxy records of vegetation change and of changes of potential vegetation drivers during the MPWP and the PPT. Stratigraphic plot including the principal curve of the original pollen data set and of the averaged stage data set, and the Local-Contribution-to-Beta-Diversity (LCBD) of each stage, all based on analyses of the Lake El'gygytgyn pollen data. It also includes environmental data of inverted globally averaged oxygen stable isotope measurements of benthic foraminifera ($\delta^{18}O_{stack}$)[22], SST of ODP1208 (mid-latitude Pacific)[30] and ODP846 (Tropical Eastern Pacific)[23], and *Sporormiella* sqrt% (herbivory indicator), *Gelasinospora* sqrt% (fire indicator) and *Glomus* sqrt% (soil disturbance/erosion indicator) relative to the vascular-plant pollen sum from Lake El'gygytgyn[21]. The names of the glacials and interglacials are indicated. Red and blue labels indicate PPT interglacials with preceding weak and strong glacials, respectively, that were included in the comparison of inter-taxa relationships (see text and Fig. 5). Dotted horizontal lines are drawn to aid visual comparisons. sqrt, square root.

LR04 stack (Fig. 3) with a mean correlation of $r=0.81$. Interglacial intensities also show a positive correlation at all core sites with the inverted LR04 stack (mean $r=0.69$). The weakest correlations are obtained for the tropical cores from the Indian Ocean (ODP722) and Atlantic (ODP662). As the interglacial temperatures at these cores are also at the temperature limit of the specific (UK[33]) proxy (Supplementary Table 1), it is unclear if this is a climatic signal or a proxy artefact.

While a systematic Plio-Pleistocene analysis, omitting pollen-based records to avoid circular reasoning, is limited to marine temperature proxies, analysing the Late Quaternary period also allows the inclusion of other climate proxy records. Here Antarctic ice cores provide independent evidence for the coherency of glacial–interglacial intensities[34]. Our comparison of stage extremes in isotope, CO_2 and methane records (Supplementary Fig. 5) show a significant correlation of the glacial intensities (mean correlation of ice-core records with the LR04 stack $r=0.52$) and a very strong correlation of the interglacial intensities ($r=0.91$). This result strongly suggests that the synchronizing factors (including greenhouse gases) are dominant over the local imprint of seasonal insolation[24] in shaping the strength of the interglacials.

Drivers of vegetation change. To separate the effects of long-term and short-term drivers on interglacial vegetation, redundancy analysis (RDA) was applied to the data set of PPT averaged interglacial pollen spectra and proxy-based environmental variables (Fig. 1). We find that variables reflecting contemporaneous interglacial climate and fire, that is, short-term ecological drivers, explain only a small portion of the variation in the pollen data when included separately in RDA (15% $P<0.1$, 18% $P<0.05$, respectively, Fig. 4 and Supplementary Table 2) compared with variables reflecting climate and vegetation type from the preceding glacial stage (42% $P<0.001$, 29% $P<0.01$, respectively). Though a large part of pollen variation is shared by all or is a combination of several variables, each variable explains a unique portion of variation as inferred from variation partitioning using a specific ensemble of RDA runs (Fig. 4 and Supplementary Figs 6 and 7). Preceding glacial climate explains the highest unique portion (18%) of pollen variation.

To test whether this result is sensitive on the choice of the climate record, RDAs of PPT interglacial pollen assemblages were repeated for each of the globally available climate records (Supplementary Table 1) with reasonable data quality. Results (Supplementary Table 3) reveal that the preceding glacial climate in all tested cases explains a significant unique portion while interglacial climate does not explain a unique portion or only a small portion. Consistent with the global coherency of the glacial–interglacial intensities, this indicates that neither the selection of a specific proxy type nor the selection of a specific site would impact the major outcome of this study.

In addition, similar RDAs were applied to the other averaged data sets. We find that the variation in the PPT glacial pollen spectra is most parsimoniously explained by the contemporary climate variable and to a lesser extent by the soil disturbance/erosion proxy, with both explaining a unique and statistically significant portion of the variation (Fig. 4). MPWP interglacial pollen are significantly correlated to climate, vegetation of

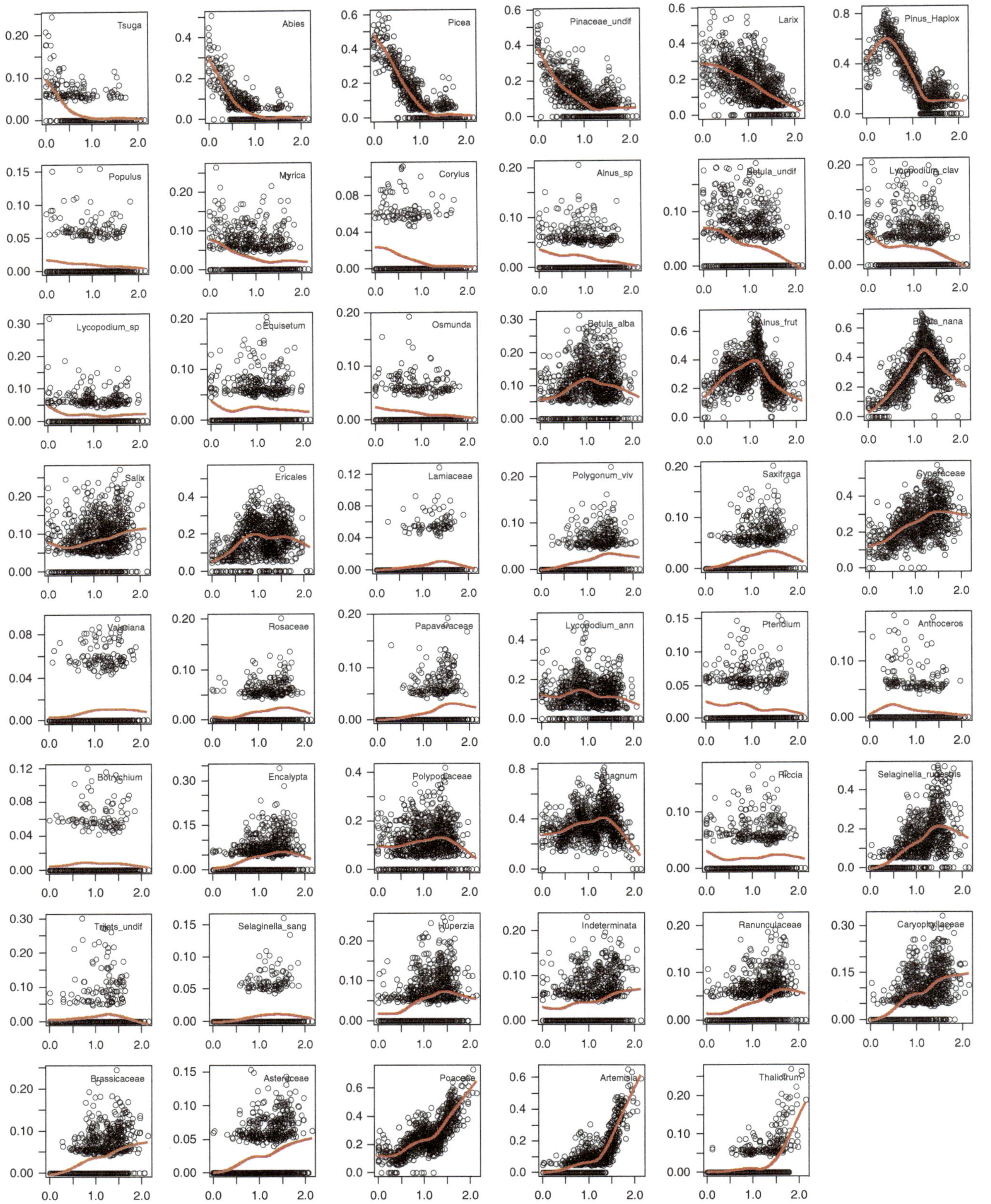

Figure 2 | Fitted response curves of each pollen or spore taxon along the principal curve. Open circles are the observed Hellinger-transformed abundance along their principal curve locations, and the solid red line is the optimized smoother from the final iteration of the principal curve. The taxa are arranged according to visual inspection of their response curves' optima (from right to left and from top to bottom). The vertical axis is the Hellinger-transformed pollen taxon abundance (%) and the horizontal axis is distance along the principal curve.

the preceding glacial, herbivory and fire when each variable (or variable set) is considered separately in RDA. However, variation partitioning shows that the major part of the explained variation is shared by all or a combination of several variables. MPWP glacial pollen spectra are most strongly related to contemporaneous glacial climate and herbivory (Fig. 4).

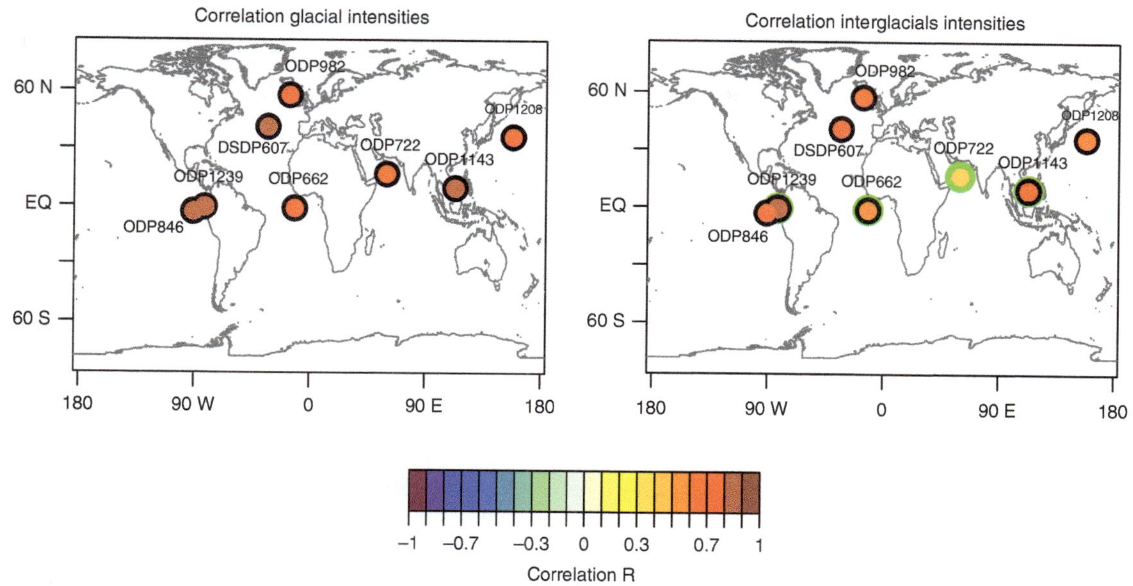

Figure 3 | Correlation of glacial and interglacial intensities of marine UK[33] temperature records with the inverted LR04 stack. All correlations are calculated during the core period (2,150–2,900 kyr ago), except ODP-982, which is evaluated for 2,500–3,500 kyr ago as the data resolution between 2150 and 2500 kyr ago does not allow determination of the minima and maxima. Black circles indicate significance at $P = 0.05$; green circles indicate records with interglacial peaks near the limit of UK[33] thermometry (>27 °C). (See Supplementary Fig.4 and Supplementary Table 1 for details and references of records.)

Differences among PPT interglacial vegetation histories. On the basis of the RDA biplots (Supplementary Fig. 6a,b) where variables for the preceding glacial are constrained to the first RDA axes and all the remaining known environmental variation is partialled out, we conclude that interglacials with preceding weak glacials and prevailing tree-tundra, such as G9, G7, G3, MIS101 and MIS93, are characterized by *Pinus* (pine) and *Picea* (spruce) forests (see also interglacial sequences in Supplementary Figs 8 and 9). Furthermore, these interglacials are closely located in the principal curve versus Local-Contribution-to-Beta-Diversity plot (Fig. 5a). Interglacials with preceding warm glacials are also characterized by similar inter-taxa relationships as inferred from the statistically significant fit of all interglacial pairs of ordination-derived taxa scores shown by Procrustes analyses and associated Protests (see Methods, Fig. 5b, see also taxa-specific residuals in Supplementary Fig. 10). This suggests that they reflect similar interglacial vegetation histories. In summary, during the PPT, interglacials with preceding warm glacial stages share similar evergreen forest vegetation, which we assume is driven by and is in equilibrium with the contemporaneous climate.

In contrast, *Pinus* and *Picea* are absent or rarely occur during interglacials with a preceding cold glacial-stage climate and extensive arctic steppe such as G1, MIS103, MIS99, MIS95 and MIS87 while deciduous shrubs (in particular *Alnus* (alder)) and wetlands (indicated by high Cyperaceae (sedge) and *Sphagnum* (bog-moss) values) are abundant (Supplementary Figs 2, 8 and 9; see also taxa-specific residuals in Supplementary Fig. 10). Some of these interglacial spectra strongly contribute to the Local-Contribution-to-Beta-Diversity when compared with spectra from other stages with similar principal curve values (Fig. 5a). Also, the inter-taxa relationships differ strongly among single interglacials as indicated by high Protest *P* values (Fig. 5b and Supplementary Figs 8 and 9). In summary, interglacials with preceding strong (cold) glacial stages lack *Pinus* and *Picea* and their vegetation characteristics are non-analogous among each other.

Discussion

With respect to PPT interglacials, we infer from our analyses of the Lake El'gygytgyn pollen record that the environmental state of the preceding glacial was more relevant for subsequent interglacial vegetation development, in particular for the presence or absence of evergreen conifers, than the contemporaneous interglacial climate, contradicting the hypothesis of a close vegetation–climate equilibrium. One alternative explanation for the observed weak relationship between proxy series for interglacial climate and interglacial pollen data would be that the analysed $\delta^{18}O_{stack}$ and mid-latitude Pacific SST do not represent the regional interglacial climate around Lake El'gygytgyn due to proxy-specific or regional-specific differences. However, our analysis shows that the strength of an interglacial is largely coherent in the PPT between all available high-quality SST records from the Northern Hemisphere (Fig. 3), and is coherent between different proxy records, including Antarctic temperature and greenhouse gases in the last 800 kyr (Supplementary Fig. 4). This is consistent with theoretical and model considerations that climate variations on long timescales are typically associated with broad spatial scales[35,36] and with evidence from climate-model simulations using orbital forcing as well as greenhouse-gas forcing[37].

Such a 'local climate' hypothesis would also be inconsistent with our finding that the climate of the preceding glacial as recorded by the $\delta^{18}O_{stack}$ and mid-latitude Pacific SST should influence the local vegetation of the interglacial. Hence, we conclude that the most parsimonious explanation for our results is that glacial climate and vegetation conditions are the main drivers of interglacial vegetation.

Accordingly, we interpret the absence of *Pinus* and *Picea* in Lake El'gygytgyn pollen spectra from interglacials following strong glacial stages to be related to three main factors: permafrost extent, distant glacial refugia, and extensive fires, plus possible interactions (Fig. 6). Today the vast continuous permafrost zone in NE Asia is covered by mono-specific larch forests with a deciduous-shrub understorey[9]. Occurrences of

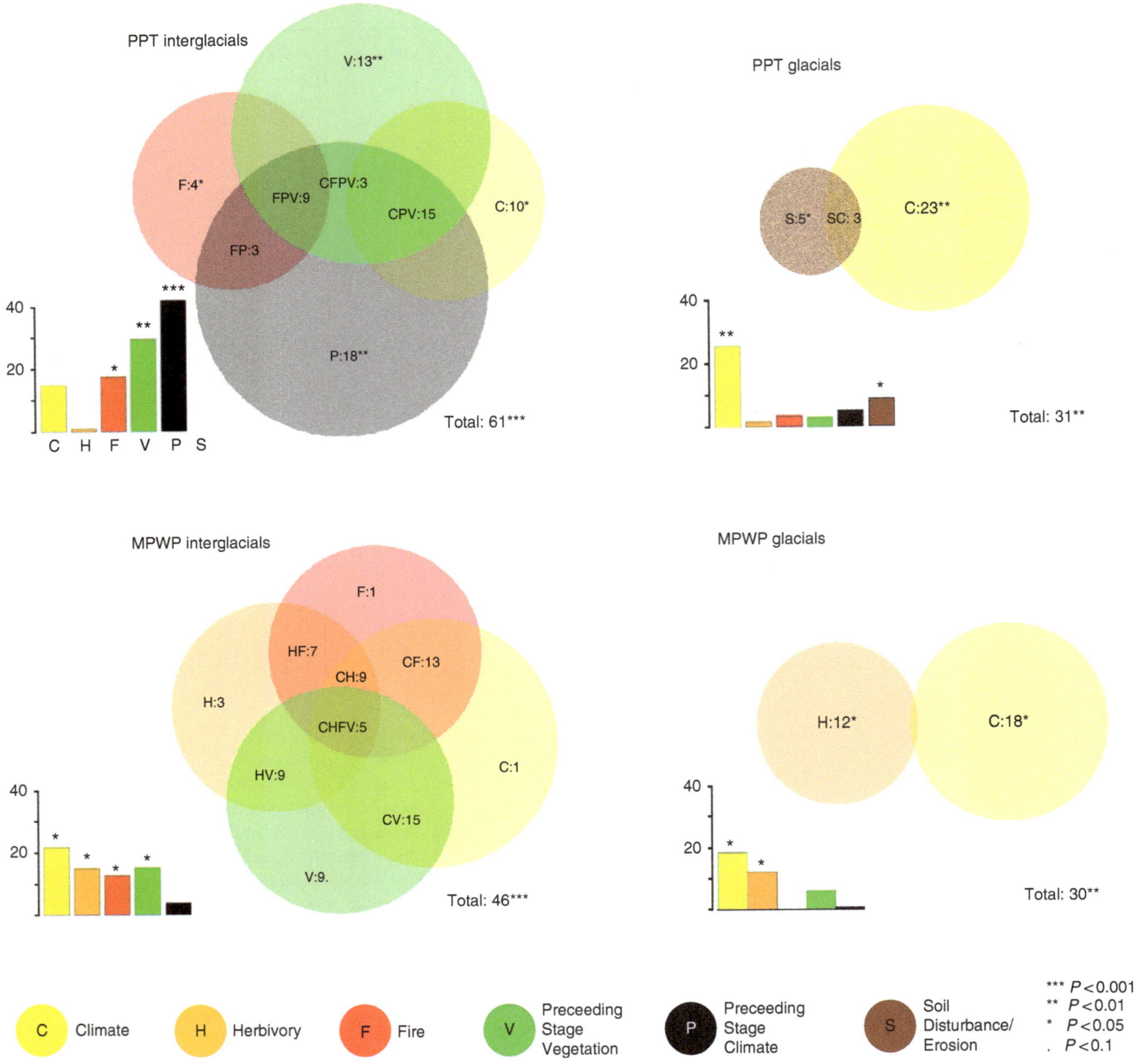

Figure 4 | Drivers of vegetation change of interglacials and glacials of the MPWP and the PPT. Bar charts indicating the variation of the pollen spectra explained by single environmental variables (climate (C), herbivory (H), fire (F), preceding-stage climate (P), preceding-stage vegetation (V) and soil disturbance erosion (S)) for each of the analysed data sets for glacials and interglacials of the PPT and of the MPWP. Venn diagrams indicate the results of variation partitioning with those variables included that separately explain a statistically significant part of the variation ($P<0.1$). Total variation explained (in %) is indicated in the lower left part of each sub-plot (see Supplementary Table 2 and Supplementary Fig. 6/7 for further detailed RDA results). Results imply that vegetation conditions of interglacials of the PPT are best explained by long-term drivers such as the climate and vegetation condition of the preceding glacial, instead of contemporaneous conditions such as climate or herbivory. In contrast, vegetation conditions of the PPT glacials and MPWP interglacials can be best explained by short-term drivers. Accordingly, we assume that vegetation conditions of interglacials with preceding cold glacials were in strong vegetation–climate disequilibria.

evergreen taxa (for example, *Picea obovata* and *Pinus sibirica*: for *Pinus pumila* see Supplementary Note 1) are restricted to the southern and western permafrost limit characterized by a summer thaw depth >1.5 m and azonal vegetation, such as on deep soils in river beds[9]. Permafrost with a shallow active layer is also assumed to limit the northward extent of *Picea* in Alaska[33]. In contrast, *Larix gmelinii* and *Larix cajanderi* can grow on soils with an active-layer depth of <40 cm (ref. 38). Thus, the question arises whether glacial-stage permafrost conditions are able to persist through strong interglacial warming and hence hinder or slow down the invasion of evergreen taxa and are therefore the key environmental link causing vegetation–climate disequilibrium. It can be assumed that permafrost during strong glacials was particularly extensive and deep because widespread continental-scale ice sheets were absent here and would not insulate soils from the very cold and harsh glacial climate[39]. Furthermore, during strong glacials, permafrost was likely more widespread due to enhanced continentality related to the absence of in-land moisture transport (that is, no snow in winter), perhaps as a result of lower sea level (60–70 m)[40] and more persistent sea-ice over the Arctic ocean[41]. Though permafrost was inferred to persist even to a few metres below the surface in

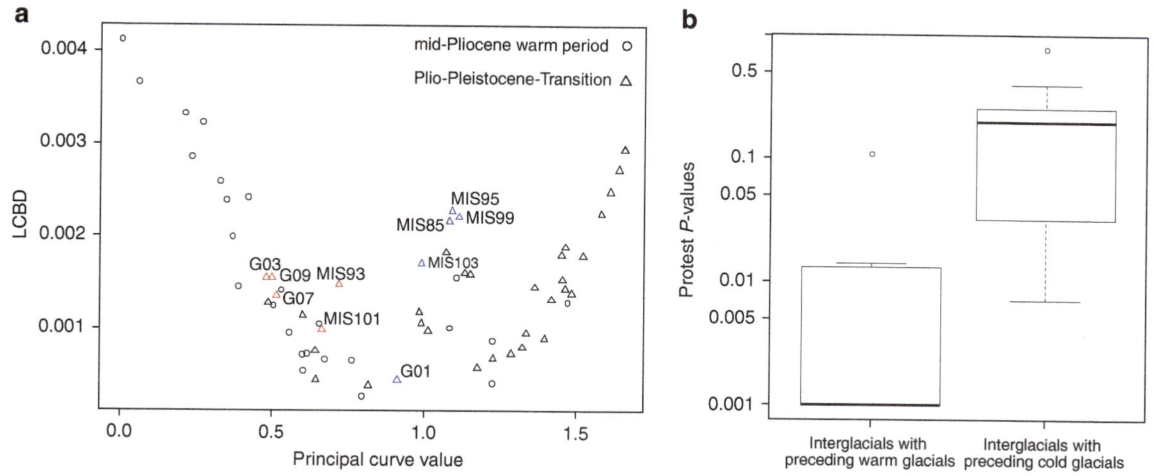

Figure 5 | Comparison of pollen-based vegetation characteristics of PPT interglacials with preceding weak glacial conditions to PPT with preceding strong glacial conditions. (**a**) Principal curve values versus Local-Contribution-to-Beta-Diversity (LCBD) values for all MPWP and PPT interglacials and glacials. Red labels denote interglacials with a preceding warm glacial (G09, G07, G03, MIS101 and MIS93), and blue labels denote interglacials with preceding cold glacials (G01, MIS103, MIS99, MIS95 and MIS85). It is clear that averaged pollen spectra of interglacials with preceding cold glacials strongly contribute to the LCBD when compared with spectra from other stages with similar principal curve values, in contrast to averaged spectra from interglacials with preceding warm glacials. (**b**) Boxplots of *P* values obtained for pairwise Procrustes comparison of ordination-derived variable scores of 13 taxa among 5 selected PPT interglacials with preceding weak (warm) glacials and among 5 selected PPT interglacials with preceding strong (cold) glacials. Whiskers represent the 1.5 time length of interquartile range. Results indicate that inter-taxa relationships among interglacials with a preceding cold glacial differ strongly among each other (that is, indicated by high Protest *P* values) while they show a significant fit among interglacials with preceding warm glacials. We assume that the vegetation histories of interglacials with preceding cold glacials were unique as a result of distant glacial refugia causing broad-scale vegetation-climate disequilibria.

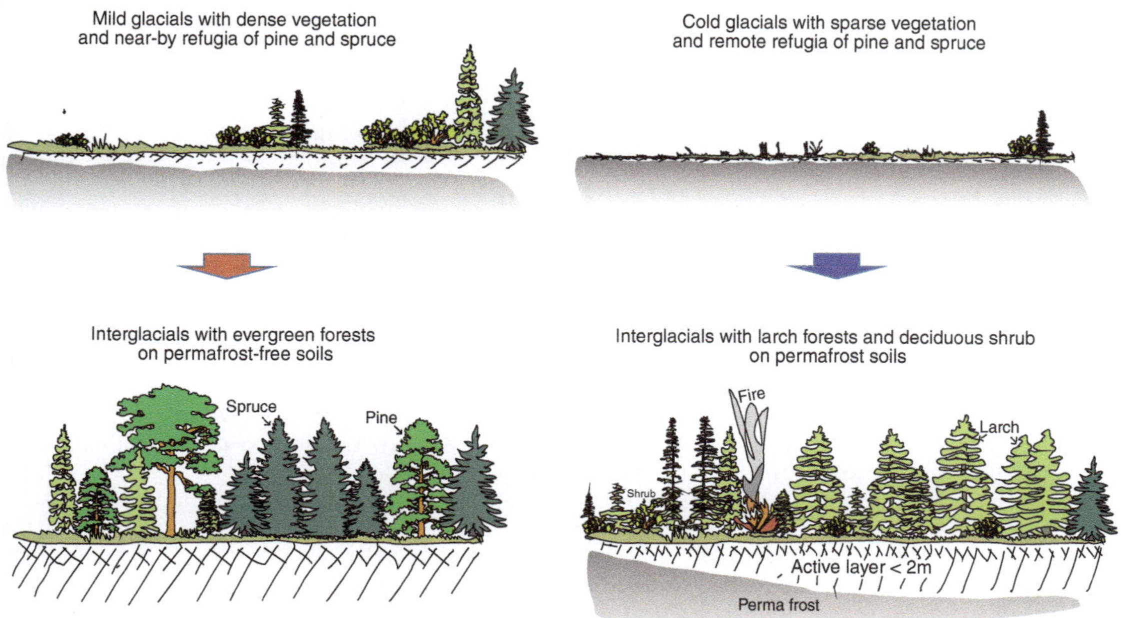

Figure 6 | Visualization of the conceptual model of the 'Glacial legacy on interglacial vegetation'. We assume that interglacial vegetation during the PPT mainly reflects conditions of the preceding glacial instead of contemporary interglacial climate. During interglacials following mild glacials, NE Asia was colonized by evergreen trees from nearby refugia. In contrast, during interglacials following cold glacials larch and deciduous shrubs expand in the area related to the combined effects of permafrost persistence and distant glacial refugia of evergreen trees, and fire. This implies that vegetation-climate disequilibrium can last for many millennia.

warmer climates during the Eemian interglacial in areas that are today located in the discontinuous permafrost zone[42], modelling results indicate that the active-layer depth responds rather directly to climate[43] while only deep permafrost may have millennial-scale lags to insolation maxima[44]. However, model studies have not been performed over long timescales and most do not include all relevant Earth-system components and their interactions. Our knowledge of permafrost response to climate change during the PPT is thus very limited.

The more southerly extensive permafrost during strong glacial stages possibly shifted the glacial refugia of *Pinus* and *Picea* farther south in addition to the severe glacial climate.

For example, the vast steppes recorded at Lake El'gygytgyn during MIS96 stretched as far south as the Lake Baikal region, which accompanied the widespread extinction of evergreen needle-leaf taxa[45]. In contrast, *Larix* (larch) probably survived in NE Russia during all glacials, by analogy to its last glacial maximum (LGM) history[46]. The pollen compositional turnover of the PPT interglacials indicates the establishment of dense larch forests with shrub *Alnus* (Supplementary Fig. 9). These forests in particular, when occurring in combination with a thick *Sphagnum* cover[47], insulate the soil from climate, thereby stabilizing the permafrost[38] and likely hinder the germination and establishment of *Picea* and *Pinus*. Such an ecosystem with a high-standing biomass in a warm, continental early interglacial climate is likely to experience frequent fires. This is indicated by the maximum *Gelasinospora* abundance coincident with the *Larix–Alnus*-type optimum in the interglacial sequences (Supplementary Fig. 9). By a study of succession in present forests from central Siberia it has been shown that the ratio of deciduous (*Larix* and short-lived shrubs) to evergreen taxa cover (*Picea* and *Pinus*) is higher at sites characterized by frequent fires[48]. Hence it is reasonable to assume that no single driver caused the expansion of deciduous shrubs and larch forests in warm interglacials but the combination of persistent permafrost, distant refugia and high fire frequency may have slowed the invasion of *Picea* and *Pinus* and thus caused long-term vegetation–climate disequilibrium (Fig. 6).

The prevailing transient vegetation types during interglacials following strong glacials have unique characteristics, which probably originated from taxon reshuffling. In contrast, during interglacials following weak glacial stages, *Pinus* and *Picea* remained in the area or in nearby refugia and invaded northern Siberia and the northern Russian Far East before shrub-rich larch forests became stabilized by feedbacks with permafrost and fire. Accordingly, vegetation–climate equilibrium, characterized by a fixed vegetation composition, was reached in a consistent manner. Hence, we conclude that disequilibrium versus equilibrium conditions differentiate interglacials with strong or weak preceding glacial states rather than a bifurcation with two stable states.

Our reasoning that permafrost presence likely contributed to the slow invasion of evergreen taxa is further supported by results from numerical analyses with the remaining averaged data sets. We find that the effect of preceding glacial drivers is much smaller during MPWP interglacials when widespread permafrost was probably absent. In contrast, contemporaneous climate during the MPWP was a main driver. For example, the warm climate supported the expansion of dark needle-leaf forests during MPWP interglacials, in particular during the two oldest recorded interglacials (MG7 and MG5) that were also among the warmest with the smallest ice-sheet extent[22,23]. Furthermore, MPWP and PPT glacial vegetation is not related to interglacial climate or vegetation conditions but is best explained statistically by contemporaneous climate. In other words, severe climate change during a glacial-to-interglacial transition causes a leading-edge disequilibrium due to slow invasion[8], while a slow die-back of long-lived trees (trailing-edge disequilibrium) is unlikely given the severe climate deterioration during an interglacial–glacial transition; the Arctic is known for its strong feedback mechanisms[13].

In addition to long-term drivers, our study provides evidence that short-term drivers can disturb the vegetation–climate relationship on glacial–interglacial timescales. Our analyses suggest that MPWP glacial and interglacial vegetation is significantly related to changes in herbivory (Fig. 4). It is assumed that large herbivores, particularly mammoths and mastodons but also deer and bison, have the potential to act as 'ecological keystones' in high-latitude vegetation[15,16,49]. The herbivore fauna of the MPWP was potentially even richer than that of MIS2 and potentially exerted its impact far into the forested areas. This is suggested by, for example, fossil records of camel ancestors on Ellesmere Island (Canadian High Arctic) formerly covered by boreal larch forests[50]. The varying herbivore abundance, as inferred from the coprophilous *Sporormiella* spore record during the MPWP, may have led to marked vegetation variation. Our results suggest that herbivory during MPWP glacials turned a deciduous shrub-tundra (less *Betula nana* type (dwarf birch) in the pollen spectra, Supplementary Fig. 6f) into an open steppe forest (more *Larix* and Poaceae (grasses)) comparable to the 'mammoth steppe' proposed for the LGM in Beringia[51]. A parallel decline of large herbivores and deciduous-shrub encroachment also occurs at the end of the late-glacial in North America[52,53] and this pattern is supported by modern exclosure experiments[53]. Accordingly, our results indicate that herbivory may represent a significant vegetation driver on glacial–interglacial timescales acting independently of climate, thereby weakening the vegetation–climate link. However, no vegetation–herbivore relationship is indicated for the PPT, which contradicts the exploitation–ecosystem hypothesis[54,55]. According to this hypothesis unproductive ecosystems can only support food-limited grazers that exert a strong control on vegetation. More productive ecosystems such as occurred during the MPWP, in contrast, should be characterized by community-level trophic cascades, which should, according to this hypothesis, result in a weaker vegetation–herbivore relationship. However, the weak vegetation–herbivore ties during the PPT can potentially be explained by a local functional extinction of the megafauna, which was previously assumed when *Sporormiella* declined to <2% of the pollen sum[53]. However, it remains unknown if our observation of generally lower herbivore densities during the PPT compared with the MPWP represents a regional peculiarity or an Arctic-wide picture. Interestingly, the two major *Sporormiella* declines, that is, at the M2/M1 and G18/G17 transitions, occur when the magnitude of late-glacial warming was particularly strong. By analogy with the climate-driven extinction of some megaherbivores during the MIS2/Holocene transition[56] we can only speculate that random processes such as extinctions during severe warming, at least partly, explain the observed herbivore density pattern around Lake El'gygytgyn. Furthermore, the potential refugia for cold-restricted plants and animals in the Asian Arctic are, compared with the North American Arctic, particularly small because the arctic coast is located further to the south and because the Bering Land Bridge was submerged during interglacials[40], which may have enhanced random effects in interglacial colonization processes.

Our results suggest that vegetation–climate disequilibrium may have lasted for millennia. With the background of our results and the severity of the LGM (MIS2) we hypothesize that today's widely distributed larch forests lacking *Picea* and *Pinus* may not be in equilibrium with contemporary climate. Support for this comes from the fact that compared with *Larix*, which colonized Yakutia during the late-glacial, evergreen taxa expanded only during the early- and mid-Holocene[57]. However, they did not significantly extend their ranges beyond the present-day limit during the Holocene thermal maximum, in contrast to *Larix*[46]. Furthermore, evergreen taxa in southern and western Siberia probably represent late-successional vegetation in present-day larch-forest regions but they only become established when fire frequency is low[48].

Vegetation records are commonly used to infer past temperature quantitatively, which is then compared with modelled or reconstructed atmospheric CO_2 concentrations to assess Earth-system sensitivity[58]. When either the modern analogues or fossil records originate from vegetation in disequilibrium with climate, the inferred climate may be misleading. As expected from our results, the PPT temperature reconstruction based on Elgygytgyn pollen data[30] poorly reflects the global SST[23] and global ice-volume[22] trends. This suggests that pollen-based quantitative climate reconstructions are less reliable than previously assumed, particularly for interglacials. However, independent regional climate records are necessary to quantify the uncertainties of pollen-based climate reconstructions. Likewise, only transient Earth-system simulations that include realistic long-term tree-population and permafrost dynamics and run over several glacial–interglacial cycles can resolve the timing and effectiveness of vegetation-related feedback mechanisms and thus provide realistic model-based estimates for Earth-system sensitivity.

With the appearance of permafrost and the functional disappearance of herbivory, the PPT at northern high latitudes represents a valuable natural experiment in a palaeoenvironmental setting with the potential to understand better the dynamics of Earth's past states when new system components are included.

Methods

Numerical analyses of original Lake El'gygytgyn pollen data. This paper uses a pollen data set from Lake El'gygytgyn[21]. Of the 1,166 samples that were analysed, 1,043 contained pollen. Of these, 844 samples with at least 150 terrestrial pollen grains and spores were included in the numerical analyses here. The pollen sum is >300 in most samples.

Principal curves[59] were derived to best summarize, in a mathematical sense, the overall pollen compositional turnover and thus to maximize the variation represented in one dimension. To increase the signal-to-noise ratio, we *a priori* decided to include only half of the original 92 reliably recorded taxa (that is, taxa recorded at least twice in the entire data set); thus, all taxa with occurrences in >74 samples were retained for the principal curve analyses. Sample scores on the first correspondence axis were used as a starting curve. The principal curve was fitted using cubic smoothing splines allowing the complexity of the individual smoothers used in the local averaging step to vary between pollen taxa, using generalized cross-validation to select the optimal degree of smoothing for each taxon. A penalty term of 1.4 was used to increase the cost of degrees of freedom in the generalized cross-validation calculations[60]. Analyses were implemented in R using the pcurve package.

Analyses of averaged interglacial and glacial pollen data. To obtain averaged interglacial and glacial data sets, principal curve minima and maxima were first assigned to marine isotope stages[22] based on the original age model[20]. MIS84 was split into two glacial stages (MIS84a/c) and an interglacial stage (MIS84b) because a full glacial–interglacial turnover is clearly reflected in the pollen data. Boundaries between glacial and interglacials were set at the point where the mean principal curve value was crossed, calculated from the minimum principal curve values for interglacials and the maximum principal curve values for glacials. Original counts of those samples that are below/above the median PC values for each glacial/interglacial were then summed for all taxa. This step was done to ensure that the averaged data sets represent the 'characteristic' pollen assemblages of each stage rather than the 'transitional' assemblages between stages. To reduce the noise in the data set, percentages were calculated including vascular-plant pollen taxa that occur in at least 12 stages with a value of at least 2% once. Finally, a data set of 32 averaged interglacial pollen spectra and another of 32 averaged glacial pollen spectra were obtained.

Principal curve analysis of the combined glacial–interglacial averaged data set was run with a similar configuration as for the original data set, only the penalty term was set to 2 to account for the lower number of samples. Local-Contributions-To-Beta-Diversity[32], that is, representing an estimation of the compositional uniqueness of a stage was likewise calculated using the combined averaged glacial–interglacial data set.

Correlation analyses of global climate data sets. We analysed the global LR04 benthic inverted $\delta^{18}O_{stack}$ (ref. 22) as well as a global collection of high-resolution SST records. We required that the full PPT (2,150–2,900 kyr ago) is covered with a mean resolution of at least 5 kyr per sample. Ten published UK[33] records and the LR04 benthic stack meet these criteria (Supplementary Table 1).

The irregular raw data of every core were low-pass filtered using a Gaussian Kernel Smoother with an effective cutoff frequency of 1/15 kyr. The assignment of the glacial minima and interglacial maxima was performed in two steps. (1) Local maxima and minima were automatically identified from the smoothed time series. (2) All records were visually compared using their published age model, and the corresponding glacial minima and interglacial maxima were manually chosen from the minima and maxima of step 1. We only interpreted time periods where a 100-kyr window did not contain a data gap >15 kyr to reduce the possibility that we missed major peaks or troughs. In some cases, where no clear assignment was possible, no assignment was made; thus most records only contain a subset of all extremes. For ODP1082 and ODP1012, only a small number of maxima and minima could be assigned, probably due to the poor data quality of the records in the time period of interest. We thus excluded both these cores from further analysis. As the ODP982 SST record does not have a high enough resolution until 2,600 kyr ago, we correlated ODP982 and LR04 on an extended time window (2,600–3,500 kyr ago) to obtain a meaningful overlap with the LR04 stack. The results are summarized in Supplementary Fig. 4.

We also investigated whether the coherency between glacial and interglacial intensities was restricted to the marine SST records. For the last 800 kyr, we analysed several proxies derived from Antarctic ice cores[34]: water isotopes related to local temperature; a methane record potentially related to low-latitude wetland dynamics; and the CO_2 record representing the global carbon cycle. We compared these records with the global benthic $\delta^{18}O_{stack}$ (ref. 22) using the same method as in the previous section (Gaussian smoothing, automatic local extremes identification and manual assignment according to the marine isotope stages[22]) (Supplementary Fig. 5). We used the same technique of comparing local extremes instead of analysing the same points in time to be consistent with our main analyses of the Plio-Pleistocene time series.

Redundancy analyses. Constrained ordination analyses were run to relate statistically the variation in the averaged glacial and interglacial pollen data sets to the proxy environmental variables. RDA was chosen over canonical correspondence analysis because of the compositional gradient lengths of the pollen data being <2 s.d. as revealed by initial detrended correspondence analyses, indicating that linear-based ordination methods are appropriate with these data sets. Because the boundary conditions strongly changed, we *a priori* separated the data of the MPWP (3,510–2,900 kyr ago, 12 glacials, 12 interglacials) from the PPT (2,900–2,150 kyr ago, 20 glacials, 20 interglacials) following definitions of the PLIOMAX group[61]. This boundary also accounts for the finding that the major decrease in atmospheric CO_2 concentrations occurred around 2,900 kyr ago (ref. 62), that the phase of interglacials having higher than present sea-level ended[40] and that ice-rafted debris markedly increased in glacial stages from MIS16 onwards[63]. Setting the boundary to that time also agrees with the results of the biome reconstruction using Lake El'gygytgyn pollen data, which indicates that MIS17 was the last interglacial that was at least temporarily dominated by cool coniferous forests[29].

The climate proxy set includes the $\delta^{18}O_{stack}$ record[22] and two alkenone-based SST records from the Pacific as no other suitable regional climate proxy data are available. Total organic carbon (TOC) values or Si/Ti ratios from Lake El'gygytgyn were not used as proxy environmental variables because terrestrial vegetation changes could have strongly impacted lake chemistry[64], thereby risking circular reasoning in our analyses. We selected the LR04 $\delta^{18}O_{stack}$ record[22], which, in addition to the temperature-sensitive ice-sheet extent, reflects the sea-level position and thus variability in continentality in northern Asia. Furthermore, the proxy set includes the high-resolution alkenone-based SST record from the northern mid-latitude ODP site 1208 in the Pacific[30]. This SST record represents the closest record to Lake El'gygytgyn and is located near to the potential glacial refugia in north-eastern Asia and has, furthermore, a high data quality. Because the SST record from ODP site 1208 does not fully cover the MPWP the alkenone-based SST from the eastern Pacific ODP846 site[23], which is well correlated to ODP site 1208 during the PPT[65], is used for analysing the MPWP. The $\delta^{18}O$ stack and SST records were interpolated to 2.5 kyr ago time slices and the maximum/minimum value for each interglacial/glacial was used. The same proxy records were used as variables to represent the preceding stage climate.

Square-root-transformed values of *Sporormiella* spore percentages (relative to the vascular-plant pollen sum), a coprophilous fungus commonly occurring in the palynomorph record, were used as a herbivore intensity proxy[31]. Likewise, percentages of *Gelasinospora* spores, which are often found at times of massive charcoal input[66,67], were used as a fire-frequency proxy. Support for this comes from the observation that the occurrence of *Gelasinospora* is significantly related to burnt soils[26], which is explained by its rapid growth rate and the resistance of its spores to high temperatures compared to other fungi. Preceding stage PC values of the combined glacial/interglacial averaged data set were used to explore long-term vegetation drivers. Furthermore, percentages of *Glomus* spores were used to indicate changes in the disturbance and erosion of soils[28].

The following sets of RDAs were run on each of the Hellinger-transformed glacial and interglacial averaged pollen data sets. Those variables (or variable sets), which when solely included in an initial RDA yielded a statistically significant relationship with the data set ($P<0.1$), were included together in variation partitioning to extract the uniquely explained variation by each single environmental variable (or variable set) and the shared explained variation. Explained variations represent adjusted r^2 values[68].

Analyses of PPT interglacial vegetation histories. We aimed to investigate whether pollen inter-taxa relationships among PPT interglacials with preceding warm glacials (that is, 'warm interglacials') are different from those among interglacials with preceding cold glacials (that is, 'cold interglacials'). For that purpose, warm interglacials (that is, having a combination of low $\delta^{18}O_{stack}$ and high SST values compared with all interglacials during the PPT, and having a minimum of 12 samples: G9, G7, G3, MIS101 and MIS93) and cold glacials (that is, having a combination of high $\delta^{18}O_{stack}$ and low SST, and having a minimum of 12 samples: G1, MIS103, MIS99, MIS95 and MIS83) were selected based on the climate proxy information and the minimum sample number. Inter-taxa relationships of a single interglacial are reflected by the species scores in an ordination. Accordingly, principal component analyses (with scaling focused on variable correlations) were performed on the original pollen samples of the respective warm or cold interglacials. Analyses were based on Hellinger-transformed pollen percentages of taxa that occur at least three times in each stage. To compare the ordination results of two interglacials the correlation of variable scores on the first four principal component analysis (PCA) axes was assessed using Procrustes analysis[69] and its associated Protest test[70]. Finally, the Protest P values gained from pair-wise comparisons among all 'warm interglacials' and among all 'cold interglacials' were presented in boxplots.

Statistical programmes used and data availability. All analyses were implemented in R version 3.03 using functions in packages analogue 0.12-0, calibrate 1.7.2, pcurve 0.6-5, vegan 2.0-10 and the function beta.div. Averaged data sets and environmental proxy data used for redundancy analyses are available at www.pangaea.de.

References

1. Shugart, H. H. & Woodward, F. I. *Global Change and the Terrestrial Biosphere* (Wiley-Blackwell, 2011).
2. Vazquez-Rivera, H. & Currie, D. J. Contemporaneous climate directly controls broad-scale patterns of woody plant diversity: a test by a natural experiment over 14 000 years. *Global Ecol. Biogeogr.* **24,** 97–106 (2014).
3. Blois, J. L., Williams, J. W., Fritzpatrick, M. C., Jackson, S. T. & Ferrier, S. Space can substitute for time in predicting climate-change effects on biodiversity. *Proc. Natl. Acad. Sci. USA* **110,** 9374–9379 (2013).
4. Svenning, J.-C. Deterministic Plio-Pleistocene extinctions in the European cool-temperate tree flora. *Ecol. Lett.* **6,** 646–653 (2003).
5. Smith, A. G. Problems of inertia and threshold related to post-glacial habitat changes. *Proc. R. Soc. London B* **161,** 331–342 (1965).
6. Prentice, I. C., Bartlein, P. J. & Webb, T. Vegetation and climate change in eastern North-America since the Last Glacial Maximum. *Ecology* **72,** 2038–2056 (1991).
7. Giesecke, T. *et al.* The pace of Holocene vegetation change—testing for synchronous developments. *Quat. Sci. Rev.* **30,** 2805–2814 (2011).
8. Svenning, J.-C. & Sandel, B. Disequilibrium vegetation dynamics under future climate change. *Am. J. Bot.* **100,** 1266–1286 (2014).
9. Tzedakis, P. C. & Bennett, K. D. Interglacial vegetation succession: a view from southern Europe. *Quat. Sci. Rev.* **14,** 967–982 (1995).
10. Cheddadi, R. *et al.* Similarity of vegetation dynamics during interglacial periods. *Proc. Natl. Acad. Sci. USA* **102,** 13939–13943 (2005).
11. Normand, S. *et al.* Postglacial migration supplements climate in determining plant species range in Europe. *Proc. R. Soc. B* **278,** 3644–3653 (2011).
12. Blonder, B. *et al.* Linking environmental filtering and disequilibrium to biogeography with a community climate framework. *Ecology* **96,** 972–985 (2015).
13. Miller, G. H. *et al.* Arctic amplification: can the past constrain the future? *Quat. Sci. Rev.* **29,** 1779–1790 (2010).
14. Vandenberghe, J. *et al.* The Last Permafrost Maximum (LPM) map of the Northern Hemisphere: permafrost extent and mean annual air temperatures, 25-17 ka BP. *Boreas* **43,** 652–666 (2014).
15. Owen-Smith, R. N. *Megaherbivores: The Influence of Very Large Body Size on Ecology* (Cambridge Univ. Press, 1988).
16. Gill, J. L. Ecological impacts of the late Quaternary megaherbivore extinctions. *New Phytol.* **201,** 1163–1169 (2014).
17. Bond, W. J. & Keeley, J. E. Fire as a global 'herbivore': the ecology and evolution of flammable ecosystems. *Trends Ecol. Evol.* **20,** 387–394 (2005).
18. Komárková, V. Vegetation type hierarchies and landform disturbance in arctic Alaska and alpine Colorado with emphasis on snowpatches. *Vegetatio* **106,** 155–181 (1993).
19. Kienast, F. *et al.* Paleontological records indicate the occurrence of open woodlands in a dry inland climate at the present-day Arctic coast in western Beringia during the Last Interglacial. *Quat. Sci. Rev.* **30,** 2134–2159 (2011).
20. Brigham-Grette, J. *et al.* Pliocene warmth, polar amplification and stepped Pleistocene cooling recorded in NE Arctic Russia. *Science* **340,** 1421–1427 (2013).
21. Andreev, A. A. *et al.* Late Pliocene and Early Pleistocene environments of the north-eastern Russian Arctic inferred from the Lake El'gygytgyn pollen record. *Clim. Past* **10,** 1017–1039 (2014).
22. Lisiecki, L. E. & Raymo, M. E. A Pliocene-Pleistocene stack of 57 globally distributed benthic $\delta^{18}O$ records. *Paleoceanography* **20,** PA1003 (2005).
23. Lawrence, K. T., Liu, Z. & Herbert, T. D. Evolution of the Eastern Tropical Pacific through Plio-Pleistocene Glaciation. *Science* **312,** 79–83 (2006).
24. Prescott, C. L. *et al.* Assessing orbitally-forced interglacial climate variability during the mid-Pliocene Warm Period. *Earth Planet. Sci. Lett.* **400,** 261–271 (2014).
25. Commandini, O. & Rinaldi, A. C. Tracing megafaunal extinctions with dung fungal spores. *Mycologist* **18,** 140–142 (2004).
26. Widden, P. & Parkinson, D. The effects of a forest fire on soil microfungi. *Soil Biol. Biochem.* **7,** 125–138 (1975).
27. Anderson, R. S., Homola, R. L., Davis, R. B. & Jacobson, G. L. Fossil remains of the mycorrhizal fungal *Glomus fasciculatum* complex in postglacial lake sediments from Maine. *Can. J. Bot.* **62,** 2325–2328 (1984).
28. Van Geel, B. in *Tracking Environmental Change Using Lake Sediments, vol. 3: Terrestrial, Algal and Silicaceous Indicators.* (eds Smol, J. P. Birks, H. J. B. & Last, W. M.) 99–119 (Kluwer, 2001).
29. Tarasov, P. E. *et al.* A pollen-based biome reconstruction over the last 3.562 million years in the Far East Russian Arcti—new insights into climate–vegetation relationships at the regional scale. *Clim. Past* **9,** 2759–2775 (2013).
30. Venti, N. L., Billups, K. & Herbert, T. D. Increased sensitivity of the Plio-Pleistocene northwest Pacific to obliquity forcing. *Earth Planet. Sci. Lett.* **384,** 121–131 (2013).
31. Isaev, A. P. *et al.* in *The Far North: Plant Biodiversity and Ecology of Yakutia.* (eds Troeva, E. I. *et al.*) 143–260 (Springer, 2010).
32. Legendre, P. & DeCaceres, M. Beta diversity as the variance of community data: dissimilarity coefficients and partitioning. *Ecol. Lett.* **16,** 951–963 (2013).
33. Lloyd, A. Ecological histories from Alaskan tree lines provide insight into future change. *Ecology* **86,** 1687–1695 (2005).
34. Masson-Delmotte, V. *et al.* EPICA Dome C record of glacial and interglacial intensities. *Quat. Sci. Rev.* **29,** 113–128 (2010).
35. Mitchell, J. M. *et al.* An overview of climatic variability and its causal mechanisms. *Quat. Res.* **6,** 481–493 (1976).
36. Jones, P. D., Osborn, T. J. & Briffa, K. R. Estimating sampling errors in large-scale temperature averages. *J. Clim.* **10,** 2548–2568 (1997).
37. Willeit, M., Ganopolski, A., Calov, R., Robinson, A. & Maslin, M. The role of CO_2 decline for the onset of Northern Hemisphere glaciation. *Quat. Sci. Rev.* **119,** 22–34 (2015).
38. Osawa, A., Zyryanova, O. A., Matsuura, Y., Kajimoto, T. & Wein, R. W. (eds) *Permafrost Ecosystems: Siberian Larch Forests* (Springer 502, 2010).
39. French, H. *The Periglacial Environment* (Wiley 478 p, 2007).
40. Miller, K. G., Mountain, G. S., Wright, J. D. & Browning, J. V. A 180-millon-year record of sea level and ice volume variations from continental margin and deep-sea isotopic records. *Oceanography* **24,** 40–53 (2011).
41. Polyak, L. *et al.* History of sea ice in the Arctic. *Quat. Sci. Rev.* **29,** 1757–1778 (2010).
42. Reyes, A. V., Froese, D. G. & Jensen, B. J. L. Permafrost response to last interglacial warming: field evidence from non-glaciated Yukon and Alaska. *Quat. Sci. Rev.* **29,** 3256–3274 (2010).
43. Jiang, Y., Zhuang, Q. & O'Donnell, J. A. Modeling thermal dynamics of active layer soils and near-surface permafrost using a fully coupled water and heat transport model. *J. Geophys. Res.* **117,** D11110 (2012).
44. Galushkin, Y. Numerical simulation of permafrost evolution as a part of sedimentary basin modelling: permafrost in the Pliocene-Holocene climate history of the Urengoy field in the West Siberian basin. *Can. J. Earth Sci.* **34,** 935–948 (1997).
45. Demske, D., Mohr, B. & Oberhänsli, H. Late Pliocene vegetation and climate of Lake Baikal region, southern East Siberia, reconstructed from palynological data. *Palaeogeogr. Palaeoclim. Palaeoecol.* **184,** 107–129 (2002).
46. Polezhaeva, M., Laxcoux, M. & Semerikov, V. L. Cytoplasmic DNA variation and biogeography of *Larix* Mill. in Northeast Asia. *Mol. Ecol.* **19,** 1239–1252 (2010).
47. Harden, J. W., Manies, K. L., Neff, J. C. & Turetsky, M. R. Effects of wildfire and permafrost on soil organic matter and soil climate in interior Alaska. *Global Change Biol.* **12,** 1–13 (2006).
48. Schulze, E.-D. *et al.* Factors promoting larch dominance in central Siberia: fire versus growth performance and implications for carbon dynamics at the boundary of evergreen and deciduous conifers. *Biogeosciences* **9,** 1405–1421 (2012).
49. Gill, J. L., Williams, J. W., Jackson, S. T., Donnelly, J. P. & Schellinger, G. C. Climatic and megaherbivory controls on late-glacial vegetation dynamics: a new high-resolution, multi-proxy record from Silver Lake, Ohio. *Quat. Sci. Rev.* **34,** 66–80 (2012).
50. Rybczynski, N. *et al.* Mid-Pliocene warm-period deposits in the High Arctic yield insight into camel evolution. *Nat. Commun.* **4,** 1550 (2013).
51. Blinnikov, M. S., Gaglioti, B., Walker, D. A., Wooller, M. J. & Zazula, G. D. Pleistocene graminoid-dominated ecosystems in the Arctic. *Quat. Sci. Rev.* **30,** 2906–2929 (2011).
52. Gill, J. L., Williams, J. W., Jackson, S. T., Lininger, K. B. & Robinson, G. S. Pleistocene megafaunal collapse, novel plant communities, and enhanced fire regimes in North America. *Science* **326,** 1100–1103 (2009).

53. Bakker, E. S. *et al.* Combining paleo-data and modern exclosure experiments to assess the impact of megafauna extinctions on woody vegetation. *Proc. Natl. Acad. Sci. USA* **113,** 847–855 (2016).
54. Oksanen, L. & Oksanen, T. The logic and realism of the hypothesis of exploitation ecosystems. *Am. Nat.* **155,** 703–723 (2000).
55. Anapuu, M. *et al.* Spatial pattern and dynamic responses of Arctic food webs corroborate the exploitation ecosystems hypothesis (EEH). *Am. Nat.* **171,** 249–262 (2008).
56. Lorenzen, E. *et al.* Species-specific responses of Late Quaternary megafauna to climate and humans. *Nature* **479,** 359–364 (2011).
57. Binney, H. A. *et al.* The distribution of late-Quaternary woody taxa in northern Eurasia: evidence from a new macrofossil database. *Quat. Sci. Rev.* **28,** 2445–2464 (2009).
58. Schmittner, A. *et al.* Climate sensitivity estimated from temperature reconstructions of the Last Glacial Maximum. *Science* **334,** 1385–1388 (2011).
59. De'ath, G. Principal curves: a new technique for indirect and direct gradient analysis. *Ecology* **80,** 2237–2253 (1999).
60. Simpson, G. L. & Birks, H. J. B. in *Tracking Environmental Change Using Lake Sediments, vol. 5: Data Handling and Numerical Techniques.* (eds Birks, H. J. B., Lotter, A. F., Juggins, S. & Smol, J. P.) 249–327 (Springer, 2012).
61. Pliomax Group http://pliomax.org/pliowiki/index.php/Welcome_to_PLIOMAX.
62. Seki, O., Foster, G. L., Schmidt, D. N. & Mackensen, A. Alkenone and boron-based Pliocene pCO_2 records. *Earth Planet. Sci. Lett.* **292,** 201–211 (2010).
63. Flesche Kleiven, H., Jansen, E., Fronval, T. & Smith, T. M. Intensification of Northern Hemisphere glaciations in the circum-Atlantic region (3.5–2.4 Ma)—ice-rafted detritus evidence. *Palaeogeogr. Palaeoclim. Palaeoecol.* **184,** 213–223 (2002).
64. Herzschuh, U. *et al.* Siberian larch forests and the ion content of thaw-lakes form a geochemically functional entity. *Nat. Commun.* **4,** 2408 (2013).
65. Nie, J. *et al.* Surface-water freshening: a cause for the onset of the North Pacific stratification from 2.75 Ma onward? *Global Planet. Change* **64,** 49–52 (2008).
66. Anderson, R. S., Ejarque, A., Brown, P. M. & Hallett, D. J. Holocene and historical vegetation change and fire history on the north-central coast of California, USA. *Holocene* **23,** 1797–1810 (2013).
67. Froyd, C. A. *et al.* The ecological consequences of megafaunal loss: giant tortoises and wetland biodiversity. *Ecol. Lett.* **17,** 144–154 (2014).
68. Peres-Neto, P., Legendre, P., Dray, S. & Borcard, D. Variation partitioning of species data matrices: estimation and comparison of fractions. *Ecology* **87,** 2614–2625 (2006).
69. Peres-Neto, P. & Jackson, D. A. How well do multivariate sets match? The advantages of a Procrustean superimposition approach over the Mantel test. *Oecologia* **129,** 169–178 (2001).
70. Jackson, D. A. PROTEST: a Procrustes randomisation test of community environment concordance. *Ecoscience* **2,** 297–303 (1995).

Acknowledgements

Arne Ramisch and Stefan Kruse contributed ideas and discussion to a very early version of the data-set analyses. We acknowledge Chris Brierley, Petra Dekens, Zhonghui Liu as well as the other data contributors for making their SST data sets publicly available and David Naafs for discussion of SST records. Cathy Jenks is thanked for her meticulous editing. This study was funded by the German Federal Ministry of Education and Research (grant no.03G0642A) and the German Research Foundation (grant nos. HE 3622/20 and ME 1169/24).

Author contributions

U.H. designed and performed the analyses with advice from H.J.B.B.; U.H. wrote the first version of the manuscript that was revised by H.J.B.B.; T.L. performed the analyses of global climate proxy data. All authors contributed to discussion and revision of the final manuscript.

Additional information

Competing financial interests: The authors declare no competing financial interests.

Uncertainty in soil data can outweigh climate impact signals in global crop yield simulations

Christian Folberth[1,2], Rastislav Skalský[1,3], Elena Moltchanova[1,4], Juraj Balkovič[1,5], Ligia B. Azevedo[1], Michael Obersteiner[1] & Marijn van der Velde[6]

Global gridded crop models (GGCMs) are increasingly used for agro-environmental assessments and estimates of climate change impacts on food production. Recently, the influence of climate data and weather variability on GGCM outcomes has come under detailed scrutiny, unlike the influence of soil data. Here we compare yield variability caused by the soil type selected for GGCM simulations to weather-induced yield variability. Without fertilizer application, soil-type-related yield variability generally outweighs the simulated inter-annual variability in yield due to weather. Increasing applications of fertilizer and irrigation reduce this variability until it is practically negligible. Importantly, estimated climate change effects on yield can be either negative or positive depending on the chosen soil type. Soils thus have the capacity to either buffer or amplify these impacts. Our findings call for improvements in soil data available for crop modelling and more explicit accounting for soil variability in GGCM simulations.

[1] Ecosystem Services and Management Program, International Institute for Applied Systems Analysis, 2361 Laxenburg, Austria. [2] Department of Geography, Ludwig Maximilian University, 80333 Munich, Germany. [3] Soil Science and Conservation Research Institute, National Agricultural and Food Centre, 82713 Bratislava, Slovak Republic. [4] School of Mathematics and Statistics, University of Canterbury, Christchurch 8140, New Zealand. [5] Department of Soil Science, Faculty of Natural Sciences, Comenius University, 84104 Bratislava, Slovak Republic. [6] European Commission, Joint Research Centre, 21027 Ispra, Italy. Correspondence and requests for materials should be addressed to C.F. (email: folberth@iiasa.ac.at).

Crop growth simulation models have become an indispensable tool for estimating future impacts of climate change on crop yield[1–6]. Projections generally indicate adverse effects of climate change on crop yield at low latitudes and mixed effects at mid and high latitudes[2,6]. Uncertainties can be traced to different spatial and temporal sources including input data, model structure and process parameterization. The use of multiple general circulation model (GCM) projections has become the norm to characterize the uncertainty in climate projections for crop impact studies[1–6]. Recently, efforts have been made to characterize the uncertainty in crop model processes and setups themselves. Forcing an ensemble of seven global gridded crop models (GGCMs) with the same set of GCM projections resulted in comparable present-day yield levels but relative climate change impacts on crop yields ranged from about − 40 to + 25% by the 2090s for a high CO_2 emission pathway[6]. Similarly, an ensemble of 27 field-scale wheat models exhibited a wide range of possible yield changes under high CO_2 and temperature change of -100 to + 100% in extreme cases[7].

Surprisingly, soils have not been recognized as a key element in agricultural climate change impact studies on the global scale. The water and nutrient storage capacity of soils enables them to sustain crop growth during periods of adverse conditions and to either buffer[8,9] or reinforce[10] impacts of climate variability. Hence, investigating the uncertainty in global soil data and its impact on global crop simulations is essential. This uncertainty encompasses at least two major components, which are the quality of global soil data in terms of accuracy and range of measured soil characteristics on the one hand and the correct spatial allocation of soil types to cropland within GGCMs on the other.

Here we focus on the second component and its relevance for interactions between climate and soil characteristics in regulating plant growth functions. Soils are generally characterized by high spatial variability. Global soil data sets usually consist of maps delineating so-called soil mapping units (SMUs), representing regions of similar topography and soil genesis, and a database containing certain characteristics of a number of soil types. Thereby, one or several soil types can be linked to each SMU[11,12]. Alternatively, such databases provide data on selected soil characteristics that can be interpolated to a high spatial resolution[13]. As GGCMs usually run at a spatial resolution of 0.5° × 0.5° (about 50 × 50 km^2 near the equator)[2,6], various soil types or combinations of soil parameters may in both cases occur within one simulation unit. To our knowledge, GGCM studies generally use the soil that is dominant with respect to coverage[1–6,14,15]. In reality, it is not known *a priori* which one of these soil types is being used for agricultural production. Farmers are likely to grow their crops either on the soil most suitable for cultivation or choose the soil based on socio-economic considerations or limitations such as land tenure[16].

A rigorous evaluation and GGCM sensitivity analysis to assess the importance of the uncertainty caused by soil data relative to climate input data and farm management is therefore crucial. The findings of recent regional studies that investigated how selection or extrapolation of soil data influences simulated crop yields are conflicting: Zhang *et al.*[17] found that finer resolution soil data improve model performance in high-resolution crop simulations in the US mid-west, with only small differences in average crop yields but large deviations in the spatial representation of yields and carbon fluxes. By contrast, Angulo *et al.*[18] found only marginal differences in simulated yields when aggregating soil data. They attributed their findings to high precipitation in the study region (Northwest Germany) and the algorithms used for estimating hydraulic parameters. Waha *et al.*[19] finally concluded that the choice of crop model (APSIM or LPJmL) and climate data are greater sources of uncertainty than soil data in crop simulations for West Africa.

Here we used the Harmonized World Soil Database[12] (HWSD; version 1.2 last updated 2012) to evaluate the impact of soil input data on yield estimates in a GGCM. Aggregating the HWSD to a 0.5° × 0.5° resolution resulted in unique combinations of up to 77 soil profiles per grid cell. The GGCM was based on the field-scale model EPIC[20] (Environmental Policy Integrated Climate) and maize was selected as a representative crop. To account for effects of soil nutrient and water supply, we simulated a total of six management scenarios, consisting of no (no-nut), business as usual (bau-nut) or sufficient (high-nut) mineral fertilizer supply, each combined with purely rainfed cultivation or sufficient irrigation water supply. Yield variability was assessed using the coefficient of variation (CV, equation (7)). With respect to meteorological data, yields may vary between climate data sets in general and inter-annually due to inter-annual climate variability. Among different soil types, yields vary due to differences in nutrient availability, soil nutrient retention capacity and water-holding capacity. Examples of processes driven by both climate and soil factors are soil hydrology, soil temperature and evapotranspiration. To disentangle the signature of climate and soil on simulated yields, we compare the inter-annual yield variability solely associated with the dominant soil type (CV_{dom}) to the variability resulting from the full range of soil types and annual yield estimates in each grid cell (CV_{tot}). In this way, we can quantify the soil-related yield signals that are lost if only the dominant soil is used. The comparison of CV_{dom} and CV_{soil}, which describes the variability of mean yields across all soil types in a grid cell, shows whether climate variability or uncertainty about the cultivation of soil types in the respective grid cells has a larger effect on yield estimates. The latter is primarily the case under agricultural low input conditions, whereas the first dominates under high input conditions and in regions with agriculturally adverse climates. We perform an assessment of the climate change impacts on simulated maize yield to highlight the importance of soil data for agricultural climate change impact studies. Soil data choices can cause differences in the magnitudes and - in extreme cases - also the direction of climate change impacts.

Results

Differences in climate- and soil-related yield variability. The yield variability over a 10-year period and the total range of possible soil types CV_{tot} is substantially higher than the solely climate-driven 10-year yield variability CV_{dom} under most crop management configurations across the different climate regions (Fig. 1a–f and Table 1). This is particularly true under irrigated conditions without any exogenous nutrient application or with present-day fertilizer supply in low input regions (Fig. 1b,d). The impact of water deficit on yields is virtually eliminated under these conditions and soil nutrient supply becomes the main driver for yield variability. The soil ensemble-driven yield variability CV_{tot} and the climate-driven yield variability CV_{dom} are nearly identical if sufficient nutrients and irrigation water are supplied (Fig. 1f). By contrast, rainfed conditions raise both CV_{dom} and CV_{tot} substantially (Figs 1a,c,e and 2a,c,e) as precipitation patterns and soil hydrologic characteristics become major factors determining yield variability. This is most apparent in (semi-)arid regions with CVs often above 150%. In the absence of fertilizer supply, CV_{tot} is higher than CV_{dom} in more than 89% of all grid cells (Table 1). Medians of CV_{tot} and CV_{dom} (Supplementary Table 2) differ by factors of about 1.2–2.5 in arid and 1.8–3.1 in non-arid climates with larger differences in the

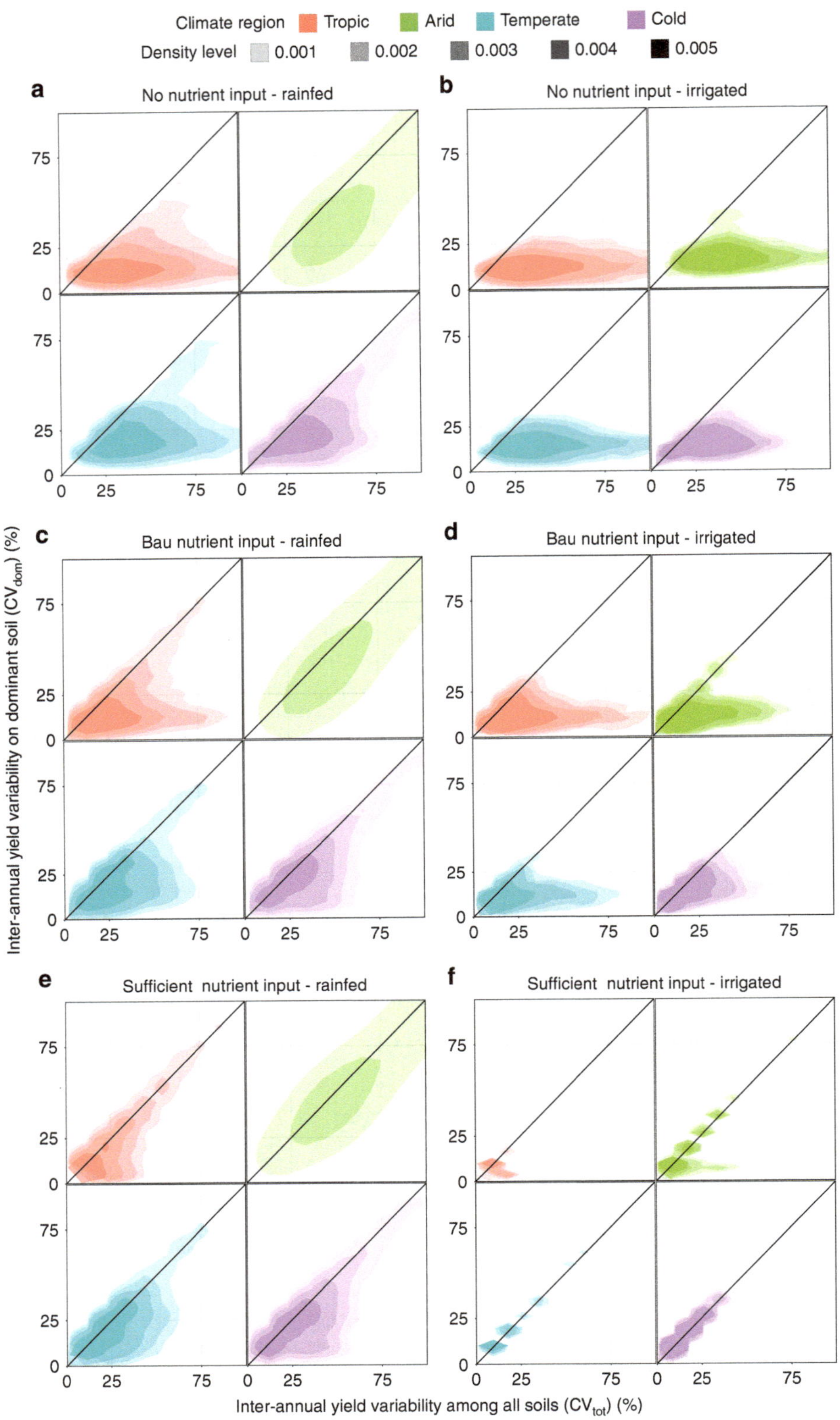

Figure 1 | Coefficient of variation of maize yields in each grid cell for all soil types compared with the dominant soil only. Each panel depicts the coefficient of variation (CV) for a different nutrient and water management scenario: (**a**) no nutrient application and rainfed water supply, (**b**) no nutrient application and sufficient irrigation water supply, (**c**) business-as-usual (bau) nutrient application and rainfed water supply, (**d**) bau nutrient application and sufficient irrigation water supply, (**e**) sufficient nutrient application and rainfed water supply and (**f**) sufficient nutrient application and sufficient irrigation water supply (see Table 3 for details). The colours in the four subpanels displayed for each crop management scenario indicate the major Koeppen-Geiger climate regions[52] (Supplementary Fig. 1 and Supplementary Table 1) tropic (red), arid (green), temperate (blue) and cold (purple). The size and shading of polygons of the same colour indicates levels of density corresponding to the bins shown in the density scale. The percentages of grid cells below the 1:1 line are shown in Table 1. Grid cells with only one reported soil type, arctic climate or without reported maize harvested area were excluded.

Table 1 | Percentage of grid cells with CV_{tot} higher than CV_{dom}.

Management*	Rainfed	Irrigated
	$CV_{tot} > CV_{dom}$	$CV_{tot} > CV_{dom}$
No-nut	89.7	96.1
Bau-nut	76.7	75.4
High-nut	61.5	28.7

*Bau-nut, business-as-usual nutrient inputs; CV, coefficient of variation; High-nut, sufficient nutrient inputs; No-nut, no nutrient inputs.

rainfed than in the irrigated scenarios. Under such low-input conditions, crop nutrients are only supplied through the weathering of soil minerals and decomposition of soil organic matter (SOM)[21] (atmospheric deposition is not taken into account due to lack of global data; see the Methods for details). The level of nutrient supply is least relevant in (semi-)arid regions where water supply and often adverse temperatures are the main factors limiting biomass production. The number of grid cells with $CV_{tot} > CV_{dom}$ decreases to $\sim$76% for the two bau-nut management scenarios (Table 1), representing present-day fertilizer application patterns with and without sufficient irrigation. This is accompanied by substantial decreases of CV_{tot} and to a lesser extent CV_{dom}, especially under irrigated conditions (Figs 1c,d and 2c,d and Supplementary Table 2). Under rainfed conditions and with sufficient nutrient inputs (high-nut), CV_{tot} is larger than CV_{dom} in more than 61% of all grid cells. The application of sufficient irrigation water and nutrients reduces the differences between CV_{tot} and CV_{dom} to a minimum, rendering them statistically insignificant (Fig. 2f). In all cases, except where both nutrient and water supplies are sufficient, we find that CV_{tot} is significantly higher than CV_{dom} (Supplementary Table 2). We attribute this to the fact that inherent edaphic and climatic conditions that drive crop yields are less relevant when water and nutrients are supplied exogenously.

Impact of area-weighting by soil extent on yield estimates. Recently, Zhang *et al.*[17], in a regional high-resolution assessment of soil organic carbon sequestration, have tested the impact of weighting EPIC simulation results by soil type extent in each grid cell. The authors found that this weighting improved simulation results significantly. We have tested the impact of this approach as well by computing an area-weighted CV_{aw} (see the Methods for details). In our global study, the area weighted yield variability CV_{aw} or yield variability derived from the dominant soils alone (CV_{dom}) often have very similar distributions (Fig. 2). They exhibit no statistically significant difference across various management scenarios and climate zones. The differences in the medians of CV_{aw} and CV_{dom} are smaller compared with the differences between one of the two and CV_{tot} (Supplementary Table 2). It needs to be stressed that area-weighting is a valuable approach if simulations refer to large coverage of grid cells as is the case in vegetation modelling or in high-resolution simulations of agricultural production in extensively cultivated regions. Globally, however, the area-weighted results do not provide a sufficient base for estimation of soil-induced yield variability in the GGCM approach. Even the soil type with the fifth largest extent has an area which is sufficiently large to account for all cultivated cropland area in >40% of all grid cells (Supplementary Fig. 2a). If only maize is taken into account, even the soil type with the 15th rank has sufficient coverage to account for the maize cultivated area in nearly 50% of all grid cells (Supplementary Fig. 2b). Therefore, it appears inevitable to take the whole range of soil types within in each grid cell into account, treating each one equally. Tentatively, this uncertainty may be bracketed by carrying out simulations for the dominant, most and least suitable soil types as was done here to highlight the importance of various subsets of soil types for model validation (see the Methods for details).

Exogenous nutrient supply and soil-related yield variability. As indicated above, the soil-induced yield variability depends strongly on nutrient management, especially if water is not a plant growth-limiting factor and temperatures are favourable for the cultivated crop. Under irrigated conditions, increasing fertilizer application rates amplify the difference in CV_{tot} among bau-nut and no-nut conditions (Fig. 3a–d). At high fertilizer application rates, this is particularly the case in tropical and temperate climates where temperatures are most favourable for maize cultivation and hence larger amounts of N can be utilized by the crop. By contrast, biomass production and therefore nutrient requirement is limited in arid and cold regions because of excessively high or low temperatures. Under rainfed conditions, such a trend is also observed albeit less pronounced than under irrigation (Fig. 3e–h). In temperate regions, the average ratio of CV_{tot} for bau-nut/no-nut increases from 0.35 when irrigated to 0.67 when rainfed (Supplementary Table 3). There are only minor differences between CV_{tot} in the no-nut and bau-nut scenarios in arid regions under rainfed conditions as yield variability is mainly climate driven (c.f. Fig. 4a,c). The mean ratio here is 0.91, but the two samples still differ significantly (Supplementary Table 3). These results emphasize the importance of soil type for crop yield estimates in GGCMs especially when it comes to assessments focusing on low input regions, which are considered the most vulnerable regarding present[16] and potential future food security[6].

Spatial patterns in soil-related yield variability. Spatial patterns indicating whether soil or weather dominate yield variability depend strongly on crop management as well (Fig. 4). If irrigation water and fertilizer are not applied (Fig. 4a), then soils dominate yield variability in nearly 51% and weather in nearly 24% of all grid cells globally (Table 2). The latter is especially evident in (semi-)arid regions such as the Great Plains, Southern Africa, North-eastern Brazil and Central to West Asia. If irrigation water is supplied but fertilizer not (no-nut-irr), then soils dominate yield variability in 81% of the maize harvest area globally (Fig. 4b and Table 2). Yield variability in this case is driven by weather in only a few areas at the edge of agricultural suitability because of temperature limitations or if soil data indicates uniform soil characteristics. By contrast, soil-related yield variability is least important under high-nutrient input conditions. With solely rainfed water supply (Fig. 4e), differences in soil hydrological characteristics remain an important factor with yield variability dominated by soil type selection in 17.7% of all grid cells and no clear dominance of soil or climate in 17.9%. In irrigated agriculture (Fig. 4f), where also the remaining differences in soil hydrological characteristics are largely eliminated, soil type selection dominates yield variability in <2% of the grid cells (Table 2), which are located in the Brazilian savannah, southern Africa and around the Himalaya. This may appear to be in contrast to Fig. 4a,b, where under irrigation conditions soils are indicated to be important in more grid cells compared with rainfed conditions. However, it can be explained by endogenous nutrient supply from soils being a major plant growth-limiting factor besides climate under low-nutrient input conditions. As the model takes only the most limiting factor on each day of the growing season into account, nutrient stress may

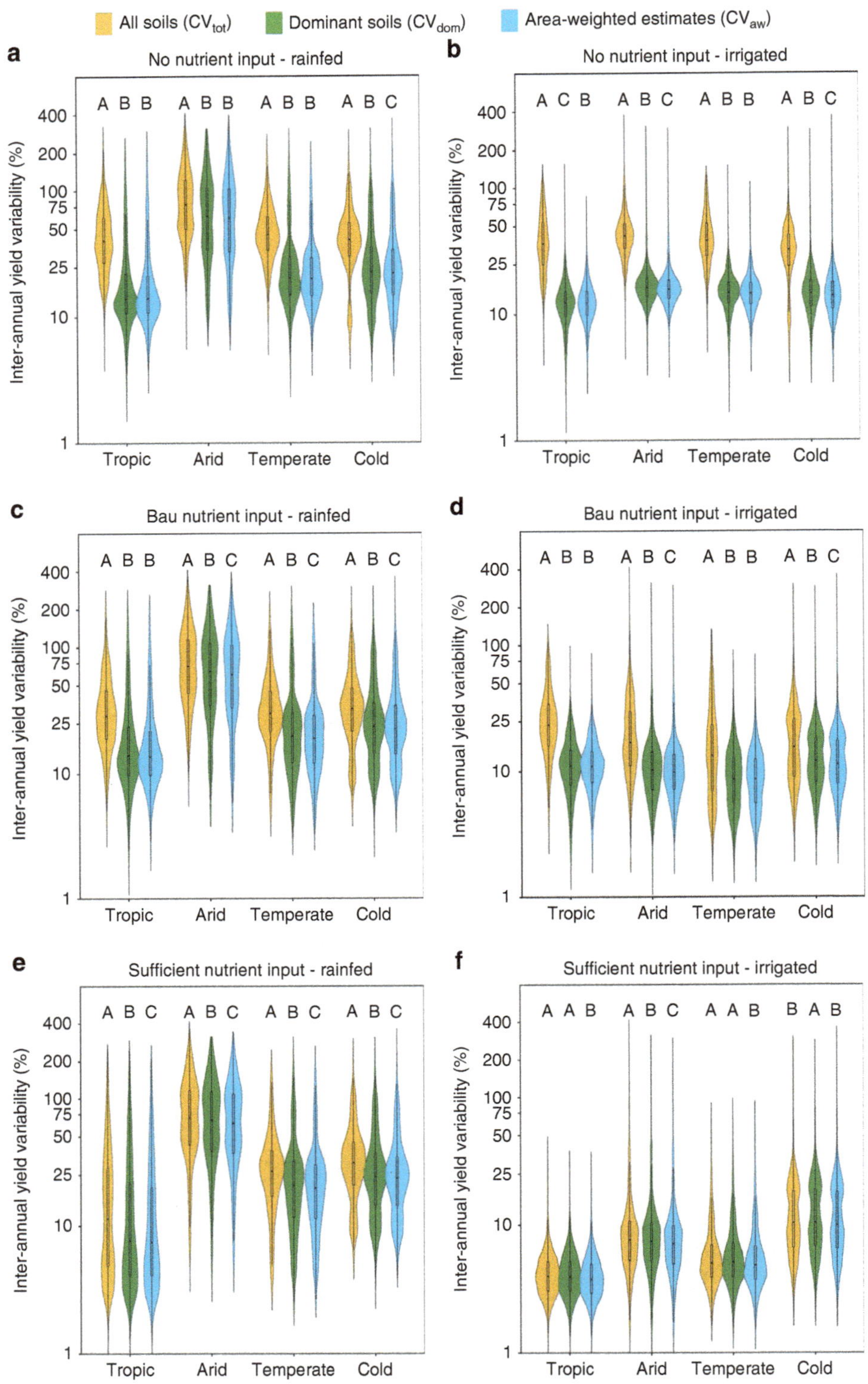

Figure 2 | Violin plots of the coefficient of variation of maize yields for various soil pools and weightings. Coefficients of variation (CV) are shown for all soil types (CV_{tot}), the dominant soil type only (CV_{dom}) or the area-weighted CV across all soil types in a grid cell based on each soil's extent (CV_{aw}). Each panel shows a nutrient and water management scenario with (**a**) no nutrient application and rainfed water supply, (**b**) no nutrient application and sufficient irrigation water supply, (**c**) business-as-usual (bau) nutrient application and rainfed water supply, (**d**) bau nutrient application and sufficient irrigation water supply, (**e**) sufficient nutrient application and rainfed water supply and (**f**) sufficient nutrient application and sufficient irrigation water supply (see Table 3 for details). The violins represent the density of values along the *y* axis and are of equal area. The capital letters A–C indicate whether the samples are significantly different according to a Tukey HSD test within each climate region and management scenario. The complete results of statistical analyses are presented in Supplementary Table 2. Grid cells with only one reported soil type, arctic climate or without reported maize harvested area were excluded.

outweigh climate-related stresses under these circumstances (see the Methods for details). The rainfed and irrigated bau-nut managements (Fig. 4c,d) provide a mixed picture because of spatially explicit information on nutrient application. Most parts of the southern hemisphere (except Brazil, Argentina, Australia and New Zealand) resemble the pattern of the no-nut scenarios (Fig. 4a,b) as they presently lack substantial fertilizer inputs (Supplementary Fig. 3). The northern hemisphere shows the

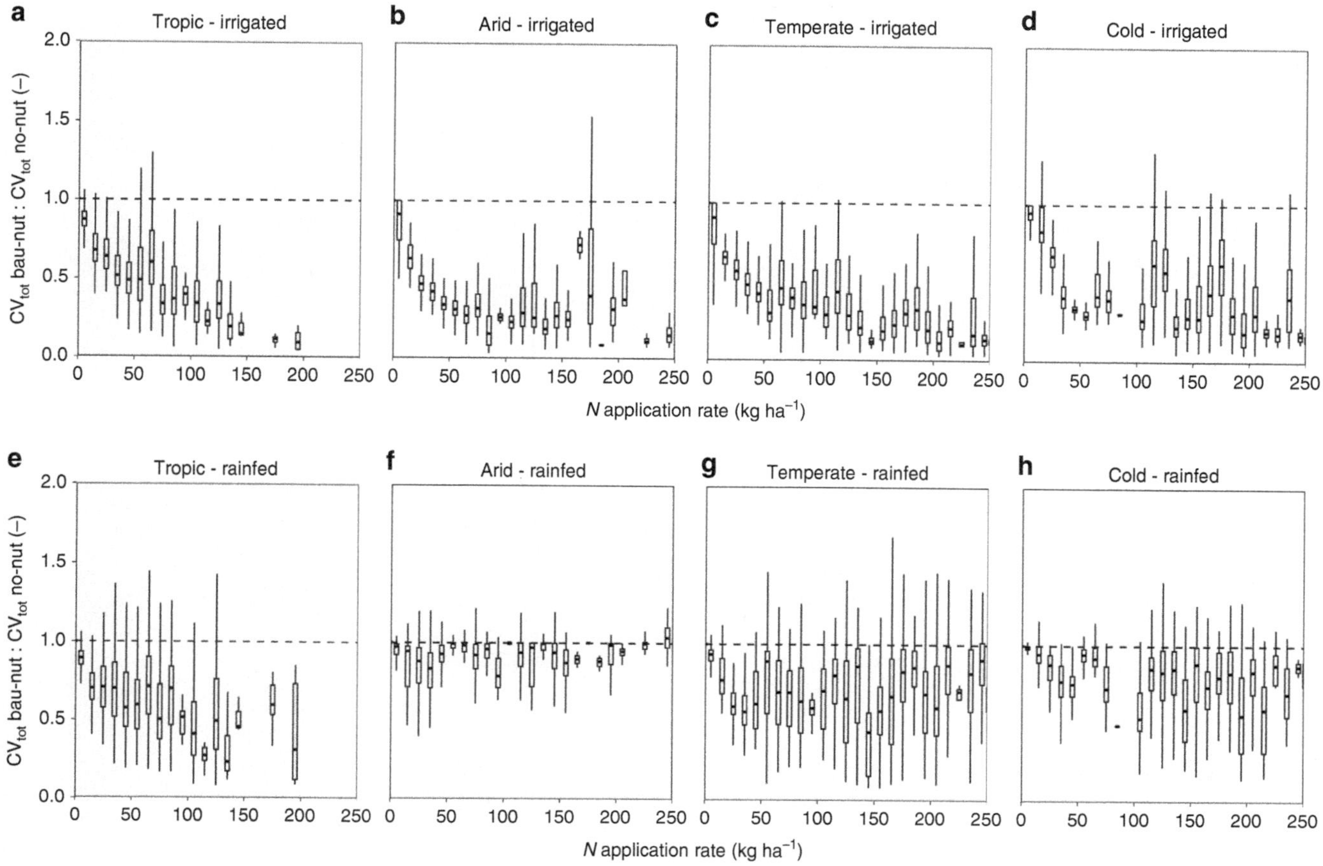

Figure 3 | Boxplots showing the ratio of the coefficient of variation in the whole soil set for the business as usual compared with the no nutrient input scenario. Ratios of the coefficient of variation for the whole soil set (CV_{tot}) for the business as usual (bau-nut) to no nutrient input (no-nut) scenarios are depicted for various Koeppen-Geiger climate regions and water management scenarios: (**a**) tropic and sufficient irrigation water supply, (**b**) arid and sufficient irrigation water supply, (**c**) temperate and sufficient irrigation water supply, (**d**) cold and sufficient irrigation water supply, (**e**) tropic and rainfed water supply, (**f**) arid and rainfed water supply, (**g**) temperate and rainfed water supply and (**h**) cold and rainfed water supply. The dashed line at intersect $y=1$ serves as a reference for CV_{tot} values in the no-nut scenario. Fertilizer application rates were binned in steps of 10 kg N ha^{-1}. The extent of the *x* axis was limited to a maximum of 250 kg N ha^{-1} for better readability. Details on the fertilizer application rates are provided in the Methods section (Table 3) and Supplementary Fig. 3. Supplementary Fig. 1 shows the major Koeppen-Geiger climate regions as defined in Supplementary Table 1[52]. CV_{tot}, the coefficient of variation for the whole soil set.

opposite, because of sufficient or even excessive fertilizer application rates in most parts such as in the United States, Europe and China[22].

The role of soils in climate change impact assessment. Projected changes in climate variables often exceed present weather variability in various regions globally and can even result in currently inexistent soil-climate combinations. This stresses the role of soils in attenuating or amplifying impacts of extreme weather or long-term climate shifts on crop yields. To highlight the importance of these climate–soil interactions, we performed an exemplary climate change impact assessment. The assessment focused on the world's major food production units (FPUs) using climate projections up to the 2050s performed with the HadGEM2-ES GCM under the RCP8.5 emission scenario (Fig. 5).

Depending on whether the highest or lowest yielding soil type is selected in each grid cell, estimated climate change impacts expressed relatively to the yield obtained under present climatic conditions can vary widely within a certain FPU. In 75% of the major FPUs, relative median yield losses are larger for the least suitable than for the most suitable soil types. This is partly due to the relative expression of the yield loss. For example, we find a nearly sixfold difference between the most (−5.5%) and the least (−32.1%) suitable soils of FPU 26 (Volga, RUS). The two subsets of soils provide yields of ~2.9 and 1.4 t ha^{-1} during the baseline period, which results in absolute yield losses of 0.15 t ha^{-1} on the highest yielding soil types and 0.45 t ha^{-1} on the lowest yielding soil types. The difference in relative yield impacts among the least and most suitable soils is hence a product of the differences in absolute baseline yields (twofold) and absolute yield losses (threefold). In any case, as large-scale agricultural climate change impact assessments often focus on relative changes in agricultural productivity[1–3,6], the specific selection of soil types for GGCM simulations can have a significant impact on the spatial interpretation of results. Besides the magnitude of the impact, even the direction (+/−) may be inversed as a function of soil type choice as is the case in FPUs 13, 19, 27 and 37.

Discussion

Soil processes modulating impacts of climate on plant growth and crop yields are highly complex and to fully disentangle them is beyond the scope of this study. To shed light on some of these processes, we have explored plant growth dynamics and associated environmental variables in two contrasting grid cells (Supplementary Discussion 1 and Supplementary Figs 4–12). The evaluation reveals that dynamic interactions of soil texture,

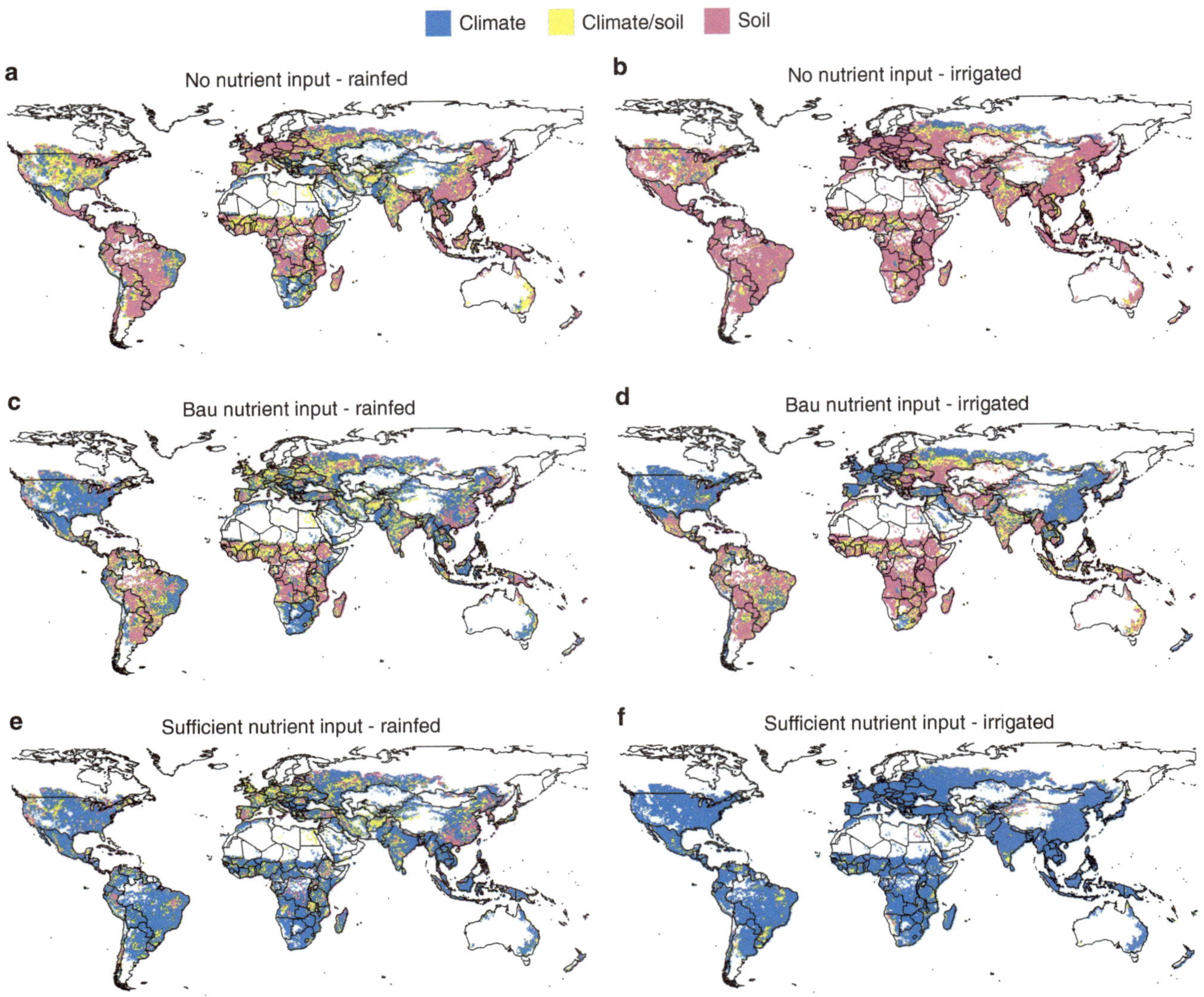

Figure 4 | Grid cells in which climate or soil are dominating yield variability. Climate variability corresponds to CV_{dom} in Fig. 1 while yield variability due to choice of soil (CV_{soil}) was computed from the 10-year means of yields from all respective soil types in each grid (see the Methods for details). Climate is considered to govern yield variability if $CV_{dom}/CV_{soil} > 30\%$ (blue colour), soil is considered to govern yield variability if $CV_{soil}/CV_{dom} > 30\%$ (magenta colour). Climate and soil are considered to govern yield variability jointly if they are in a range of $\pm 30\%$ of the respective larger value (yellow colour). Each panel represents a nutrient and water management scenario with (**a**) no nutrient application and rainfed water supply, (**b**) no nutrient application and sufficient irrigation water supply, (**c**) business-as-usual (bau) nutrient application and rainfed water supply, (**d**) bau nutrient application and sufficient irrigation water supply, (**e**) sufficient nutrient application and rainfed water supply, and (**f**) sufficient nutrient application and sufficient irrigation water supply (see Table 3 for details). Percentages of grid cells falling into either one of the categories are shown in Table 2. Grid cells with only one reported soil type, arctic climate or without reported maize harvested area were excluded.

Table 2 | Percentages of grid cells in which yield variability is largest for CV_{soil}, CV_{dom} or not dominated by either one as displayed in Fig. 4.

Management*	Rainfed			Irrigated		
	Soil	Climate/soil	Climate	Soil	Climate/soil	Climate
No-nut	50.6	25.7	23.7	81.0	13.0	6.0
Bau-nut	33.0	24.5	42.6	45.4	19.4	35.3
High-nut	17.7	17.9	64.5	1.4	4.4	94.3

*Bau-nut, business-as-usual nutrient inputs; CV, coefficient of variation; High-nut, sufficient nutrient inputs; No-nut, no nutrient inputs.

precipitation and plant water requirement can result in unexpected climate change impact responses. If precipitation decreases moderately, climate change can cause a more severe impact on crop yields in soils with lower water-holding capacity that are drained rapidly, whereas a soil of finer texture may buffer the potential deficit by storing water for longer periods. However, very low precipitation levels can also result in lower plant-available water in fine-textured soils. For such soils, only the very fine pore-space may be filled with water that is not readily accessible for plants or the fine texture can cause higher run-off[10].

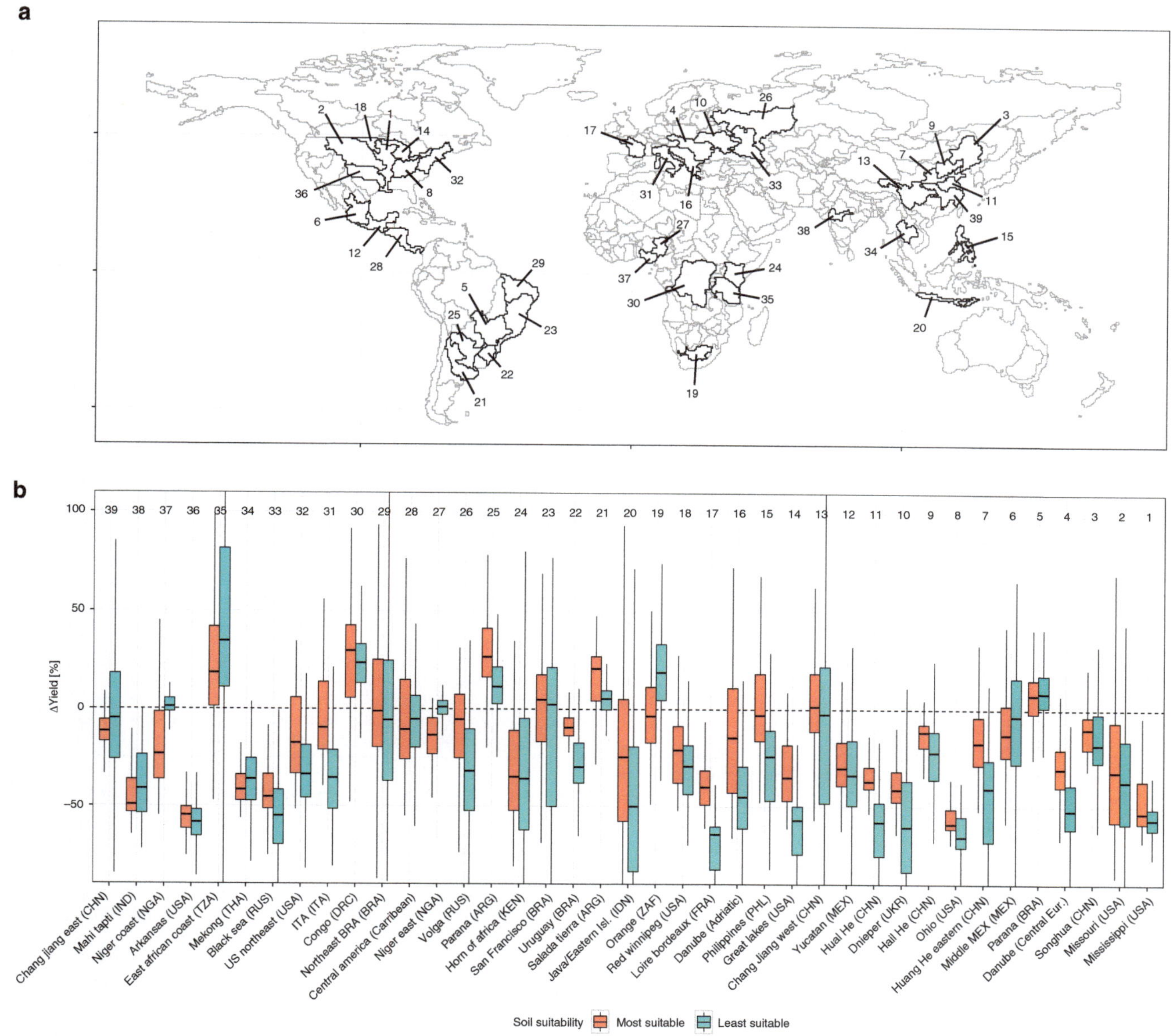

Figure 5 | Projected percentage change in maize yields by the 2050s in the 39 largest food production units (FPUs). (**a**) FPUs as defined by intersecting water basins with major administrative boundaries[53]. (**b**) Relative change in maize yields for the presently most or least suitable soil type in each grid cell for the rainfed business-as-usual nutrient input scenario (bau-nut-rf; Table 3). The climate projection is based on HadGEM2-ES for RCP8.5 and is compared here with present-day climate data (see the Methods for details). Each of the selected FPUs comprises more than 10^6 ha of harvested maize area. Ranking in terms of decreasing cultivated maize area is indicated by the order from right (largest) to left (smallest).

This would result in a more vulnerable crop while a coarse textured soil can result in higher levels of plant available water. Besides these water-related processes, the nutrient supply of a soil can also attenuate or amplify climate change impacts by modulating the plant's sensitivity to climate variables (see the Methods for details) as has been discussed extensively above with respect to the nutrient management scenarios. Temperature, solar radiation and atmospheric CO_2 concentration affect biomass accumulation independently from soils and soil processes in the model.

At the global scale, changes in maize yields vary from -26% on the most suitable soil to -27% on the dominant soil and -31% on the least suitable soil in the climate change scenario (not shown). This results in an absolute uncertainty range of 5%. A recent assessment based on yield estimates from 14 GGCMs forced with projections from 5 GCMs showed that the estimated change in maize yields may vary from -20 to $+15\%$ among crop models around the year 2050, whereas the GGCM used here (GEPIC) had an absolute uncertainty range of $\sim 15\%$ when driven by the 5 GCMs[6]. This indicates that the largest source of uncertainties in agricultural climate change impact assessments lies in crop models themselves followed by uncertainties in GCM projections. Nevertheless, soil characteristics and data have an undeniable impact on uncertainty in estimates of crop yields under climate change. Thus, for the GGCM used here, the magnitude of the range in observed impacts due to the selection of soil data amounted to about a third of the uncertainty that was associated with the five GCM projections. Quantifying this relative importance in GGCM × GCM ensembles will require the direct inclusion of various soil data in future ensemble runs.

Further sources of uncertainty in global crop yield simulations are the correct spatial allocation of cropland, crop management and crop rotations besides varying algorithms applied by different crop growth simulation models. Several of these uncertainties are presently being addressed within the GGCM intercomparison project[23]. Porwollik *et al.* (in preparation), for example, found that depending on the cropland data set used for aggregation, global average maize yield estimates differ only slightly, but can range widely in single countries. A recent study based on the GEPIC model estimated that rotation of maize with a leguminous shrub and soil conservation practice can buffer adverse changes in precipitation under climate change as compared with conventional intensification[5]. Although such uncertainties are presently addressed individually, the contributions of single factors to overall model uncertainty and potential interactions will have to be investigated in multi-dimensional experiments.

Similar effects can be expected for other crop models with detailed representation of soil processes[6,24], such as APSIM, DSSAT, DNDC or Century and for certain ecosystem and hydrologic models. The importance of soil data quality and allocation will increase in the future as the range of soil processes taken into account in field-scale models and GGCMs is constantly expanding[24,25]. Considering that yield estimates from bio-physical crop models often form the basis for subsequent agro-economic (climate change impact) assessments[26], uncertainties and errors originating from soil data will propagate throughout the whole assessment chain from bio-physical to economic impacts and further to policy recommendations. This amplifies the need for developing new strategies on how to deal with soil-related uncertainty in GGCM simulations.

As any other soil mapping product, the HWSD itself has intrinsic uncertainties and quality limitations. For example, substantial differences in the characteristics and locations of soil types can be observed when comparing global databases like HWSD to finer-scale regional soil data[27]. Such soil mapping uncertainties can be further amplified when used in models as has been shown, for example, by Hendriks *et al.*[27] and Lin *et al.*[28] One way to avoid this issue is the use of soil profile information together with spatially represented covariates in digital soil mapping products to infer directly quantitative soil information at a given spatial resolution[13].

Indeed, on-going efforts to improve the quality of basic soil data, for example, by increasing the coverage of soil samples and combining them with remote sensing (for example, http://globalsoilmap.net[29]) are essential. Yet, our results show that the correct spatial allocation of these soil data to present cropland is at least equally important. One project targeting the regional matching of soil data and cropland by merging ground and remote sensing data is presently carried out by the Africa Soil Information Service (http://africasoils.net). As long as there are no global high-resolution input data that match all required inputs (climate, soil, cropland and management) per grid, however, the full ensemble of contrasting soil types present in each location should be taken into account in GGCM simulations. Alternatively, the role of soils and their variability in buffering or amplifying future climate impacts on crops should at least be bracketed by simulating the most and least suitable soils under cultivation. Another key ingredient for improved climate impact assessments is the collection and GGCM representation of crop management practices. This and better characterizing soils and their variability in GGCMs will allow for identifying farm-level adaptation options tailored to the range of marginal to near-optimal conditions experienced in the field.

Methods

Model description. We used the geographic information system (GIS)-based global crop modelling framework GEPIC[30] based on the field-scale model EPIC v. 0810 (refs 20,25) for simulating crop production. Maize was used as a representative crop to model as it globally accounts for the largest production volume and the second largest harvested area. Concordantly, it is widely simulated with GGCMs[1–6,14]. GEPIC reads large-scale input data sets and runs EPIC for each grid cell of a defined area treating each grid cell as a field with specific climate, soil, topography and management. The spatial resolution applied here was 0.5° × 0.5°, which is presently the norm in global crop model simulations[6]. The last 10 simulation years after a spin-up period of 30 years were used for evaluation of the simulation results. EPIC uses the energy-to-biomass conversion approach for estimating potential biomass increase at a daily time step. The potential increase is subsequently adjusted by the major stress factor out of nutrients, water, temperature and aeration to an actual biomass gain. Besides crop growth, soil processes ranging from organic matter cycling to wind erosion are simulated. Soils in the EPIC model provide rooting space for plants and serve as the medium storing, cycling and providing nutrients and water. After crop uptake, drainage, leaching and other losses to the environment, both can be replenished by fertilizer and irrigation water application. To account for this, we simulated six management scenarios with varying levels of water and nutrient supply as shown in Table 3. The scenarios with ample irrigation and fertilization are assumed to eliminate water- and nutrient-related stresses on plant growth. The parameterization applied here has been specified in Folberth *et al.*[31] and Rosenzweig *et al.*[6] using mainly default model parameters. More detailed descriptions of model routines and further input data are provided at the end of the Methods section.

Input data for climate and topography. Climate data from the WFDEI.GPCC database[32] were provided through the ISI-MIP and AgGRID projects (http://isi-mip.org; http://agmip.org/ag-grid) at a resolution of 0.5° × 0.5°. WFDEI.GPCC is based on temperature and solar radiation from ERA-interim[33] and precipitation and snowfall from GPCC[34]. Climate projections for HadGEM2-ES RCP8.5 were provided by ISI-MIP[35] and are based on runs for the CMIP5 project[36]. Elevation and slope were derived from the digital elevation model GTOPO30 (ref. 37).

Soil database processing and evaluation. The HWSD in its version 1.2 (ref. 12) was used for preparing soil input data for GEPIC. The HWSD is presently the most up-to-date global soil database, bringing together data from several national and regional soil assessments. Thereby, 45,000 SMUs were defined by local experts as homogenous soil regions to which certain soil types were attributed (Supplementary Fig. 13). Depending on the size of the SMU and the given soil heterogeneity, these can be up to ten different soil types with varying fractions of coverage. Exceptions are China and Greenland among other smaller units, for which only one soil type has been reported per SMU.

The mapping raster of the HWSD was aggregated from its native resolution of 30 arcsec to a 0.5° × 0.5° grid complying with the present state-of-the-art in global crop modelling. Thereby, all soil types from all SMUs in each grid cell were attributed to the respective grid cell. Their share was calculated based on the coverage of the SMU within the grid cell and the reported share of the soil type in the SMU (Supplementary Fig. 14). This resulted in up to 77 soil profiles in each grid (Supplementary Fig. 15), with the dominant soil type covering between <10%

Table 3 | Nutrient and water management scenarios used in the study.

Fertilizer scenario*	Abbreviation	Max. irrigation vol. (mm a^{-1})	N (kg ha^{-1} a^{-1})	P (kg ha^{-1} a^{-1})
Business as usual	Bau-nut-irr	2,000*	Variable†	Variable†
Business as usual	Bau-nut-rf	—	Variable†	Variable†
No nutrient inputs	No-nut-irr	2,000*	—	—
No nutrient inputs	No-nut-rf	—	—	—
Sufficient nutrients	High-nut-irr	2,000*	Max. 500	100
Sufficient nutrients	High-nut-rf	—	Max. 500	100

*The upper limit of irrigation assumes sufficient supply of irrigation water, which is presently the norm in global crop simulations 6 due to lack of data for spatial crop-specific application volumes.
†Business as usual refers to fertilizer application rates around the year 2000 (see the Methods for details).

and 100% of each simulated grid cell (Supplementary Fig. 16). Owing to the small extent of SMUs in China, this approach resulted also here in various combinations of soils per grid cell because of intersecting SMUs (Supplementary Fig. 17 and Supplementary Fig. 15b).

The aggregation resulted also in smaller extents of the dominant soil type in each grid cell as shown in Supplementary Fig. 16. The dominant soil type in each grid cell would still be sufficient for covering all cropland in >80% of all grid cells (Supplementary Fig. 2a). For the subsequent soil types, this figure decreases continuously. However, even up to the fifth soil type more than 50% of all cropland could be allocated on the respective soil type in grid cells that have this number of soil types reported. This highlights that using the dominant soil type in each grid cell is not justified by the extent of cultivated areas in each grid cell. A similar and even more pronounced pattern can be observed if only the harvested area for maize is taken into account (Supplementary Fig. 2b).

Crop nutrient management scenarios. Six nutrient and water management scenarios were simulated (Table 3). The business-as-usual (bau-nut) nutrient management scenario represents the norm in most GGCMs[6] and is based on data representative for 'around the year 2000'. Reported global planting dates and growing season lengths were obtained from the database by Sacks *et al.*[38]. Fertilizer application rates for N and P were provided by the GGCM intercomparison coordination team[23]. They are based on global, spatially explicit and crop-specific fertilizer application rates by Mueller *et al.*[39] combined with nutrients embedded in manure. The application rates for N and P in this scenario are displayed in Supplementary Fig. 3 and range from 1 (parts of sub-Saharan Africa) to 372 (Egypt) $kg\,N\,ha^{-1}\,a^{-1}$ and 0 (parts of sub-Saharan Africa) to 150 (New Zealand) $kg\,P\,ha^{-1}\,a^{-1}$. Although this management scenario is important as a reference for comparisons with other studies, the global imbalance in agricultural nutrient application rates limits the spatial comparability of soil- and climate-related yield variability.

Hence, two additional nutrient management scenarios were run for evaluating the effect of soil nutrient contents on crop yields (no-nut) or the effect of soil texture and water supply (high-nut). No nutrients were supplied in the no-nut scenario in order to capture the most pronounced effects of endogenous soil nutrient supply in the model. Although such conditions are rare globally, they are prevalent in parts of sub-Saharan Africa, West Asia and South America (Supplementary Fig. 3a and b). For the high-nut scenario, a maximum amount of $500\,kg\,N\,ha^{-1}\,a^{-1}$ was set and the fertilizer was applied automatically by the model based on the plant nutrient deficit in order to virtually eliminate nutrient limitations for plant growth. As shown in Supplementary Fig. 18 and Supplementary Table 4, computed optimal N application rates were mostly <225–$280\,kg\,N\,ha^{-1}\,a^{-1}$, which corresponds to present application levels in parts of Europe, the United States of America and China, and values $<170\,kg\,N\,ha^{-1}\,a^{-1}$ were found for 50% of all grid cells. Phosphorus was applied at a rigid level of $100\,kg\,P\,ha^{-1}\,a^{-1}$ at planting, which corresponds to common practice in present high-input regions and ensures sufficient P supply for the plant. Although this causes an oversupply in all grid cells, there are no interactions with other soil functions in the model and hence the purpose of eliminating P limitations for plant growth is fulfilled.

Crop water management scenarios. Irrigation water supply was either turned off in all grid cells in order to mimic rainfed only agriculture or irrigation water was supplied in sufficient amounts. The latter was used for evaluating the effect of soil nutrient contents and soil texture on maize yields alone if water supply from the soil—depending on precipitation patterns and soil hydrologic characteristics—is eliminated as a limiting factor for plant growth. Automatic irrigation takes place in the model if plant water stress limits potential biomass increase by ≥10% on a given day during the growing season. The model then applies sufficient water to level out plant water stress on this specific day. The total annual allowable amount was set to 2,000 mm in order to fully eliminate water stress.

Supplementary Fig. 3c depicts an example of irrigation volumes applied by the model in scenario high-nut-irr. The actually applied volumes were $<440\,mm\,a^{-1}$ in nearly all grid cells and $<155\,mm\,a^{-1}$ in 50% of all grids (Supplementary Table 4), the highest value found was $1{,}020\,mm\,a^{-1}$. Owing to this optimization, the irrigation volumes applied by the model can be considered equal to plant water requirement supplementary to precipitation.

Model performance evaluation. Model performance in reproducing observed yields around the year 2000, the time period for which reported yields and crop management are representative, was evaluated at the grid cell (Supplementary Fig. 19) and national levels (Supplementary Fig. 20). National average yields based on separate irrigated and rainfed simulations for each grid cell were calculated as

$$Y_{av,c} = \frac{\sum_{g=1}^{m}\left[Y_{i,g}\times A_{i,g} + Y_{r,g}\times A_{r,g}\right]}{\sum_{g=1}^{m}\left[A_{i,g} + A_{r,g}\right]} \quad (1)$$

where $Y_{av,c}$ is the national average yield in country c, $Y_{i,g}$ is yield under irrigated conditions in grid cell g, $Y_{r,g}$ is yield under rainfed conditions in grid cell g, $A_{i,g}$ is irrigated area in grid cell g and $A_{r,g}$ is rainfed area in each grid cell g. Average yields per grid cell (Supplementary Fig. 19) were calculate using the same approach without aggregation at the country level. Rainfed and irrigated maize harvest areas per grid cell were adopted form the MIRCA2000 data set[40]. Reported yields were derived from a data set based on national or subnational statistics[41]. The respective grid-cell-specific estimates were then produced for most suitable soils (equation (5)), least suitable soils (equation (6)), dominant soils and soils producing the estimate closest to the reported values.

Table 4 | Evaluation of model performance for various soil subsets at the grid cell level.

Soil subset	r^2	ME	MSE	MAE
Most suitable	0.29 (P<.0001)	0.46	6.07	1.87
Least suitable	0.29 (P<.0001)	−1.16	6.52	1.76
Dominant	0.36 (P<.0001)	−0.25	4.93	1.55
Closest to the observed	0.54 (P<.0001)	−0.40	3.34	1.07

MAE, mean absolute error; ME, mean error; MSE, mean squared error.

Among the most suitable, least suitable and dominant soil type, the dominant provides the best correlation with reported yields under present-day reported management conditions at the grid cell level (Supplementary Fig. 19a–c and Table 4). Using the most suitable soil type, results in a lower intercept. When selecting the yield in each grid cell that is closest to the reported value (Supplementary Fig. 19d), R^2 improves significantly and slope and intercept improve slightly. The mean absolute error (MAE) follows a similar pattern (Supplementary Fig. 21a). The magnitude of the MAE is at the grid cell level overall lowest at low-yield levels and highest at high-yield levels. The lowest error occurs for the minimum yielding soil types and soil types with the closest yield subsets at low yields. The lowest error at high yields was found for the latter and the maximum yielding soil types.

Pairwise t-tests were applied to compare the MAE of yield estimates between the various subsets of soils, and all the differences were found to be statistically significant ($P<0.0001$). There was not much practical difference in the performance of the models based on most and least suitable soils, respectively, apart from the fact that the former tended to overestimate where the latter tended to underestimate. The estimates of the model, based on the soil producing the yield closest to the observed, understandably had the highest correlation with the observed, smallest MAE and smallest mean squared error (MSE). The validation shows that model performance also strongly depends on which soil type is selected in each grid cell. However, the approach of selecting the soil providing the yield closest to the reported corresponds to a calibration and must be treated with care. It ignores uncertainties originating from management input data like growing seasons, fertilizer inputs, extents of irrigation areas and volumes and planted cultivars as well as uncertainties in model algorithms as, for example, the estimation method for potential evapotranspiration. Also the often coarse resolution of the reported yields data set not only limits the quality of the validation at the grid cell level but would also impact such a calibration approach. It also has to be taken into account that a global validation of crop yields mainly reflects management conditions that vary in the case of maize strongly between countries with high agricultural intensification mainly in the Northern hemisphere and countries with prevailing low-input agriculture.

The agreement between simulated and reported national average yields is far better in terms of statistical coefficients (Supplementary Fig. 20). The maximum yielding soil type causes a systematic overestimation of low yields, which is evident in the high intercept (Supplementary Fig. 20a), whereas the low slope for the lowest yielding soil type is caused by an overestimation of yields in high-yielding countries (Supplementary Fig. 20b). The dominant soil type shows at better agreement in terms of coefficients with reported yields (Supplementary Fig. 20c). Selecting the best matching yield in each grid cell also improves the agreement at the national scale massively (Supplementary Fig. 20d). The MAE at the national scale (Supplementary Fig. 21b) for the minimum yielding soil type follows mostly the pattern at the grid cell level, except for the highest yield bin, in which simulated yields match well. The MAE for the maximum yielding soil type in contrast decreases with increasing reported yields, which is due to the over-estimation of yields in low-yielding countries. The MAEs for the dominant and best matching soil types show a similar pattern with constantly comparably low errors.

Yield estimates across time and soil types. Different subsets of crop yield estimates were analysed to identify and compare different sources of variability. The whole set of possible yields within a grid cell is represented as

$$YD_{n,t} = \begin{bmatrix} YD_{1,1} & \cdots & YD_{1,10} \\ \vdots & \ddots & \vdots \\ YD_{LN,1} & \cdots & YD_{LN,10} \end{bmatrix} \quad \text{with } LN \in \{1,2..77\} \text{ and } t \in \{1,2,..10\} \quad (2)$$

where ($YD_{n,t}$) is the crop yield on a given soil type n in year t and LN describes the number of soil types in a given grid cell. These can range between 1 and 77. n also identifies the ranking of soil types according to their aerial coverage with $n=1$ standing for the dominant soil type. t refers to the simulation years and ranges from 1 to 10. Hence,

$$\mathrm{YD}_{n,t} = [\mathrm{YD}_{1,1} \ \cdots \ \mathrm{YD}_{1,10}] \quad (3)$$

describes the annual yields on the dominant soil type ($n = 1$) in each grid cell and

$$\overline{\mathrm{YD}}_n = [\overline{\mathrm{YD}}_1 \ \cdots \ \overline{\mathrm{YD}}_{\mathrm{LN}}] \quad (4)$$

refers to the 10-year means of yields across all soil types (n) up to the number of soil types (LN) in each grid cell.

Identification of most and least suitable soils. The most or least suitable soil types were defined as maximum or minimum yielding soil types within a grid cell. They were identified by moving through the yield estimates for each grid cell from the dominant to the least abundant soil and selecting the soil type providing the highest or lowest yield according to

$$\mathrm{YD}_{\max} = \max(\overline{\mathrm{YD}}_1 \ldots \overline{\mathrm{YD}}_{\mathrm{LN}}) \quad (5)$$

and

$$\mathrm{YD}_{\min} = \min(\overline{\mathrm{YD}}_1 \ldots \overline{\mathrm{YD}}_{\mathrm{LN}}) \quad (6)$$

where YD_n is the array of 10-year mean yields for all soil types in a grid as shown in equation (4).

Arithmetic and area-weighted coefficient of variation. The coefficient of variation (CV, %) as a measure for yield variability was calculated as

$$\mathrm{CV} = \frac{S}{\overline{X}} \times 100 \ \% \quad (7)$$

where S is the standard deviation and $\overline{X}$ is the mean of a sample. Thereby, the combination of equation (7) and the yield sample in equation (2) is termed $\mathrm{CV}_{\mathrm{tot}}$, the combination of equation (7) and yield sample in equation (3) is $\mathrm{CV}_{\mathrm{dom}}$, and the combination of equation (7) and the yield sample in equation (4) is $\mathrm{CV}_{\mathrm{soil}}$.

An area-weighted $\mathrm{CV}_{\mathrm{aw}}$ based on the CVs analogue to the yields in equation (4) was calculated as

$$\mathrm{CV}_{\mathrm{aw}} = \sqrt{\frac{\sum_{i=1}^{\mathrm{LN}} (\sigma_i^2 * \mathrm{fr}_i^2)}{\sum_{i=1}^{\mathrm{LN}} \mathrm{fr}_i^2}} \Big/ \frac{\sum_{i=1}^{\mathrm{LN}} (\overline{\mathrm{YD}}_i * \mathrm{fr}_i)}{\sum_{i=1}^{\mathrm{LN}} \mathrm{fr}_i} \quad (8)$$

where σ_i is the standard deviation of yields on soil type i, fr_i is the fraction of coverage by soil type i, YD_i is the mean yield for soil type i and LN is the number of soil types in the respective grid cell.

Statistical evaluation. The statistical significance of differences between yield subsets in Fig. 2 was tested by an analysis of variance combined with Tukey's honest significant difference (HSD) test at $P = 0.05$. Tukey's HSD tests whether the means of two samples are significantly different from each other, which is indicated by different letters. Yields were log-transformed for both tests to achieve a normal distribution. Details of the evaluation results are provided in Supplementary Table 2. Kendall's tau-test in a modified form that accepts ties within a sample was used for evaluating the correlation between the ratios of $\mathrm{CV}_{\mathrm{tot}}$ for the bau-nut and no-nut management scenarios and fertilizer application rates in various climate regions (Fig. 3 and Supplementary Table 3). This test evaluates a rank correlation between two non-parametric samples, which is provided if $tau \neq 0$ and $P < 0.01$.

Relevant routines of the EPIC model. The subsequently outlined routines are an excerpt of the model processes relevant for interpreting the results presented herein. Their descriptions are based on the original documentation of the EPIC model, which has been made publicly available by the developers at http://epicapex.tamu.edu/files/2015/05/EpicModelDocumentation.pdf.

Phenologic development of the crop takes place according to the heat unit (HU) approach. Daily HU are calculated as

$$\mathrm{HU}_k = \frac{T_{\max,k} - T_{\min,k}}{2} - T_{\mathrm{b}} \quad (9)$$

where HU_k is the HUs accumulated on day k [°C], $T_{\max,k}$ and $T_{\min,k}$ are the maximum and minimum temperatures on the day (°C) and T_{b} is the base temperature (°C) of a specific crop. Maturity is reached when total accumulated HU are equal to potential HUs (°C), the sum of the daily HU based on long-term climate data and reported growing seasons provided by Sacks *et al.*[38]

EPIC estimates potential biomass increase ΔB_{p} on each day according to

$$\Delta B_{\mathrm{p}} = 0.001 \times \mathrm{BE} \times \mathrm{PAR} \quad (10)$$

where ΔB_{p} (t ha^{-1}) is biomass gain, BE [(kg ha^{-1})/(MJ m^{-2})] is the biomass-energy-conversion coefficient and PAR [MJ m^{-2}] is intercepted photosynthetic active radiation depending on leaf area index (LAI) and solar radiation. Actual biomass is subsequently obtained by correcting ΔB_{p} for the maximum stress out of nutrients, water, temperature, aeration, salinity (see further below).

At maturity, crop yield is calculated by multiplying total above-ground biomass with a water stress-adjusted harvest index (HIA*). Many grains like maize are most sensitive to water stress during flowering, when major yield components are determined[42] and barrenness of flowers can cause massive yield losses[43]. The water-stress adjusted HI (HIA*) is estimated from simulated potential HI (HIA; depending on HU accumulation) and a defined minimum HI (HIA_0) according to

$$\mathrm{HIA}^* = (\mathrm{HIAHIA}_0) \times \frac{\mathrm{WUR}}{\mathrm{WUR} + \exp(6.13 - 0.0883 \times \mathrm{WUR})} + \mathrm{HIA}_0 \quad (11)$$

where WUR is the simulated water use ratio. WUR is estimated at harvest as

$$\mathrm{WUR} = 100 \times \frac{\sum_{i=1}^{K} U_i}{\sum_{i=1}^{K} E_{\mathrm{P}i}} \quad (12)$$

where U_i [mm d^{-1}] is the actual and $E_{\mathrm{P}i}$ [mm d^{-1}] the potential plant water use rate for day i. K is the total number of days of the growing season. The exponential parameters in equation (11) are set to result in 0.05 × (HIA-HIA_0) when WUR = 0.10 and 0.90 × (HIA-HIA_0) when WUR = 0.50. Hence, there is little reduction in HIA* if the ratio of U to E_{P} is greater than 0.5.

Above-ground biomass growth is constrained mainly by water, nutrients (N and P), temperature and aeration stress. The major stress on each day of the growing season limits biomass accumulation by a fraction ranging from 0 to 1. The sum of the daily values for each stress factor over the growing season is referred to as 'stress days'. The stresses are computed as follows:

The effect of water deficit or water stress (WS) on plant biomass production is based on the concept that drought stress is proportional to the transpiration reduction[44]. It is calculated as

$$\mathrm{WS}_i = \frac{\sum_{l=1}^{M} u_{l,i}}{\mathrm{EP}_i} \quad (13)$$

where WS_i is the amount of water stress on day i [–], l is a given soil layer [–], M is the total number of soil layers [–], $u_{l,i}$ is the plant available water in layer l on day i [mm], and EP_i is the potential ET on day i (mm). In addition, water deficit has an impact on HI as described above. Temperature stress (TS) occurs on a given day i if the average air temperature TG is above the optimum temperature (TO) or below the base temperature (TB) according to

$$\mathrm{TS}_i = \sin\left(\frac{\pi}{2} \times \frac{\mathrm{TG}_i - \mathrm{TB}_i}{\mathrm{TO}_i - \mathrm{TB}_i}\right) \quad (14)$$

or if the average daily temperature exceeds TO by 50%. Nutrient stresses (N stress (NS) and P stress (PS)) vary nonlinearly between optimum or excessive supply and 50% of the optimum supply when stress is 100% (ref. 45). First, a scaling factor SNS (here for *NS*) on a given day i is calculated as

$$\mathrm{SNS}_i = 200 \times \left(\frac{\mathrm{UN}_i}{\mathrm{cNB}_i \times B_i} - 0.5\right) \quad (15)$$

where UN_i is the N uptake on day i (kg ha^{-1}), cNB_i is the optimum N concentration in biomass on day i (kg kg^{-1}) and B_i is the total plant biomass on day i (kg ha^{-1}). This factor is then used for estimating the actual nitrogen stress NS according to

$$\mathrm{NS}_i = \frac{\mathrm{SNS}_i}{\mathrm{SNS}_i + \exp(3.52 - 0.026 \times \mathrm{SNS}_i)} \quad (16)$$

The calculation of PS follows the same pattern. Aeration stress occurs if the soil pore space approaches water saturation. A crop-specific saturation factor SAT for day i is estimated as

$$\mathrm{SAT}_i = 100 \times \frac{\frac{\mathrm{SW1}_i}{\mathrm{PO1}_i} - \mathrm{CAF}_j}{1 - \mathrm{CAF}_j} \quad (17)$$

where SW1 is the soil water content (mm) on day i in the top 1 m of the soil profile, PO1 is the pore volume (mm) on day i in the top 1 m of the soil profile, and CAF is the critical aeration factor [–] of crop j, which can vary between 0 and 1. If $\mathrm{SAT}_i > 0$, aeration stress (AS) is subsequently calculated as

$$\mathrm{AS}_i = 1 - \frac{\mathrm{SAT}_i}{\mathrm{SAT}_i + \exp(5.10 - 0.82 \times \mathrm{SAT}_i)} \quad (18)$$

Further plant growth constraints occur through limitations of root growth, whereas EPIC selects that major stresses out of soil strength, aluminium toxicity and temperature stress to limit root development on a given day. Soil strength depends on bulk density according to

$$\mathrm{SS}_{l,i} = \frac{\mathrm{BD}_{l,i}}{\mathrm{BD}_{l,i} + \exp(bt_1 + bt_2 + \mathrm{BD}_{l,i})} \quad (19)$$

where $\mathrm{SS}_{l,i}$ is the soil strength [–] of layer l on day i, $\mathrm{BD}_{l,i}$ is the bulk density (g cm^{-3}) of layer l on day i and bt_1 and bt_2 are coefficients based on the soils sand content. Aluminium toxicity stress (ATS) depends on the amount of aluminium set free at a certain pH and the aluminium sensitivity of the crop. The amount of aluminium (AL0) a crop j can tolerate is estimated according to

$$\mathrm{AL0}_j = 10 + 20 \times (\mathrm{ALT}_j - 1) \quad (20)$$

where ALT_j is an aluminium tolerance index that can vary between 1 (= highly sensitive) and 5 (= very tolerant). ATS on day i is then estimated as

$$ATS_i = \frac{100 - ALS_{l,i}}{100 - AL0_j} \quad (21)$$

where $ALS_{l,i}$ is the amount of dissolved aluminium in soil layer l on day i.

SOM and organic nitrogen cycling follow the approach of the CENTURY model[46]. As described in Izaurralde *et al.*[47], SOM is split into several pools with varying exchange and turnover rates: standing dead residue and roots, metabolic and structural litter, slow humus, passive humus and microbial biomass. C, N and P may leave the system through erosion, leaching and volatilization. Fluxes between different pools depend on soil and crop management, soil hydrology, temperature and depth within the profile. The C/N ratio has in addition an impact if microbial processes are involved.

Out of various options for calculating water erosion in EPIC, the MUSLE approach was selected as it has been adapted to small watersheds, which have most similarity with single agricultural fields. The algorithm is based on rainfall kinetic energy, soil erodibility, crop management, erosion control practice, slope length and steepness, and soil coarse fragment content. Daily wind erosion is estimated taking into account soil erodibility, surface roughness, vegetative cover, mean unsheltered travel distance of wind across the field and duration of wind greater than threshold velocity (here with a default of 6 m s^{-1}). Atmospheric deposition as an opposite process was not taken into account. Recent studies have shown that atmospheric loading with and deposition of N and P strongly depend on emissions from fossil fuel combustion and atmospheric transport processes[48–50]. Resulting deposition rates are presently not available in a form suitable for global crop models, but should be included in the future.

The Hargreaves method[51] was used for calculating potential evapotranspiration. Potential soil evaporation depends on potential evapotranspiration and soil cover. Actual soil evaporation was estimated from the top 20 cm of the soil profile.

References

1. Fader, M., Rost, S., Müller, C., Bondeau, A. & Gerten, D. Virtual water content of temperate cereals and maize: present and potential future patterns. *J. Hydrol.* **384,** 218–231 (2010).
2. Liu, J. *et al.* A global and spatially explicit assessment of climate change impacts on crop production and consumptive water use. *PLoS ONE* **8,** e57750 (2013).
3. Balkovič, J. *et al.* Global wheat production potentials and management flexibility under the representative concentration pathways. *Glob. Planet. Chang* **122,** 107–121 (2014).
4. Elliott, J. *et al.* The parallel system for integrating impact models and sectors (pSIMS). *Environ. Model. Softw* **62,** 509–516 (2014).
5. Folberth, C. *et al.* Effects of ecological and conventional agricultural intensification practices on maize yields in sub-Saharan Africa under potential climate change. *Environ. Res. Lett.* **9,** 044004 (2014).
6. Rosenzweig, C. *et al.* Assessing agricultural risks of climate change in the 21st century in a global gridded crop model intercomparison. *Proc. Natl Acad. Sci. USA* **111,** 3268–3273 (2014).
7. Asseng, S. *et al.* Uncertainty in simulating wheat yields under climate change. *Nat. Clim. Chang* **3,** 827–832 (2014).
8. Knapp, A. K. *et al.* Consequences of more extreme precipitation regimes for terrestrial ecosystems. *BioScience* **58,** 811–821 (2008).
9. Wang, E., Cresswell, H., Xu, J. & Jiang, Q. Capacity of soils to buffer impact of climate variability and value of seasonal forecasts. *Agric. Forest Meteorol* **149,** 38–50 (2009).
10. Fernandez-Illescas, C. P., Porporato, A., Laio, F. & Rodriguez-Iturbe, I. The ecohydrological role of soil texture in a water-limited ecosystem. *Water Resour. Res.* **37,** 2863–2872 (2001).
11. FAO. *Digital Soil Map of the World.* (FAO, 1995).
12. FAO/IIASA/ISRIC/ISS-CAS/JRC. *Harmonized World Soil Database (version 1.2)* (FAO, 2009).
13. Hengl, T. *et al.* SoilGrids1km—Global soil information based on automated mapping. *PLoS ONE* **9,** e105992 (2014).
14. Stehfest, E., Heistermann, M., Priess, J. A., Ojima, D. S. & Alcamo, J. Simulation of global crop production with the ecosystem model DayCent. *Ecol. Model.* **209,** 203–219 (2007).
15. Van der Velde, M., Bouraoui, F. & Aloe, A. Pan-European regional-scale modelling of water and N efficiencies of rapeseed cultivation for biodiesel production. *Glob. Chang. Biol* **15,** 24–37 (2009).
16. Hazell, P. & Wood, S. Drivers of change in global agriculture. *Phil. Trans. R. Soc. B* **363,** 495–515 (2008).
17. Zhang, X. *et al.* Multi-scale geospatial agroecosystem modeling: a case study on the influence of soil data resolution on carbon budget estimates. *Sci. Total Environ.* **479,** 138–150 (2014).
18. Angulo, C. *et al.* 'Fingerprints' of four crop models as affected by soil input data aggregation. *Eur. J. Agron.* **61,** 35–48 (2014).
19. Waha, K., Wang, E., Huth, N. & Carberry, P. How model and input uncertainty impact maize yield simulations in West Africa. *Environ. Res. Lett.* **10,** 024017 (2015).
20. Williams, J. R. *The EPIC Model.* (ed Singh, V. P.) (Computer Models of Watershed Hydrology, Water Resources Publications, 1995).
21. Lal, R. Soil carbon sequestration impacts on global climate change and food security. *Science* **304,** 1623–1627 (2004).
22. Foley, J. A. *et al.* Solutions for a cultivated planet. *Nature* **478,** 337–342 (2011).
23. Elliott, J. *et al.* The Global Gridded Crop Model Intercomparison: data and modeling protocols for Phase 1 (v1.0). *Geosci. Model Dev* **8,** 261–277 (2015).
24. Boote, K. J., Jones, J. W., White, J. W., Asseng, S. & Lizaso, J. I. Putting mechanisms into crop production models. *Plant Cell Environ.* **36,** 1658–1672 (2013).
25. Izaurralde, R. C., McGill, W. B. & Williams, J. R. *Development and Application of the EPIC Model for Carbon cycle, Greenhouse gas Mitigation, and Biofuel Studies.* (eds Liebig, M. A., Franzluebbers, A. J. & Follet, R. F.) (Managing Agricultural Greenhouse Gases, 2012).
26. Nelson, G. C. *et al.* Climate change effects on agriculture: Economic responses to biophysical shocks. *Proc. Natl Acad. Sci. USA* **111,** 3274–3279 (2014).
27. Hendriks, C. M. J., Stoorvogel, J. J. & Claessens, L. Exploring the challenges with soil data in regional land use analysis. *Agr. Syst.* **144,** 9–21 (2016).
28. Lin, H., Wheeler, D., Bell, J. & Wilding, L. Assessment of soil spatial variability at multiple scales. *Ecol. Model.* **182,** 271–290 (2005).
29. Sanchez, P. A. *et al.* Digital soil map of the world. *Science* **325,** 680–681 (2009).
30. Liu, J. A GIS-based tool for modelling large-scale crop-water relations. *Environ. Model. Softw* **24,** 411–422 (2009).
31. Folberth, C., Gaiser, T., Abbaspour, K. C., Schulin, R. & Yang, H. Regionalization of a large-scale crop growth model for sub-Saharan Africa: model setup, evaluation, and estimation of maize yields. *Agr. Ecosyst. Environ.* **151,** 21–33 (2012).
32. Weedon, G. P. *et al.* The WFDEI meteorological forcing data set: WATCH Forcing Data methodology applied to ERA-Interim reanalysis data. *Water Resour. Res.* **50,** 7505–7514 (2014).
33. Dee, D. P. *et al.* The ERA-Interim reanalysis: configuration and performance of the data assimilation system. *Q. J. R. Meteorol. Soc* **137,** 553–597 (2011).
34. Schneider, U. *et al.* GPCC's new land surface precipitation climatology based on quality—controlled *in situ* data and its role in quantifying the global water cycle. *Theor. Appl. Climatol.* **115,** 15–40 (2013).
35. Hempel, S., Frieler, K., Warszawski, L., Schewe, J. & Piontek, F. A trend-preserving bias correction—the ISI-MIP approach. *Earth Syst. Dynam* **4,** 219–236 (2013).
36. Taylor, K. E., Stouffer, R. J. & Meehl, G. A. An overview of CMIP5 and the experiment design. *B. Am. Meteorol. Soc* **93,** 485–498 (2012).
37. USGS. *GTOPO30 Digital Elevation Model* (U.S. Geological Survey EROS Data Center, 1997).
38. Sacks, W. J., Deryng, D., Foley, J. A. & Ramankutty, N. Crop planting dates: an analysis of global patterns. *Glob. Ecol. Biogeogr* **19,** 607–620 (2010).
39. Mueller, N. D. *et al.* Closing yield gaps through nutrient and water management. *Nature* **490,** 254–257 (2012).
40. Portmann, F. T., Siebert, S. & Döll, P. MIRCA2000—global monthly irrigated and rain-fed crop areas around the year 2000: a new high-resolution dataset for agricultural and hydrological modeling. *Glob. Biogeochem. Cycles* **24** GB 1011 (2010).
41. Monfreda, C., Ramankutty, N. & Foley, J. A. Farming the planet. Part 2: Geographic distribution of crop areas, yields, physiological types, and net primary production in the year 2000. *Glob. Biogeochem. Cycles* **22** GB 1022 (2008).
42. Doorenbos, J. & Kassam, A. H. *Yield Response to Water* (Irrigation and Drainage Paper 33, 1979).
43. Gaiser, T., de Barros, I., Sereke, F. & Lange, F.-M. Validation and reliability of the EPIC model to simulate maize production in small-holder farming systems in tropical sub-humid West Africa and semi-arid Brazil. *Agr. Ecosyst. Environ.* **135,** 318–327 (2010).
44. Hanks, R. J. *Yield and Water Use Relationships: An Overview.* (eds Taylor, H. M., Jordan, W. R. & Sinclair, T. R.*Limitations to Efficient Water Use in Crop Production* (American Society of Agronomy, Crop Science Society of America, Soil Science Society of America (1983).
45. Jones, C. A. A survey of the variability in tissue nitrogen and phosphorus concentrations in maize and grain sorghum. *Field Crops Res* **6,** 133–147 (1983).
46. Parton, W. J., Ojima, D. S., Cole, C. V. & Schimel, D. S. *A General Model for Soil Organic Matter Dynamics: Sensitivity to Litter Chemistry, Texture and Management* (Quantitative Modeling of Soil Forming Processes, SSSA Special Publication 39, Soil Science Society of America (1994).
47. Izaurralde, R. C., Williams, J. R., McGill, W. B., Rosenberg, N. J. & Quiroga Jakas, M. C. Simulating soil C dynamics with EPIC: model description and testing against long-term data. *Ecol. Model.* **192,** 362–384 (2006).

48. Schulz, M., Prospero, J. M., Baker, A. R., Dentener, F. & Ickes, L. Atmospheric transport and deposition of mineral dust to the ocean: Implications for research needs. *Environ. Sci. Technol.* **46,** 10390–10404 (2012).
49. Liu, X. *et al.* Enhanced nitrogen deposition over China. *Nature* **494,** 459–462 (2013).
50. Wang, R. *et al.* Significant contribution of combustion-related emissions to the atmospheric phosphorus budget. *Nat. Geosci.* **8,** 48–54 (2014).
51. Hargreaves, G. H. & Samani, Z. A. Reference crop evapotranspiration from temperature. *Applied Engr. Agric* **1,** 96–99 (1985).
52. Peel, M. C., Finlayson, B. L. & McMahon, T. A. Updated world map of the Koeppen-Geiger climate classification. *Hydrol. Earth Syst. Sci.* **11,** 1633–1644 (2007).
53. Rosegrant, M. W. *et al. International Model for Policy Analysis of Agricultural Commodities and Trade (IMPACT): Model Description* (International Food Policy Research Institute, 2008).

Acknowledgements
This project was funded by the European Research Council Synergy grant ERC-2013-SynG-610028 IMBALANCE-P. C.F. was partly supported by a Research Fellowship of the Center for Advanced Studies of LMU Munich. We thank Joshua Elliott from the Global Gridded Crop Model Intercomparison (GGCMI) project for processing climate input data and the GGCMI and ISI-MIP project teams for providing various input data used in this study.

Author contributions
C.F., R.S., M.V., J.B., M.O. and L.B.A. designed the research; C.F. and R.S. compiled and prepared the data; C.F. and E.M. performed the data analyses; C.F. and M.V. wrote the manuscript; all authors interpreted results and commented on the paper.

Additional information

Competing financial interests: The authors declare no competing financial interests.

Permissions

All chapters in this book were first published in NC, by Nature Publishing Group; hereby published with permission under the Creative Commons Attribution License or equivalent. Every chapter published in this book has been scrutinized by our experts. Their significance has been extensively debated. The topics covered herein carry significant findings which will fuel the growth of the discipline. They may even be implemented as practical applications or may be referred to as a beginning point for another development.

The contributors of this book come from diverse backgrounds, making this book a truly international effort. This book will bring forth new frontiers with its revolutionizing research information and detailed analysis of the nascent developments around the world.

We would like to thank all the contributing authors for lending their expertise to make the book truly unique. They have played a crucial role in the development of this book. Without their invaluable contributions this book wouldn't have been possible. They have made vital efforts to compile up to date information on the varied aspects of this subject to make this book a valuable addition to the collection of many professionals and students.

This book was conceptualized with the vision of imparting up-to-date information and advanced data in this field. To ensure the same, a matchless editorial board was set up. Every individual on the board went through rigorous rounds of assessment to prove their worth. After which they invested a large part of their time researching and compiling the most relevant data for our readers.

The editorial board has been involved in producing this book since its inception. They have spent rigorous hours researching and exploring the diverse topics which have resulted in the successful publishing of this book. They have passed on their knowledge of decades through this book. To expedite this challenging task, the publisher supported the team at every step. A small team of assistant editors was also appointed to further simplify the editing procedure and attain best results for the readers.

Apart from the editorial board, the designing team has also invested a significant amount of their time in understanding the subject and creating the most relevant covers. They scrutinized every image to scout for the most suitable representation of the subject and create an appropriate cover for the book.

The publishing team has been an ardent support to the editorial, designing and production team. Their endless efforts to recruit the best for this project, has resulted in the accomplishment of this book. They are a veteran in the field of academics and their pool of knowledge is as vast as their experience in printing. Their expertise and guidance has proved useful at every step. Their uncompromising quality standards have made this book an exceptional effort. Their encouragement from time to time has been an inspiration for everyone.

The publisher and the editorial board hope that this book will prove to be a valuable piece of knowledge for researchers, students, practitioners and scholars across the globe.

List of Contributors

K. Van Tricht, S. Lhermitte, I.V. Gorodetskaya and N.P.M. van Lipzig
KU Leuven, Department of Earth and Environmental Sciences, Celestijnenlaan 200E, Leuven 3001, Belgium

J.T.M. Lenaerts, B. Noël and M.R. van den Broeke
Institute for Marine and Atmospheric research Utrecht—Utrecht University, Utrecht 3584CC, The Netherlands

T.S. L'Ecuyer
Atmospheric and Oceanic Sciences, University of Wisconsin-Madison, Madison, Wisconsin 53706, USA

D.D. Turner
National Severe Storms Laboratory, NOAA, Norman, Oklahoma 73072, USA

Sanne Muis, Jeroen C.J.H. Aerts and Philip J. Ward
Institute for Environmental Studies (IVM), Vrije Universiteit Amsterdam, 1081 HV Amsterdam, The Netherlands

Hessel C. Winsemius
Deltares, 2600 MH Delft, The Netherlands

Martin Verlaan
Deltares, 2600 MH Delft, The Netherlands
Mathematical Physics, Electrical Engineering, Mathematics and Computer Science (EEMCS), TU Delft, 2628 CD Delft, The Netherlands

Wei-Jun Cai and Yuening Ma
School of Marine Science and Policy, University of Delaware, Newark, Delaware 19716, USA
Department of Marine Sciences, University of Georgia, Athens, Georgia 30602, USA

Kenneth D. Hoadley, D. Tye Pettay and Mark E. Warner
School of Marine Science and Policy, University of Delaware, Newark, Delaware 19716, USA

Yongchen Wang and Brian M. Hopkinson
Department of Marine Sciences, University of Georgia, Athens, Georgia 30602, USA

Yohei Matsui, Justin H. Baumann, Stephen Levas and Andréa G. Grottoli
School of Earth Sciences, The Ohio State University, Columbus, Ohio 43210, USA

Ye Ying
Ocean College, Zhejiang University, Hangzhou 310058, China

Qian Ding
Department of Marine Sciences, University of Georgia, Athens, Georgia 30602, USA
Ocean College, Zhejiang University, Hangzhou 310058, China

Xinping Hu
Department of Marine Sciences, University of Georgia, Athens, Georgia 30602, USA
Department of Physical and Environmental Sciences, Texas A&M University - Corpus Christi, Corpus Christi, Texas 78412, USA

Xiangchen Yuan
Department of Marine Sciences, University of Georgia, Athens, Georgia 30602, USA
Key Laboratory of Marine Bio-resources Sustainable Utilization, South China Sea Institute of Oceanology, Chinese Academy of Sciences, Guangzhou 510301, China

Verena Schoepf
School of Earth Sciences, The Ohio State University, Columbus, Ohio 43210, USA
ARC Centre of Excellence for Coral Reef Studies, School of Earth and Environment and UWA Oceans Institute, University of Western Australia, Crawley,Western Australia 6009, Australia

Hui Xu
Department of Marine Sciences, University of Georgia, Athens, Georgia 30602, USA
Ocean College, Zhejiang University, Hangzhou 310058, China
School of Mechanical Engineering and Automation, Zhejiang Sci-Tech University, Hangzhou 310023, China

Chenhua Han
Department of Marine Sciences, University of Georgia, Athens, Georgia 30602, USA
Ocean College, Zhejiang University, Hangzhou 310058, China
Key Laboratory of Marine Ecosystem and Environment, State Oceanic Administration (SOA) Second Institute of Oceanography, Hangzhou 310012, China

Todd F. Melman
Reef Systems Coral Farm, New Albany, Ohio 43054, USA

Michael L. Griffiths
Department of Environmental Science, William Paterson University, Wayne, New Jersey 07470, USA

Alena K. Kimbrough and Michael K. Gagan
Research School of Earth Sciences, Australian National University, Canberra, Australian Capital Territory 0200, Australia

Russell N. Drysdale
School of Geography, University of Melbourne, Parkville, Victoria 3010, Australia
EDYTEM, UMR CNRS 5204, Universite´ de Savoie, 73376 Le Bourget du Lac, France

Julia E. Cole
Department of Geosciences, University of Arizona, Tucson, Arizona 85721, USA
Department of Atmospheric Sciences, University of Arizona, Tucson, Arizona 85721, USA

Kathleen R. Johnson
Department of Earth System Science, University of California, Irvine, California 92697-3100, USA

Jian-Xin Zhao
School of Earth Sciences, University of Queensland, Brisbane, Queensland 4072, Australia

Benjamin I. Cook
NASA Goddard Institute for Space Studies, New York, New York 10025, USA

John C. Hellstrom
School of Earth Sciences, The University of Melbourne, Parkville, Victoria 3010, Australia

Wahyoe S. Hantoro
Research Center for Geotechnology, Indonesian Institute of Sciences, Bandung 40135, Indonesia

M. Tedesco
Lamont-Doherty Earth Observatory, Columbia University, New York, New York 10964, USA
NASA Goddard Institute of Space Studies, New York, New
York 10025, USA

T. Mote
University of Georgia, Athens, Georgia 30602-2502, USA

X. Fettweis
University of Liege, Liege 4000, Belgium

E. Hanna
University of Sheffield, Sheffield S10 2TN, UK

J. Jeyaratnam and J.F. Booth
The City College of New York, New York, New York 10031, USA

R. Datt
Lamont-Doherty Earth Observatory, Columbia University, New York, New York 10964, USA
The City College of New York, New York, New York 10031, USA
The Graduate Center of the City University of New York, New York, New York 10016, USA

K. Briggs
University of Leeds, Leeds LS2 9JT, UK

A.C. Nölscher, J. Kesselmeier and J. Williams
Air Chemistry and Biogeochemistry Department, Max Planck Institute for Chemistry, Mainz 55128, Germany

A.M. Yañez-Serrano and S. Wolff
Air Chemistry and Biogeochemistry Department, Max Planck Institute for Chemistry, Mainz 55128, Germany
Clima e Ambiente (CLIAMB), Instituto Nacional de Pesquisas da Amazônia (INPA), Avenue André Araújo 2936, Manaus, Amazonas CEP 69083-000, Brazil

A. Carioca de Araujo
Embrapa Amazônia Oriental, Empresa Brasileira de Pesquisa Agropecuaria, Belem, Pará CEP 66095-100, Brazil

J.V. Lavrič
Biogeochemical Systems Department, Max Planck Institute for Biogeochemistry, Jena 07745, Germany

Sev Kender and Melanie J. Leng
Centre for Environmental Geochemistry, School of Geography, University of Nottingham, Nottingham NG7 2RD, UK
British Geological Survey, Keyworth, Nottingham NG12 5GG, UK

Erin L. McClymont
Department of Geography, Durham University, Durham DH1 3LE, UK

Aurora C. Elmore
Department of Geography, Durham University, Durham DH1 3LE, UK
Department of Earth Sciences, University of Cambridge, Cambridge CB2 3EQ, UK

Henry Elderfield
Department of Earth Sciences, University of Cambridge, Cambridge CB2 3EQ, UK

Dario Emanuele
DST—Università del Sannio, Via dei Mulini 59a, Benevento 82100, Italy

Chundi Hu
Institute of Earth Climate and Environment System, Sun Yat-sen University, 135 West Xingang Road, Guangzhou, Guangdong 510275, China
School of Atmospheric Sciences, Nanjing University, Nanjing, Jiangsu 210023, China

Qigang Wu
School of Atmospheric Sciences, Nanjing University, Nanjing, Jiangsu 210023, China

Zhenning Li, Kaiqiang Deng and Tuantuan Zhang
Institute of Earth Climate and Environment System, Sun Yat-sen University, 135 West Xingang Road, Guangzhou, Guangdong 510275, China
School of Atmospheric Sciences, Sun Yat-sen University, Guangzhou, Guangdong 510275, China

Junwen Chen
School of Atmospheric Sciences, Sun Yat-sen University, Guangzhou, Guangdong 510275, China

Song Yang
Institute of Earth Climate and Environment System, Sun Yat-sen University, 135 West Xingang Road, Guangzhou, Guangdong 510275, China
School of Atmospheric Sciences, Sun Yat-sen University, Guangzhou, Guangdong 510275, China
Guangdong Province Key Laboratory for Climate Change and Natural Disaster Studies, Sun Yat-sen University, Guangzhou, Guangdong 510275, China

Chengyang Zhang
Institute of Earth Climate and Environment System, Sun Yat-sen University, 135 West Xingang Road, Guangzhou, Guangdong 510275, China
Climate Center, Guangxi Meteorological Bureau, Nanning, Guangxi 530022, China

E. Freeman, L.C. Skinner and D. Hodell
Godwin Laboratory for Palaeoclimate Research, Department of Earth Sciences, University of Cambridge, Cambridge CB2 3EQ, UK

C. Waelbroeck
LSCE/IPSL, Laboratoire CNRS-CEA-UVSQ, Domaine du CNRS, baˆt 12 91198 Gif-sur-Yvette, France

Kirsten Fahl, Gregor Knorr, Frank Niessen, Catalina Gebhardt, Laura Jensena and Jens Matthiessen
Alfred Wegener Institute (AWI) Helmholtz Centre for Polar and Marine Research, Am Alten Hafen 26, Bremerhaven 27568, Germany

Ruediger Stein and Wilfried Jokat
Alfred Wegener Institute (AWI) Helmholtz Centre for Polar and Marine Research, Am Alten Hafen 26, Bremerhaven 27568, Germany
Department of Geosciences (FB5), University of Bremen, Klagenfurter Strasse 4, Bremen 28359, Germany

Achim Kopf
Department of Geosciences (FB5), University of Bremen, Klagenfurter Strasse 4, Bremen 28359, Germany
MARUM-Center for Marine Environmental Sciences, University of Bremen, Leobener Strasse, Bremen 28359, Germany

Michael Schreck
Arctic Research Centre, Korea Polar Research Institute, 26 Songdomirae-ro, Yeonsu-gu, Incheon 406-840, Korea

Matthias Forwick
Institute of Geology, University of Tromsø—The Arctic University of Norway, P O Box 6050
Langnes, Tromsø 9037, Norway

Michael Kaminski
Geosciences Department, College of Petroleum Engineering & Geosciences, King Fahd University of Petroleum & Minerals, Dhahran 31261, Saudi Arabia

Gerrit Lohmann
Alfred Wegener Institute (AWI) Helmholtz Centre for Polar and Marine Research, Am Alten Hafen 26, Bremerhaven 27568, Germany
MARUM-Center for Marine Environmental Sciences, University of Bremen, Leobener Strasse, Bremen 28359, Germany

Yannick Donnadieu
Laboratoire des Sciences du Climat et de l'Environnement, LSCE-IPSL, CEA-CNRS-UVSQ, Université Paris-Saclay, 91191 Gif sur Yvette, France

Emmanuelle Pucéat, Mathieu Moiroud and Jean-François Deconinck
Laboratoire des Sciences du Climat et de l'Environnement, LSCE-IPSL, CEA-CNRS-UVSQ, Université Paris-Saclay, 91191 Gif sur Yvette, France
Biogéosciences Dijon, Université Bourgogne-Franche-Comté, UMR CNRS 6282, Dijon 21000, France

François Guillocheau
Géosciences Rennes, Université de Rennes, UMR CNRS 6118, Rennes 35042, France

Xinyu Wen and Shaowu Wang
Department of Atmospheric and Oceanic Sciences & Laboratory for Climate and Ocean-Atmosphere Studies, School of Physics, Peking University, Beijing 100871, China

Zhengyu Liu
Department of Atmospheric and Oceanic Sciences & Laboratory for Climate and Ocean-Atmosphere Studies, School of Physics, Peking University, Beijing 100871, China
Department of Atmospheric and Oceanic Sciences & Center for Climatic Research, Nelson Institute for Environmental Studies, University of Wisconsin-Madison, WI 53706, USA

Jiang Zhu
Department of Atmospheric and Oceanic Sciences & Center for Climatic Research, Nelson Institute for Environmental Studies, University of Wisconsin-Madison, WI 53706, USA

Jun Cheng
Polar Climate System and Global Change Laboratory, Nanjing University of Information Science and Technology, Nanjing 210044, China

Øivind Hodnebrog, Gunnar Myhre, Jana Sillmann and Bjørn H. Samset
Center for International Climate and Environmental Research-Oslo (CICERO), P.O. Box 1129 Blindern, N-0318 Oslo, Norway

Piers M. Forster
School of Earth and Environment, University of Leeds, Leeds LS2 9JT, UK

A.W. Jacobel, J.F. McManus, R.F. Anderson and G. Winckler
Department of Earth and Environmental Sciences, Columbia University, New York, New York 10027, USA
Lamont-Doherty Earth Observatory, Palisades, New York 10964, USA

Jacob N.W. Howe and Alexander M. Piotrowski
Department of Earth Sciences, University of Cambridge, Cambridge CB2 3EQ, UK

Taryn L. Noble
Department of Earth Sciences, University of Cambridge, Cambridge CB2 3EQ, UK
Institute for Marine and Antarctic Studies (IMAS), University of Tasmania, Hobart, Tasmania 7001, Australia

Stefan Mulitza
MARUM-Center for Marine Environmental Sciences, University of Bremen, Leobener Strasse, D-28359 Bremen, Germany

Cristiano M. Chiessi
School of Arts, Sciences and Humanities, University of São Paulo, Av. Arlindo Bettio 1000, CEP03828-000 São Paulo SP, Brazil

Germain Bayon
Institut Français de Recherche pour l'Exploitation de la Mer (IFREMER), Unité de Recherche Géosciences Marines, F-29280 Plouzané, France

Hongmei Li, Tatiana Ilyina, Wolfgang A. Müller and Frank Sienz
Max Planck Institute for Meteorology, Bundesstrae 53, 20146 Hamburg, Germany

R.L. Thompson
Norsk Institutt for Luftforskning (NILU), Kjeller, Norway

P.K. Patra and T. Saeki
Department of Environmental Geochemical Cycle Research, Japan Agency for Marine Earth
Science and Technology (JAMSTEC), Yokohama 236-0001, Japan

F. Chevallier and P. Ciais
Laboratoire des Sciences du Climat et l'Environnement (LSCE, CEA-CNRS-UVSQ), Gif-sur-Yvette, France

T. Shirai, M. Ishizawa, S. Maksyutov and T. Machida
National Institute for Environmental Studies (NIES), Tsukuba 305-8506, Japan

R.M. Law and T. Ziehn
Oceans and Atmosphere, Commonwealth Scientific and Industrial Research Organisation (CSIRO), 3195 Aspendale, Australia

I.T. van der Laan-Luijkx
Department of Meteorology and Air Quality, Environmental Sciences Group Wageningen University (WU), 6708 PB Wageningen, The Netherlands

W. Peters
Department of Meteorology and Air Quality, Environmental Sciences Group, Wageningen University (WU), 6708 PB Wageningen, The Netherlands
University of Groningen, Centre for Isotope Research, 9747 AG Groningen, The Netherlands

A. Ganshin
Department of Upper Atmospheric Layers Physics, Central Aerological Observatory (CAO), Moscow 141700, Russia
National Research Tomsk State University (TSU), 634050 Tomsk, Russia

R. Zhuravlev
Department of Upper Atmospheric Layers Physics, Central Aerological Observatory (CAO), Moscow 141700, Russia
National Research Tomsk State University (TSU), 634050 Tomsk, Russia
Department of Atmospheric Physics and Microwave Diagnostics, Institute of Applied Physics of the Russian Academy of Sciences, Nizhny Novgorod 603950, Russia

T. Maki
Atmospheric Environment and Applied Meteorology Research Department, Meteorological Research Institute (MRI), Tsukuba 305-0052, Japan

T. Nakamura
Japan Meteorological Agency (JMA), Global Environment and Marine Department, Tokyo 100-8122, Japan

B. Poulter
Institute on Ecosystems and Department of Ecology, Montana State University (MSU), 59717 Bozeman, Montana, USA

J.G. Canadell
Global Carbon Project, Commonwealth Scientific and Industrial Research Organisation (CSIRO), 2601 Canberra, Australia

Bryn Hubbard and David W. Ashmore
Centre for Glaciology, Department of Geography and Earth Sciences, Aberystwyth University, Aberystwyth SY23 3DB, UK

Martin O'Leary, Ian Rutt, Suzanne Bevan, Bernd Kulessa, Peter Kuipers Munneke and Adrian Luckman
Glaciology Group, Department of Geography, Swansea University, Swansea SA2 8PP, UK

Jean-Louis Tison and Morgane Philippe
Laboratoire de Glaciologie, Département Géosciences, Environnement et Société, Université Libre de Bruxelles, Bruxelles 1050, Belgium

Daniela Jansen
Alfred Wegener Institut, Helmholtz-Zentrum für Polar- und Meeresforschung D-27568, Bremerhaven, Germany

Adam Booth
School of Earth and Environment, University of Leeds, Leeds LS2 9JT, UK

Heidi Sevestre
Department of Arctic Geology, University Centre in Svalbard, N-9171 Longyearbyen, Norway

Wei Liu and Shang-Ping Xie
Climate, Atmospheric Sciences, and Physical Oceanography (CASPO), Scripps Institution of Oceanography, University of California, San Diego, 9500 Gilman Drive, La Jolla, California 92093, USA

Jian Lu
Pacific Northwest National Laboratory, 902 Battelle Blvd, Richland, Washington 99354, USA

Antoine Crémière, Shyam Chand, Terje Thorsnes and Simone Sauer
Marine Geology, Geological Survey of Norway, Postal box 6315 Sluppen, 7491 Trondheim, Norway
CAGE—Centre for Arctic Gas Hydrate, Environment and Climate, Department of Geology, UiT the Arctic University of Norway, 9037 Tromsø, Norway

Aivo Lepland
Marine Geology, Geological Survey of Norway, Postal box 6315 Sluppen, 7491 Trondheim, Norway
CAGE—Centre for Arctic Gas Hydrate, Environment and Climate, Department of Geology, UiT the Arctic University of Norway, 9037 Tromsø, Norway
Institute of Geology, Tallinn University of Technology, 19086 Tallinn, Estonia
Department of Geology, University of Tartu, Ravila 14A, 50411 Tartu, Estonia

Tõnu Martma
Institute of Geology, Tallinn University of Technology, 19086 Tallinn, Estonia

Diana Sahy, Daniel J. Condon and Stephen R. Noble
British Geological Survey, Keyworth, Nottingham NG12 5GG, UK

Harald Brunstad
Lundin Norway AS, 1366 Lysaker, Norway

Gerald A. Meehl, Aixue Hu and Haiyan Teng
National Center for Atmospheric Research, 3090 Center Green Drive, Boulder, Colorado 80301, USA

Wenhao Dong, Yanluan Lin, Jonathon S. Wright, Yuanyu Xie, Yong Luo, Wenyu Huang, Jianbin Huang, Yiran Peng and Fanghua Xu
Ministry of Education Key Laboratory for Earth System Modeling, Center for Earth System Science, and Joint Center for Global Change Studies (JCGCS), Tsinghua University, Room S813, MengMinwei Science Building, Qinghuayuan 1, Beijing 100084, China

Yi Ming
Geophysical Fluid Dynamics Laboratory, Princeton/NOAA, Princeton, New Jersey 08540-6649, USA

Bin Wang
Ministry of Education Key Laboratory for Earth System Modeling, Center for Earth System Science, and Joint Center for Global Change Studies (JCGCS), Tsinghua University, Room S813, MengMinwei Science Building, Qinghuayuan 1, Beijing 100084, China
State Key Laboratory of Numerical Modeling for Atmospheric Science and Geophysical Fluid Dynamics, Institute of Atmospheric Physics, Chinese Academy of Sciences, Beijing 100029, China

Lei Wang and Lide Tian
Key Laboratory of Tibetan Environment Changes and Land Surface Processes, Institute of Tibetan Plateau Research, Chinese Academy of Sciences (CAS), and the CAS Center for Excellence in Tibetan Plateau Earth Sciences, Beijing 100101, China

Olivier Dangles
Institut de Recherche pour le Développement (IRD), UMR EGCE, IRD-247 CNRS-UP Sud-9191, Avenue de la Terrasse, Bâtiment 13, 91198 Gif-sur Yvette, France
Pontificia Universidad Católica del Ecuador, Escuela de Ciencias Biológicas, Avenida 12 de Octubre, Quito 170150, Ecuador

Sophie Cauvy-Fraunié
Institut de Recherche pour le Développement (IRD), UMR EGCE, IRD-247 CNRS-UP Sud-9191, Avenue de la Terrasse, Bâtiment 13, 91198 Gif-sur Yvette, France
Pontificia Universidad Católica del Ecuador, Escuela de Ciencias Biológicas, Avenida 12 de Octubre, Quito 170150, Ecuador
IRSTEA, UR MALY, 5 rue de la Doua, Villeurbanne 69100, France

Patricio Andino and Rodrigo Espinosa
Pontificia Universidad Católica del Ecuador, Escuela de Ciencias Biológicas, Avenida 12 de Octubre, Quito 170150, Ecuador

Dean Jacobsen
Pontificia Universidad Católica del Ecuador, Escuela de Ciencias Biológicas, Avenida 12 de Octubre, Quito 170150, Ecuador
Freshwater Biological Laboratory, Department of Biology, University of Copenhagen, Universitetsparken 4, 2100 Copenhagen, Denmark

Roger Calvez
Institut de Recherche pour le Développement (IRD), UMR G-EAU, 361 Rue Jean-François Breton, BP 5095, 34196 Montpellier, France

N.C. Surawski
CSIRO Agriculture, GPO Box 1700, Canberra, Acton 2601, Australian Capital Territory, Australia
Energy, Environment and Water Research Center (EEWRC), The Cyprus Institute, PO Box 27456, Nicosia 2121, Cyprus (N.C.S.)

A.L. Sullivan and S.H. Roxburgh
CSIRO Land and Water, GPO Box 1700, Canberra, Acton 2601, Australian Capital Territory, Australia

P.J. Polglase
CSIRO Land and Water, GPO Box 1700, Canberra, Acton 2601, Australian Capital Territory, Australia
Fenner School of Environment & Society, Australian National University, Linnaeus Way, Acton, Australian Capital Territory 2601, Australia (P.J.P.)

C.P. Mick Meyer
CSIRO Oceans and Atmosphere, 107–121 Station Street, Aspendale, Victoria 3195, Australia

Thomas Laepple
Periglacial Research Section, Alfred Wegener Institute, Helmholtz Centre for Polar and Marine Research, Telegraphenberg A43, 14473 Potsdam, Germany.

Ulrike Herzschuh
Periglacial Research Section, Alfred Wegener Institute, Helmholtz Centre for Polar and Marine Research, Telegraphenberg A43, 14473 Potsdam, Germany
Institute of Earth and Environmental Sciences, Faculty of Sciences, University Potsdam, 14479 Potsdam-Golm, Germany

H. John B. Birks
Department of Biology, University of Bergen, Postboks 7803, 5020 Bergen, Norway
Bjerknes Centre for Climate Research, Bergen, Norway
Environmental Change Research Centre, University College London, London WC1E 6BT, UK

Martin Melles
Institute of Geology and Mineralogy, Faculty of Mathematics and Natural Sciences, University of Cologne, 50674 Köln (Cologne), Germany

Andrei Andreev
Institute of Geology and Mineralogy, Faculty of Mathematics and Natural Sciences, University of Cologne, 50674 Köln (Cologne), Germany
Institute of Geology and Petroleum Technologies, Faculty of Natural Sciences, Kazan Federal University, Kazan 420008, Russia

Julie Brigham-Grette
Department of Geosciences, University of Massachusetts Amherst, Amherst, Massachusetts 01003-9297, USA

Ligia B. Azevedo and Michael Obersteiner
Ecosystem Services and Management Program, International Institute for Applied Systems Analysis, 2361 Laxenburg, Austria

Christian Folberth
Ecosystem Services and Management Program, International Institute for Applied Systems Analysis, 2361 Laxenburg, Austria
Department of Geography, Ludwig Maximilian University, 80333 Munich, Germany

Rastislav Skalský
Ecosystem Services and Management Program, International Institute for Applied Systems Analysis, 2361 Laxenburg, Austria
Soil Science and Conservation Research Institute, National Agricultural and Food Centre, 82713 Bratislava, Slovak Republic

Elena Moltchanova
Ecosystem Services and Management Program, International Institute for Applied Systems Analysis, 2361 Laxenburg, Austria
School of Mathematics and Statistics, University of Canterbury, Christchurch 8140, New Zealand

Juraj Balkovič
Ecosystem Services and Management Program, International Institute for Applied Systems Analysis, 2361 Laxenburg, Austria
Department of Soil Science, Faculty of Natural Sciences, Comenius University, 84104 Bratislava, Slovak Republic

Marijn van der Velde
European Commission, Joint Research Centre, 21027 Ispra, Italy

Index